Flame-Retardant Polymeric Materials

CONTRIBUTORS

S. M. Atlas
Bronx Community College of the City University of New York
New York, New York

R. H. Barker
Department of Textiles
Clemson University
Clemson, South Carolina

L. Benisek
International Wool Secretariat
Technical Center
Ilkey, Yorkshire, England

Wen-Hsuan Chang
Coating and Resins Division
PPG Industries, Incorporated
Allison Park, Pennsylvania

Robert T. Conley
Department of Chemistry
Wright State University
Dayton, Ohio

C. P. Fenimore
General Electric Research and Development Center
Schenectady, New York

Kurt C. Frisch
Polymer Institute
University of Detroit
Detroit, Michigan

R. E. Fruzzetti
Research Division
Goodyear Tire and Rubber Company
Akron, Ohio

E. L. Lawton
Fibers Department
Monsanto Research and Development Corporation
Research Triangle Park, North Carolina

Menachem Lewin
Fiber Institute
Jerusalem, Israel

H. F. Mark
Polytechnic Institute of New York
Brooklyn, New York

Eli M. Pearce
Polytechnic Institute of New York
Brooklyn, New York

Daniel F. Quinn
Department of Chemistry
Wright State University
Dayton, Ohio

Sidney L. Reegen
Polymer Institute
University of Detroit
Detroit, Michigan

T. H. Rogers
Research Division
Goodyear Tire and Rubber Company
Akron, Ohio

Ronald Blair Ross
Coating and Resins Division
PPG Industries, Incorporated
Allison Park, Pennsylvania

Roger L. Scriven
Coating and Resins Division
PPG Industries, Incorporated
Allison Park, Pennsylvania

Stephen B. Sello
J. P. Stevens & Company
Technical Center
Garfield, New Jersey

C. J. Setzer
Fibers Department
Monsanto Research and Development Corporation
Research Triangle Park, North Carolina

S. W. Shalaby
Research Center
Ethicon Corp.
Somerville, N. J.

Flame-Retardant Polymeric Materials

Edited by

Menachem Lewin

Israel Fiber Institute

S. M. Atlas

*Bronx Community College
of the City University of New York*

and

Eli M. Pearce

*Department of Chemical Engineering
Polytechnic Institute of New York*

SPRINGER SCIENCE+BUSINESS MEDIA, LLC

Library of Congress Cataloging in Publication Data
Main entry under title:

Flame-retardant polymeric materials.

Includes bibliographical references and index.
1. Fire resistant polymers. I. Lewin, Menachem. II. Atlas, Sheldon M. III. Pearce.
Eli M.
TH1073.F58   668   75-26781
ISBN 978-1-4684-2150-7   ISBN 978-1-4684-2148-4 (eBook)
DOI 10.1007/ 978-1-4684-2148-4

© 1975 Springer Science+Business Media New York
Originally published by Plenum Press, New York in 1975
Softcover reprint of the hardcover 1st edition 1975

Preface

Flammability has been recognized as an increasingly important social and scientific problem. Fire statistics in the United States (Report of the National Commission on Fire Prevention and Control, "America Burning," 1973) emphasized the vast devastation to life and property—12,000 lives lost annually due to fire, and these deaths are usually caused by inhaling smoke or toxic gases; 300,000 fire injuries; 11.4 billion dollars in fire cost at which 2.7 billion dollars is related to property loss; a billion dollars to burn injury treatment; and 3.3 billion dollars in productivity loss. It is obvious that much human and economic misery can be attributed to fire situations. In relation to this, polymer flammability has been recognized as an increasingly important social and scientific problem. The development of flame-retardant polymeric materials is a current example where the initiative for major scientific and technological developments is motivated by sociological pressure and legislation. This is part of the important trend toward a safer environment and sets a pattern for future example. Flame retardancy deals with our basic everyday life situations—housing, work areas, transportation, clothing and so forth—the "macroenvironment" capsule within which "homosapiens" live. As a result, flame-retardant polymers are now emerging as a specific class of materials leading to new and diversified scientific and technological ventures.

From the humble beginnings of flame-retardance treatments of existing polymers, the field is now developing into the design and engineering of new heat-resistant molecules, polymers, and commodities which are inherently flame-retardant. It is an interdisciplinary development and involves several scientific, engineering, legal, medical, and sociological consequences.

The objective of this series is varied. In many cases it is to give an up-to-date summary of the state of the art in flame-retarding polymeric materials so as to be an aid to those involved in solving these problems. Interpretation as to mechanism and conjecture about future approaches

v

has been encouraged on the part of the authors. Since polymer degradation is the precursor to flammability, suitable importance has also been placed on this area. There will be occasional chapters which also deal with a specific test method, some of which may have historical importance. In the first volume, we have included a general article on the development of the oxygen index test because of its broad interest and significance. Future issues will also be concerned with similar subject areas.

We are hopeful that through these means, meaningful solutions to a number of the flammability problems and their subsequent positive social ramifications will be accomplished.

M. Lewin
S. M. Atlas
E. M. Pearce

Contents

CHAPTER 3

Flame Retardance of Protein Fibers

L. BENISEK

CHAPTER 4

Flame-Retardant Polyethylene Terephthalate Fibers

E. L. LAWTON AND C. J. SETZER

CHAPTER 5

Flame Retardance of Rubbers

T. H. ROGERS, JR. AND R. E. FRUZZETTI

CHAPTER 6

Retardation of Combustion of Polyamides

ELI M. PEARCE, S. W. SHALABY, AND R. H. BARKER

CHAPTER 10

Flame-Retardant Organic Coatings

WEN-HSUAN CHANG, ROGER L. SCRIVEN, AND RONALD BLAIR ROSS

1

Combustion of Polymers and Its Retardation

H. F. Mark, S. M. Atlas, S. W. Shalaby, and E. M. Pearce

1. Introduction

The rapidly growing demands for polymers and the increasing public awareness of their potential as fire hazards have revived the old problem of polymer flammability and made it a pressing challenge to our modern technology. With our present capabilities the initial goal in this respect should be the flame retardation of polymers and not their flame proofing, especially when organic systems are being considered. Upon examining the different means for achieving this goal, a number of guidelines have been established. The ideal flame-retardant polymer system has (1) a high resistance to ignition and flame propagation, (2) a low rate of combustion, (3) a low rate and amount of smoke generation, (4) low combustibility and toxicity of combustion gases, (5) retention of reduced flammability during use, (6) acceptability in appearance and properties for specific end-uses, and (7) little or no economic penalty. Therefore, a flame-retardant

H. F. Mark · Polytechnic Institute of Brooklyn, Brooklyn, New York. *S. M. Atlas* · Bronx Community College of the City University of New York, New York. *S. W. Shalaby* · Chemical Research Center, Allied Chemical Corporation, Morristown, New Jersey. *E. M. Pearce* · Polytechnic Institute of New York, Brooklyn, New York.

treatment, in addition to being formulated from efficient, economic chemicals, should require no unusual processing conditions, must be applicable in commercial equipment, reproducibly, with no effect on other processing steps, and must be durable under all use conditions. All of these requirements dictate the type of evaluation necessary for flame-retardant systems.

Flame retardation is essentially an interruption of the burning process. The burning process requires a heat source which causes degradation and decomposition with subsequent ignition and combustion. The degradative processes can be thermal or thermal-oxidative and can be exothermic or endothermic. Certain types of flame retardants lead to endothermicity as their mode of retardancy. Decomposition leads to various modes of fragmentation, including regeneration of monomers. These polymer fragments are quite susceptible to oxidation and, thus, combustion. The products of combustion depend on polymer structure and exact burning conditions. Products such as CO, CO_2, HCl, NO_2, organic acids, and NH_3 would be typical. Hydrogen cyanide is produced from a number of nitrogen-containing polymers. The toxicity levels of some of these products become quite important and are difficult to discern because of synergistic effects on each other, as well as the physiological effects of a diminishing oxygen condition.

Approaches to the flame retardation of a polymeric material are normally developed on the basis of information regarding (1) chemical and physical properties of the polymer, (2) phases of the polymer degradation, (3) combustion of the polymer and its dependence on the nature of the degradation products, and (4) chemical and physical properties of the flame-retardant compositions and their interaction with the polymeric substrate. The evaluation of the flame retardancy of many polymer systems and end-use conditions is usually rather difficult and, thus, a number of semiquantitative test methods have been developed to provide flammability data which may relate to an actual fire situation. Among these methods, the limiting oxygen index (LOI) is a widely used test.

2. Brief Description of the Phases of Polymer Degradation and Combustion

Fruitful analysis of the process of polymer combustion and its retardation requires certain knowledge of a number of other pertinent processes. Of these, the thermal and thermal-oxidative degradation of a polymeric material are relevant to its combustion. The combustion of a polymeric substrate in an oxidizing atmosphere is largely dependent upon the formation of combustible volatile products. On the other hand, the retardation of combustion is affected by the evolution of noncombustible vapors and certain physical and chemical changes in the condensed phase in the presence of a heat source.

2.1. Thermal Degradation of Polymers

Studies on the thermal degradation of a polymer in an inert atmosphere usually precede those of the more complex thermal-oxidative degradation in an oxidizing environment. Information on the production of combustible volatiles, such as monomers, carbon monoxide, hydrocarbons, and hydrogen, and noncombustible gases, including carbon dioxide, hydrogen halides, ammonia, and water vapor, can be obtained by studying the thermal degradation of a particular polymer. This information may help one to make an initial judgment as to the general behavior of this polymer in the presence of a flame. A polymer such as polystyrene, polymethyl methacrylate, or polyoxymethylene, which undergoes a facile thermal depolymerization to produce its combustible monomer, is considered to be less flame resistant than polyvinyl chloride, which affords the noncombustible hydrogen chloride at the initial stages of its thermal decomposition.

2.2. Thermal-Oxidative Degradation of Polymers

Although the thermal degradability of a polymer is pertinent to its combustibility, more practical information can be gathered by studying its thermal-oxidative degradation in atmospheres similar to those encountered in an actual fire situation. The rate of thermal-oxidative degradation of a polymer depends on the prevailing temperature of the environment. At temperatures between 25 and 100 C, this process is usually referred to as oxidation, since limited degradation occurs under these conditions. Occasionally, the polymer performance at the 25–100 C temperature range can be extrapolated to higher temperatures comparable to those developed in the combustion zone of burning polymers.

Most practical objects made of organic polymers—a sheet of paper, a nylon stocking, a woolen sweater, a polyester–cotton shirt or a rubber glove —are reasonably stable under ambient environmental conditions. Also, if such materials are exposed to temperatures between 80 and 100 C in air for a period of a few hours, no visible or readily recognizable change occurs. However, closer analysis reveals that a minute amount of oxidation, resulting in the formation of hydroperoxyl groups (—O—OH), does occur under these conditions. Sensitive methods for the analysis of free radicals are now available and can be used to determine quantitatively the presence of hydroperoxides at a concentration of 10 ppm or less. These techniques have permitted the study of the kinetics of the hydroperoxyl formation, through the insertion of an oxygen molecule into a carbon–hydrogen bond.

$$\text{>}C^*\text{—H} + O_2 \longrightarrow \text{>}C^*\text{—O—O—H} \tag{1}$$

The formation rate of an —O$_2$H group depends to a large degree on the type of atoms or groups which are bonded to the starred carbon atom, C*. Thus, in the presence of electronegative groups, e.g., CN, COOR, or Cl, the rate of hydroperoxide formation increases, but electropositive groups, such as —CH$_3$ or —C$_2$H$_5$, tend to slow hydroperoxidation.

The hydroperoxyl groups which exist in a polymeric material undergo slow, progressive degradation or aging at ambient temperatures. They dissociate, according to equation (2), into free radicals.

$$\text{>}C\text{—O—O—H} \longrightarrow \text{>}C\cdot + \cdot O\text{—O—H} \quad \text{or} \quad \text{>}C\text{—O}\cdot + HO\cdot \tag{2}$$

In turn, these radicals initiate chain reactions in the course of which —C—C— bonds are severed and new free radicals are formed. This process can be counteracted by chain transfer agents such as amines, phenols, or halogenated organic molecules. These are capable of interrupting the chain depropagation through the formation of stable free radicals in a chain transfer process. Hence, hydrogen-donating molecules such as those designated as ∿SH are referred to as chain stabilizers and are commonly used to delay substantially the aging of polymers.

$$\sim\!\!\sim SH + \text{—}C\cdot \longrightarrow \text{—}CH + \sim\!\!\sim S$$
$$\sim\!\!\sim S + \cdot S\sim\!\!\sim \longrightarrow \sim\!\!\sim S\text{—}S\sim\!\!\sim$$
(stable)

At ambient temperatures, the steady-state concentration of free radicals in a polymeric material is very low (about 10^{-7} mol %), and relatively small amounts of a stabilizer suffice to protect the object against deterioration over practical periods of time (months and years). However, in the case of combustion, the material does not remain at ambient temperatures, and its temperature rises owing to the added heat. Therefore, an object such as a sheet of paper or a nylon stocking, at a temperature of 50–100°C in air, may undergo accelerated hydroperoxidation, and free radicals will form along the polymer backbone according to reactions (1) and (2). This may then lead to the degradation of the polymer by the cleavage of carbon–carbon bonds as shown in reaction (3).

$$
\begin{array}{ccc}
\overset{\displaystyle H}{\underset{\displaystyle H}{|}} & \overset{\displaystyle H}{\underset{\displaystyle H}{|}} & \overset{\displaystyle H}{\underset{\displaystyle H}{|}} \\
\mathcal{WW}-C-C-C-\mathcal{WW} & \longrightarrow & \mathcal{WW}-C\ +\ CH_2{=}CH-\mathcal{WW} \\
\end{array}
\tag{3}
$$

The rate of formation of hydroperoxyl groups in the presence of oxygen at a temperature of about 100°C is determined, to a great extent, by the polymer structure. Thus, polypropylene undergoes hydroperoxidation almost 100 times faster than polyethylene at that temperature, owing to the presence of tertiary hydrogens along the backbone of the former polymer.

The formation of active free radicals on the polymer backbone can be successfully interrupted by chain transfer agents. This may slow or inhibit the chain reactions leading to the chain scission and degradation of the polymer between 50 and 100°C. At a relatively high temperature range, such as 150–180°C, the chain transfer agents may act as stabilizers. Many polymers are processed and/or used at such temperature ranges, and amines, phenols, and halogen compounds are used as thermal-oxidative stabilizers. The noticeable volatility of many amine and phenol stabilizers at 150–180°C and their high cost limit their use and encourage the application of halogen-based systems. Examples of the halogen compounds which can also be considered as the first line of defense in flame retardation are shown in Table 1. These stabilizers are capable of interfering with the earliest preignition processes, namely, formation of active chain radicals, chain degradation, and the generation of combustible volatile products. Below a maximum service temperature of about 180°C, most polymeric materials can be sufficiently stabilized to survive and to perform their duties as pipes, bottles, filters, or gaskets. However, under conditions characteristic for combustion, more heat will be added, the temperature will rise continually, and a number of different events will occur.

In a polymeric material which is present in air at a temperature range between 180 and 250°C, the rate of the chain degradation increases sharply owing to the accelerated formation of active free radicals on the main chain under these severe conditions. The chain transfer agents (stabilizers) will no longer be capable of slowing these reactions and several undesirable conditions develop:

1. The chain reaction in the presence of more oxygen is exothermic and contributes to a further rise in temperature.
2. The chemical decomposition of the polymer starts to produce volatile, readily flammable products.
3. Most polymers soften and some even melt at that temperature range, which leads to a deformation of the object and, in many cases, to an increase of the surface area exposed to oxygen.

TABLE 1

Examples of Commercial Flame Retardants

Trade name	Composition or structure	Supplier
(a) *Halogen-rich, organic flame retardants*		
Chlorowax	Chloriated paraffin	Diamond Shamrock
DER	Brominated epoxy resin	Dow Chemical
Tetrabromobisphenol A	$C(CH_3)_2(C_6H_2Br_2OH)_2$	Great Lakes Chemicals
Tribromophenol	$C_6H_2Br_3OH$	Great Lakes Chemicals
Phosphorus tribromide	PBr_3	Great Lakes Chemicals
Polytribromstyrol	$\sim\sim CH(C_6H_2Br_3)CH_2 \sim\sim$	Kalk (W. Germany)
HET	Chlorendic acid	Hooker Chemicals
Firemaster PHT4	Tetrabromophthalic anhydride	Michigan Chemicals
Firemaster BP4A	Tetrabromobisphenol A	Michigan Chemicals
Firemaster 5BT	Pentabromotoluene	Michigan Chemicals
BP-10	Decabromobiphenyl	Michigan Chemicals
Aroclor 5442 & 5460	Chlorinated triphenyl	Monsanto Chemical
Hexabromobenzene	C_6Br_6	Michigan Chemicals
Hexachlorocyclopentadiene	C_5Cl_6	Hooker Chemical
Dechlorane 510 & 4070	$(C_5Cl_6)_2$	Hooker Chemical
EB-80	Pentabromoethylbenzene	Great Lakes Chemicals
(b) *Miscellaneous flame retardants*		
Antimony oxide	Sb_2O_3	Hooker Chemicals
Antimony sulfide	Sb_2S_3	Associated Lead (UK)
Antimony trichloride	$SbCl_3$	Stauffer Chemical
Sodium antimonate	$NaSbO_3$	Chemetron
Phosphonitrilic chloride (trimer)	$(\!-\!\overset{\displaystyle Cl}{\underset{\displaystyle Cl}{\vert}}P\!=\!N\!-\!)_3$	El-Monte, Millmaster Onyx
Phosphonitrilic chloride (polymer)	$+\overset{\displaystyle Cl}{\underset{\displaystyle Cl}{\vert}}P\!=\!N\!+_n$	El-Monte, Millmaster Onyx
Diammonium phosphate	$(NH_4)_2HPO_4$	Stauffer Chemical
Firebrake ZB	$2ZnO(B_2O_3)_3(H_2O)_{3.5}$	U.S. Borax
Zinc borate	$Zn(BO_2)_2$	Kraft Chemical
C-30 BF	$Al_2O_3 \cdot 3H_2O$	Alcoa
Ammonium bromide	NH_4Br	Allied Chemical Dow Chemical
Molybdenum oxide	MoO_3	Climax Molybdenum
Molybdenum sulfide	MoS_3	Climax Molybdenum
Fyrol 32-B	$(BrCH_2CH_2Br—CH_2—O)_3PO$	Stauffer Chemical
TPP	Triphenylphosphite, $P(O—C_6H_5)_3$	Weston
Triphenylphosphine oxide	$OP(C_6H_5)_3$	M & T
Thiourea	$(NH_2)_2C{=}S$	Allied Chemical

There exist several means to counteract these events and to establish a second line of defense against further progress of the combustion process. These include:

1. Incorporating into the polymer one or more ingredients which decompose in the temperature range 180–250 C and can develop noncombustible gases such as H_2O, hydrogen halides, CO_2, and/or NH_3: If one adds a mixture of NH_4Br and $(NH_4)_2CO_3$, HBr forms at about 220 C. This acts as a very efficient radical-chain terminator. The CO_2, H_2O, and NH_3 are very desirable diluents for combustible decomposition products such as CH_2O, CH_3—CHO, and low-molecular-weight hydrocarbons. In addition to being a gas-phase diluent, the water released from flame retardants containing hydrated alumina is associated with an energy loss due to the endothermic dehydration and evaporation processes.
2. Adding to the polymer certain inorganic, finely powdered fillers such as carbon black, silica, alumina, or limestone: Such fillers increase the thermal conductivity of the system and slow the rate of temperature increase. They also raise the melt viscosity and retard or prevent the deformation of the degrading polymeric object.
3. Mixing the polymer with a finely powdered composition of ingredients which are capable of forming glassy coatings around the decomposing polymer mass. This lowers the rate of diffusion of gases between the condensed and gas phases. Thus, both the accessibility of oxygen to the burning site and the fugacity of the volatile flammable decomposition products will be lowered. The glasslike coating may also increase the heat-conductivity and eventually prevent the molten polymer from dripping, by the proper dissipation of heat. Useful additives of this type include certain borates, phosphates, and silicates, such as $(NH_4)_3BO_3$, $(NH_4)_4B_2O_5$, $(NH_4)_3PO_4$, and the corresponding Na salts.

3. Analysis of the Combustion Process and Pertinent Aspects of Flame Retardation

Knowledge of the phases of polymer degradation and their relation to combustion should be available prior to any attempt to retard the combustion process. However, this alone does not allow one to develop a realistic approach to the flame ratardation of polymers, and the understanding of some basic features of flame retardancy and the composition of flame retardants are obvious necessities.

3.1. Basic Features of Flame Retardancy

To allow for the generation of noncombustible vapors in the gas phase, one may use, for instance, flame retardants which are capable of generating hydrogen halides or water vapor. Halogenated flame retardants are usually capable of producing hydrogen halides upon interacting with the organic polymer substrate. Water can be incorporated in the flame retardant as a water of crystallization, as in carbonates, borates, and phosphates.

Typical flame retardants which are capable of participating in the second line of defense against continued flame reproduction, by one or more of the three processes described earlier, are listed in Tables 2, 3, and 4.

In the absence of a flame retardant in a burning polymer, there exist two possibilities: (1) if no further heating occurs and/or the access to oxygen is interrupted, the burning may stop at this stage and there will be burnt patches in a carpet, a hole in a car seat, or some similar damage; or, (2) if more heat is added to a burning non-flame-retarded polymer and more oxygen becomes available, the combustion will continue with a temperature increase up to and above 400°C, with the development of an open flame. The object probably will be lost at that stage for all practical purposes, and the problem will now be to localize the fire and to prevent its spreading. Accordingly, a glassy coating of the hot material and/or a char is a desirable feature in polymer flame retardation. In addition, the conversion of an appreciable fraction of the polymer into a char which coats the surface prevents further access of oxygen and forms a solid, immobile residue instead of a fluid burning mass. For this purpose, one could add to the flame-retarding composition certain compounds, such as phosphoric acid

TABLE 2

Examples of Glass-Forming Flame Retardants

Trade name	Composition	Supplier
Aerotex	Mixture of borates and phosphates	American Cyanamid
Arko-AS	Mixture of borates and phosphates	Arkansas Co.
Fi-Retard	Mixture of borates and phosphates	Arkansas Co.

TABLE 3

Flame Retardants Which Produce Nonflammable Gases

Composition	Gaseous products
$NH_4Br + ZnSO_4 \cdot 6H_2O + 2(NH_4)_3PO_4$	$9H_2O + 7NH_3 + SO_3 + ZnO + HBr + P_2O_3$
$ZnCO_3 + (NH_4)_2CO_3 \cdot 4H_2O) + (NH_4)Br$	$5H_2O + HBr + 3NH_3 + ZnO$

TABLE 4

Examples of Flame-Retardant Fibers

Name	Main constituent(s)
Carbon black	Carbon
Glass fibers	SiO_2
Slag wool	SiO_2, CaO, Al_2O_3
Asbestos	$3MgO \cdot 2SiO_2$
Rock wools, e.g., Rocksil	$CaMg(CO_3)_2$
Metal wools, e.g., steel wool	Fe

TABLE 5

Char-Forming and Halogen-Containing Systems in Building Compositions

Trade name	Composition	Supplier
Aerotex NDC	Mixture of phosphates	American Cyanamid
CZC	Chromated zinc chloride	DuPont
Firemaster HHP	Sb_2O_3	Michigan Chemical
Oncor 23A	$Sb_2O_3-(SiO_2)_n$	National Lead Co.
F 1309	Sb_2O_3–tricresylphosphate	Podell Ind. Inc.

salts (preferably of zinc or other heavy metals), which are known to catalyze the char formation of certain organic matrices. Inorganic fillers, as mentioned above, will also have some favorable action during this last stage of the combustion process. Typical examples of flame retardants which contain char-forming components are shown in Table 5.

3.2. Additional Features of Flame-Retarding Compositions

The brief and somewhat oversimplified anatomy of polymer combustion discussed above should be complemented by a few remarks concerning (1) the cooperation of the different individual ingredients of a flame-retardant composition with each other, and (2) the incorporation of the protective agents into the polymer chain.

The first set of remarks is related to the fact that certain organic compounds which contain F, Cl, or Br are either volatile or readily decomposable at 200–400°C. This can be associated with the high fugacity of these compounds at temperatures characteristic of the early stages of combustion or during the preignition processes. Hence, these halogen moieties become unavailable at later stages (at higher temperatures) when their chain-stopping action is surely needed. One, therefore, may incorporate into the flame-retarding systems certain ingredients which can combine chemically with the halogen components and increase their residence time in the burning mass. A typical example of this situation has been observed in the Sb_2O_3–halogen flame retardants. While Sb_2O_3 is of little value alone as a flame retardant, it displays very pronounced "synergism" when used in conjunction with halogen compounds. Another favorable relationship exists between zinc salts and bromine compounds. In this system

$$ZnCO_3 + 2(NH_4)Br \longrightarrow CO_2 + H_2O + 2NH_3 + ZnBr_2$$

three nonflammable gases are produced and the halogen remains in the condensed phase because of the high thermal stability of the $ZnBr_2$. Phosphates and silicates of Zn, Ti, and other heavy metals are used to attain similar synergistic effects.

Another important set of factors relate to the stability of the flame-retarding compositions in certain environments to which the polymeric system is normally exposed. For instance, if textile goods must be flame retarded, the agents have to withstand laundering and dry-cleaning conditions and, thus, resist migration to the liquid phase. Inorganic fillers would, in many cases, meet these requirements, with the exception of leachable ones. In addition, several types of flame-retarding resin formers have been developed and some of them are quite successful for practical purposes. Forerunners of these preparations were halogenated phenolic resins and epoxides; newer types include tetrakismethylolphosphonium halides, e.g., THPC,

$$[P(CH_2OH)_4]^+Cl^-$$

Many halogenated phenolic compounds may advantageously be cured onto textile goods with the aid of THPC and various methylolic compounds which have been developed in the course of the permanent or durable press techniques. Another interesting additive is the tris(1-aziridinyl)phosphine oxide

$$O{=}P\left[N{\overset{\displaystyle CH_2}{\underset{\displaystyle CH_2}{\Big|}}}\right]_3$$

which is a good resin former as such and can be readily used in conjunction with halogenated aliphatic and aromatic compounds. However, the health problem associated with the carcinogenic aziridinyl moiety in this additive will definitely limit its use.

In addition to the commercial polymers containing flame-retarding additives, a number of new high-performance, high-temperature polymers such as poly-*m*-phenylene isophthalamide, polydiphenyl ether pyromelliti-mide, and polybenzimidazoles show improved flammability as a result of their high content of aromatic rings and thermal stability.

3.3. Polymer Properties Which Affect the Heating and Combustion Processes

It has already been mentioned that in the initial stages of oxidation (during aging or combustion), the chemical structure of a polymer plays an important role in controlling the rate of this reaction. However, there are a number of bulk properties, such as those associated with the glass transition temperature (T_g), melting temperature (T_m), decomposition temperature (T_d), specific heat (C_p), and thermal conductivity (k), which are of particular significance at the advanced stages of combustion. At these stages, the temperature of the polymer reaches higher levels, and the processes of heat generation and convection begin to control the course of combustion. A brief discussion of these properties and their relevance to the polymer flammability is contained in the following paragraphs.

At the three critical temperatures, T_g, T_m, and T_d, a polymer undergoes phase transitions which are associated with significant changes in physical properties such as thermal conductivity, modulus, viscosity, and density.

3.3.1. Glass Transition Temperature (T_g)

At this temperature, the amorphous phase of a polymer passes from a relatively hard state into a somewhat softer and usually rubbery condition. Above this temperature, bottles start to buckle, pipes sag, and plates warp. Table 6 shows the T_g values for a number of important polymers.

3.3.2. Melting Temperature (T_m)

At its T_m, a crystalline polymer changes abruptly from a hard solid into a mobile liquid of relatively lower viscosity (a few thousand to ten thousand poises), with diminished mechanical properties. Table 7 shows the melting temperatures of a few important polymers.

3.3.3. Decomposition Temperature (T_d)

This is the temperature range associated with the decomposition of the polymer in the presence of oxygen, and it depends strongly on the chemical nature of the polymer and to a lesser extent, on its specific shape (rod, plate, film, fiber, web, or sponge). The decomposition temperature

TABLE 6

Glass Transition Temperatures (T_g) of Various
Polymers

Polymer	T_g, °C
Polyethylene	−120
Polypropylene	−22
Polybutylene	−25
Polybutadiene	−80
Polyvinyl fluoride	−20
Polyvinyl chloride	85
Polyvinylidene chloride	−20
Polystyrene	95
Polyformaldehyde	−80
Nylon 6	50
Nylon 6-6	55
Polyethylene terephthalate	80
Poly-4,4'-isopropylidene-diphenylene carbonate	149
Teflon (polytetrafluoroethylene)	125
Polydimethyl siloxane	−125
Polydiphenyl-ether sulfone	250

TABLE 7

Melting Temperatures (T_m) of Various Polymers

Polymer	T_m, °C
Low-density polyethylene	110
High-density polyethylene	130
Polypropylene (isotactic)	175
Nylon 6	215
Nylon 6-6	260
Polyethylene terephthalate	260
Teflon (polytetrafluoroethylene)	330
Nomex (poly-*m*-phenylene isophthalamide)	380

signals the beginning of a relatively rapid development of numerous low-molecular-weight products, most of which are flammable. Most decomposition processes are exothermic in the presence of sufficient oxygen, capable of self-propagation, and can be considered as a significant heat source for attaining flame reproduction. Decomposition temperatures of several common polymers are listed in Table 8.

In addition to T_g, T_m, and T_d, material constants such as the specific heat and heat conductivity are important for the rate of temperature rise and distribution in polymeric systems. Specific-heat and heat-conductivity

TABLE 8

Decomposition Temperatures (T_d) of
Various Polymers

Polymer	T_d. C
Polyethylene	340 -440
Polypropylene	320 -400
Polyvinyl acetate	215 -315
Polyvinyl chloride (PVC)	200 -300
Polyvinyl fluoride (PVF)	370 -470
Polytetrafluoroethylene (Teflon)	500 -550
Polystyrene	300 -400
Polymethyl methacrylate	180 -280
Polyacrylonitrile	250 -300
Cellulose acetate	250 -310
Cellulose	280 -380
Nylon 6	300 -350
Nylon 6-6	320 -400
Polyethylene terephthalate	280 -320

TABLE 9

Specific Heats (C_p) of Various Materials

Polymer	C_p. cal/g C
Polyethylene	0.55
Polypropylene	0.46
Teflon (polytetrafluoroethylene)	0.25
Polyvinyl chloride (PVC)	0.25
Polyvinyl fluoride (PVF)	0.30
Polystyrene	0.32
Styrene-butadiene copolymers (SBR)	0.45
Acrylonitrile-butadiene-styrene terpolymers (ABS)	0.35
Ceelulose acetate	0.40
Nylon 6	0.38
Nylon 6-6	0.40
Polyethylene terephthalate	0.30
Phenol formaldehyde polymers	0.40
Epoxy resins	0.25
Polyimide (poly-4,4-biphenylene pyromellitimide)	0.27

values of a number of commercial polymers are summarized in Tables 9 and 10, respectively.

Depending on their chemical structure, different polymers have different tendencies to react with oxygen at elevated temperatures prior to ignition. Table 11 lists temperatures at which flash ignition and self- or

TABLE 10

Thermal Conductivities (k) of Various Polymers

Polymer	k, cal/sec cm^2 × 10^4 at 1 C/cm
Polyethylene (low density–high density)	8.0 –12.5
Polypropylene	2.8
Polytetrafluoroethylene (Teflon)	6.0
Polyvinyl chloride (PVC)	3.0 - 6.0
Polystyrene	2.0 – 3.5
Styrene-butadiene copolymers (SBR)	3.5 – 4.0
Acrylonitrile-butadiene-styrene terpolymers (ABS)	4.5 – 8.0
Cellulose acetate	4.0 – 8.0
Nylon 6	6.0
Nylon 6-6	6.0
Polyethylene terephthalate	3.5 – 4.5
Phenol formaldehyde polymers	3.0 – 6.0
Epoxy resins	4.0 – 5.0
Polydimethyl siloxane (silicone)	3.5 – 4.5

TABLE 11

Ignition Temperatures of Various Polymers

Polymer	Self-ignition, C	Flash ignition, C
Polyethylene	350	340
Polypropylene	550 (570)[a]	520
Polytetrafluoroethylene (Teflon)	580	560
Polyvinyl chloride (PVC)	450	390
Polyvinyl fluoride (PVF)	480	420
Polystyrene	490	350
Styrene-butadiene copolymers (SBR)	450	360
Acrylonitrile-butadiene-styrene terpolymers (ABS)	480	390
Polymethyl methacrylate	430	300
Polyacrylonitrile (PAN)	560	480
Cotton	400	210
Cellulose acetate	470	340
Nylon 6	450 (530)[a]	420
Nylon 6-6	530 (532)[a]	490
Polyethylene terephthalate	480 (450)[a]	440
Wool	600	
Rayon	420	

[a] Other reported values.

autoignition of different commercial polymers occur. It can be seen that different behaviors can be recorded for some similar polymers.

In summary, one can classify the individual components of a flame-retarding system as follows:

1. chain transfer agents which slow the free-radical chain reactions.
2. ingredients which develop nonflammable gases—H_2O, CO_2, NH_3—and keep the flash point of the decomposition gases high.
3. compounds which are capable of forming glassy coatings on the burning mass and hinder the access of oxygen.
4. materials which act as catalysts for char formation.
5. molecules which reduce the volatility of important components of the flame-retarded system and retain them in the burning mass up to higher temperatures.
6. resin formers which act as coupling agents between the substrate and the flame-retarding agent.

At least five speculative modes of action have been proposed for fire retardation.

1. *Gas Theory*: Generation of noncombustible gases which dilute the flame oxygen supply and tend to exclude oxygen from the polymer surface.
2. *Thermal Theory*: Radicals or molecules from retardant degradation react endothermally with flame species or substrate species; retardant decomposes endothermally.
3. *Chemical Theory*: Retardant degrades into free-radical acceptors which interfere with flame chain reactions.
4. *Coating Theory*: Nonvolatile char or liquid barrier is formed which minimizes the oxygen diffusion to the condensed phase and also reduces the heat transfer from flame to the polymer.
5. *Physical Interaction Theory*: Finely divided particles or solid interfaces which may form endothermally and lower the net heat of combustion or reduce flame propagation by altering the course of gas-phase reactions and lead to less reactive radicals.

It is interesting to note that these proposed theories have their obvious place in the comprehensive analysis of combustion processes and do not contradict, but rather complement, each other.

4. Semiquantitative Evaluation of Polymer Flammability: The Limiting Oxygen Index (LOI)

In an attempt to establish a simple, semiquantitative test method for evaluating the ignitability and burning behavior of the different polymers, Fenimore and Martin developed the limiting oxygen index. This is a

TABLE 12

Limiting Oxygen Index (LOI) Values of Different Polymers

Polymer	LOI
(a) *Polymers tested as bars*	
Polyethylene	17.4
Polypropylene	17.5–18
Polymethylene oxide	15.3
Polyethylene oxide	15.0
Polyphenylene oxide	28.5
Poly(2,6-dimethyl)phenylene oxide	30.5
Poly(2,6-diphenyl)phenylene oxide	33.7
Polystyrene	18.3
Polybutadiene	18.3
Polyisoprene (rubber)	18.3
Polymethyl methacrylate	17.3
Nylon 6	23 –26
Nylon 6-6	24 –26
Polyethylene terephthalate	22 –26
Poly-bisphenol terephthalate	37.6
Polyvinyl carbazole	21.6
Polyvinyl alcohol	21.6
Polyvinyl chloride	38 –45
Polyvinylidene chloride (Saran®)	60
Polyvinylidene fluoride (Kynar®)	43.7
Polytetrafluoroethylene (Teflon®)	95.0
Polychlorotrifluoroethylene	95.0
Polyethylene-co-chlorotrifluoroethylene (Halar®)	60
Polyarylene carbonates	26 –29
Polyarylene sulfones	29 –31
Poly-3,3-bis(chloromethyl)-1,3-epoxy alkane (Pentone®)	23.2
(b) *Polymers tested as fabrics*	
Cellulose	18 –20
Cellulose acetate	18.5
Polypropylene	18
Polyacrylonitrile	19.9
Polyethylene terephthalate	20 –21
Wool	25.0
Nylon 6	20 –21.5
Nylon 6-6	20 –21.5
Chlorinated modacrylics (vinyl chloride acrylonitrile copolymers)	29 –30
Poly-*m*-phenylene isophthalamide (Nomex®)	28 –30
Polybenzimidazole	40 –42
Carbon/graphite	56 –64
Kynol® (phenol formaldehyde polymer)	35

measure of the minimum concentration of oxygen in an oxygen–nitrogen atmosphere that is necessary to initiate and support a flame for ≥ 3 min:

$$\text{LOI} = \frac{\text{volume of } O_2}{\text{volume of } O_2 + \text{volume of } N_2} \times 100$$

It is used here just to illustrate the relative flammability, as measured by a specific test, of a number of polymeric materials. Typical LOI values of different polymers are shown in Table 12. The data in Table 12 indicate that polyethylene and polypropylene can burn in a mixture of N_2 and O_2, whose oxygen volume percent is less than that known for air (21%). On the other hand, Teflon can only burn in an atmosphere of almost pure oxygen.

Despite the fair reproducibility of the LOI test in evaluating the flammability of many systems, one should realize that: (1) the oxygen accessibility to the burning polymer in actual fires is usually less than that in the LOI test; (2) both the air (or gas) velocity and temperature about a burning object in a real fire are usually higher than those in the LOI test; (3) the LOI is dependent on a number of factors which cannot be controlled in actual fire situations and these include sample geometry, orientation of sample with respect to the flame, the air (or gas) temperature around the sample, burning time, fluidity and dripping of molten polymer, formation of char or similar barriers, wicking effect of fillers; and thus (4) the LOI test must be complemented with other test methods for a rigorous evaluation of flame-retardant systems.

2

Technology and Test Methods of Flameproofing of Cellulosics

Menachem Lewin and Stephen B. Sello

1. Introduction

Cellulose and cellulosic products are considered flammable because they are readily ignited and rapidly consumed after ignition. The idea of imparting flame resistance to cellulose dates back many centuries, the first significant recorded attempt being made in England in 1735, when a patent was granted to Jonathan Wyld for a flame-retardant mixture consisting of alum, ferrous sulfate, and borax. Early references on Flameproofing are listed in J. E. Ramsbottom.[1]

In 1821 Gay-Lussac[2] published the first systematic investigation of the use of flame retardants. In this paper he disclosed the development of a flame-resistant finish for linen and jute using a mixture of ammonium phosphate, ammonium chloride, and borax. He also established that the most effective flame-retardant salts were those which have either a low melting point and are capable of coating the material with a glassy layer or those which decompose on heating into nonflammable vapors.

In 1912 Perkin[3] attained durable flame resistance by precipitating stannic oxide within the fiber. However, the afterglow was severe and consumed the textile.

Menachem Lewin · Israel Fiber Institute, Jerusalem, Israel. *Stephen B. Sello* · J. P. Stevens & Company, Incorporated, Technical Center, Garfield, New Jersey.

Cellulosic fibers still amount to over half of the fibers used in most countries, and therefore the reduction of the flammability of cellulosic products is of great importance.

The terms used in connection with the flammability of materials are very often confusing, and we must definitely distinguish between the terms "flameproof" and "flame resistant." While a flameproof material entirely resists fire or flame without any appreciable change in physical or chemical properties, a flame-resistant material does not support combustion and does not glow after removal of the source of ignition, but does show some changes in physical and chemical properties.

When cellulose is heated to the temperature of decomposition, it yields volatile, flammable gases, as well as liquid and tarry products which may also volatilize and ignite, leaving a char consisting mainly of carbon. The slow oxidation of this char is responsible for the afterglow, which is as great a fire hazard as the flaming of the volatile products. The flame-retarding treatment essentially reduces the proportion of the volatile products to the amount of char formed. An efficient flame-retarding finish therefore must satisfy two requirements: (1) reduce the formation of flammable tar and gaseous products, and (2) prevent the afterglow of the increased amount of char.

Substantial chemical add-ons are generally required to attain a satisfactory flame-retarding effect. Consequently, the flame-retarding treatment of cellulose is expensive and generally impairs the physical properties of the substrate. Although many processes are documented in the scientific and patent literature to reduce the flammability of cellulose and cellulosic products, their usage has been limited because of the high costs and undesirable side effects.

For practical purposes, it is of great importance that the flame-resistant effect be durable under all conditions encountered; e.g., textiles must not only withstand water, but in many cases also repeated launderings and dry cleanings. Since several water-soluble inorganic salts are used as flame-retardant chemicals, the question of the permanence of the treatment must be considered as of main importance and can be categorized as:

1. Nondurable: Flame retardants (generally water-soluble inorganic salts) which are easily removed by water, rain, or perspiration.
2. Semidurable: Treatments which resist leaching, but lose their effectiveness after a limited number of launderings.
3. Durable: Flame-retardant finishes which withstand leaching, laundering, and dry cleaning for the life of the material.

2. Nondurable Flame Retardants

In the past, as in the present, great interest has been focused upon water-soluble chemicals as flame-retarding agents. These evidently can impart only temporary protection, since the effect of the treatment is destroyed not only by laundering, but also by rain and perspiration. Periodic reprocessing is thus necessary to maintain flame retardancy.

The deposition of any nonflammable substance onto the cellulose substrate in sufficient amount will suppress the propagation of flame. Since organic materials are commonly considered flammable, mostly inorganic salts and acids have been suggested as flame retardants, and most of the common inorganic salts have been investigated for this purpose.[4] In practice, only a few very efficient agents, or mixtures of such, are used which are capable of imparting a high degree of resistance to both afterflaming and afterglowing. These two characteristics are attained by different mechanisms, and many effective flame retardants fail to reduce the afterglow. Very few inorganic compounds, e.g., ammonium salts of phosphoric acid, are able to suppress effectively both the flame propagation and afterglow.

The amount of flame-retarding chemicals required depends not only on their efficiency, but also on the structure of the substrate to be treated; e.g., a heavier-weight, tight-weave cotton fabric needs less flame-retarding chemical in relation to the weight of the fabric to be treated than a light-weight, loose-weave fabric.

Cellulosic materials treated with water-soluble inorganic salts must be dried carefully, since fast drying might cause crystallization of the chemicals on the surface and drying at too high a temperature might result in the decomposition of ammonium salts by loss of ammonia.

The nondurable, water-soluble inorganic flame retardants can be divided into three main groups.[5]

2.1. Group I

The retardants in this group melt at relatively low temperatures and subsequently resolidify in the form of a solid foam produced by the evolution of decomposition products. The solid foam serves as a barrier between the flame and the substrate.

The most important examples of this group are boric acid and its salts. Boric acid itself imparts only moderate levels of flame retardancy, but applied in large amounts it prevents afterglow. Its sodium salt, borax $(Na_2B_4O_7 \cdot 10H_2O)$, imparts better protection against flame propagation, but does not suppress the afterglow.

A mixture of boric acid with its sodium salt is much more effective than either component alone. For example, an add-on of approximately 34% borax is required to prevent the burning of a cotton fabric, and boric acid itself is ineffective in any amount applied. A 7:3 mixture of borax and boric acid is an effective flame retardant with as little as approximately 5% add-on, although a 20% add-on level is needed to decrease substantially the time of afterglow. Satisfactory prevention can be attained, however, only by adding ammonium phosphate to the borax–boric acid mixture; the deposition of a 7:3:5 borax–boric acid–primary ammonium phosphate mixture in 10% or higher concentration, for instance, entirely suppresses the afterglow.

A solution of ammonium sulfate, ammonium phosphate, boric acid, and borax is suggested for the nondurable flame-retardant treatment of cellulosic textiles, papers, or porous materials.[6] The flame-retarding treatment of cellulosic fabrics and papers with a mixture of borax, boric acid and urea is disclosed by Knight[7] and, with the addition products of alkalifluoride and boric acid or boric acid anhydride, by Clare and Deyrup.[8] Organic derivatives of boric acid, e.g., boroxoles, are suggested by Hoffman[9] as flame retardants for cellulose.

$$\text{where } R = \phi\text{—}C_2H_5 \text{ or } \phi CH\text{—}CH_3$$
$$\qquad\qquad\qquad\qquad\qquad |$$
$$\qquad\qquad\qquad\qquad\qquad R$$

The effectiveness of the flame retardants based on boric acid and borax depends on the satisfaction of the following requirements:[5]

1. One component, at least, must decompose at low temperature (80–200°C) with the evolution of nonflammable gaseous products (H_2O, CO_2, NH_3, SO_2, etc.).
2. The other component must melt near the temperature at which gas evolution begins. The mixture should form a uniform melt which solidifies as decomposition progresses.
3. The solidified foam must be stable at temperatures as high as 500°C.

2.2. Group II

The flame retardants of this group consist of inorganic acids, acidic salts, and salts capable of releasing acids on heating. The importance of furnishing free acid groups at the time of combustion is illustrated by the relative effectiveness of orthophosphoric acids and their sodium salts. When the acid anhydride is balanced by an equivalent amount of alkali oxide in the residue, the salt does not exhibit flame-retardant properties.

2.2.1. Sulfamic Acid and Its Salts

Sulfamic acid (H_2NSO_2OH, a strong monobasic acid), and especially its ammonium salt have gained important roles in the flame-retarding treatment of cellulose, even though they are ineffective in suppressing afterglow.

In addition, the sulfamates, and especially the amine salts of sulfamic acid, have a softening effect upon cellulose which is independent of humidity. The ethylamine salt is most widely used because of its lower price.

The treatment of cellulosic products with sulfamates does not affect strength properties if the drying temperature does not exceed 70–100°C, so that esterification of the cellulose does not occur, and the treated product is not exposed to strong light for an extended period of time. In order to avoid the yellowing effect of light on cellulosic materials treated with sulfamates, buffering agents (e.g., trisodium phosphate or sodium acetate) or reducing agents (sodium hydrosulfite or sulfite) must be added as stabilizers to the sulfamate solution.

Chicopee Manufacturing Corp. disclosed[10] the treatment of cellulose and regenerated cellulose with a mixture of ammonium sulfamate and dicyandiamide. This combination, it is claimed, imparts good flame retardancy without affecting the hand and physical properties of the substrate.

Gutman and Herbst[11] suggest the treatment of cellulosic textiles for flame retardancy with a mixture of sulfamide, formaldehyde and urea

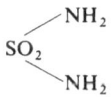

in the presence of a small amount of sulfamic acid.

2.2.2. Phosphoric Acid and Its Salts

Ammonium phosphate and diammonium phosphate are the most widely used nondurable flame retardants. These salts decompose into ammonia and phosphoric acid on heating. The phosphoric acid loses water and is converted to pyrophosphoric acid at approximately 200°C. At higher temperature the pyrophosphoric acid loses more water and is converted to metapyrophosphoric acid, which can form a glassy melt.

The flame-retarding effectiveness of phosphoric acid, its ammonium salts, and of monosodium phosphate ($NaH_2PO_4 \cdot H_2O$) can be attributed to their acidity. Disodium phosphate ($Na_2HPO_4 \cdot 12H_2O$) and trisodium phosphate ($Na_3PO_4 \cdot 12H_2O$), on the other hand, are ineffective as flame retardants. Their very moderate effect, if any, may be attributed to the volatilization of crystal water at the temperature of ignition.

Phosphoric acid itself and the monosodium and ammonium salts of phosphoric acid are also very effective in suppressing the afterglow; significantly less phosphorus (approximately 0.5%) is required to suppress the afterglow than to impart self-extinguishing properties. This small amount of phosphorus catalyzes the oxidation of the carbon char to carbon monoxide instead of carbon dioxide during pyrolysis. The lower exothermicity of the former reaction is not sufficient to maintain the continued oxidation of the char.[12,13]

Michelitsch disclosed[14] the flame-retarding treatment of cellulosic textiles with a mixture of diammonium phosphate, ammonium sulfate, and a neutral acid acceptor such as hexamethylene tetramine, the latter serving to increase the lifetime of the finished fabric. Aarons, Baumgartner, and English suggest[15] the flame-retarding treatment of cellulose with a mixture of the ammonium salts of inorganic acids (e.g., ammonium sulfate, diammonium or monoammonium phosphate), a neutral or slightly alkaline acid acceptor (e.g., dicyandiamide, urea, hexamethylene tetramine, etc.) and a water-soluble boric compound.

2.2.3. Metallic Salts

Easily decomposable metallic salts and, primarily, metallic halides can be considered as flame retardants for cellulosic materials. The flame-retarding effect of the metallic salts can be attributed to the release of acid at the temperature of ignition, but the possibility of releasing acid even on storage must be considered.

Ciba[16] suggested the single-step treatment of cellulose with $ZnCl_2$ and urea. However, although $ZnCl_2$ is very effective in suppressing flame propagation and afterglow, it is generally not used as flame retardant because of its hygroscopic and acidic character.

Kurlychek[17] suggested the application of water-insoluble zinc soaps in molten or emulsion form, followed by impregnation with sodium stannate, then with aluminium sulfate, and finally with soap. In this way flame resistance and water repellency can be simultaneously attained.

The treatment of cellulosic substrates with metal ureides (reaction products of metal chlorides with urea) are disclosed by Vallernaud-Barnier-Bauer.[18] Aluminum, copper, chromic, zirconium, and stannic chlorides are suggested for the preparation of metal ureides. The treatment of cellulosic textiles with metal ureides results not only in flame-retardant, but also in water-repellent properties.

2.3. Group III

The flame retardants of this group are inorganic compounds which decompose or sublime on heating, producing large amounts of nonflammable gases or vapors. Carbonates, halides, ammonium salts, and highly hydrated salts are characteristic members of this group.

Ammonium salts decompose on heating, and therefore play an important role as flame retardants. Ammonium phosphates, sulfate, and sulfamate have been mentioned under group II because their effectiveness is based partially on the release of NH_3 at the temperature of decomposition, but mostly on formation of free acids, which are capable of altering the course of degradation of the cellulose substrate into a less dangerous direction.

The ammonium halides exhibit no foaming tendencies on heating alone, and although they decompose to ammonia and the corresponding halogen acid at approximately 350°C, their flame-retardant properties seem to be more related to their ease of sublimation.[5]

Ammonium halides are suggested as flame retardants for cellulose acetate. British Celanese disclosed[19] the flame-retarding treatment of cellulose acetate with the aqueous or alcoholic solution of urea and one of the following ammonium salts: ammonium bromide, chloride, borate, or diammonium phosphate.

Besides the ammonium salts, the halides of organic amines are suggested by Wesson and Olpin[20] for the flame-retarding treatment of cellulose acetate. Cotton and regenerated cellulose also can be rendered flame resistant by treating with 15% ethylene diammonium bromide from aqueous solution.

Lincoln and Campbell suggested[21] the treatment of cellulose with a neutral or slightly alkaline solution of alkanolamine phosphate (phosphoric acid neutralized with mono-, di-, or triethanolamine). Satisfactory flame retardancy can be attained without affecting the hand of the cellulosic fabric.

3. Semidurable Flame Retardants

Cellulosic materials treated with semidurable flame retardants are required to withstand not only leaching in water, but also a limited number of launderings.

The most obvious means of attaining semidurable flame resistance is the application of insoluble salts. It must be taken into consideration that the flame-retarding effect of the simple inorganic salts is based on their capability of decomposing in heat and releasing strong acid or alkali which is responsible for the reduction of flame propagation. Generally these thermally unstable salts of weak acid/strong base or of strong acid/weak base are very soluble in water. On the other hand, the water-insoluble inorganic salts generally do not easily decompose on heating. The bonding energies which resist ionization and solution also resist the thermal decomposition of the molecular structure. This effect is illustrated by the flame-retarding efficiency of ammonium phosphate compared with the ineffectiveness of calcium phosphate.

It is thus evident that flame-retarding effectiveness is being limited to the insoluble salts of amphoteric cations or anions, e.g., the phosphates or borates of tin, zinc, aluminum, or the stannates, tungstates, aluminates, etc. The easily reducible metal oxides and hydroxides are also capable of catalytically altering the course of the thermal decomposition of cellulose and combine water-insolubility with flame-retarding properties. Such compounds are, for example, stannic, ferric, plumbic, titanic, chromic, zinc, cerium, bismuth, tungsten, arsenic, and silicon oxides.

Ramsbottom and Snoad[22] studied systematically the flame-retarding effectiveness of water-insoluble inorganic compounds and concluded that the flame-retardant property might result from the soluble salts adsorbed on the insoluble precipitate and not entirely removed by washing. The following tabulation shows the minimum add-on required to attain flame retardancy comparable with that of cellulose treated with a 10% 1:1 borax–boric acid solution:

Compound	Add-on, %
Ferric hydroxide	25
Antimony oxychloride	30
Stannic oxide, hydrated	40
Titanic hydroxide	40
Bismuth trioxide, hydrated	40
Zinc stannate	40
Aluminum borate	59
Lead peroxide	60
Cerium hydroxide	69
Aluminum hydroxide	70
Chromic hydroxide	91
Silica, hydrated	100
Aluminum silica	100
Magnesium silica	116
Magnesium ammonium phosphate	125

Metallic hydroxides, although producing flame resistance, generally enhance the afterglow. Insoluble phosphates and borates do not exhibit flame-retardant properties, but are effective in suppressing the afterglow. Tungstic acid functions similarly to boric and phosphoric acid in suppressing the afterglow, and its combination with stannic, ferric, or aluminum hydroxide reduces both the flame propagation and afterglow.

The precipitation of the water-insoluble inorganic compounds can be carried out by several techniques.

The precipitation of metastannic acid can be attained by the alkaline hydrolysis of a stannic salt or by the acidification of alkali stannate. Stannic chloride is usually hydrolyzed with ammonia:

$$SnCl_4 + 4NH_4OH \longrightarrow 4NH_4Cl + H_2O + H_2SnO_3$$

The acidification of alkali stannate is usually attained with ammonium sulfate, ammonium, zinc, or aluminum acetate, or acetic acid:

$$Na_2SnO_3 + (NH_4)_2SO_4 \longrightarrow Na_2SO_4 + 2NH_3 + H_2SnO_3$$

Ferric and zinc hydroxides are usually precipitated from the corresponding chlorides, sulfates, or acetates with ammonium hydroxide. Aluminum hydroxide precipitate is obtained by the acidification of an aluminate or by the hydrolysis of the acetate.

Tungstic acid is precipitated from sodium tungstate by acidification, and chromic hydroxide from chromic sulfate by alkaline hydrolysis:

$$Na_2WO_4 + 2CH_3COOH \longrightarrow 2CH_3COONa + WO_2(OH)_2$$

$$Cr_2(SO_4)_3 + 6NH_4OH \longrightarrow 2Cr(OH)_3 + 3(NH_4)_2SO_4$$

Often mixtures of several flame-retarding chemicals are used. It is very difficult to determine in which form these compounds are actually present on the fiber. For example, by applying stannic chloride and sodium tungstate, we must consider not only the formation of stannic and tungstic acid, but also the formation of stannic tungstate. The actual compositions of the mixed hydroxides formed within the cellulosic fiber depend on concentration and pH at the location where the precipitation occurs.

The chemicals used in a double-bath procedure are not substantive materials, so it must be taken into consideration that the chemical in the second bath is removed from solution by reacting with the chemical already deposited onto the fiber from the first bath. Therefore a gradual decrease in chemical concentration in the second bath can be expected. The concentration in the second bath should be kept constant to ensure the presence of only a minimum amount of unreacted water-soluble chemical on the fiber.

Usually 25–30% weight increase is required to attain adequate flame retardancy. It is preferable to attain the high add-on by applying the components from concentrated solutions rather than from dilute solutions by multiple impregnations.

The impregnation of cellulosic fabrics is generally carried out on padding machines. Excess solution is removed by squeezing the fabric through two or three rollers. The fabric impregnated with the first chemical can be immersed in the second chemical solution without intermediate drying or after partial or complete drying. The latter method gives the most reliable results because the chemical add-on can be carefully controlled.

A classic example of the two-bath technique is the Perkin process.[23] The cellulosic substrate is first impregnated with the aqueous solution of sodium stannate. After intermediate drying, it is immersed in aqueous ammonium sulfate solution and dried. The fabric is washed to remove the excess of water-soluble chemicals.

The strength of the interbonding forces between the cellulose and the metallic oxides determines the durability of the treatment. Both the hydrated metallic oxides and cellulose are surrounded by water molecules held in place by hydrogen bonds. When dehydrated, the free hydroxyl groups of the cellulose molecules replace the water molecules, and secondary valences form between the metallic oxide and cellulose molecules.

Generally the antiglow constituents are more sensitive to pH changes than the flame retardants. The antiglow components ionize more readily and lose their effectiveness on alkaline soaping:

$$Zn_3(PO_4)_2 + 6OH^- \longrightarrow 3Zn(OH)_2 + 2PO_4^{3-}$$

$$WO_2(OH)_2 + 2OH^- \longrightarrow WO_4^{2-} + 2H_2O$$

The incorporation of resins into the double-bath procedure improves the durability of the flame-retardant properties. The binders retard the removal of the precipitated metallic oxides by emulsification. The number of resins which can be used as binders is limited, however, because many of them enhance the afterglow. Urea-, melamine-, and phenol-formaldehyde are the most suitable binders and small amounts, which do not cause significant stiffening, are effective.

Efforts have been made to simplify the two-bath procedure by applying the two components capable of forming the insoluble precipitate from a single bath. Hopkinson[24] disclosed the application of the two chemicals from a single bath using an organic solvent of sufficient hydrophilic character to enable the wetting of the cellulosic fiber and penetration of the chemicals, and of low enough dielectric constant to hold the inorganic salt in molecular form and prevent its precipitation. Among suitable solvents are ethylene and diethylene glycol monomethylether, glycerol, acetamide, triethanolamine, or tricresyl phosphate. Water-insoluble borates, phosphates, and tungstates of zinc, aluminum, magnesium, and manganese are formed *in situ* in the fiber by first treating it with the water-soluble salts from organic solvent, drying, and finally immersing in water.

This process has several advantages when compared with the aqueous double-bath process. In the organic-solvent process both active ingredients are uniformly deposited onto the fiber. The resin, if desired, can be added directly to the organic solution.

Another example of this method is impregnation with a mixture of zinc chloride and borax in glycol solution, followed by exposing the treated cellulosic material to water to form zinc borate precipitate. The immersion in water can be substituted by exposure to humid atmosphere.

Frisch[25] describes an interesting application of metal phosphate. The process is based on the solubility of the tertiary orthophosphate of zinc in aqueous ammonia under certain conditions. When cellulose impregnated with such a solution is dried, an insoluble shiny film consisting of zinc ammonium oxide phosphate $[Zn_3O(NH_4PO_4)_2]$ is formed.

3.1. Titanium and Antimony Compounds

Several flame-retarding procedures have been developed based on this principle of precipitating insoluble products *in situ* on cellulosic fiber. Among these procedures, those processes which are based on titanium and antimony compounds have special importance.

Panik, Sullivan, and Jacobsen[26] studied the treatment of cotton with titanium compounds. Very good flame resistance has been attained without significantly impairing the esthetic and physical properties of the cellulosic substrate. The treatment does not impart protection against afterglow and is not durable. The incorporation of the antimony salt into a titanium salt solution, however, imparts improved durability and protection against afterglow. Panik, Sullivan, and Jacobsen treated cotton fabric with a mixture of titanium chloride acetate and antimony chloride. After partial drying, the cellulosic material is treated with ammonia to hydrolyze the metallic salts to oxides. Finally the unreacted soluble salts are removed by rinsing. The optimum Sb_2O_3/TiO_2 weight ratio is 1:1.3. These titanium–antimony compounds impart good flame retardancy at relatively low (15%) add-on.[27] The TiO_2 functions to absorb the alkali liberated at the temperature of ignition and alters the course of combustion to produce a lesser amount of flammable tars. The Sb_2O_3 improves the fixation of TiO_2 to the cellulose and increases its capacity to absorb alkali. The afterglow can be further reduced by adding silicate to the alkaline bath. Sodium silicate melts at the temperature of combustion and the melt coats the fiber.

The National Lead Co. possesses several patents[28-31] disclosing the flame-retarding treatment of native and regenerated cellulosic products, acetyl cellulose, wood, paper, etc., with a combination of titanium and antimony compounds. The titanium compounds are generally used in the form of titanium chloride formate, acetate, or propionate. These titanium compounds are obtained by reacting titanium tetrachloride with formic, acetic, or propionic acid and are applied in conjunction with antimony trichloride. The cellulosic substrate impregnated with the aqueous solution containing the titanium and antimony compounds is immersed after drying into an alkaline solution to form a transparent gel within the fiber.

The simultaneous application of titanium compounds in the form of basic chlorides with valences of $3+$ and $4+$, in conjunction with antimony chloride, is disclosed by Beacham and Panik.[31] The gelatinization takes place upon drying the treated textile. The main function of $TiCl_3$ in this process is to swell the cellulose. It is also possible to use $TiCl_3$ without the presence of Ti^{4+}. In this case the Ti^{3+} must be partially or entirely oxidized *in situ* on the cellulosic substrate.

Du Pont owns several patents for the flame-retarding treatment of cellulose with titanium and antimony compounds. Dills[32] disclosed the treatment of textiles with an aqueous–isopropanol solution of $TiOCl_2$, $SbCl_3$, and $ZnCl_2$, followed by drying and neutralization in Na_2CO_3 solution. The Sb/Ti atom ratio must not be greater than 2.

Riches[33] described the treatment of wood, paper, and regenerated cellulose with the aqueous solution of titanium, antimony, and/or zirconium salt. As a new approach, the free or latent acids are partially or entirely neutralized with gaseous ammonia, and finally the gelatinization is completed by rinsing in Na_2CO_3 solution.

Generally the antimony compound (e.g., Sb_2O_3) is dissolved in the aqueous solution of the titanium compound (e.g., $TiCl_4$):

$$4TiCl_4 + Sb_2O_3 + H_2O \longrightarrow 4TiOCl_2 + 2SbCl_3 + 2HCl$$

Such a hydrochloric acid-containing solution remains stable even after dilution, and the treated cellulose must be exposed to alkali to re-form Sb_2O_3.

Du Pont markets, under the trade name Erifon, a flame-retarding composition which consists of an aqueous solution of titanium and antimony salts and also a small amount of free HCl. Because of the hydrolytic cleavage of the titanium and antimony salts, the pH of the Erifon solution is less than 1. The components of the solution swell the cellulose and penetrate the fiber. After neutralization they form bonds with the cellulosic substrate. The minimum amount of solid Erifon required to attain flame retardancy is 11%, based on the dry weight of the substrate. Generally, for practical purposes, 13–16% metallic oxide must be fixed to attain a satisfactory level of flame retardancy. However, it is difficult to attain good penetration into the cellulose fiber, and the excess of Erifon precipitates on the surface of the fibers upon neutralization in the form of white metallic oxides.

The surface deposition of these oxides, which do not contribute to the flame resistance, can be avoided by diluting Erifon with water-soluble volatile alcohols instead of water. After impregnation with the alcoholic Erifon solution, the cellulose substrate is dried for 15 min at room temperature to volatilize the alcohol and is finally neutralized in soda-ash solution. It is important to remove the residual alkali with acetic acid. The use of phosphates, especially acid phosphates, oxyacids, or acid fluorides in conjunction with Erifon must be avoided because they react with its components and adversely affect the flame retardancy.

There are several procedures documented in the literature which use antimony compounds without titanium salts. Jordan and O'Neil[34] describe a flame-retarding treatment with a mixture of KH_2SbO_4 and $FeSO_4$, the effect being based on the deposition of the corresponding metallic oxides.

White[35] disclosed a flame-retarding treatment of cellulose with the aqueous solution of antimony trifluoride and a peptizing agent (e.g., sodium, potassium, or ammonium fluoride). After drying, antimony trioxide is precipitated *in situ* on the fiber by alkali treatment. This procedure imparts good flame retardancy, but the afterglow is objectionable.

3.2. Metallic Oxide and Halogenated Organic Binder Compositions

The deposition of metallic oxides *in situ* within the cellulosic fiber via a two-bath procedure imparts good flame retardancy, but the durability of the finish is limited and generally the protection against afterglow is unsatisfactory. These disadvantages can be minimized by using non-flammable binders in conjunction with the metallic oxides. Halogen-containing organic compounds, such as polyvinyl chloride, polyvinyl chloroacetate, chlorinated paraffins, etc., are the most suitable. These substances are effective because they form a coating and prevent oxygen from reaching the flammable substrate. The presence of metallic oxides catalyzes the release of hydrochloric acid in the vicinity of the temperature of combustion.

The most effective systems are the combinations of antimony trioxide with chlorinated organic compounds. Although each alone does not impart satisfactory flame-retardant properties to cotton, the combination of the two systems is very efficient. Metallic oxides, such as antimony pentoxide, tin, zinc, or manganese oxide, are significantly less effective in reducing flammability than antimony trioxide. However, a mixture of equal amounts of antimony pentoxide and trioxide is very effective, and a 1:2 mixture of antimony trioxide and zinc oxide imparts less flame retardancy, but is more effective in suppressing the afterglow.

Since the effectiveness of the halogenated compounds depends on the amount of acid liberated, only those compounds are effective which can easily release the respective halogen acid. Therefore the halogen content of the organic compound, its volatility at the temperature of combustion, and the nature of the halogen bond (e.g., aromatic or aliphatic) determine the effectiveness. The highly chlorinated paraffins and polyvinyl chloride are very effective because they are not volatile and their chlorine content is high.

Neoprene, aniline hydrochloride or hydrobromide, ammonium chloride, ethylene diamine hydrochloride, the chlorinated paraffins, and polyvinyl chloride readily liberate halogen acids and consequently are very effective. Aniline sulfate, which liberates sulfuric acid on pyrolysis, is significantly less effective. Similarly, chlorinated diphenyls are only very moderately effective in conjunction with antimony compounds because chlorine on double-bonded carbon is thermally quite stable. Only 3% of its chlorine content is readily released, while in PVC or in highly chlorinated paraffins 40–60% of the chlorine content is thermally available.[5]

The effective flame-retarding species is actually the $SbCl_3$ or $SbOCl$, which forms from the interaction of Sb_2O_3 and chlorinated paraffin or polyvinyl chloride at the temperature of pyrolysis.[36] The metallic oxides and halogenated compounds must be used in definite ratio to obtain optimum effectiveness, and therefore at least one Cl atom is required for each Sb atom. It is preferable to use an excess of chlorinated compound because the amount of hydrochloric acid liberated is lower than the theoretical amount. The optimum ratio is independent of the amount of flame retardant used.

The paraffins with 40% chlorine content impart soft hand, but their effectiveness as acid releasers is insufficient and the durability of the treatment is limited. The highly chlorinated paraffins (chlorine content 70%) are effective acid releasers and the treatment is durable, but they cause substantial stiffening. Sometimes the mixture of paraffins with high and low chlorine content is used with the objective of attaining a durable treatment without excessive stiffening.

Addition of a glow-retardant agent, such as a salt of boric or phosphoric acid, to these compositions is useful in further reducing the afterglow. In order to attain a semidurable effect such compounds must be water-insoluble, thus limiting the number of compounds that can be used. Zinc borate or phenyl diamidophosphate is more effective in suppressing afterglow than phosphated polyvinyl alcohol or zinc, tin, copper, or iron phosphates.

In order to attain satisfactory protection against flame propagation and afterglow, the following three components should be present in an effective flame-retarding composition:

1. Metallic oxide, preferably antimony trioxide.
2. Compounds capable of releasing hydrochloric acid.
3. Compounds capable of releasing phosphoric or boric acid.

The mixture of antimony trioxide, chlorinated polymer, and glow retardant can be applied from solvent suspension; e.g., a suspension of antimony trioxide, polyvinyl chloride–polyvinyl acetate copolymer, zinc borate, and triphenyl phosphate in methyl ethyl ketone is effective when applied in such a concentration as to give 30% solid on the weight of cellulosic substrate. The durability of the fixation of the glow-retardant agent and metallic oxide depends on the binder used. High-molecular-weight PVC/PVA copolymers impart good durability, but their solubility is low and their stiffening effect is substantial. The stiffness can be reduced by replacing the PVC/PVA copolymer with chlorinated paraffins (42% Cl content), although the flame-retarding effectiveness is simultaneously reduced.

In an oil-in-water emulsion, ammonium oleate is used as emulsifier and methylcellulose as stabilizer. The addition of glow retarders, e.g., zinc borate or phenyl diamidophosphate, to the mixture reduces the stability of the emulsion.

Water-in-oil type emulsions are generally preferred since their stability is better than that of oil-in-water emulsions. They can be prepared to the desired viscosity and can also penetrate the fibers more easily. The emulsifiers used in water-in-oil emulsions should be soluble in organic solvents. Generally esters or amino-condensate-type compounds are used as emulsifiers. In both types of emulsion the oil phase is generally a petroleum fraction, although it can be replaced with nonflammable halogenated solvents like dichloroethane or carbon tetrachloride.

The application of emulsions is simpler than that of suspensions. Emulsions can be applied with conventional padding machines, while doctor blades or printing machines are required to apply suspensions of high viscosity.

The antimony oxide–chlorinated polymer combination is the most widely used system for the "FWWMR" finish, an abbreviation used for fire, water, weather, and mildew resistance. These finishes generally also contain plasticizers, coloring pigments, fillers, stabilizers, and fungicides. Chemical add-ons of up to 60% are required to attain satisfactory flame resistance. The finish withstands 4–5 yr of outdoor exposure. It is suitable for heavy-weight outdoor cellulosic fabrics, such as military tents, but is not suitable for apparel or interior decorating fabrics because of the adverse effect of the treatment on hand and color.

Cellulose treated with antimony trioxide in conjunction with chlorinated paraffin is damaged by exposure to light or heat.[37] After 4 hr exposure at 130°C temperature, a substantial amount of HCl is released from the chlorinated paraffin. It is thus desirable to incorporate an acid acceptor (e.g., calcium carbonate) into the treating bath to reduce the acid damage which might occur at elevated temperatures. Calcium carbonate, however, does not give protection against the effect of light.

The simultaneous application of flame-retardant metallic oxides with halogen-containing binders in conjunction with glow-retardant agents is the basis of many patents. Quehl owns a series of patents[38] which are based on the flame-retarding treatment of cellulose with metallic oxides in conjunction with halogenated polymers. These patents are the basis of the semidurable Aflaman flame retardants. These are compositions consisting of antimony compounds and organic chloro compounds which impart a simultaneous flame-retardant and water-repellent effect. The Aflaman N products are dispersions which contain only a small amount of organic solvent and can be diluted with water. The Aflaman L products are soluble in chlorinated solvents.[39-41]

Broatch[42] claimed a durable flame-retardant treatment of cellulose by treating the substrate with an aqueous dispersion of antimony ortho-phosphate, chlorine-containing vinyl polymer, and stabilizer (e.g., carboxy-methyl cellulose) followed by heat curing. Tricresyl phosphate is recommended as softener for the formulation containing PVC, and the durability of the flame-retarding treatment can be improved by incorporating resins into the formulation.

Protein condensation products, e.g., condensate of urea and formaldehyde with casein, have also been suggested as binders[43,44] for the antimony oxide–chlorinated paraffin combination. The degrading effect on cellulose of the HCl released from the antimony trioxide–chlorinated polymer composition on exposure to light is only minimal in the presence of protein condensation products.

Timmons[45] disclosed the fixation of insoluble antimony compounds with numerous resins. Polymers such as polyester, polystyrol, or polyvinyl derivatives are suggested by Lurie.[46]

The flame-retarding treatment of cellulosic fibers using copper antimonate dispersed in an organic halogen compound is claimed by Giordano and Straka.[47] Halogenated paraffins, polyvinyl chloride, chlorinated rubber, etc., are suitable halogen compounds. The copper antimonate also imparts bacteriostatic properties.

4. Flame-Resistant Cotton Flote

The treatment of cotton flote (a resin-treated cotton-batting product used in automobiles, furniture, mattresses, insulation, etc.) is an important field for the application of nondurable or semidurable flame-retardant chemicals. The flame retardants[48-50] used in the treatment of this product should be inexpensive; applicable from aqueous system; compatible with resins used in the processing of cotton flote; durable to mild leaching, high humidity, and high temperature; odorless after drying, and effective in suppressing afterglow.

Antimony trioxide in conjunction with polyvinylidene chloride and melamine imparts adequate flame resistance. Because the chemicals in this composition are insoluble in water, the flame retardants are resistant to high humidity and saturation.

Ammonium salts of phosphoric acid are very effective, but show a propensity toward ion exchanging when the treated cotton flote is saturated with tap water. Organic salts of phosphoric acid, such as propyl ammonium phosphate, are less susceptible to ion exchange. Urea–phosphate or dicyandiamide–phosphate complexes remain effective in suppressing flame propagation and afterglow, even after exposure to high humidity or saturation with water, if the flame-retardant concentration exceeds 5%. The combination of borate and PVC also yields satisfactory protection against both flame propagation and afterglow.

5. Durable Flame Retardants

5.1. Phosphorylation

Phosphoric acid and the ammonium salts of phosphoric acid are very effective in inhibiting the combustion of cellulose. Since the protection afforded by the deposition of acid is temporary in nature, attempts were made to bind these compounds directly to the cellulose to obtain a durable effect.

The heat curing of cellulose with an acidic substance in the presence of a buffering agent and swelling medium is a suitable condition for cellulose esterification. The heat treatment of cellulose with phosphoric acid in the presence of certain nitrogenous compounds thus leads to the formation of cellulose phosphate with excellent flame-resistant properties.

In practice, the cellulose is impregnated with the aqueous solution of the acid and the nitrogenous compound. After drying, the cellulose is cured at an elevated temperature to accomplish the esterification, and finally the unreacted components are removed by rinsing. Naturally, heat curing cellulose with a nonvolatile strong mineral acid for prolonged periods of time at the temperatures required for esterification causes very substantial degradation of the cellulose. By using a large excess of certain organic nitrogen compounds, the degradation can be reduced to an acceptable level. Moreover, the presence of the nitrogenous compound favors the esterification reaction. The esterification actually takes place in the molten solution of the nitrogenous compound containing the acid or acid salt. The high temperature is also suitable for the evaporation of water released in the esterification reaction.

In the reaction of cellulose with acidic phosphates or phosphoric acid in the presence of organic nitrogen compounds, the formation of mono-substituted esters predominates. However, the formation of disubstituted products by esterification of two hydroxyl groups on the same anhydro-glucose unit or on neighboring cellulose chains cannot be excluded. Mono-substitution may take place on any of the three hydroxyl groups available in the anhydroglucose unit, but the primary hydroxyl is most readily esterified.

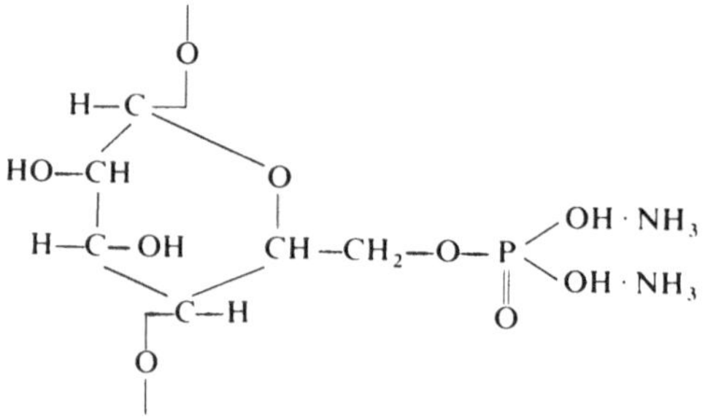

The reactivity of the substrate determines the extent and uniformity of the esterification reaction. By using Kier boiled, desized, mercerized cotton, the yield of phosphorylation is close to 100% under optimum reaction conditions.

The analysis of cellulose rendered flame retarding with the simple urea–phosphoric acid system indicates that an average degree of substitution (DS) of 0.16, corresponding to about one ester group for each six anhydroglucose units, is required to attain self-extinguishing properties. The esterified positions are probably not evenly distributed but are preferentially located in the more reactive, accessible portions of the cellulose fiber. The nitrogen content of the urea–phosphate-treated cellulose corresponds to two atoms of nitrogen per atom of phosphorus, and the structure shown can be assigned.

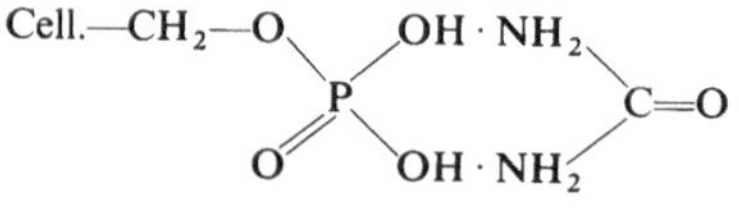

5.1.1. Acid Component

Reid and Mazzeno[51] have reviewed the reagents which have been used to phosphorylate cellulose. These include ammonium phosphate, urea–phosphoric acid, phosphorus tri- and oxychloride, monophenyl phosphate, phosphorus pentoxide, and chlorides of partially esterified phosphoric acids. Phosphoric acid can be used in the form of meta-phosphoric, orthophosphoric, pyrophosphoric, polyphosphoric, ortho-phosphorous, or pyrophosphorous acid. It appears that the form of the original acid in the higher or lower valence state has no effect on the phosphorylation reaction, since the esterification takes place at 150–200°C, where the acid is not present necessarily in the same form as it was applied.

Substituted acids can be employed for cellulose esterification provided the substituting radical does not reduce the rate of the esterification or the hydrophilic character of the cellulose. Substituents of high carbon content generally impair the effectiveness of the acid. The flame resistance of cellulose esterified with ethyl-, phenyl-, or boron-substituted phosphoric acids is inferior. Halogen-substituted phosphoric acids cause greater cellulose degradation in the course of curing or storage.

The nitrogen-substituted products of phosphoric acid are as satisfactory phosphorylating agents as the free acids. Diammonium phosphate, urea–orthophosphate, or guanidine–pyrophosphate are as effective as the free acid. Amido acids, such as phosphamic or the so-called nitride hexaphosphoric acid, and many other nitrogen-containing phosphoric acids of presently unknown structure also are effective.

The metallic salts of phosphoric acids are much less effective than the nitrogen-containing derivatives mentioned above. Monosodium phosphate imparts fair flame resistance, but disodium and trisodium phosphates or calcium and magnesium phosphates do not appreciably reduce flammability.

Gallagher[52] used sodium hexametaphosphate with a heat-curing technique to produce cellulose phosphate esters with high strength retention. Such reaction products, which contained more than 1.6% phosphorus, were insoluble in cupriethylene diamine, indicating an apparent crosslinking of the cellulose.

5.1.2. Nitrogen Component

The phosphate esters obtained by the esterification of cellulose have ionic character, the free-acid groups being neutralized by the nitrogen base. The ionized salts are capable of exchange with metallic ions and, for example, by immersion in hard water, the calcium salt of the cellulose phosphate may be formed.

$$\text{Cell.}-CH_2-OP \overset{/\,O-NH_4}{\underset{/\!/\,\diagdown O-NH_4}{}} + Ca^{2+} \longrightarrow \text{Cell.}-CH_2-O-P \overset{/\,O\diagdown}{\underset{/\!/\,\diagdown O\diagup}{}} Ca + 2NH_4^+$$

Such a replacement of the ammonium groups with alkali, alkaline earth, or other metal ions or with hydrophobic groups, to form salts which do not decompose to release free acid at the temperature of combustion, reduces or even entirely destroys the flame-retarding effectiveness. The flame-retarding property can be regenerated, however, by treatment of the ion-exchanged cellulose in mineral acid or ammonium salt-containing bath.

Laundering of phosphorylated cellulose with ionic detergents or alkali salts also leads to reduction of the flame-retarding properties through this ion-exchange mechanism. Although the hydrolytic stability of the phosphate ester of cellulose is high, strong alkaline washing might also lead to partial saponification of the cellulose ester. Naturally, if the loss of flame resistance is caused by saponification, the flame resistance cannot be regenerated by aftertreatment with mineral acids or ammonium salts.

The reaction of cellulose with urea–phosphoric acid has been thoroughly studied by Nuessle *et al.*[53] It was shown that below 170°C, an acidic monoammonium cellulose phosphate is formed. At temperatures higher than 170°C or at prolonged periods of curing, the ammonium salt loses water.

$$\text{Cell.--CH}_2\text{--O} \diagdown \diagup \text{O--NH}_4 \qquad \xrightarrow{-2\text{H}_2\text{O}} \qquad \text{Cell.--CH}_2\text{--O} \diagdown \diagup \text{NH}_2$$
$$\text{P} \qquad\qquad \text{P}$$
$$\text{O} \diagup \diagdown \text{O--NH}_4 \qquad\qquad\qquad \text{O} \diagup \diagdown \text{NH}_2$$

The cellulose amidophosphate formed does not exhibit ion-exchange properties and withstands mild alkaline launderings, but is saponified by strongly alkaline washes.

Besides amide formation, insolubilization by polycondensation reactions or the use of more complex amines or amides limits the rate of ion exchange because of the increased molecular size of the side group.

The organic nitrogen compounds used might undergo condensation, since the curing temperatures are generally above the dissociation and condensation temperatures. For example, on prolonged heat curing, urea might condense with itself to form biuret

$$2\text{NH}_2\text{--CO--NH}_2 \longrightarrow \text{NH}_2\text{CO--NH--CO--NH}_2 + \text{NH}_3$$

or with cellulose to form cellulose carbamate,

$$\text{NH}_2$$
$$|$$
$$\text{Cell.--O--C}{=}\text{O}$$

The reaction medium must be a solvent for the acid and a swelling agent for the cellulose. It must be liquid above 140°C and inert toward cellulose. At the temperature of esterification it should not decompose to form highly acidic or alkaline decomposition products.

Weak nitrogen bases with strong hydrogen-bonding groups satisfy these requirements. The esterification reaction proceeds well in the presence of formamide, acetamide, urea, biuret, dicyandiamide, and melamine. Substituted derivatives of these amides are also suitable, provided the substituents do not affect the hydrophilic character or the hydrogen bonding capability of the molecule. Halogen, alkyl, or aryl substituents unfavorably affect these properties, as demonstrated by the unsatisfactory performance of methyl urea. The introduction of amino or hydroxyl groups should alter the weak basic character of the media, e.g., guanidine (iminourea) can be regarded as unsuitable because of its strongly alkaline character.

Strongly acidic conditions are the most favorable for the esterification reaction, but the danger of hydrolytic degradation of cellulose is great under these conditions. In the alkaline range, oxidative degradation of cellulose may take place. Therefore it is necessary to buffer the system carefully to carry out the esterification at an acceptable rate with minimum degradation of cellulose. Such favorable conditions can be approximated by esterifying cellulose in the presence of an excess of urea. Such a buffer system is almost independent of the temperature, and substantial changes can be made in the treating bath without appreciably altering the pH of the solution. By using guanidine as a buffer, similar variations cause substantial changes in pH and the solution can become definitely alkaline.

A buffering system with a slightly higher pH can be attained by using a mixture of urea and guanidine carbonate. Similar effects can be obtained with other guanidine derivatives like dihydroxyguanidine, guanylurea, aminoguanidine, and biguanide. Since these chemicals are definitely alkaline, an excess of urea is necessary for an adequate buffering system. The resulting cellulose phosphates are the salts of these strong bases and are less susceptible to ion exchange.

The treatment of cellulose with urea–phosphoric acid cannot be carried out without cellulose degradation, which is caused by the reaction of cellulose with the acid and occurs in the first minutes of heat curing. Such degradation can be reduced by incorporating formaldehyde into the treating solution. In such a system the formation of urea–formaldehyde resin and also the cross-linking of cellulose must be considered. However, formaldehyde also reduces the resistance of phosphorylated cellulose to ion exchange.

The extent of the ion-exchange character of phosphorylated cellulose depends on the amount of urea used. The higher the urea content, the greater the stability of the cellulose phosphate against ion exchange. The resistance of the cellulose phosphate to ion exchange is proportional not only to the amount of urea present in the impregnating bath, but also to the extent of the conversion of urea to biuret, cyanuric acid, and melamine.

5.2.3. Phosphorylation Procedures

Besides urea–phosphoric acid, many other combinations of organic nitrogen compounds and inorganic acids are suggested for the flame-retarding treatment of cellulose, but generally more drastic heat curing is required. The higher curing temperature or the longer curing period might cause more severe fiber damage.

Many of the phosphorylated cellulose substrates which show unsatisfactory resistance to ion exchange contain the required amount of phosphate, which is on the order of 9–10%. The permanence of the flame-retarding effect is independent of the phosphate content, provided it is at least 6–7%, and depends instead upon the degree of salt formation or the extent of polymerization.

The urea–phosphoric acid procedure is the subject of several patents. After World War II a series of patents based on urea–phosphoric acid treatment of cellulose was issued to Bancroft & Sons Co.[54,55] claiming durable flame retardancy without significant cellulose degradation (Ban-Flam process).

According to the studies of Davis, Findlay, and Rogers,[56] phosphorylation starts at 130°C, and in practice the upper limit of the curing temperature is approximately 175°C. Curing time is about 2 hr at 130°C, but only 5 min at 175°C. Diammonium pyrophosphate is formed, which serves as the esterifying agent in the presence of urea. The cellulose pyrophosphate formed is converted to the orthophosphate in washing.

In order to attain self-extinguishing properties at least 3% phosphorus content is required. The concomitant strength loss is at least 20% and increases with increasing degree of phosphorylation. Further fiber damage results from the formation of hydrocellulose.

Ciba[57] claimed that the fixation of the nitrogen components in the fiber can be improved by formaldehyde aftertreatment. The cellulosic substrate is treated with an aqueous solution containing urea–phosphoric acid condensate neutralized with ammonia and a melamine–formaldehyde precondensate using a pad–heat-cure technique. After rinsing in cold water, the cellulosic substrate is treated with an aqueous solution of formaldehyde and ammonium chloride at 90°C.

Berger[58] disclosed a flame-retarding treatment with a mixture of the reaction product of urea with ortho-, meta-, or pyrophosphoric acid and methylol melamines. The ratio of urea–phosphoric acid to methylol melamine may range from 3:1 to 6:1, with at least 6% weight gain being required to attain adequate flame resistance. If the phosphorylated cellulose is aftertreated with trimethylol melamine using a pad–heat-cure procedure, an improvement in durability can be attained. Such a fabric can be washed with soap at boil without reducing the flame retardancy, provided the washed fabric is aftertreated with an ammonium salt solution.

The replacement of urea with other nitrogen compounds is the subject of many patents. Among them, Loukomsky[59] disclosed the durable flame-retardant treatment of cellulose with a combination of guanyl-melamine salts and pyrophosphoric acid. The cellulosic substrate is first impregnated with guanylmelamine hydrochloride solution (obtained by the reaction of dicyandiamide with anhydrous hydrochloric acid at 100–150°C) and then with pyrophosphoric acid. Thus water-insoluble guanylmelamine pyrophosphate is precipitated *in situ* within the fiber. The fixation of this insoluble precipitate can be further improved by resin aftertreatment.

Fluck and Moretti[60] described a flame-retarding treatment of native and regenerated cellulose using a combination of diammonium phosphate, hexamethylene tetramine, and dicyandiamide. Preferably 10–15% weight increase is required to attain adequate flame resistance. The durability of the treatment to laundering is limited.

O'Brien[61] studied the durable flame resistance and other physical properties of cotton and rayon treated with cyanamide and phosphoric acid. The single-step pad–heat-cure (curing 3–5 min at 132–154°C) procedure imparted exceptionally good durability to home launderings (approx. 50 washes) done in soft water. The flame retardancy of the treated fabric laundered in hard water was restored by soaking in 5% NH_4Cl solution, indicating that the formation of calcium salt as well as loss of nitrogen were responsible for the limited durability of flame resistance. The Pyroset CP process of American Cyanamid is based on this principle.

A suggestion has been made[62] to improve the durability of such a finish by incorporating halogenated alkyl phosphate (e.g., a mixture of chloroethyl phosphates) into the phosphoric acid–cyanamide impregnating solution. After drying and curing at elevated temperature, the treated viscose rayon contained 3.2% phosphorus and 2.5% nitrogen and withstood several home launderings.

Japanese researchers[63] have suggested the replacement of urea with ammonia or with organic amines and the substitution of phosphoric acid with a polymerized-type phosphoric acid in the urea–phosphoric acid process. The polymerized acid is termed metaphosphoric acid $(HO \cdot PO_3)_n$, in which $n = 1$ to 6 or higher. However, if $n < 3$, the fabric has an objectionable yellowing, which decreases as the degree of the condensation of metaphosphoric acid increases. Metaphosphoric acid must be present in the form of its ammonium or organic amine (mono-, di-, or trisubstituted methyl, ethyl, or propylamine or corresponding alkanolamine) salt. Certain quaternary ammonium compounds (such as tetramethyl, -ethyl, or -propyl ammonium hydroxides) can also be used in conjunction with the metaphosphoric acid. The P/N ratio should be 1:0.5–2.0. If the P/N atom ratio is higher than 1:0.5, the degradation of the cellulosic substrate is severe.

The cellulosic fabric treated with the amine salt of metaphosphoric acid is cured at 155°C for 2 min. It is claimed as an additional feature that fabric so treated is resistant to soiling.

Phosphoramides also phosphorylate cellulose in high yield.[87] While unsubstituted phosphoric triamide (PA) reacts with cellulose almost quantitatively and, on washing, most of the P–N bonds cleave to yield cellulose phosphate, substituted phosphoramides, e.g., *NN'N''*-trimethylphosphoramide, react in significantly lower yields.[348a]

Although cellulose phosphorylated with PA exhibits some limited durability, the maintenance of flame resistance after multiple launderings in hard water cannot be considered sufficient. The slight loss of phosphorus is insufficient to account for loss of flame retardancy. Laundering in soft water does not destroy the flame retardancy of cellulose phosphate. Thus ion exchange with calcium and/or magnesium is undoubtedly responsible for the lack of durability. After hard-water launderings, treatment in acid or ammonia is required to reestablish the flame-retardant properties.

Several approaches were attempted to block the ionic P—OH groups by complex formation, esterification, or amidation in order to eliminate the ion-exchange characteristics. Of the active organic compounds, after-treatment with 8% APO appeared the most effective. The low crease recoveries obtained on the phosphorylated cotton indicated that little or no cross-linking of the cellulose occurred during the esterification. Thus, the main problem in blocking the P—OH groups appears to be a decreased reactivity of the remaining acidic groups of the phosphorylating agent.[348a]

5.2. Sulfation

Braconnot[64] dissolved linen in cold concentrated sulfuric acid and obtained a highly degraded cellulose sulfate as a product. Since then many improved sulfation procedures have been reported, although most of these cause excessive cellulose degradation.

Moede and Curran[65] sulfated cellulose with a triethylamine–sulfur trioxide complex with only moderate fiber damage.

Cellulose can also be sulfated with sulfamic acid:

$$\text{Cell.—OH} + \text{H}_2\text{N—SO}_2\text{OH} \longrightarrow \text{Cell.OSO}_3\text{NH}_4$$

The nitrogen-substituted products of sulfamic acid, i.e., the ammonium or organic amine salts of sulfamic acid, are suitable sulfating agents, as is the free acid itself.

In its reactions involving sulfation, sulfamic acid resembles the SO_3–tertiary amine complexes and can be regarded as an NH_3/SO_3 complex. While the SO_3–tertiary amine complexes can be used for sulfation in aqueous alkaline medium, sulfamic acid and its salts must be employed in anhydrous medium at elevated temperature.

The previously discussed phosphorylation of cellulose proceeds in yields close to 100% under optimum conditions, but the yields of sulfation are generally low (approximately 30%), and the esterification reaction proceeds in acceptable yields only under severe curing conditions (6–10 min at 180–190°C).

Although the ammonium salt of cellulose sulfate exhibits excellent flame-retardant properties at approximately 2% sulfur content, the afterglow of sulfated cellulose is severe and may consume the fabric.

Morton and Ward[66] have described a procedure for flame-retarding cellulose by esterifying it with phosphoric or sulfuric acid or with the corresponding ammonium salts in the presence of cyanamide.

The nitrogen compounds play a role in the sulfation procedure similar to that in the phosphorylation process. In the presence of suitable nitrogen-containing additives (e.g., dicyandiamide), the sulfation proceeds at a lower curing temperature, in better yields, and with reduced fiber damage. Stiffening of the fabric is generally not a problem in sulfation procedures.

5.3. Combined Sulfation–Phosphorylation

The phosphate and sulfate esters of cellulose exhibit excellent flame-resistant properties and withstand multiple home launderings without saponification of the ester linkage. However, the undesirable characteristics of the cellulose esters—impairment of the durability of the flame resistance by the ion-exchange properties of cellulose phosphate and the severe afterglow of cellulose sulfate—hindered the acceptance of esterification as a durable flame-retarding treatment. The combination of sulfation–phosphorylation procedures has been reported[348a,b] as a feasible approach to obtain durable flame-retardant cotton with minimal afterglow. Since the main purpose of the addition of phosphorus is only to suppress the afterglow, the undesirable ion-exchange properties of cellulose phosphate might be tolerated.

Treatment of cellulose with a mixture of ammonium sulfamate (AS), urea derivatives, and diammonium phosphate (DAP) is ineffective in suppressing the afterglow because the cellulose esterification with DAP is impeded in the presence of AS. However, the phosphorylation of cellulose with PA proceeds in high yield even in the presence of sulfamate, and furthermore the PA functions as an amide in promoting the sulfation of cellulose with AS. The incorporation of methylolated amide-type cross-linkers, such as DMDHEU and especially bis(methoxymethyl)uron, is effective in minimizing hydrophilicity, strength losses, and discoloration.

The ion exchange of sulfated–phosphorylated cotton with sodium ions in soft-water laundering decreases the effectiveness of the phosphate groups in preventing afterglow, while the alkali earth ions do not impair it. The success of the combined process depends upon obtaining a well-defined P/S ratio as well as a minimum amount of S within narrow tolerances in the modified cellulose. It appears that a minimum of 1.7–1.8% S and a P/S weight ratio of 1.3–2.0 are the optimum requirements to attain flame resistance without afterglow in water with high Ca and high Na content.

Decreasing this ratio to 1.1 results in an afterglow of approximately 30 sec in the charred area.

5.4. Mesylation and Tosylation

The chemical modification of cellulose with methane sulfonyl chloride and *p*-toluene sulfonyl chloride in the presence of a tertiary amine, e.g., pyridine, has been studied by Pacsu and Schwenker.[67,68] The mesyloxy groups in the C-6 position were replaced with halogen:

$$Cell.OH + ClSO_2CH_3 \xrightarrow{\text{pyridine}} Cell.OSO_2CH_3$$

$$Cell.OSO_2CH_3 + NaX \longrightarrow Cell.X + CH_3SO_3Na$$

$$(X = I, Br, Cl, F)$$

It has been demonstrated that the substitution of the primary alcohol group with a mesyl group or with a halogen atom (Br, I) significantly reduces the formation of levoglucosan, increases the amount of dehydration products, and imparts permanent flame-retardant properties. Among the mesyl/halogen cellulose derivatives, the bromo and iodo derivatives are more effective in suppressing the flammability than the mesyl cellulose. The effectiveness of chloro/mesyl cellulose is comparable to that of mesyl cellulose, while the fluoro derivative is more flammable than mesyl cellulose. The afterglow of these cellulose derivatives is objectionable, but can be suppressed by phosphorylation, for instance, by treating mesyl cellulose with dialkyl phosphoryl chloride in pyridine medium.

5.5. Phosphoric and Phosphorous Acid Derivatives

The inorganic salts of phosphoric acid have previously been shown to play an important role in the flame-retardant finishing of cellulose. The organic derivatives of phosphoric acid have also been suggested as flame retardants for cellulose.

5.5.1. Phosphoric and Phosphorous Acid Esters

Phosphoric acid esters such as tricresyl phosphate or triphenyl phosphate are well-known softeners for plastics and, although they are effective in reducing flammability, they are not widely used because of their toxicity.

The esters of phosphoric acids with polyvalent alcohols have been suggested as flame retardants especially in conjunction with polyalkylene imines.[69,70] The cellulosic substrate is pretreated with 8–12% polyethylene imine solution, dried, and subsequently impregnated with the phosphoric acid ester. Specifically, dipentaerythritol hexaorthophosphate was suggested as the ester. Aftertreatment with polyalkoxymethyl melamine is said to improve the durability of the flame-retarding finish.

Weaver, Frick, and Reid[71] suggested the treatment of cellulose with an aqueous solution of the ammonium salt of the phosphoric acid ester of a polyhalogen propanol, e.g., bis(2,3-dibromopropyl)phosphate, in conjunction with melamine–, urea–, or guanidine–formaldehyde precondensate. Miles and Delasanta[72] described the treatment of cotton fabrics and garments from chlorinated solvents with tris(2,3-dibromopropyl)phosphate in conjunction with polyvinyl chloride, polyvinyl acetate, or their copolymer. Fabric finished this way remained flame resistant after 15 accelerated launderings.

Redfarn[73] disclosed the incorporation of tris(2,3-dibromopropyl)-phosphate into the spinning solution of cellulose triacetate. Courtaulds, Ltd.[74,75] has suggested the incorporation of this compound into viscose solution. Such an addition to the dope before extrusion avoids the need for a later flame-retardant treatment of the viscose yarn or fabric.

Among the phosphoric acid esters, the triallyl ester gained special importance because it can be relatively easily polymerized and the double bonds can be brominated.[76]

Bromo triallyl phosphate (BAP) polymer is obtained by the telomerization of a polybrominated methane with an unsaturated organic phosphate.

Triallyl phosphate and bromoform are used as the most readily available compounds of these classes. They can be substituted with other compounds such as carbon tetrabromide and trimethallyl phosphate. This reaction proceeds in emulsion by a free radical polymerization mechanism producing a stable emulsion of a polymer or telomer. The polymer retains some unsaturated groups and can be further polymerized by heat curing after application to the cellulosic substrate to produce a flame-retardant finish durable to at least 12 launderings.

Similar polymers, but containing chlorine as the halogen, are also suitable for use as flame retardants for cellulose.[77] Such polymers are attained by the reaction of triallyl phosphate with a halomethane, especially carbon tetrachloride or methylene chloride, in the presence of organic peroxide.

Predvoditelev, Nifantev, and Rogovin[78] have reported the synthesis of cellulose phosphites by the action of monomethylphosphite on cellulose. Yuldashev, Muratova, and Askarov[79] reported the transesterification of cotton cellulose by the reaction of phosphorous acid esters with chlorocellulose. The phosphorus-containing derivatives of cellulose thus produced possess reduced flammability and strong bactericidal activity. They have been obtained by the reaction of chlorocellulose with both the neutral and acid esters of phosphorous acid, such as di- and trialkyl phosphites, although esterification proceeds more readily with the acid than with the neutral esters. The degree of modification depends on the chlorine content of the chlorocellulose and on the reactant ratios, reaction time, and reaction temperature.

The flame-retardant treatment of cellulose with organic phosphites in conjunction with chlorinated paraffins and naphthalenes is reported by Truhlar.[80]

Durable flame-retardant cellulose derivatives have been prepared by Tesoro and Sello[81,82] by treating cotton with tris(2-chloroethyl)phosphite in DMF medium. The transesterification product can isomerize to a hydrolytically stable intermediate capable of further reaction, as shown in the following equations:

$$\text{Cell.OH} + \text{P(OC}_2\text{H}_4\text{Cl})_3 \longrightarrow \text{Cell.OP(OC}_2\text{H}_4\text{Cl})_2$$

$$\longrightarrow \underset{\underset{\text{C}_2\text{H}_4\text{Cl}}{|}}{\text{Cell.O—P(O)OC}_2\text{H}_4\text{Cl}} \xrightarrow{\text{P(OC}_2\text{H}_4\text{Cl})_3} \underset{\underset{\text{C}_2\text{H}_4\text{P(O)(OC}_2\text{H}_4\text{Cl})_2}{|}}{\text{Cell.OP(O)OC}_2\text{H}_4\text{P(O)(OC}_2\text{H}_4\text{Cl})_2}$$

The chemical stability of the bound phosphorus in this product confirms the occurrence of the isomerization reaction.

Trialkyl phosphites have been used also for the Arbuzov reaction of haloacetamidomethyl cellulose to obtain flame-retarding cellulose phosphate esters.[81,83]

5.5.2. Phosphoramides

The alkaline hydrolytic stability of organophosphorus compounds containing phosphorus bonded to nitrogen or carbon was studied by Jones and Noone.[84] They concluded that compounds containing P–N bonds, such as phosphoramides, can be more easily hydrolyzed by alkali than compounds containing P–C bonds. Alkali-resistant, flame-retardant agents should preferably have the P atom bonded through C atom to electron-donating substituents.

Phosphoramide, $\text{P(O)(NH}_2)_3$, by itself is too susceptible to hydrolysis.[85] The introduction of substituents on the amido groups, however, increases the resistance of the phosphorus–nitrogen bond to hydrolysis.

Malowan[86] disclosed the flame-retarding treatment of cotton with a phosphoramide made from phosphoryl chloride and ammonia, followed by heating at 110–150°C. Suggested formulas for this material are

$$
\begin{array}{c}
O \\
\parallel \\
P-ONH_4 \\
HN \qquad NH \\
| \qquad | \\
O-P \qquad P-O \\
H_2N \quad N \quad NH_2 \\
H
\end{array}
$$

or

$$
H_2N-\overset{\overset{\displaystyle O}{\parallel}}{\underset{\underset{\displaystyle NH_2}{|}}{P}}-NH\left[\overset{\overset{\displaystyle O}{\parallel}}{\underset{\underset{\displaystyle NH_2}{|}}{P}}-NH\right]_x\overset{\overset{\displaystyle O}{\parallel}}{\underset{\underset{\displaystyle NH_2}{|}}{P}}-NH_2
$$

Although the exact chemical structure cannot be formulated, the

$$
\begin{array}{c}
O \\
\parallel \\
-P-NH- \\
| \\
NH_2
\end{array}
$$

group is an important part of the molecule. On heating under anhydrous conditions, this product loses ammonia and becomes acidic.

Nielsen[87] treated cotton fabric with phosphoramide using a heat-curing technique. It was shown that phosphoramide reacted with cellulose, and the alternate possibility of self-condensation was excluded. The samples with 3.5% P content resisted 20 launderings. The treated cotton also had excellent antifungal properties. Although the reacted cellulose exhibited ion-exchange character, the effectiveness of the phosphorus moiety as a flame-retardant agent was not impaired. It appeared that the —P(O)—NH— structure of the phosphorylamide was retained as a chain of uncertain length. For every four phosphorus–nitrogen units, there were found to be approximately two ammonium groups, one hydroxyl, and one amido group. The imido nitrogen accounts for the unique properties with respect to ion exchange compared with the simple cellulose phosphate.

Monsanto[88] disclosed the incorporation of the reaction product of phosphoroxychloride and ammonia (N/P atom ratio 2:1 to 3:1) into the aqueous spinning solution of regenerated cellulose. The flame resistance of the rayon yarn is claimed to be excellent.

Frick et al.[89] investigated phosphoramides as flame retardants for cotton; they were partially substituted with low-molecular-weight substituents to give resistance to hydrolysis. These amides were converted to methylol derivatives by reaction with formaldehyde. By applying these methylolated phosphoramides in conjunction with other N-methylol compounds, durable flame retardancy could be attained. The best finish was obtained by applying N,N',N''-trimethyl-N,N',N''-tris(hydroxymethyl)-phosphoramide, $P(O)[N(CH_3)CH_2OH]_3$, in conjunction with trimethylol melamine using an acid-catalyzed heat-curing technique.

By coreacting N,N',N''-trialkyl phosphoramide with trimethylol melamine (TMM) *in situ* in the fiber, excellent durable flame resistance can also be attained.[349a,b]

There is a large difference in efficiency between the phosphoramide used by itself and the phosphoramide which has been cross-linked with a melamine resin. The enhancement of the flame-retardant activity in this case results from physical rather than chemical effects. Phosphorus analyses on chars recovered from the burning of samples containing the low-molecular-weight, volatile phosphoramide alone indicate that a considerable quantity of the phosphoramide is lost from the fabric during burning. When the system is cross-linked with the melamine resin, however, its volatility is decreased and almost all of the phosphorus remains in the fabric after burning. The chemical action on the cellulose substrate is probably the same whether the melamine resin is present or not. However, when the resin is present, more of the reagent is in contact with the cellulose at the time of pyrolysis.[350]

Maeder[90] suggested the reaction product of diethyl phosphoramide and chloral as a flame-retardant agent:

$$(C_2H_5O)_2P(O)NH_2 + OHC\!-\!CCl_3 \longrightarrow (C_2H_5O)_2P(O)NHCHOHCl_3$$

Courtaulds Ltd.[91] disclosed the flame-retarding treatment of cellulose acetate by incorporating bis(2-chloroethyl)phosphoramide, $(ClCH_2CH_2O)_2$-$P(O)NH_2$, into the spinning solution or impregnating the extruded fiber.

Glade[92] disclosed the simultaneous flame-retardant and water-repellent treatment of a cellulosic substrate using melamine–formaldehyde condensate in conjunction with phosphoramides of the following generic formula:

$$\underset{X}{\overset{RO}{\diagdown}}\!\!P\!\!\overset{\diagup O}{\underset{\diagdown NHR'}{}}$$

R = alkyl, alkoxyalkyl, or haloalkyl with 1–4 C atoms
R' = H or alkyl with 1–4 C atoms
X = a second RO or a second NHR' group

Frick *et al.*[89] investigated the conversion of partially substituted phosphoramides to imides. The thermal deamination of N,N',N''-trimethylphosphoramide did not yield the expected monomeric N.N′-dimethyl-phosphoramidic imide,

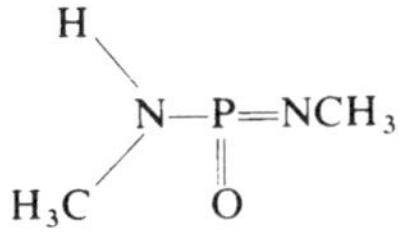

$$
\begin{array}{c}
H \\
\diagdown \\
\quad N-P=NCH_3 \\
\diagup \quad \parallel \\
H_3C \quad\quad O
\end{array}
$$

but it gave a polymeric product. By applying this product to cotton substrate from aqueous alkaline solution with heat-curing techniques, good flame resistance was obtained originally, but the finish was not durable.

The thermal deamination of N,N′,N″-trimethylthiophosphoramide gave a dimeric product which was a less effective flame retardant than the oxygen analog and showed limited durability only after excessive heating. Under these conditions the fiber damage was substantial.

5.5.3. Aziridines

Tris(1-aziridinyl)phosphine oxide (APO) and tris(1-aziridinyl)-phosphine sulfide (APS) are the most important compounds of this category.

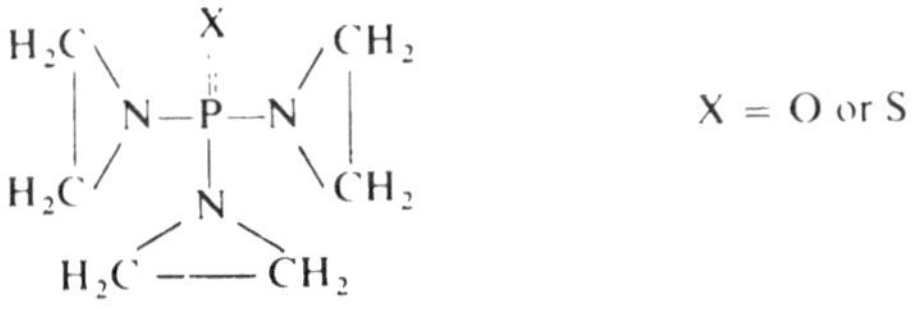

$$
\begin{array}{ccc}
H_2C \diagdown & X & \diagup CH_2 \\
\mid \quad \diagdown & \parallel & \diagup \quad \mid \\
\mid \quad N-P-N & & \mid \\
H_2C \diagup & \mid & \diagdown CH_2 \\
& N & \\
H_2C & \diagdown & \\
& \text{---} CH_2 &
\end{array}
\qquad X = O \text{ or } S
$$

They are both highly reactive compounds which are obtained by reacting POCl$_3$ or PSCl$_3$ with ethylene imine using a tertiary amine as an acid acceptor.

$$
\begin{array}{c}
\quad\quad\quad H_2C \diagdown \\
PCl_3 + 3 \quad\quad \mid \quad NH \longrightarrow \quad P\!\!\left[\!\!\begin{array}{c} \diagup CH_2 \\ N \quad\mid \\ \diagdown CH_2 \end{array}\!\!\right]_3 \\
\parallel \quad\quad\quad H_2C \diagup \quad\quad\quad \parallel \\
X \quad\quad\quad\quad\quad\quad\quad\quad X
\end{array}
$$

$$X = O \text{ or } S$$

APO can be reacted with cellulose by an acid catalyzed heat-curing technique. Drake and Guthrie[93] reported the cross-linking of cotton with APO in the presence of zinc fluoroborate catalyst. The impregnated fabric was cured at 140°C for 4 min.

Under the conditions of acid-catalyzed heat curing, both chemical reaction with cellulose and polymerization may take place.[13]

$$
n\ \begin{array}{c} H_2C\text{---}CH_2 \\ \diagdown\ \diagup \\ CH_2\quad N \\ |\quad \diagdown \\ |\quad N\text{---}P{=}O \\ CH_2\quad \diagup \\ \diagdown\ N \\ H_2C\text{---}CH_2 \end{array}
\ \longrightarrow\
\left[\begin{array}{c} \quad\quad N\text{---}CH_2\text{---}CH_2 \\ H_2C\diagdown\quad\quad \diagdown \\ \quad N\text{---}P{=}O \\ H_2C\diagup\quad \diagup \\ \quad N \\ H_2C\text{---}CH_2 \end{array}\right]_n
\xrightarrow{\text{Cellulose}}
$$

$$
\left[\begin{array}{c} CH_2\quad N\text{---}CH_2\text{--}CH_2 \\ |\quad \diagdown \\ |\quad N\text{--}P{=}O \\ CH_2\quad \diagup \\ \diagup N \\ H_2C\text{---}CH_2 \end{array}\right]_{n-1}
\ +\
\begin{array}{c} CH_2\quad N\text{---}CH_2\text{---}CH_2\text{---} \\ |\quad \diagdown \\ |\quad N\text{---}P{=}O \\ CH_2\quad \\ \quad NH \\ \quad | \\ \quad CH_2 \\ \quad | \\ \quad CH_2\text{---}OCell. \end{array}
$$

Miles, Hoffman, and Meiola[94] studied the effectiveness of catalysts. Viscose films were treated with APO in the presence of various acid catalysts, and after curing (140°C for 5 min) the condensation products were studied by infrared spectroscopy. Best results were attained with diammonium phosphate catalyst. The stability of the treating bath and the durability of the flame-retardant finish were excellent.

APO-treated cotton turns yellow on hot ironing subsequent to hypochlorite bleaching,[95] although neither the hypochlorite bleaching nor scorching alone causes yellowing. It is believed that in the bleaching process some primary amino groups are formed by the hydrolysis of the N–P–N bond in APO and alcohol groups are formed and oxidized to aldehyde as a result of the cleavage of the aziridine rings. Combination of the primary amine and aldehyde groups leads to the formation of Schiff bases which decompose on heating to yellow products. Therefore it is preferable to use sodium perborate or sodium persulfate rather than sodium hypochlorite for the bleaching of APO-finished cotton.

By reacting APO or APS with polyfunctional amines, amides, or with N-methylol amido compounds in acidic, alkaline, or neutral aqueous solution, phosphorus-containing polymers are obtained.[96] The cellulosic substrate is treated with the aqueous solution of the monomers or with the aqueous dispersion of a partially polymerized product. Melamine and urea are the preferred amides and ethylene diamine is the preferred amine. For example, by applying a mixture of APO and ethylene diamine in equimolar ratio (pH = 12) with a pad–cure technique, good durable flame retardancy is attained without substantial fiber damage.

Chance, Drake, and Reeves[97] suggested the flame retarding of cellulosic substrate with aziridine derivatives (in combination with polyols) of the following formula:

$$H_2C \diagdown \quad X \quad \diagup CH_2 \qquad X = O \text{ or } S$$
$$N\text{-}P\text{-}N$$
$$H_2C \diagup \quad Y \quad \diagdown CH_2 \qquad Y = -N \diagup{CH_2} \diagdown{CH_2} \quad or \quad N \diagup{CH_3} \diagdown{CH_3}$$

One or two hydroxyl groups must be present for each aziridine group; thus, ethylene glycol, triethanolamine, diethanolamine, or pentaerythritol are suitable coreactants. After curing at 100–170°C for 10–2 min, excellent flame retardancy can be attained. The same authors[98] disclosed the flame-retarding treatment of cotton with phosphorus-containing aziridinyl–polyol polymers. These polymers are obtained by heating APO or APS with polyols in aqueous solution.

The reactions of modified cellulose with APO or APS are disclosed by Reeves *et al.* A flame-retardant, cross-linked cellulose product is obtained by reacting APO or APS with carboxyalkyl cellulose.[99]

$$\overset{O}{\overset{\|}{-P}}-N-CH_2-CH_2-O-CO(CH_2)_x-OCell.$$

They also reported the reaction of APO and APS with phosphonoalkyl cellulose.[100]

The simultaneous application of APO and thiourea[101,102] with a conventional pad–heat-cure technique yields a very effective flame-resistant finish which exhibits good durability in laundering and dry cleaning. The ordinary thiourea can be replaced by other thioureas such as

$$SC \diagup{NH_2} \diagdown{NHCH_3} \qquad SC \diagup{NH_2} \diagdown{NHCH_2OH} \qquad SC \diagup{NHC_2H_5} \diagdown{NHC_2H_5} \qquad SC \diagup{NHCH_2OH} \diagdown{NHCH_2OH}$$

The replacement of thiourea by urea leads to objectionable fumes in the course of heat curing.[103] In general the APO/thiourea molecular ratio ranges from 0.5 to 2.0. Such a mixture reacts with cellulose without the presence of catalyst, but the reaction is slow and it is preferable to use $Zn(BF_4)_2$ catalyst. It is believed that in the course of heat curing, APO and thiourea yield phosphorus-, nitrogen-, and sulfur-containing polymers which chemically combine with the hydroxyl groups of the cellulose.

LeBlanc and Symm[104] disclosed the flame-retarding treatment of cellulosic substrate with a composition comprised of a polyaziridinyl phosphoramide and a sulfamide derived from ammonia or a primary alkyl amine.

Kassack, Malz, and Lober[105] suggested the flame-retarding treatment

of cellulose by coapplying phosphorus-containing polyazirinyl derivatives
with phosphoramides in the presence of an acidic catalyst such as ammonium
chloride. The suggested aziridine derivatives are APO or compounds with
the following structures:

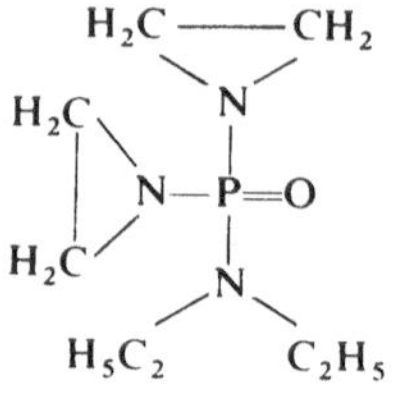

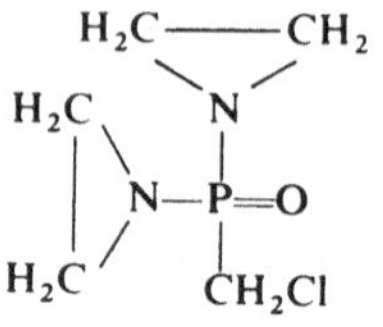

$$
\begin{array}{cc}
\text{(structure 1)} & \text{(structure 2)}
\end{array}
$$

and the suggested phosphoramide is N,N-dialkylphosphoramide.

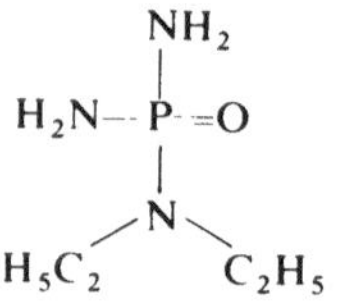

$$
\begin{array}{c}
NH_2 \\
| \\
H_2N-P=O \\
| \\
N \\
H_5C_2 \qquad C_2H_5
\end{array}
$$

Similarly, Steinhauer[106] disclosed the flame-retarding treatment of
cellulose with 10–35% aqueous solution containing APO and a phosphoramide obtained by reacting P_2O_5 with ammonia or ethylene diamine or by
reacting $POCl_3$ with ammonia.

The disadvantages of the APO-type finishes are the toxicity of the
compounds and the high loss in tensile strength and yellowing of the treated
fabric.[107]

5.6. Phosphonitrilic Halides and Their Derivatives

Phosphonitrilic chloride is obtained by reacting phosphorus pentachloride with ammonium chloride in an inert solvent (e.g., tetrachloroethane)
at 140°C.

$$nPCl_5 + nNH_4Cl \longrightarrow (PNCl_2)_n + 4nHCl$$

The tri- and tetramers are the most readily formed polymers:

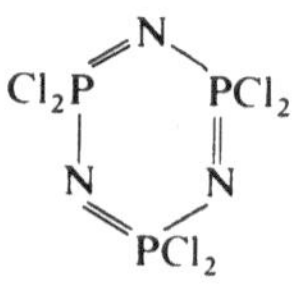

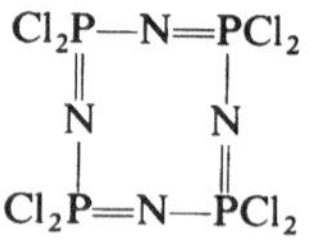

$$
\begin{array}{cc}
\text{(triphosphonitrile structure)} & \text{(tetraphosphonitrile structure)}
\end{array}
$$

These polymers are very good flame-retarding agents, but since they are
water-insoluble, they must be applied from organic solvent, e.g., benzene or
alcohol.

By reacting phosphonitrilic chloride with dry ammonia in carbon tetrachloride, a water-soluble product is obtained which can be used for the flame-retarding treatment of cellulose.[108] The cellulose substrate is impregnated with the aqueous solution of the phosphonitrilic polymer and then cured at a temperature above 100°C.[109,110]

The halogen atoms of phosphonitrilic halides can be reacted with sodium alcoholates to obtain the corresponding esters. Hamalainen[111] disclosed the flame-retarding treatment of cellulose with the reaction product of polymeric phosphonitrilic halide and polyhalogen alcohols, especially 2,3-dibromopropanol.

By reacting polymeric phosphonitrilic halide with sodium allyl alcoholate, the unsaturated ester is obtained. The allyl ester (PNE) can

$$(H_2C=HC-H_2CO)_2P \overset{N}{\underset{N}{\overset{\displaystyle \diagdown}{\bigcirc}}} P-(OCH_2CH=CH_2)_2$$

$$(OCH_2CH=CH_2)_2$$

then be reacted[112] with liquid bromine in benzene. The resulting 2,3-dibromopropyl ester gives a durable flame-retardant finish on cellulosic substrates.

The bromoform adduct of the allyl ester[112,113] was prepared by a procedure similar to that described[114] for the preparation of a flame retardant by the telomerization of triallyl phosphate with bromoform. PNE and bromoform are heated in aqueous emulsion in the presence of potassium

$$2n\mathrm{CHBr_3} + \left[NP \overset{OCH_2-CH=CH_2}{\underset{OCH_2-CH=CH_2}{\diagup}} \right]_n \longrightarrow \left[-N=P \overset{OCH_2-CHBr-CH_2-CHBr_2}{\underset{OCH_2-CHBr-CH_2-CHBr_2}{\diagup}} \right]_n$$

persulfate catalyst. The resulting polymer is also an effective, durable flame retardant.

Phenylaminophosphonitrilic polymers[115] obtained by reacting phenyl phosphorus tetrachloride with ammonium chloride are also suggested for the flame-retarding finish of cotton.

Permanently flame-retardant rayon has been obtained by incorporating liquid phosphonitrilate polymer of the following generic formula into the viscose solution:

$$\left[-N=P \overset{OR}{\underset{OR'}{\mid}} - \right]_n$$

where R and R' are the same or different alkyl or alkenyl radicals.

A liquid mixture of trimer, tetramer, and higher cyclic polymers is pumped into the viscose line supplying the spinnerettes. The polymer is thus dispersed readily and evenly throughout the fiber and only slightly impairs the tensile strength, appearance, and general physical properties. Since the flame-retardant rayon differs only slightly from ordinary rayon, it can be processed in the normal fashion.

5.7. Phosphines and Phosphine Oxides

Phosphines, phosphine oxides, and phosphine sulfides are not widely used for the flame-retarding treatment of cellulosic substrates.

The preparation of methylol phosphine compounds is disclosed in several patents of Farbwerke Hoechst A.G. Tris(hydroxymethyl)phosphine, $P(CH_2OH)_3$, is obtained by reacting 1 mol PH_3 with 3 mol CH_2O in the presence of cadmium, iron, or mercury catalyst.[117]

A flame-retardant finish based on tris(hydroxymethyl)phosphine (THP), methylolurea (MU, 1.5 mol formaldehyde per mole of urea), and trimethylol malanine (TMM), was developed at the Southern Regional Research Center of the United States Department of Agriculture using $MgCl_2$–citric acid catalyst.[351a] Better yield of insolubilization, and thus better durable flame resistance, has been attained by using a lactic acid–$Al_2(OH)_5Cl$ catalyst.[351b]

Alkylamino phosphines are suggested as flame retardants by Coates and Hoye.[118] They are obtained by reacting hydroxymethylphosphonium salts with primary and secondary amines

$$CH_3P^+(CH_2OH)_3Cl^- + 2(C_2H_5)_2NH \longrightarrow$$

$$CH_3—P[CH_2N(C_2H_5)_2]_2 + CH_2O + HCl + 2H_2O$$

or by reacting phosphine with primary or secondary amine and formaldehyde:

$$PH_3 + 3C_6H_5NH_2 + 3CH_2O \longrightarrow P(CH_2NH—C_6H_5)_3 + 3H_2O$$

Tris(hydroxymethyl)phosphine oxide (THPO), $O{=}P(CH_2OH)_3$, is obtained by the oxidation of THP.[119] In general the reaction of THPO with bifunctional molecules yields relatively low-molecular-weight, water-soluble polymers. THPO shows low reactivity toward esterification with carboxylic acids, but it readily reacts with anhydrides and acyl halides to form esters and adds to diisocyanates to yield polyurethanes. THPO exothermally reacts with urea to yield a product of the following structure:

$$+NH—CONH—CH_2P(O)—CH_2)_x$$
$$\underset{\displaystyle CH_2OH}{|}$$

Nitrogen-containing tertiary phosphine oxides are suggested by Rauhut and Semsel[120] as flame retardants. They are prepared by reacting elementary phosphorus with unsaturated amides in alkaline aqueous medium. By using acrylamide, the phosphine oxide $(H_2NCOCH_2CH_2)_3P{=}O$ is obtained.

The methylolated derivative of this carboxamide[352] and that of the corresponding carbamate, $P(O)CH_2CH_2OCONH_2$,[353] were evaluated as durable flame-retarding agents for cotton fabrics. Since they do not co-condense with trimethylol melamine in any appreciable yield, the two components can be employed in any combination to attain an optimum N/P atom ratio. A phosphorus content of 1.6–1.8% attained in cotton fabric with the phosphine oxides alone is insufficient to impart self-extinguishing properties. At this phosphorus level additional amide nitrogen to give an N/P ratio of at least 6, must be supplied.

The compound 3(dimethylphosphinyl)propionamide[354]

$$CH_3 \diagdown \!\!\!\!\!\!\!\!\!\!\!\! \underset{CH_3 \diagup \ \overset{\|}{O}}{P} \ CH_2CH_2CONH_2$$

was prepared by reacting acrylamide with dimethylphosphine oxide. The methylolated derivative of this carboxamide reacted with cellulose under acid-catalyzed, heat-curing conditions produces a flame-resistant fabric.

Tris(hydroxymethyl)phosphine sulfide (THPS) is suggested as a flame retardant by Reuter and Jakob[121] and is prepared from THP by reacting it with sulfur. THPS can be used as flame retardant in conjunction with aminoplast resins.

5.8. Phosphonium Salts

5.8.1. Tetrakis(hydroxymethyl)phosphonium Salts

The breakthrough in flame-retardant finishes[122] came in 1953 when tetrakis(hydroxymethyl)phosphonium chloride was first suggested at the Southern Regional Research Center of the U.S. Department of Agriculture for the flame-retarding treatment of cellulose substrates. It is now one of the most widely used phosphorus compounds imparting durable flame resistance to cellulosic substrates. Its abbreviation, THPC, is now a registered trademark of Hooker Chemical Co.

The compound, a crystalline solid soluble in water and lower alcohols, was first described in 1921 by Hoffman.[123] It is prepared from phosphine, formaldehyde, and hydrochloric acid at room temperature in 90%

yield.[124]

$$PH_3 + 4CH_2O + HCl \longrightarrow (HOCH_2)_4P^+Cl^-$$

It is reductive in character and more reactive than THPO.[125]

The methylol groups in THPC react with amines to form insoluble polymers.[126]

$$(HOCH_2)_4P^+Cl^- + RNH_2 \longrightarrow [(RNCH_2)_3P = O]_n$$
$$\text{(polymer)}$$

The phosphine oxide structure present in the final polymeric product is hydrolytically very stable.

Reeves, McMillan, and Guthrie[127] first suggested the flame-retarding treatment of aminated cellulose with THPC:

$$2Cell.OCH_2CH_2NH_2 + (HOCH_2)_4P^+Cl^- \longrightarrow$$

$$Cell.OCH_2CH_2NHCH_2P(O)CH_2NHCH_2CH_2OCell.$$
$$|$$
$$CH_2OH$$

This reaction proceeds very easily; impregnation of the aminated cellulose with aqueous THPC solution and subsequent drying at 100°C is sufficient to obtain a durable flame-resistant cellulose derivative.

THPC also reacts with many other chemicals containing active hydrogens such as *N*-methylol compounds, phenols, and polybasic acids[128,129] to form insoluble polymers in the cellulosic substrate.

5.8.1.1. THPC-Amide Finish—Heat Cure. The condensation of THPC with polyfunctional amides such as urea is shown in the following equation:

$$(HOCH_2)_4P^+Cl^- + NH_2CONH_2 \longrightarrow [(HOCH_2)_3P^+CH_2NHCONH_2]Cl^- + H_2O$$

or

$$(HOCH_2)_2P(O)CH_2NHCONH_2 + HCl + CH_2O + H_2O$$

The cocondensation of THPC with polymethylol melamine is schematically illustrated on the following page.

The water-soluble condensation product still contains halide ion, but only traces of halogen can be found in the fully condensed heat-cured product.[130] The condensation proceeds faster with melamine and its derivative than with urea.

THPC itself is reactive toward cellulose, but the reaction is slow. The treatment of unmodified cellulose with THPC alone yields cross-linked product insoluble in cuene and imparts some flame resistance, but the durability of the effect is limited. The coapplication with aminoplast resins ensures better and more durable flame-retarding properties.[103]

The THPC-amide finish[103,124,126,131-137] and some modifications of it are commercially used for the durable flame-retarding treatment of heavy- and medium-weight cotton fabrics.

$$(HOCH_2)_4P^+Cl^- + HOCH_2NH-C \overset{N}{\underset{N}{\diagup}} \overset{}{\diagdown} C-NHCH_2OH \longrightarrow$$

(with melamine ring bearing NHCH_2OH)

$$-NH\left[CH_2-P^+(CH_2OH)-CH_2-NH-C\underset{}{\overset{N}{\diagup}}C-NH\right]-CH_2- \overset{-O_2}{\longrightarrow}$$

$$-NH\left[-CH_2P(O)-CH_2-NH-C\underset{}{\overset{N}{\diagup}}C-NH\right]-CH_2-$$

The treatment of cellulosic substrates with a composition consisting of THPC, polymethylol melamine, urea, and triethanolamine with pad-heat-cure technique[138] imparts a flame-resistant finish which is durable to commercial and home launderings and dry cleanings. A chemical add-on of 18–20% is required to attain satisfactory flame and glow resistance. In the course of polycondensation, HCl is released from the THPC, which might lead to severe fiber damage. However, the incorporation of urea into the pad bath not only improves the yield of insolubilization, but also serves to reduce the fiber damage. Besides the THPC–melamine cocondensation, a THPC–urea condensation to form a polymeric product must also be considered.

THPC, polymethylol melamine, and urea condense in aqueous solution within several hours even at room temperature to form a viscous solution or gel. The adjustment of the pH of the pad solution to approximately 7 (with triethanolamine) significantly increases the stability of the aqueous pad bath by hindering any premature condensation of the ingredients. Triethanolamine also assists in protecting the fibers from excessive tendering during the curing stage by acting as an acid acceptor in neutralizing the HCl that is liberated.[103]

Although the THPC–amide process is very effective, its primary limitations[126] are stiffness of the treated fabric and reduction in its breaking strength. There is slight increase in crease recovery, indicating that some cellulose cross-linking occurs under the acidic heat-curing conditions. Miles, Hoffman, and Delasanta[139] suggest the aftertreatment of THPC–TMM-finished cellulosic textiles with dilute mineral acid at 75–100°C to increase the breaking strength.

Another disadvantage of the THPC-amide process is that the ingredients of the impregnating composition are prone to migrate through the fabric during the drying and curing stage, leading to unevenly flame-resistant and locally excessively tendered fabric. In order to minimize the migration and stiffening, the impregnated fabric must be air-dried at relatively low temperature ($\sim 105°C$) to avoid premature polymer formation on the fiber surface.

The THPC-amide formulation has been modified in several ways with the objective of reducing cost and improving the physical and esthetic properties of the flame-resistant cellulosic substrate. For instance, the replacement of triethanolamine with inorganic alkali (such as NaOH, Na_2CO_3, or $NaHCO_3$) is claimed[140,141] to give a more stable padding liquor and to permit a greater latitude in curing conditions. The cellulosic substrate is impregnated with the polymerizable composition consisting essentially of THPC; a cyclic copolymerizable nitrogen-containing resin, such as trimethylol melamine or dimethylol alkylene urea; a copolymerizable carbamic acid derivative, such as urea, thiourea, dicyandiamide, etc.; and an inorganic alkali as stabilizer. The fabric is then dried and cured at a suitably high temperature for a relatively short period to induce enough copolymerization of the ingredients so that they withstand leaching in aqueous ammonium hydroxide aftertreatment, which is effective as the final stage of curing.

The Southern Regional Research Center of the U.S. Department of Agriculture studied THPC finishes using several nitrogen-containing compounds in an attempt to obtain new flame-retardant finishes for cotton which would be cheaper and/or produce better properties. One property which is of importance in flame-retardant cotton finishes is that of durable press. Some of the nitrogen compounds which have been studied recently in conjunction with THPC are guanazole, monomethylglycoluril, and dimethylol cyanoguanidine.

Cotton fabric was treated with THPC and dimethylol cyanoguanidine (DMCG) to impart flame resistance and excellent improvement in dry-wrinkle recovery. At 17% add-on, the treated fabric had conditioned wrinkle-recovery angles greater than 290°. It also passed the vertical flame test after 50 home-laundry cycles.[355]

Recent work has shown that the reaction of THPC with urea can be sufficiently catalyzed by sodium phosphate salts, thus other reactive nitrogen compounds are not needed.[356] A 1:1 THPC/urea mole ratio was selected as optimum. Mono- and dibasic phosphate catalysts gave excellent durability; soluble tribasic phosphate imparted insufficient durability. Based on the durability of flame resistance and on the strength retention of the treated cotton fabric, Na_2HPO_4 was selected as catalyst in a concentration 3–6% to obtain optimum results. The pH should be adjusted to a level of 6.0 and curing should be 1–2 min at 160°C.

The rapid polymerization of the THPC-amide finish and the release of formaldehyde generally cause extreme stiffening. The addition of 1.0–1.4 mol of a sulfite or bisulfite per mole of THPC reduces the stiffening by acting as a formaldehyde acceptor.[142] Sodium, potassium, ammonium sulfites and bisulfites, triethanolamine sulfites, etc., are suggested for this purpose.

Hooker Chemical Co.[143,144] also suggests modified formulations containing antimony oxide and organic chloro compounds, such as polyvinyl chloride, chlorinated paraffin, or a mixture of the two. The chlorinated polymers and the antimony oxide are less effective flame retardants than the THPC-amide and thus, by partially substituting the THPC-amide with these ingredients, a higher total chemical add-on is required to attain durable flame-resistant properties comparable to those obtained with THPC-amide only.

The combination of the THPC-amide process with the bromoform adduct of triallyl phosphate (BAP) is a further modification suggested by the Southern Regional Division of the U.S. Department of Agriculture.[145] BAP emulsion can be directly added to the THPC-amide formulation. However, BAP is not quite as effective a flame retardant as the THPC-amide finish.[126] The THPC-amide monomeric finish penetrates into the fiber, while the BAP polymeric finish coats the fiber surface, thus making possible a wider range of application of the THPC-amide finish.

The bromine-containing phosphonitrilates, such as 2,3-dibromopropylphosphonitrilate or the bromoform adduct of allylphosphonitrilate are good flame retardants and can be added in aqueous emulsion to the THPC-amide finish.[126,146] The bromoform–phosphonitrilic allyl ester adduct (BPNE) increases the flame-retardant effectiveness of the THPC-amide finish, and thus the total chemical add-on has not to be increased. The polymeric BPNE finish, similar to the polymeric BAP finish, does not penetrate into the fiber, but coats the surface of the fibers.

5.8.1.2. THPC-Amide Finish—Ammonia Cure. Methylol phosphorus polymers with free hydroxymethyl groups present can be insolubilized with ammonia. This discovery is the basis of the flame-retarding procedure suggested by Reeves and Guthrie.[147] The cellulosic substrate is first impregnated with a water-soluble precondensate obtained by reacting

THPC with urea, melamine, methylolated melamine, etc. After drying it is treated with ammonia to obtain an insoluble polymer with higher nitrogen content. Drying before the ammonia treatment is necessary to minimize leaching of the precondensate into the ammonia solution. However, without high-temperature curing, the leaching cannot be entirely avoided; therefore the impregnated and dried fabric is first exposed to ammonia vapor.[148-150] The insolubilization may be completed in the ammonia-gas chamber or by a subsequent aftertreatment with aqueous ammonia solution. The THPC–urea precondensate used in this procedure is marketed under the trade name Proban. The Proban finish imparts good, durable flame retardancy with less stiffening and tendering of the cellulosic fabric than with a heat cure.

Several other procedures have been suggested to avoid the undesirable side effects (migration, stiffening, tendering) of the THPC-amide heat-cure finish.

Proban, Ltd.[151] pretreats the cellulosic substrate with an aminoplast resin using acid-catalyzed heat curing. Ciba[152] pretreats the cellulosic textile with polymethylol melamine using H_2O_2-catalyzed wet fixation (self-condensation). An insoluble resin is formed in and around the fiber. The cellulosic fabrics pretreated with the aminoplast resins are then impregnated with THPC (neutralized with triethanolamine), dried, and immersed in aqueous ammonium hydroxide solution. The pretreatment with the aminoplast resins increases the level of flame retardancy and prevents leaching in the aqueous ammonia solution. If wet fixation of polymethylol melamine is used, the stiffening and tendering of the cellulosic fabric are minimal.

5.8.1.3. THPC–APO. THPC reacts rapidly with alkyleneimines yielding polymeric products:

$$(HOCH_2)_4P^+Cl^- + CH_2{-}CH_2 \longrightarrow (HOCH_2)_2P(O)CH_2OCH_2CH_2NH_2 + CH_2O + HCl$$
$$\underset{NH}{\diagdown \diagup}$$

or

$$\begin{array}{c} | \\ -N{-}CH_2 \\ \diagdown \\ \qquad\qquad P(O)CH_2OCH_2CH_2N{-} \\ \diagup \qquad\qquad\qquad\quad | \\ -N{-}CH_2 \\ | \end{array}$$

One of the most effective and durable flame retardants for a cellulosic substrate is the tetrakis(hydroxymethyl)phosphonium chloride/tris(1-aziridinyl)phosphine oxide (THPC–APO) finish.[153] This polymer imparts very durable flame-resistant finishes at relatively low resin add-on and is satisfactory for use on cotton fabrics, with the exception of tightly constructed, lightweight fabrics weighing less than about 3 oz/yd^2.[137] Satisfactory flame resistance can be attained with lower chemical add-ons (8–12%) for heavier fabrics, but lighter fabrics require higher add-ons (15–18%), which cause stiffening and weakening.

The THPC–APS finish gives results comparable to those of THPC–APO. APS is more stable than APO and can be stored longer, but is less soluble in water, thus limiting the concentration which can be used for application, and fabrics treated with it initially have a disagreeable odor.[137]

The durable, flame-retarding treatment of cellulosic substrates with phosphorus-containing aziridines (such as APO or APS) in conjunction with methylol phosphorus compounds [such as THPC or tris(hydroxymethyl)phosphine oxide (THPO)], or with the reaction product of methylol phosphorus compounds and methylol melamine, are disclosed by Drake, Reeves, and Chance.[154] The treatment essentially consists of impregnating the fabric with the aqueous solution of the ingredients, followed by drying at relatively low temperature and then by curing at approximately 160°C for 3 min to form a highly cross-linked polymer within the fibers. The so-treated cellulosic substrate, especially cotton fabric, is not only flame and glow resistant, but rot resistant as well.

The THPC–APO-finished textiles, when fully cured, are physiologically inert, but adequate precautions must be taken to prevent the chemicals, especially the APO, from coming in contact with the skin. Persons working with the material must carefully avoid direct contact because repeated exposure may cause skin sensitization and induce respiratory allergic reactions.[137]

5.8.1.4. Miscellaneous THPC Polymers. Polymers prepared by reacting THPC or THPO with monomeric aliphatic, alicyclic, or aromatic diisocyanates and their application as flame retardants to cellulose or cellulosic derivatives (cyanoethylated, carboxyethylated, aminoethylated cellulose) are suggested by the U.S. Department of Agriculture.[155]

Cotton fabric can be rendered durably flame retardant by treatment with aqueous precondensates of THPC with tris(carbamoylethyl)phosphine (TCEP) or tris(carbamoylethyl)phosphine oxide (TCPO).[156] The preferred mole ratio of THPC to TCEP or TCPO is 4:1. The fixation is preferably completed by exposing the treated textile to ammonia gas.

5.8.1.5. THPOH Finishes. Reactive methylol phosphorus compounds can be made from THPC by reacting it with alkali hydroxide in a 1:1 mole ratio[126,157]:

$$(HOCH_2)_4PCl + NaOH \longrightarrow (HOCH_2)_4POH + NaCl$$

$$(HOCH_2)_4POH \Longrightarrow (HOCH_2)_3P + CH_2O + H_2O$$

$$(HOCH_2)_3P + CH_2O \Longrightarrow HOCH_2OCH_2P(CH_2OH)_2$$

The product of the THPC–NaOH reaction is not, however, adequately characterized and might be considered as the equilibrium mixture of tetrakis(hydroxymethyl)phosphonium hydroxide (THPOH) and tris-(hydroxymethyl)phosphine (THP).[158] There is also evidence[159] that the solution consists of THP and the hemiacetal of THP.

When reacting THPC with NaOH or other base, the pH of the solution must not exceed 7.5–7.8. Since THPOH can be converted to the less reactive tris(hydroxymethyl)phosphine oxide (THPO) by an excess of base, it is very important to keep the pH of the solution below 7.8.[126]

THPOH reacts with amides which contain 2 or more NH groups to produce insoluble polymers. In the polymerized product, the phosphorus is present in the same form (phosphine oxide) as the polymer produced from THPC and amides.

A durable flame-retardant finish based on THPOH, trimethylol melamine, and urea has been developed and applied to cotton fabric with a conventional pad–dry–heat-cure procedure.[160] This formulation is very similar to that used in THPC-amide finishing, the THPOH being substituted for THPC on a molar basis. The impregnated fabric is heat cured at 150°C for 3 min. The properties of THPC-amide and the THPOH-amide heat-cured cellulose, however, are different. The THPOH-amide-treated samples retain 80–90% of their original breaking strength, and the increase in stiffness is only moderate. They exhibit less tendency to yellow when exposed to hypochlorite bleach than the THPC-amide heat-cured cellulosic products.

THPOH–trimethylol melamine–NH$_3$ finishes[161] have been designed to attain durable flame retardancy at a lower cost by using lower phosphorus and higher nitrogen levels. The cellulosic substrates impregnated with an aqueous solution containing THPOH, urea, and TMM in 2:4:1 mole ratio are cured with ammonia vapor after partial drying. The ammonia cure is followed by a heat cure. This process requires less chemical add-on and causes less strength loss and stiffening than the THPOH–TMM pad–dry–heat-cure procedure.

THPOH reacts with ammonia at room temperature, and it was this observation which led to the THPOH–NH$_3$ flame-retardant finish.[157,162] The procedure essentially consists of impregnating the cellulosic substrate with aqueous THPOH solution and, after partial drying to about 10–20% moisture content, exposing it to ammonia vapor. The reaction of ammonia with THPOH produces water, which combines with ammonia to form ammonium hydroxide. Since ammonium hydroxide reacts with THPOH, imparting water-soluble products instead of insoluble polymer, the formed water and ammonium hydroxide must be removed from the gas chamber. The THPOH–NH$_3$ procedure gives good flame retardancy without strength loss and stiffening. The P/N ratio is 3.5:1 in the fabric treated with this process.[126] Because of the high phosphorus content required to attain self-extinguishing properties, this process is more expensive than the THPC-amide or THPOH-amide finishes. The durability of the flame retardancy is also lower.

Copper salts were found to stabilize THPOH–NH_4OH solutions by forming a complex, thereby making it possible to apply THPOH to cellulosic substrate from a single bath without the use of gaseous ammonia.[163] The minimum mole ratio of cupric nitrate to THPOH is 1:4. The process consists of padding the textiles with THPOH–NH_4OH solution stabilized with $Cu(NO_3)_2$ and curing immediately at 155°C for 2–5 min.

The advantages of this copper complex procedure are stability of the solutions for at least 20 days and the avoidance of the use of gaseous ammonia. However, the so-treated cellulosic product has a bluish-green color due to the copper present.

5.8.2. Other Phosphonium Salts

By reacting THP with ethylene oxide and acetic acid,[164]

$$(HO—CH_2)_3P + CH_2—CH_2 + CH_3COOH \longrightarrow [(HOCH_2)_3P^+CH_2CH_2OH]CH_3COO^-$$
$$\underset{O}{\diagdown\diagup}$$

or by reacting THP with alkyl halides[165]

$$(HOCH_2)_3P + CH_3I \longrightarrow [(HOCH_2)_3P^+CH_3]I^-$$

quaternized phosphonium salts are obtained which can be used as flame retardants.

The reaction of 1 mol of epihalohydrin and 2 mol of THP yields a mixture of mono-, di-, and triphosphonium compounds. The treatment of cellulosic substrate with the reaction product of epihalohydrin and THP in conjunction with APO imparts durable flame-retardant properties.[166] After partial drying to 7–20% residual moisture content, the impregnated cellulosic substrate is exposed to ammonia vapor or treated with ammonium carbonate or ammonium acetate solution containing 5–20% neutral salt. The ammonia aftertreatment renders the flame-retarding ingredients entirely insolubilized.

5.9. Phosphinic Acid and Its Derivatives

Petrov, Sopikova, and Nifantev[167] reported the esterification of cellulose with alkyl or aryl phosphinic acid anhydrides. Cellulose phosphinates containing 1.4–10.3% phosphorus have been obtained by reacting cellulose with methyl and phenyl phosphinic acid anhydrides. At a phosphorus content of about 4%, the modified cellulose becomes flame resistant.

Kiselev, Aksenova, and Kutsenko[168] reported the esterification of cellulose with the chloroanhydrides of dialkylphosphinic acids. Dimethyl, diethyl, dipropyl, and dibutyl phosphinates of cellulose have been prepared by treating cotton fiber with the chloroanhydrides of the appropriate acid. It is shown that at temperatures of 100–110°C the esterification can be carried out without preliminary activation of the fiber; at lower temperatures activation is necessary. The rate of reaction gradually decreases as the size of the alkyl group increases. Cellulosic fiber esterified with dialkylphosphinic acids to a phosphorus content of 7–8% does not burn or glow in air; at lower phosphorus content the modified fiber burns but extinguishes itself. Fiber modified by ethylphosphinic acid is rendered incombustible at lower phosphorus contents (3.5–4.0%).

5.10. Phosphonic Acid Derivatives

Besides the phosphoric and phosphorous acid derivatives and phosphonium salts, the phosphonates are playing an increasingly important role in the durable flame-retarding treatment of cellulose.

While in the phosphoric acids only P–O bonds are present, in the phosphonic acids one of the organic moieties is attached directly to the P atom through a P–C bond.

$$
\begin{array}{ccc}
\text{OH} & & \text{OH} \\
| & & | \\
\text{HO—P—OH} & & \text{R—P—OH} \\
\| & & \| \\
\text{O} & & \text{O} \\
\text{Phosphoric acid} & & \text{Phosphonic acid}
\end{array}
$$

Phosphonic acid is obtained by the oxidation of primary phosphines, e.g., with nitric acid.

$$
\text{R—PH} \xrightarrow{\text{ox}} \text{R—P}{\overset{\displaystyle \text{OH}}{\underset{\displaystyle \text{OH}}{=\!\!=\!\text{O}}}}
$$

The phosphonic acids have two reactive OH groups which can be esterified or converted to amides. The organic group attached directly to the P atom may be polymerizable (e.g., vinyl group) or may contain reactive moieties.

The alkaline hydrolytic stability of the P–C bond is significantly greater than that of the P–N bond.[84] Therefore, whenever a high level of hydrolytic resistance is required from the flame-retardant finish, the phosphonates are possible candidates, but the stability of the ester linkage should be considered.

The treatment of cellulose with chloromethyl disodium phosphonate yields a phosphorus-containing cellulose ether. Chloromethyl phosphonic acid dichloride can also be used for the etherification.

$$Cell.OH + ClCH_2P\overset{ONa}{\underset{O\quad ONa}{\diagdown}} \overset{OH^-}{\longrightarrow} Cell.OCH_2P\overset{ONa}{\underset{O\quad ONa}{\diagdown}}$$

$$Cell.OH + ClCH_2POCl_2 \overset{OH^-}{\longrightarrow} Cell.OCH_2P\overset{ONa}{\underset{O\quad ONa}{\diagdown}}$$

Drake, Reeves, and Guthrie[169] disclosed the preparation of phosphonomethyl cellulose of reduced flammability by impregnating the cellulose fiber with the sodium salt of chloromethyl phosphonic acid in alkaline medium using the pad–cure technique (curing 5–30 min at 140–160°C). Flame-resistant product is attained by converting the alkali salt of phosphonomethylated cotton to the free acid with mineral acid and then neutralizing it with ammonium hydroxide. By this technique, phosphonomethyl cellulose with a maximum of 5% P content has been obtained.

Polymers from dialkylene chloromethyl phosphonate[170]

$$ClCH_2-P-(OR)_2 \atop O \qquad \text{where R = alkenyl (allyl or methallyl) group}$$

or from dialkylene-β-cyanoethane phosphonate[171,172]

$$NC-CH_2CH_2-P(OR)_2 \atop O \qquad \text{where R = alkenyl (allyl or methallyl) group}$$

are suggested for the durable flame-retarding treatment of cellulose. Dialkylene cyanoethane phosphonate is obtained by reacting dialkylene phosphite with acrylonitrile. The cellulosic substrate is impregnated with the dialkylene cyanoethane phosphonate solution containing persulfate initiator. By heating the treated fabric to 76–78°C, the polymerization of the allyl groups is induced and insoluble polymer is formed.

Toy and Rattenbury[173] suggested the use of partially brominated dialkylene cyanoethane phosphonate polymers for the flame-retarding treatment of cellulosic substrates. They are prepared by partial polymerization (approximately 50% conversion) of diallyl or dimethallyl chloromethyl phosphonate, followed by saturation of the remaining double bonds by addition of bromine. The brominated phosphonate polymer is then insolubilized *in situ* on the cellulosic substrate by heat curing.

Polymeric β-(carboxyalkyl)alkane phosphonates,

$$ROOCCHR'—CH_2P(O)(OR'')_2$$

$$R = CH_3, C_2H_5$$
$$R' = CH_3, H$$
$$R'' = CH_2—CH=CH_2$$
$$\text{or } CH_2—C=CH_2$$
$$\qquad\quad |$$
$$\qquad\quad CH_3$$

have also been suggested as flame retardants.[174] They are prepared by reacting acrylic or methacrylic acid esters with diallyl or dimethallyl phosphite. The partially polymerized products are applied to cellulose from emulsion and further polymerization occurs at elevated temperatures *in situ* within the fiber.

Dialkyl epoxyphosphonates[175] are also suggested as flame retardants for cellulosic substrates. They are prepared by reacting α-haloaldehydes with dialkyl phosphites with subsequent dehydrohalogenation:

$$ClCH_2—CHO + HOP(OR)_2 \longrightarrow ClCH_2—CHP(O)(OR)_2 \xrightarrow{-HCl} CH_2—CH—P(O)(OR)_2$$
$$\qquad\qquad\qquad\qquad\qquad\qquad\quad |\qquad\qquad\qquad\qquad \diagdown\!\!\diagup$$
$$\qquad\qquad\qquad\qquad\qquad\qquad\quad OH\qquad\qquad\qquad\qquad\ \ O$$

Vinyl phosphonic acid and its esters[176] may also be used as flame retardants for cellulosic products. Vinyl phosphonic acid is obtained by the hydrolysis of vinyl phosphonic dichloride. The latter is prepared by reacting vinyl phosphonic acid dialkyl esters with phosphorus pentachloride:

$$CH_2=CHP(O)(OR)_2 + 2PCl_5 \longrightarrow CH_2=CHP(O)Cl_2 + 2POCl_3 + RCl$$

Vinyl phosphonic acid dichloride itself can also be reacted directly with cellulose.

Bis(2-haloethyl)vinyl phosphonate and bis(2-halopropyl)propenyl phosphonate[177] have been proposed as flame-retarding agents for paper and cellulosic textiles.

Schiffner and Lange[178] studied the flame-retarding treatment of cellulose with chloromethyl phosphonic acid and urea, and in comparing it with the phosphoric acid–urea procedure, it was found that higher reagent concentration is required to attain self-extinguishing properties with the chloromethyl phosphonic acid process. The concomitant strength losses were less severe than in the phosphoric acid–urea process, especially if formaldehyde-containing solutions were used.

Halogenated polyphosphonate (Phosgard C-22R, Monsanto)

$$XCH_2CH_2O\overset{\overset{O}{\|}}{P}\!-\!-\!OCH\!-\!-\!\overset{\overset{O}{\|}}{P}\!-\!O\!-\!-\!\overset{\overset{CH_3}{|}}{CH}\!-\!-\!\overset{\overset{O}{\|}}{P}\!-\!(OCH_2CH_2X)_2$$

$$\overset{|}{XCH_2CH_2}\quad \overset{|}{CH_3}\quad \overset{|}{OCH_2CH_2X}$$

$$X = Cl$$

is a water-insoluble product obtained by the reaction of phosphorus trichloride, ethylene oxide, and trialkyl phosphite. The amination of this halogenated polyphosphonate yields water-soluble polyaminophosphonates in which up to 5 chlorine atoms are replaced by amine residues[179] by reacting Phosgard with ammonia, hydrazine, diethanolamine, ethylene diamine, or guanidine. The N/P atom ratio of the aminated products ranged from 1.6 to 5.0, depending on the amine used in the preparation and on the extent of conversion. The flame-retarding effectiveness of the aminated polyphosphonates increases with increasing nitrogen content, demonstrating the contribution of nitrogen to the flame-retarding effectiveness of the organophosphorus compound.

Insolubilization of a polyaminophosphonate ($X = -HNCH_2CH_2NH_2$, reaction product of Phosgard C-22R and ethylene diamine) by condensation with a suitable coreactant, e.g., tris(1-aziridinyl)phosphine oxide (APO), yields an effective durable flame-retardant finish without substantially impaired fabric properties. The N/P ratio of APO is closely similar to that of Phosgard aminated with ethylene diamine, and thus the N/P ratio of the insolubilized finish does not differ from that of either of its components. The insolubilization proceeds in high yield by curing the treated cellulosic substrate at 160–190°C for 2–6 min. The durability of the flame-retardant finish obtained by this process is excellent in laundering and severe alkaline scouring.

A hydrolytically stable phosphonate derivative is obtained by the Arbuzov reaction of haloacetamidomethyl cellulose with trialkyl phosphite.[81] The reaction rate of haloacetamidomethyl cellulose with trialkylphosphite decreases in the order: iodo > bromo > chloro.

$$\text{Cell.OH} + HOCH_2NHCOCH_2X$$
$$\downarrow [H^+]$$
$$\text{Cell.OCH}_2NHCOCH_2X$$
$$\downarrow (RO)_3P$$
$$\text{Cell.OCH}_2NHCOCH_2P(O)(OR)_2$$

$$X = Cl, Br, I$$
$$R = -CH_3, -C_2H_5, -C_2H_4Cl$$

The new cellulose phosphonate ester derivatives exhibit excellent chemical stability. Self-extinguishing properties are attained at approximately 2% P content. Organophosphorus structures, in which the N/P atom ratio is unity, are not the most efficient flame retardants. The after-treatment of dialkyl phosphonoacetamidomethyl cellulose with trimethylol melamine by a wet-fixation procedure dramatically improves the flame-retardant properties.

Phosphonate ester derivatives of cellulose were prepared also by reacting cellulose with N-hydroxymethyl-3-(dialkyl phosphono)propionamide[180-182]:

$$\text{Cell.OH} + \text{HOCH}_2\text{NHCOCH}_2\text{CH}_2\text{P(O)(OR)}_2 \xrightarrow[\Delta]{\text{H}^+} \text{Cell.OCH}_2\text{NHCOCH}_2\text{CH}_2\text{P(O)(OR)}_2$$

$$R = \text{alkyl or haloalkyl}$$

The methylol group is responsible for the chemical fixation of this compound to cellulose. No cellulose cross-linking occurs, and the alkyl ester groups are not involved in transesterification.[183]

The intermediate dialkyl phosphonopropionamide[184,185] is prepared by reacting acrylamide with dialkyl phosphite in the presence of alkaline catalyst:

$$\text{H}_2\text{NCOCH}=\text{CH}_2 + \text{HOP}\begin{matrix} \diagup \text{OR} \\ \diagdown \text{OR} \end{matrix} \xrightarrow[\text{dioxane}]{\text{CH}_3\text{ONa}} \text{H}_2\text{NCOCH}_2\text{CH}_2\text{P(O)(OR)}_2$$

Methylol amides of a series of dialkyl phosphonocarboxylic acids were studied by Aenishänslin et al.[181] regarding flame-retarding effectiveness, durability of the flame retardancy, and the physical and aesthetic properties of the finished cellulosic substrate. They showed that the N-methylol compounds of dimethyl and diethyl phosphonopropionamide and the corresponding 1-methyl propionamides are the most suitable flame retardants for cotton:

$$(\text{RO})_2\text{P(O)(CH}_2\underset{\underset{\text{X}}{|}}{\text{C}}\text{HCONHCH}_2\text{OH)}$$

$$R = \text{CH}_3 \text{ or } \text{C}_2\text{H}_5$$
$$X = \text{H or CH}_3$$

It is preferable to apply N-methylol dialkyl phosphonopropionamide (marketed by Ciba under the name Pyrovatex CP) in conjunction with aminoplast resins in order to increase the nitrogen content of the treated cellulose and the effectiveness of the flame retardant. Among the aminoplast resins, polymethylol melamine is very efficient. This combination can be applied from a single bath by acid-catalyzed pad–dry–cure technique or the acid-catalyzed application of Pyrovatex CP can be followed by the application of trimethylol melamine using a wet-fixation technique. With the two-step system, cotton or rayon can be modified to exhibit excellent durable flame-retardant properties without appreciable increase in fabric stiffness.[182] since it is possible by this method to introduce trimethylol melamine even in high concentration without fabric stiffening and thus attain self-extinguishing properties at low-phosphorus and high-nitrogen content. By using this technique, satisfactory flame-retardant properties have been attained on cotton and rayon at 0.8 and 1.0% phosphorus content, respectively. If a higher degree of flame resistance is required, larger amounts of flame-retardant agents must be used, resulting in excessive stiffening in the single-step acid-curing procedure.

The amounts of phosphorus and nitrogen required for self-extinguishing properties depend on the orientation of the cellulose fiber. Ramie, which is known to be the most highly oriented natural cellulosic fiber, needs the lowest, and spun rayon, which has a low degree of orientation, requires the highest reagent concentration to attain an adequate level of flame retardancy.[183]

The flame-retarding effectiveness of the *N*-methylol derivatives of diethyl-, bis(2-chloroethyl)-, and bis(2-bromoethyl)phosphonopropionamides

$$(RO)_2P(O)CH_2CH_2CONHCH_2OH$$

$$R = -C_2H_5, -C_2H_4Cl, -C_2H_4Br$$

has been compared,[182] and a further synergistic effect was noted for halogen-containing reagents. Bromine is more effective than chlorine in enhancing the flame-retarding efficiency of the phosphorus- and nitrogen-containing reagents.

N-hydroxymethyl derivatives of carbamates[357,358] and carbamate-substituted carboxamide[359] containing phosphorus in dialkyl phosphorus moieties have been reported as flame retardants for cellulose.

The carbamate analog of Pyrovatex CP, $(RO)_2P(O)CH_2CH_2OCONH-CH_2OH$, was disclosed[357,358] as a flame retardant for cotton and a phosphonocarbamate of the following structure[353]

$$(C_2H_5O)_2P(O)CH_2N\left[CH_2CH_2OCON{\overset{H}{\underset{CH_3}{\big<}}}\right]_2$$

was prepared by reacting *O,O'*-diethyl-*N,N*-bis(2-hydroxyethyl)aminomethyl phosphonate with methyl isocyanate. The methylol derivatives of the carbamate were applied to cellulosic substrates as flame retardants in conjunction with trimethylol melamine.

Phosphonates, suitable for the flame-retarding treatment of cellulosic substrates have been obtained by the condensation of phosphines and formaldehyde with amides.[186] Similar products were prepared from guanidine sulfate or dicyandiamide with formaldehyde and phosphine. Sanderson[187] reported the preparation of flame retardants by reacting guanidine, formaldehyde, and dialkyl phosphite.

The copolymerization of vinyl phosphonates, e.g.,

$$CH_2{=}CH{-}\overset{\overset{\displaystyle O}{\|}}{\underset{\displaystyle OCH_2CH_2Cl}{P}}{-}O{\left(CH_2CH_2O{-}\overset{\overset{\displaystyle O}{\|}}{\underset{\displaystyle CH{=}CH_2}{P}}{-}O\right)}_x CH_2CH_2Cl$$

where $x = 1$–20, with *N*-methylol acrylamide via free-radical catalysis *in situ* on the fiber is a viable approach to attain durable flame resistance.[360]

Petersen[188] studied the α-ureido alkylation of nucleophilic phosphorus compounds and arrived at a series of acyclic and cyclic urea–phosphorus compounds. In addition to phosphine, suitable nucleophilic PH compounds

$$-N\overset{\overset{\displaystyle O}{\|}}{\underset{\displaystyle |}{\underset{\displaystyle C}{\diagdown}}}NH{-} \;+\; H{-}\overset{\overset{\displaystyle R}{|}}{C}{=}O \;+\; H{-}P\overset{\diagup R}{\diagdown R} \longrightarrow$$

$$-N\overset{\overset{\displaystyle O}{\|}}{\underset{\displaystyle |}{\underset{\displaystyle C}{\diagdown}}}N{-}\underset{\displaystyle |}{\overset{\overset{\displaystyle R}{|}}{C}}H{-}P\overset{\diagup R}{\diagdown R} \;+\; H_2O$$

PH–acid components:

$$PH_3 \qquad P_2H_6 \qquad H_2P{-}R$$

$$H{-}P\overset{\diagup R}{\diagdown R} \qquad HO{-}P\overset{\diagup OR}{\diagdown OR} \;\rightleftharpoons\; H{-}\overset{\overset{\displaystyle O}{\uparrow}}{P}\overset{\diagup OR}{\diagdown OR}$$

are provided by diphosphine, primary and secondary phosphines, and also by dialkyl phosphites.

The phosphorus links up with the nitrogen atom of an acyclic or cyclic urea through the carbon atom of an aldehyde to form a carbon–phosphorus bond.

5.11. Phosphorus-Containing Triazines

The synergistic role of nitrogen in enhancing the flame-retardant characteristics of cellulosic substrates with organophosphorus chemicals has been reported in several publications.[81,179,189] The contribution of nitrogen to the flame-retarding efficiency of organophosphorus chemicals depends on the source of nitrogen.[190] Cellulose derivatives containing amide, amine and nitrile substituents (Cell.OCH$_2$CH$_2$CONH$_2$, Cell.OCH$_2$CH$_2$NH$_2$, Cell.OCH$_2$CH$_2$CN), of comparable structure have been aftertreated with phosphorus chemical (THPOH), and it was found that, while the amido and amino nitrogen enhance, the nitrile nitrogen adversely affects the flame-retarding properties. Among the amide nitrogens, melamine and melamine derivatives have been found to be the most effective in enhancing flame-retardant properties.[81,182,191]

Several approaches are documented to introduce phosphorus moieties into the triazine molecules and to use the phosphorus-containing triazine derivatives for the flame-retarding treatment of cellulosic substrate. Moreau and Chance reported[192] the flame-retarding treatment of cotton with phosphorus-containing triazine derivatives obtained by the Arbuzov reaction of chlorotriazines with trialkyl phosphites.

The application of the methylolated derivatives of **ADPT** and **DAPT** in conjunction with trimethylol melamine imparts good flame retardancy, but the loss in phosphorus content is appreciable in laundering.

Better hydrolytic stability can be expected from triazine derivatives containing phosphonate ester moieties.

The condensation of tris(hydroxymethyl)phosphine (THP) with *N*-methylol compounds (e.g., trimethylol melamine) in the presence of acid catalyst yields precondensates which are suitable for the flame-retarding treatment of cellulosic substrates.[193]

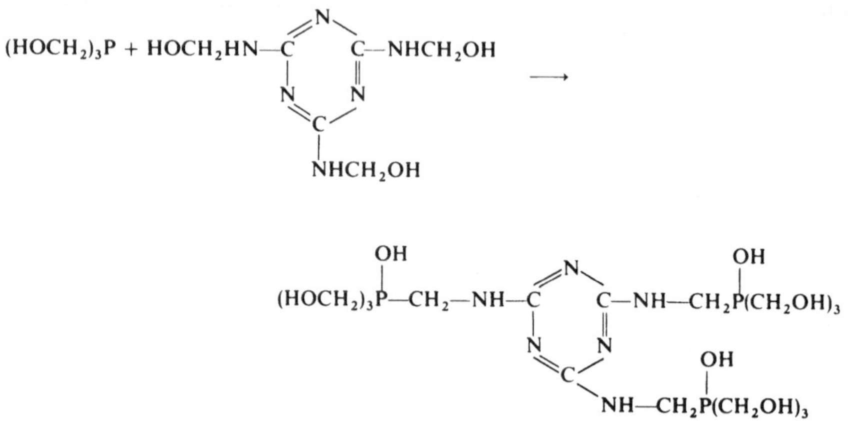

$$(HOCH_2)_3P + HOCH_2HN-C\underset{\underset{C}{\underset{|}{NHCH_2OH}}}{\overset{N}{\diagup\diagdown}}C-NHCH_2OH \longrightarrow$$

$$(HOCH_2)_3\overset{OH}{\overset{|}{P}}-CH_2-NH-C\underset{\underset{C}{\underset{NH-CH_2\overset{OH}{\overset{|}{P}}(CH_2OH)_3}}}{\overset{N}{\diagup\diagdown}}C-NH-CH_2\overset{OH}{\overset{|}{P}}(CH_2OH)_3$$

By condensing polymethylol melamines (penta- or hexamethylol melamine) with 3-dialkyl phosphonopropionamide in 1:4 mole ratio in benzene in the presence of acid catalyst (e.g., *p*-toluene sulfonic acid), a water-soluble product is obtained in which 4–5 methylol groups are replaced with phosphonate ester residues.[194] By applying this precondensate in conjunction with trimethylol melamine to cotton fabric, using acid-catalyzed heat-cure or wet-fixation technique, excellent durable flame retardancy can be attained, but the stiffening of the textile is excessive.

The reaction product of hexamethylol melamine and dibutyl phosphine[186] imparts durable flame resistance to cotton when applied in

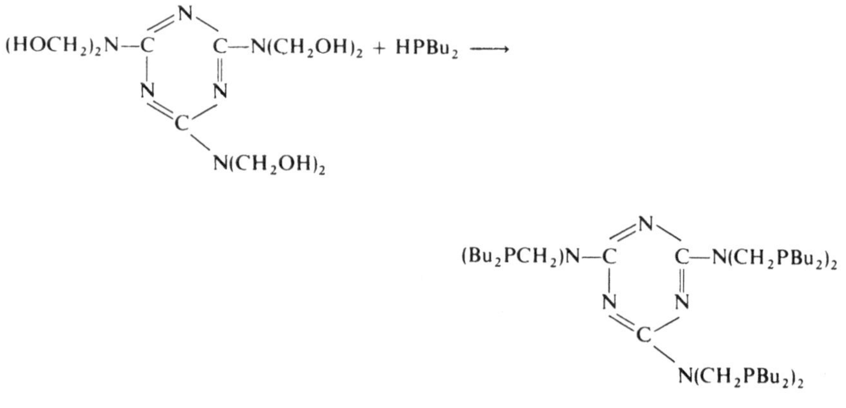

$$(HOCH_2)_2N-C\underset{\underset{C}{\underset{N(CH_2OH)_2}}}{\overset{N}{\diagup\diagdown}}C-N(CH_2OH)_2 + HPBu_2 \longrightarrow$$

$$(Bu_2PCH_2)N-C\underset{\underset{C}{\underset{N(CH_2PBu_2)_2}}}{\overset{N}{\diagup\diagdown}}C-N(CH_2PBu_2)_2$$

conjunction with trimethylol melamine.

Anionic guanamine phosphonates of the following structure have been prepared by Schuller[195]:

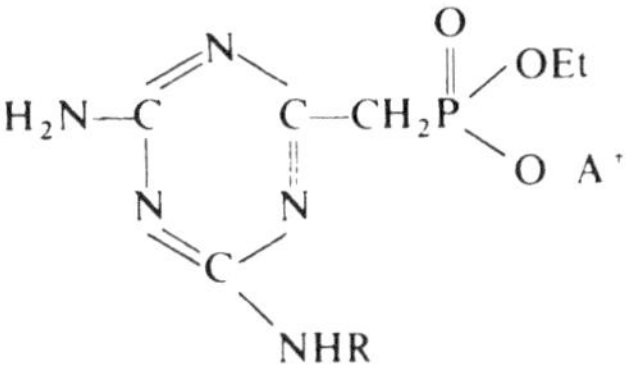

Triazine derivatives contining phosphonate ester groups of the following generic formula[196]:

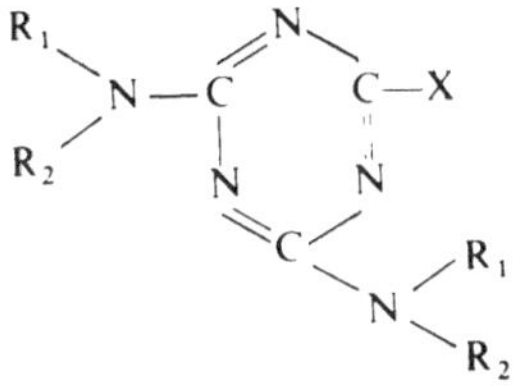

$R_1 R_2 = $ —H

X = —$CH_2P(O)(OR)_2$.

　　—$C_2H_4P(O)(OR)_2$.

　　—$N[C_2H_4P(O)(OR)_2]_2$.

　　—$NHC_2H_4P(O)(OR)_2$.

　　—$SC_2H_4P(O)(OR)_2$.

R = —CH_3 or —C_2H_5

were prepared by the following sequence of reactions:

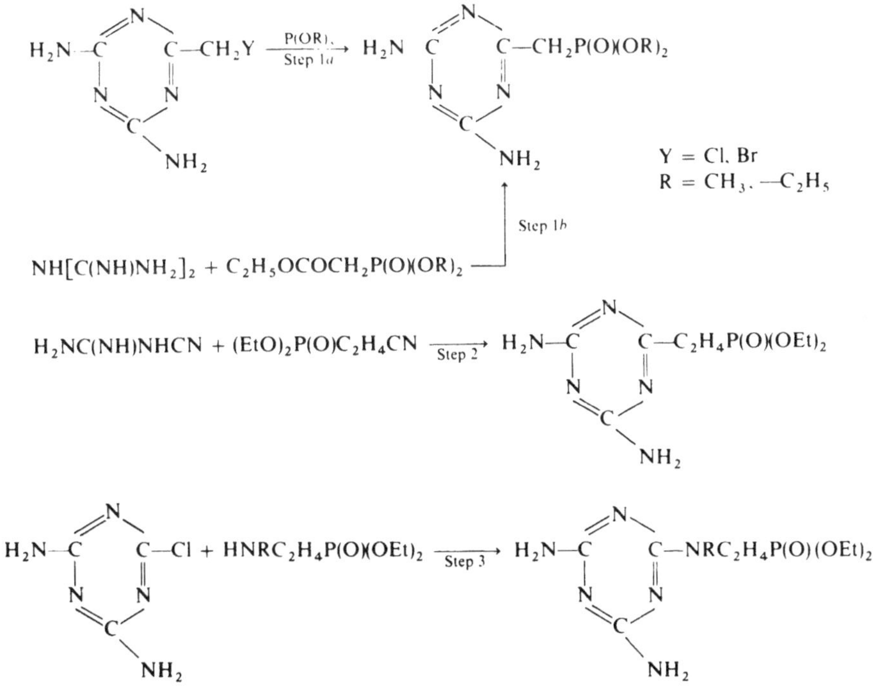

$Y = Cl, Br$

$R = CH_3, —C_2H_5$

$NH[C(NH)NH_2]_2 + C_2H_5OCOCH_2P(O)(OR)_2$ ——

$H_2NC(NH)NHCN + (EtO)_2P(O)C_2H_4CN \xrightarrow{Step\ 2} H_2N—C\ \ C—C_2H_4P(O)(OEt)_2$

$R = H$ or $C_2H_4P(O)(OEt)_2$

The flame-retarding treatment of cotton has been systematically studied by applying the methylolated phosphorus-containing triazine derivatives alone and in conjunction with trimethylol melamine using acid-catalyzed heat-curing and H_2O_2-catalyzed wet-fixation technique. In these structures the N/P ratio ranges from 3:1 to 6:1.

The synthesized compounds are not the most efficient flame retardants, and therefore it is preferable to combine the phosphonate ester derivatives of triazines with polymethylol melamine. In this manner excellent durable flame retardancy can be attained.

Hexahydro-s-triazine phosphonates were prepared by reacting 1,3,5-triacryloylhexahydro-s-triazine with dimethyl phosphite. The phosphonate obtained was applied to cellulose and to other polymeric substrates to render them self-extinguishing.[361]

6. Halogens as Flame Retardants

Halogens, particularly bromine compounds, occupy an important position today among the fire-extinguishing and flame-retardant agents. The effectiveness of these compounds is relatively high. It has been shown, for instance, that in order to reduce the burning velocity of a stoichiometric hexane–air laminar flame, 4.05 molecules of carbon dioxide or of nitrogen are needed per molecule of hexane as compared to 1.54 molecules of chlorine and 0.33 of bromine.[197]

The mechanism of the inhibitory action of halogen compounds is believed to be based on the interaction of the halogen with some of the reactive moieties of the flame itself. According to Rosser,[198,199] the active species is the hydrogen halide, which reduces the concentration of the free radicals OH and H considered responsible for the propagation of the flame:

$$OH + CO \longrightarrow CO_2 + H$$
$$H + O_2 \longrightarrow OH + O$$
$$RH + Br \longrightarrow HBr + R$$
$$Y + HBr \longrightarrow Br + HY$$

In this scheme RH denotes the fuel and Y refers to active radicals such as H, OH, CH_3.

The hydrogen bromide is either formed directly by the thermal decomposition of the bromine-containing compound, as in the case of ammonium bromide, or as in the case of hexabromobenzene, bromine atoms are first produced in the pyrolysis and later act to regenerate hydrogen bromide according to the above formulas. The activity of the bromine compounds thus depends greatly on the mode of their decomposition and on the dissociation energies of the bonds between the bromine and the carbon or other atoms to which the bromine is linked. Since the bond dissociation energies of chlorine-containing compounds is generally greater than that of their bromine analogs, the higher activity observed in many cases for the latter is understandable.[200]

6.1. Activity of Bromine and Chlorine Compounds

Several comparisons of the relative effectiveness of bromine and chlorine derivatives were published in recent years, particularly with a view to their use in the preparation of flame-retardant polyesters. Pape, Nulph, and Nametz[201] found 13%_0 bromine to be as effective as 22% chlorine when comparing tetrahalophthalic anhydrides. Schneider, Pews, and Herring[202] while basically confirming these results, found that at least 30%_0 aromatically bound chlorine is needed to get self-extinguishing properties in a polyester.

In the case of ethylene oxide adducts of tetrahalobisphenol, the bromine derivative was found to be twice as effective as the chlorine derivative. As for the influence of structure, they state that aliphatic bromine

$$X = Br, Cl$$

compounds are approximately 1.5 times more effective than the aromatic compounds.[202]

In another comparison carried out on rigid polyurethane foam samples,[200] it was found that hexachlorohexane is 3 times more effective than hexachlorobenzene.

When similar comparisons were carried out on a cellulosic substrate, i.e., chromatographic-quality paper strips, such differences between the aromatic and aliphatic structures are not observed.[203] As can be seen from Table 1, 19.2% Br in tris(tribromophenyl)phosphate and 17% Br in tris(2,3-dibromopropyl)phosphate are needed to prevent flame propagation on the paper strip in the vertical direction. Since the percentage of phosphorus is higher in the second derivative and assuming that it exerts an additive activity, it appears that in the case of paper the aromatic compound is at least as effective as the aliphatic.

The difference, however, between the flame-retardant activity of chlorine and bromine, as exemplified by comparing ammonium bromide to ammonium chloride, is more pronounced in the case of cellulose. At a 45° testing angle, the bromine compound is 3 times more effective[203] (see Table 1).

TABLE 1

Flammability Limits of Bromine Compounds on Cellulose[a(203)]

Compound	Angle	% Compound	% Br	% Cl	% P	% N
Tris(2,3-dibromo-	90	24.8	17		1.1	
propyl)phosphate	45	20.5	14.1		0.91	
	0	8.6	5.9		0.38	
Tris(tribromophenyl)-	90	27.7	19.2		0.83	
phosphate	45	19.3	13.4		0.58	
	0	7.8	5.4		0.23	
Hexabromocyclo-	90	34.5		All test strips burned.		
dodecane	45	31.4	23.4			
	0	12.7	9.4			
Pentabromoethane		Too volatile for quantitative data, but left appreciable carbonaceous residue; at 0° some flame inhibition at 20% add-on.				
2,4,6-Tribromophenol	0	30.7		All test strips burned. Very little carbonaceous residue.		
2,4,6-Tribromoaniline	0	31.6		All test strips burned. Very little carbonaceous residue.		
Ammonium bromide	90	8.8	7.2			1.3
	45	7.0	5.7			1.0
	0	3.4	2.8			0.49
Ammonium chloride	90	26.6		All test strips burned.		
	45	21.6		14.5		5.7
	0	5.7		3.8		1.5
1,4-Dimethyl-5-	90	All test strips burned at highest add-on achieved (not				
iminotetrazolinium	45	measured).				
bromide	0	21.0	8.0			7.6

[a] Chromatographic grade paper strips, 12 × 0.5 cm, were used in the tests, 8 strips for each concentration of the tested compound. The strips were soaked in water or directly in formamide solutions of the compounds in special narrow test tubes, for 3 min, and blotted between filter papers. In order to avoid inhomogeneous distribution in the compound owing to chromatographic migration, the strips were flash dried in a hot air stream, and then oven dried to constant weight. Eight concentrations of each compound were tested starting with a 40% solution and diluting at 5% intervals. Each strip was individually ignited by a match while firmly clamped at a predetermined angle in a draft-free box. Three angles were used: 0°, 45°, and 90°.

6.2. Synergism

The flame-retardant activity of bromine-containing compounds was found to be greatly enhanced by a number of compounds considered as synergists. A particularly effective synergist in the case of the rigid polyurethane foam and hexabromobenzene is antimony oxide. Ammonium borate and ammonium oxalate showed only limited activity, while boron oxide, ammonium carbonate and bicarbonate, ammonium sulfate, stannous oxide, stannous orthophosphate, and stannic oxide and phosphate did not show any synergistic effect.[200] A similar activity of Sb_2O_3 was found in the cases of nylon and polyester fibers,[204] polystyrene foam,[205] epoxy and polyester resins,[206,207] and polyolefins,[208] but no reports of such a synergism seem to have been published on cellulosics.

Another group of synergists for bromine, recently discovered, are the peroxides[209] and free-radical initiators: hydroperoxides, quinone imines, dibenzyl compounds, hydroizones,[209] and N-chloro and N-nitroso compounds.[210] A striking effect was obtained when adding dicumyl peroxide along with 1,1,2,2-tetrabromoethane to polystyrene foam.[209] The effect was explained by an extension of the resident time of the halogen atoms in the flame resulting from reactions between the halogen and polymer fragments formed in the pyrolysis in which halogen-containing intermediates are formed. On the other hand, the free radicals may accelerate the decomposition of the compound and thus enhance its activity and enable effectiveness at lower concentrations.

The information available on the mode of activity and on the practicality of the application of this group of synergists is scarce. Little information is available on the activity of these synergists in the case of cellulosics. It is also not known what the extent of the synergistic effect will be under testing methods other than those described for plastics, particularly methods in which high ignition temperatures, relatively long testing times, and high air drafts are applied. Studies on the application of synergists of this type to cellulosics might prove to be of considerable theoretical and practical interest.

The possibility of using phosphorus derivatives as synergists with bromine compounds has been considered by several investigators.[200,203,211] The addition of 0.5% P as an organic molecule along with 15% bromine in the form of tetrabromophthalic anhydride was found to accelerate greatly the self-extinguishment of a polyester resin.[200]

Results of combustion tests carried out on Southern pine wood blocks impregnated with various amounts of phosphorus and chlorine and of phosphorus with bromine dissolved in a 50:50 creosote–oil mixture are shown in Figs. 1 and 2.[211] Addition of 2.66% Br or 9% Cl to 0.8% P-containing mixtures decreases the weight loss from 60–70% of the control and 40–55% of 0.8% P-containing blocks to 7–15% only. This effect seems to be more than additive and suggests a synergism.

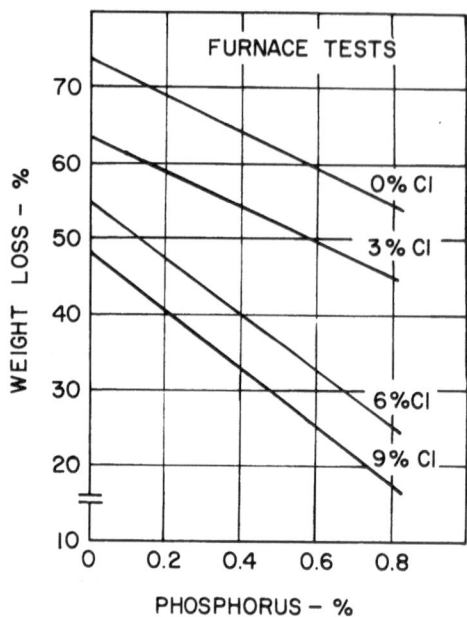

FIGURE 1. Combustion furnace test, showing effect of chlorine (as pentachlorophenol) on the weight loss of Southern pine blocks treated with a 50:50 creosote–oil mixture containing various amounts of phosphorus (as triaryl phosphate).[211] (From: *Forest Prod. J.*, **9**(10), Oct. 1959; Figs. 6 and 7, p. 328.)

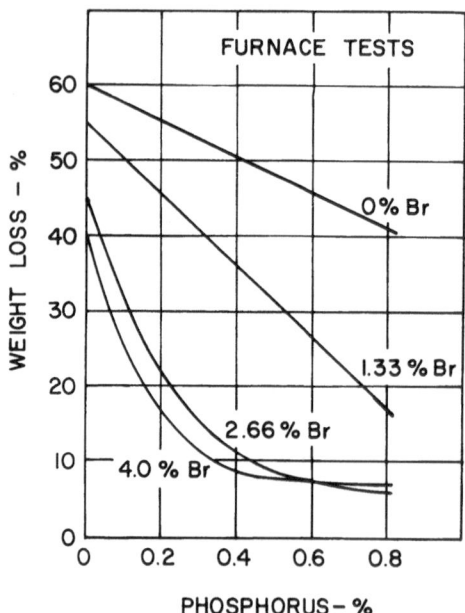

FIGURE 2. Combustion furnace tests showing effect of bromine as 2,4,6-tribromoaniline on the weight loss of Southern pine blocks treated with a 50:50 creosote–oil mixture containing various amounts of phosphorus (as triaryl phosphate).[211] (From: *Forest Prod. J.*, **9**(10), Oct. 1959; Figs. 6 and 7, p. 328.)

The bromine–phosphorus synergism is, however, not confirmed for polyurethane systems in which the addition of phosphorus to the bromine derivatives yields an additive activity only.[200] Similarly, the role of the phosphorus in tris(2,3-dibromopropyl)phosphate is unclear. Different relationships are obtained in the case of paper. As seen in Table 1, tris-(2,3-dibromopropyl)phosphate and tris(tribromophenyl)phosphate are both more effective in flame retardance than hexabromocyclododecane.

It is likely that the main contribution of the phosphorus is in decreasing the volatility of the compounds. Tribromophenol and tribromoaniline are more volatile than hexabromocyclododecane and did not exhibit any flame-inhibiting properties under the test conditions used (Table 1). On the other hand, pentabromoethane, being even more volatile than the last 3 compounds, showed some flame inhibition. It appears, therefore, that the activity of these compounds depends on the volatility as well as on the C–Br bond strength. The stability of the aromatic C–Br bonds in tribromo-phenol and tribromoaniline allows these compounds to evaporate before they can decompose and participate in the flame reactions. The higher effectiveness of pentabromoethane, in spite of its higher volatility, was due to the weaker C–Br bonds in the aliphatic structure.

The high flame-retardant activity of ammonium bromide compared to that of tris(dibromopropyl)phosphate is noteworthy. It may be attributed to the relative ease of the formation of the hydrogen bromide rather than to a synergistic effect between bromine and nitrogen. The much lower performance of 1,4-dimethyl-5-iminotetrazolinium bromide confirms this view. It appears to be the result of a higher thermal dissociation energy to the free base and the hydrogen bromide, which must reach the gas phase at the appropriate time in the decomposition of the combustible substrate in order to be effective.

The contribution of the volatility to the activity of the flame retardants seems to explain the different effectiveness of 2,4,6-tribromoaniline in the case of paper[203] from that of timber.[211] The presence of this compound, together with the triaryl phosphate in the case of the timber, could have, however, brought about the formation of less volatile phosphorus- and bromine-containing intermediates in the high temperature of the flame; this would explain the apparent synergistic effect. On the other hand, there is a pronounced difference in the mode of combustion between a thin paper strip and a block of timber in the tests applied. In the latter case the time of combustion is longer, the temperature is higher, and the relative contribu-tions of the volatility and C–Br bond strength of the halogen-containing flame-retardant compound would tend to favor a higher effectiveness. These conditions would allow the formation of the active bromine radicals before the volatilization of the undecomposed compound; this would not be the case with paper. Furthermore, the thickness of a timber sample which has been thoroughly impregnated with the flame retardant ensures a continuous supply of the compound during the combustion test.

The possibility of reactions occurring between bromine and phosphorus was investigated[212] by applying to cotton the combination flame retardant based on the bromoform adduct of triallyl phosphate and THPC-amide.[145] Only 8% of the originally added bromine was found in the char of the pyrolyzed fabric as compared to 95% of the phosphorus.[212] No bromine was found in the pyrolysis char of a fabric impregnated with 14% bromine in the form of tris(2,3-dibromopropyl)phosphate.[212] In both cases volatilization of the bromine during combustion did not markedly influence the amount of the phosphorus in the char, and it was concluded that no apparent reaction between bromine and phosphorus occurs during combustion.

6.3. Brominated Lignin and Cellulose

It appears from the above considerations that bromine-containing polymers with relatively low volatility might prove to be effective flame retardants. This would explain the high flame-retardant effectiveness of the bromine in brominated wood products.[213-216] The flame-retardant compound is bromolignin formed *in situ* by brominating timber or lignin-containing pulp with elementary bromine or with the chlorine–bromide system. In Fig. 3, the influence of the bromine content of the timber speci-

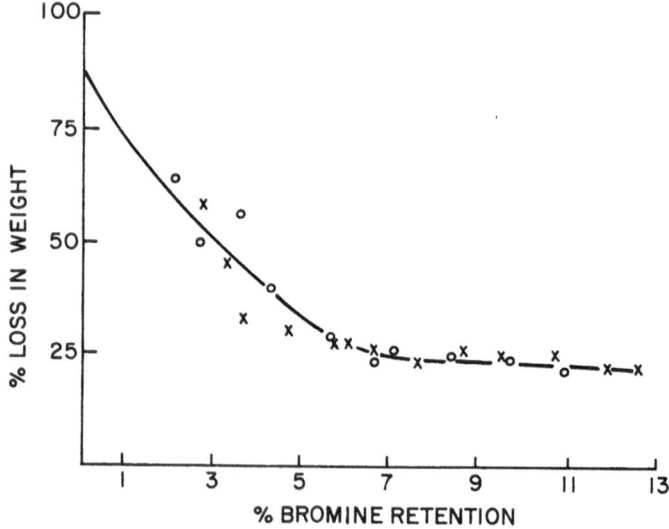

FIGURE 3. Effect of % Br in brominated picea excelsa on final loss in weight in the fire tube test.[216]

mens on the weight loss in the Fire Tube Test[217] is shown. At 5–7% bromine, the weight loss decreased from 88% in the untreated timber to 25%. Corresponding decreases in weight loss were reported for timbers impregnated with 13–14% diammonium phosphate.[217] The same brominated samples show a sharp decrease in afterglow time and in the temperature of the flue gases from the combustion.

Unlike the case of tris(dibromopropyl)phosphate and the bromoform adduct discussed above, substantial amounts of bromine were found in the char fraction, as well as in the tar and aqueous fractions obtained on pyrolyzing brominated wood powder at 250°C for 3 hr in vacuum.[218] The presence of the bound bromine in the char suppressed its afterglow and reflects the lower rate of decomposition of lignin and bromolignin compared to cellulose and hemicellulose. The presence of bound bromine in 6-bromomesyl cellulose[67,68] did not prevent the samples from afterglowing upon combustion. Schwenker and Pacsu[67] do not report data on the bromine content of the char, and it is possible that during the pyrolytic decomposition of the cellulose the bromine was essentially volatilized and, in order to protect the samples from afterglowing, phosphorus had to be added to the system. Percentages of 4–5% of bromine were, however, sufficient to prevent the brominated mesyl cellulose from flaming in the match test.[67] Addition of bromine to cellulose by reacting it with methylol derivatives of dihalocyanoacetamide[219] yields similar results to the direct bromination.

$$NC-\underset{\underset{Br}{|}}{\overset{\overset{Br}{|}}{C}}-CON(CH_2OH)_2$$

The bromine-containing flame retardants have not yet been applied on a sizable scale to cellulosics. The activity of bromine compounds is seen in the light of the foregoing discussion to vary with the substrate and with the performance variables and testing methods specified for it. Specific compounds might have to be tailor-made for particular applications if high effectiveness and low cost are required. Of basic importance for textiles is the durability of the flame retardant-treated fabrics against repeated laundering, dry-cleaning, light, and bleaching agents. The bromine compounds thus have to be insolubilized on the fabric or made to react with it. In both cases the strength properties, stiffness, abrasion resistance, and dye fastness have to be kept within acceptable limits. These are the current requirements for all flame retardants. Essentially, the relatively low add-on of an effective bromine compound needed for a high degree of flame retardance might facilitate the preservation of the fabric properties. It is thus to be expected that interest in bromine-based flame retardants, possibly in conjunction with phosphorus and nitrogen, will be sustained in the future.

7. Flame-Retardant Treatment of Wood, Board, and Paper

7.1. Flame-Retardant Requirements

Wood-base materials, such as timber, insulation board, hardboard, and particle board, are flammable. They can be ignited by a variety of fire sources, and once ignited the flame spreads either across their surface or from one surface to another, until it reaches the "flashover point" when the fire becomes general throughout the room. After that point the further spread of the fire to adjacent rooms in a building is limited only by the "fire resistance" of the structure[219] in question. While spread of flame and ignition are properties of the material, fire resistance depends on, in addition to combustibility, the strength properties under heat, the dimensions, and the thermal conductivity. Fire resistance of a heavy timber structure may thus be higher than that of uncombustible materials owing to its low heat conductivity enhanced by the insulating layer of charcoal formed on its surface in a fire, its high load-bearing qualities, and its low coefficient of expansion.

The importance of flame-retardant treatments of timber in raising the ignition temperature and lowering spread of flame characteristics is greater than in enhancing its fire resistance, which can be improved equally well by other means, such as changes in design and dimensions.[220,221] It has been shown that heavy impregnation of solid wood walls with flame-retardant chemicals decreased the rate of flame spread by a factor of 8 as compared with untreated timber, but the increase in failure time was only by 25%. Only by rendering wood completely incombustible, i.e., preventing its pyrolytic decomposition, could one produce a fire-resistant timber. However, the known flame retardants to date can prevent ignition, after-flaming, and afterglow, and direct the pyrolysis reactions to the production of high amounts of char and water and low amounts of tar and nonflammable gases, but they cannot prevent the pyrolysis itself.

At present flame-retardant treatments of wood-base materials therefore aim at preventing ignition and reducing the rate of flame spread. In the case of fiberboard, particle board, plywood, and most timber products these are the main requirements. Flame-retarded wood products are at present used mainly in interior partitions, decorative and insulating materials, roof framing, and, where timber replaces noncombustible construction, in doors and exterior wood scaffoldings.

Flame-retardant treatment of paper and paperboard is of importance in packaging, decoration, electrical applications, and structural and utility uses. It has not achieved widespread application. The disposal of treated paper presents serious problems which may possibly be overcome by organized reuse.

7.2. Treating Processes

Flame retardance of wood can be achieved basically in two ways: impregnation treatments and surface coating. The present chapter does not deal with coatings.

The impregnation treatments are carried out in pressure vessels as practiced by the wood preservative industry.[222.223] A vessel previously loaded with the wood is evacuated. The treatment solution is admitted and pressure is applied for a predetermined time, depending on the wood species. The wood is dried in specially constructed kilns by a relatively slow process. Complete impregnation of timber is possible only for permeable wood species or for sapwood. For less permeable species and for heartwood, incision is practiced. In all cases, it is aimed at a penetration of 1/2 in. on both faces. The average dry-flame retardant retention is approximately 3 lb/ft^3, but for maximum protection, and in special cases, retentions exceeding 5 lb/ft^3 are being used. It is very difficult to obtain completely homogeneous penetration of all surfaces of all pieces of timber. The retention of the chemical on a batch of timber is an average of the retentions of the individual specimens in the batch.

It recently has been found possible to increase the retention by precompressing the wood specimen across its grain between rollers. A reduction in thickness of the specimens by 5–20% brought about an increase in retention of chemical upon subsequent impregnation from 8.2% to 11%.[368] Impregnations with flame-retarding chemicals by subjecting the panels to compressional wave energy ranging from sonic to ultrasonic frequencies have also been described.[373]

An adequate retention of chemicals was recently demonstrated by using high-velocity jet impregnations. Experiments[371] with a number of flame-retarding chemicals on 5/8-in. thick Douglas fir and on particle board showed that high flame-spread protection can be obtained by this impregnation technique.

The main chemicals used as flame retardants are salts: diammonium and monoammonium phosphate, zinc chloride, ammonium sulfate, borax and boric acid, and ammonium chloride. In a comparison study carried out on Douglas fir plywood samples impregnated with these salts, it was found that $(NH_4)_2HPO_4$ is most effective in retarding flame spread, followed by $ZnCl_2$, $(NH_4)_2SO_4$, and $NaBO_3$. However, both $(NH_4)_2HPO_4$ and $ZnCl_2$ increased the smoke level index in the 8-ft tunnel tests at dry retentions higher than 2 lb/ft^3, while the other chemicals reduced the value with increasing retentions.[224-226] Some formulations of these salts used commercially are described in the literature.[227] The cost of the treated wood is in the range of \$70–100/1000 board feet above the untreated wood. A list of the accepted flame-retardant treatments for lumber and plywood is published by the U.S. Department of Defense.[228] The Minalith formulation is made up of 10 parts $(NH_4)_2HPO_4$, 60 parts $(NH_4)_2SO_4$, 10 parts borax, and 20 parts boric acid. The Pyresote formulation is made up of 35 parts each of $ZnCl_2$ and $(NH_4)_2SO_4$, 25 parts H_3BO_3, and 5 parts $Na_2Cr_2O_7$. The salts are dissolved in water at 12–15% concentrations.

Ammonium polyphosphate liquid fertilizers (11% N, 37% P_2O_5) at a retention of 4–6 lb/ft³ of wood alone or in equal mixture with ammonium sulfate were found to be effective with regard to flame spread, fuel contributed, and smoke developed. These formulations were considered to be economically advantageous compared to the currently used phosphate formulations.[363] The use of mixtures of $NH_4H_2PO_4$ and NH_4HF_2 in the presence of dichromate and a wetting agent for impregnation was also suggested.[364] A leach-resistant impregnating composition consisting of a mixture of arsenic pentoxide dihydrate, zinc oxide, phosphoric acid, potassium dichromate, and copper sulfate was also recently described.[365]

The treatment of fiberboards is done either by treating the pulp before the sheet formation or by spraying or soaking the wet pressed mat before drying. Hardboards have been treated by pressure impregnation after hot pressing. The use of urea–phosphate[369,372] and of mixtures of $(NH_4)_2HPO_4$ $(NH_4)_2SO_4$[370] for such impregnation was suggested. For paper, the addition of the flame retardant as a beater additive or in the sizing stages is possible.

7.3. Properties of Flame-Retardant Wood

In the above-mentioned flame-retardant treatments, essentially no chemical reaction takes place between the wood and the salts, and little is known on interactions between the salts themselves.[224] The properties of the salts influence the behavior of the wood. Most of the applied salts are highly hygroscopic. Ammonium sulfate deliquesces above 81% and monoammonium phosphate above 93% relative humidity. Boric acid, on the other hand, is unaffected by water vapor; borax is efflorescent at normal humidities. The treated wood is thus sensitive to relative humidity. Migration of salts to the surface will take place on increasing relative humidity, and exudation and loss of flame retardant and fire protection will occur.[229] Furthermore, a deleterious effect on dimensional stability, paint coatings, and corrosivity may be expected. The addition of corrosion inhibitors, such as sodium dichromate, in small amounts strongly reduces the corrosion of the treated timber to various metals at ambient conditions, but does not prevent electrolytic corrosion taking place on the metal surfaces by exuded salts at high relative humidities.[224] It has been suggested recently that the strength properties of timber should be tested in the wet state because of the hygroscopic nature of the treated timbers.[230] In plywood a reduction in shear strength proportional to the amount of flame retardant was found.[230] Although the inherent lack of homogeneity of the treated timbers makes it difficult to determine strength losses resulting from treatment, decreases of 5–8% in modulus of elasticity and of 10–17% in modulus of rupture as compared to untreated specimens are reported.[224]

The presence of salt crystals in wood was found to have an abrasive effect on the cutting tools used in machining treated timbers and prompted the use of tools topped with tungsten–carbide or similar abrasive-resistant materials. The presence of salts was found to interfere with the gluing of treated timbers both in the manufacture of plywood[231] and of structural timber assemblies.[224] While the bondings obtained are satisfactory for interior use, they are inadequate for exterior use and for structural assemblies.

The most important drawback of the treatments currently in use is the leachability of the salts on outside exposure. The leachability, in addition to the salt migration, renders the treatments nondurable and limits their application. Considerable effort has been exerted in recent years to overcome this drawback both by modifying the existing treatments, and by developing new ones.

7.4. Recent Developments

7.4.1. Sealing Processes

The use of an organic chlorinated sealer was suggested to reduce the leachability of chromated zinc chloride-treated wood.[232] The precipitation of insoluble double salts within the wood was described in several disclosures. $ZnNH_4PO_4$ soluble in excess ammonia was impregnated into the wood, which was then heated to remove the ammonia and precipitate the compound. A similar system involves the composition $2ZnO \cdot ZnCl_2 \cdot 2NH_3$. A two-stage process involving impregnation with a soluble zinc salt following precipitation with a base was also disclosed.[233,234] Insolubilization of borax by precipitating manganese borate with $MnCl_2$ inside the impregnated timber was suggested[235] in order to decrease the leachability. A double impregnation, first with a solution of $(NH_4)_2HPO_4$ and then with a solution of $MgSO_4$ to precipitate $MgNH_4PO_4$ on the wood, was also described.[362] Zinc chloropropyl phosphate and zinc bis(chloropropyl)phosphate solubilized with NH_4OH as well as other water-soluble heavy-metal salts of acid phosphate esters were suggested recently for the impregnation of wood to produce a weather-resistant flame-retardant effect.[389] All the above sealing treatments were, however, found to be subject to leaching.

Sodium silicate, together with sodium borate, has been used as an impregnation solution for wood. The silicate is precipitated in the wood with CO_2 under a pressure of 500 psi. The leachability is said to be reduced and the flame-retardant partially fixed.[236] No reduction in strength is observed.

Fire-resistant, as distinct from flame-retardant, plywood is claimed to have been made by impregnating fiber mats with a sodium silicate solution, then drying and pressing for 5–6 hr at 20 atm. The pressed sheet is coated with a phenol–resorcinol–HCHO adhesive and heat-pressed together with a layer of chips sprayed with a urea–formaldehyde binder. The sheet is then used as the inner layer in plywood. The temperature at the back of the sheet after heating for 30 min with a gas burner was only 135°C.[237]

For the treatment of red cedar shingles, in which the nonleachability is of basic importance, the sealing technique was likewise suggested. A mixture of trimethyl thiophosphate and triphenyl phosphite in a bakelite-type reaction product of epichlorhydrin and 4,4'-dihydroxydiphenylmethane, in the presence of a silicone resin as fluidity modifier and triethylenetetramine as hardener, impregnated into red cedar shingles and dried at 100°C was used for flame retardance. The flame retardant remains embedded in the cured resin which decreases leachability.[238]

A durable flame-retardant treatment for red cedar involves impregnation with PCl_3, $PSCl_3$, and a vinyl chloride–vinyl acetate copolymer binder at a retention of 11.7%.[374] The use of impregnating baths containing mixtures of melamine–formaldehyde resins and strong inorganic acids such as H_3PO_4, HBr, H_2SO_4, and H_2NSO_3H for sliced veneer and filter papers at a 10–15% add-on was described.[378] In another procedure an aqueous emulsion containing inorganic phosphates, a wax, and poly(vinyl acetate) for the impregnation of paper and corrugated board is used.[379]

An acrylic resin in conjunction with urea or dicyandiamide was used as a sealing agent in conjunction with NH_4Br in an impregnation bath for paper with resulting good elasticity and tearing strength.[380] An impregnation of paper with poly(vinyl bromide) was disclosed.[381] Another application of bromine for flame retardance of paper involved the use of the product obtained from reacting of brominated oils, e.g., cottonseed oil, castor oil, soybean oil, or methyl 9,10-dibromostearate, with $P(OCH_3)_3$ or $P(OC_2H_5)_3$ at 150–210°C for 5–6 hr.[382]

Polyethylene was also used as a sealing agent in a process for the production of building panels. Wood waste previously impregnated with hexachlorobutadiene is pressed with a mixture composed of polyethylene and a number of flame-retardant compounds, e.g., antimony oxide, chlorinated paraffin, and $MgSiO_3$.[390]

The application of antimony oxide as a flame retardant to paper was carried out using the sealing technique. A suspension of antimony oxide in polyvinyl chloride latex, in the presence of carboxymethyl cellulose and a fatty acid soap, was mixed with pulp and precipitated by the addition of aluminum sulfate. The paper produced from the stock was flame resistant.[239] Butadiene rubber was also used as bonding agent.[240]

An emulsion of antimony trioxide in vinylidene chloride copolymer (1:1 ratio) has been developed as a flame retardant for paper, paperboard, and nonwovens with the trade name Fire-Tard. It is applied by wet-end addition, surface sizing, or coating. The treated paper retains its flame-retardant properties upon leaching. The flame-retardant properties are good, and the strength is said to increase.[241] The chlorine-containing copolymer performs as both the sealing agent and fire retardant in these cases.[239,241]

A similar approach to flame retardance of fiberboard is the addition to the slurry of cellulosic fibers of such noncombustible elements as glass fibers, magnesium oxide, aluminum sulfate, water glass, diammonium phosphate, starch,[241a] kaolin,[366] and zeolite,[367] and then mold pressing. The cellulosic fibers constitute 50–75% of the final board obtained.[241a] The flame retardant is thus embedded in the structure in sufficient amounts to protect the cellulosic fibers against fire.

7.4.2. Reactive Flame Retardants

A considerable number of durable flame-retardant treatments based on a chemical interaction between the substrate and the applied compound were studied. Such interactions are possible with cellulose as well as with the lignin in the lignocellulosic material.

7.4.2.1. Compounds Interacting With Cellulose. Most of the systems based on interactions of phosphorus derivatives with cellulose were described in Section 5 above for textiles. The patent literature, however, recommends the use of many of these compounds on paper and fiberboard, as well as timber. In the following, several treatments will be briefly described.

Partial phosphorylation is apparently achieved by impregnating wood with phosphoric acid[242] in the presence of dicyandiamide and formaldehyde. The treated wood is claimed to be nonhygroscopic, resistant to leaching, and less subject to swelling. A similar composition with the addition of urea and the use of monoammonium phosphate was disclosed by Stumpf.[243] Additions of dextrin and lactic acid to the above formulation are included in a later patent[244] designed for the fire retardance of boards. Another composition for fiberboard includes pentaerythritol instead of dicyandiamide.[245] Mixtures of monoammonium phosphate with phosphoric acid and urea[246] or with carbamide[247] are suggested for the impregnation of hardboard. The absorption is 40% of the weight of the board. The board is then cured at 100–140°C for 1–3 hr.

A stable, leach-resistant, flame-retardant, fungistatic treatment based on phosphorylation involves the use of a solution composed of melamine, dicyandiamide, phosphoric acid, and a 37% formaldehyde solution in ratios of 0.25:0.75:1:3. The solution is stated to have a pot life exceeding 14 days and to be applicable by impregnation, surface treatments, or as an adhesive to red cedar shingles and shakes, plywood, and particle board, without affecting their esthetic quality. It is fixed by heat curing.[377]

A partial phosphorylation probably also takes place when an impregnating solution composed of H_3PO_4, $H_2NC(NH)NHCN$, and a thermosetting resin is used.[383] Another phosphorylating treatment applied to pulp and to polyvinyl alcohol fibers involves impregnation with an aqueous ammonium polyphosphate and heating in xylene at 139°C. This treatment was claimed to cause less degradation of the cellulose than other phosphorylating treatments.[384]

The phosphorylation processes appear to be more effective when applied to hardboard and particle board due to the relatively high temperatures of pressing. The application of a partly polymerized mixture of THPC and THPO with melamine resins or with acrylamide[248] is performed by impregnating the wood and heat treating it. The impregnation of pulp sheets with APO [tris(l-aziridinyl)phosphine oxide] in the presence of a boron–trifluoride–amine complex catalyst and subsequent curing has also been suggested[249] and increased tensile and burst strengths, along with flame retardancy of the paper, were claimed. In this treatment, the cellulose is partially cross-linked. A cross-linked, flame-retardant wood and paper is also obtained by impregnating with cyclic ethylene vinyl phosphonate.[386]

The use of alkyl or aryl halogenated phosphites or phosphonates for wood impregnation was found to be particularly effective.[387] The use of bis(2-chloroethyl)vinyl phosphonate and bis(2-chloroisopropyl)isopropenyl phosphonate together with copolymers of an alkyl acrylate with N-methylolacrylamide or N-butoxymethylacrylamide for the impregnation of paper and cellulosic textiles was also recently disclosed.[388]

Flame-retardant wet-strength papers are made by applying a cationic colloidal solution of methylolated tris[2-(4,6-diamino-S-triazine-2-yl)ethyl]-phosphite oxide as a heater additive to the paper stock[250]:

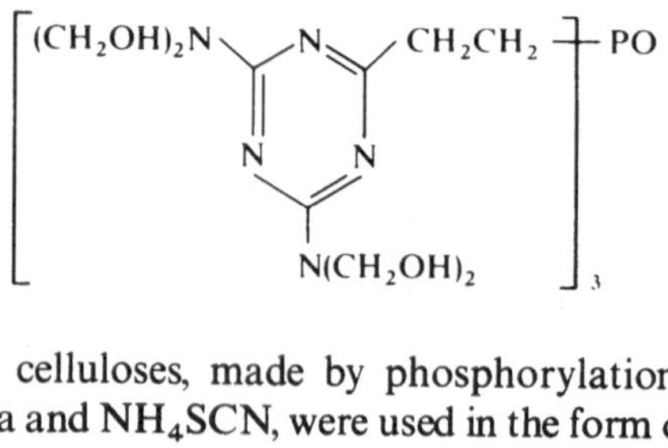

Phosphorylated celluloses, made by phosphorylation in molten urea or in a mixture of urea and NH_4SCN, were used in the form of the ammonium salt as dispersants or binders for the preparation of flame-retardant papers having improved mechanical properties.[385]

Radiation polymerization *in situ* of the wood was investigated in order to obtain a nonleachable flame-retardant effect.[251] Wood was impregnated *in vacuo* with flame-retardant unsaturated organophosphorus compounds: bis(2-chloroethyl)vinyl phosphate, dimethyl methallyl phosphonate, triallyl phosphate, trimethallyl phosphite, diallylchloromethylphosphonate and diallyl phosphite. Cross-linking monomers, such as divinyl benzene, ethylene glycol dimethacrylate, were also used. When an even distribution of the compound in the wood is achieved, flame retardance and leaching resistance are obtained after irradiation with a cobalt source. Particularly good results were obtained on western white pine.

The use of tris(chlorotribromopropyl)phosphite for impregnation of lumber, plywood, and particle board, followed by application of mixtures of styrene and methyl methacrylate and irradiation with a high-energy radiation source was recently suggested.[375] Other formulations which include acrylonitril and styrene were claimed to produce higher dimensional stability and weather resistance compared to methyl methacrylate.[376]

7.4.2.2. Reactions With Lignin. Compared to the cellulose, little attention has been directed to the lignin component of the wood. Use of lignin as the basic substrate from which a flame-retardant compound is produced *in situ* is of interest provided the reaction can be carried out without excessive degradation of the cellulose component.

Since it was established that elementary bromine dissolved in water does not degrade cellulose[252,253] to any significant extent, it is possible to brominate wood with elementary bromine.[213,214] A low yield of approximately 30–33% substituted bromine, however, is obtained. Bromination of primarily fiberboard-grade pulp containing lignin is therefore being carried out by suspending the pulp in a sodium bromide solution and introducing gaseous chlorine or chlorine water in excess into the slurry.[213,253a] The bromine yield obtained in the rapid bromination (30–60 sec) is over 90%, and the hydrochloric acid formed is later neutralized by calcium hydroxide addition. The brominated pulp is dewatered and formed into insulation board or pressed into hardboard. A 1.4:1 ratio of chlorine to bromine is used. A lignin oxidation reaction consumes 42% of the added chlorine and 2–7% of chlorine is used in the chlorination of the lignin. There is no noticeable degradation of the cellulose as evidenced by viscosity-DP determinations, and the strength properties of sheets made from brominated pulp do not differ from the unbrominated sheets.[214]

The bromination of timber is carried out by the vacuum-pressure method in autoclaves with a sodium bromate–sodium bromide 60:40 water solution with the addition of sulfuric acid and a buffering salt.[216,253a] The main bromination reaction occurs after the treatment solution is pressed into the timber. The pressure applied is ~ 150 psi, and the final pH of the brominated timber is 3.5–4.5.

A small percentage of sodium dichromate is added to the treatment solution in order to eliminate the corrosivity of the treated timber. Retentions of 6–8% of bromine yield a high degree of fire retardance stable to prolonged severe leaching, weather exposure, and storage. The strength properties, paintability, and workability are not significantly changed by the treatment, but swelling and water absorption are decreased. The brominated wood was found to be durably stable against wood-attacking fungi.[216]

The application of halogen and phosphorus was also suggested by first brominating lignin-containing fibers and thereafter phosphorylating with Et_3PO_3.[253] Percentages of 2.9% Br and 2.6% P appear to be sufficient for flame retardance of hardboard. Relatively long reaction times and low phosphorus yields are obtained, and the application to timber is difficult.

A lignin–phosphorus ester was prepared by reacting lignin with PCl_3 and applied as a flame retardant for paper and textiles.[254] A chlorinated lignosulfonic acid containing about 13% combined chlorine, obtained by chlorinating sulfite waste liquor, was suggested as a flame-retardant and fungicidal agent for wood,[254a] especially for impregnating pit props. The impregnation solution contains 6–10% of chlorosulfonic acid. The treatment is not durable although it is stated that even under excessively humid conditions leaching out is very slow. Urotropine or thiourea is added in order to reduce corrosivity of metal containers.

7.4.3. Solvent Treatments

Impregnation of wood with flame-retardant compounds dissolved in low-boiling solvents is of particular interest since it might emable the elimination of the slow and costly kiln-drying operation. Several such procedures were disclosed.

A 15% solution of $(MeO)_2POH$ in benzene was used for the impregnation of timber with the vacuum-pressure technique. The excess solvent is removed by vacuum evaporation, and in a second stage dry gaseous ammonia at a pressure of 12 psi is admitted to the vessel for 30 min. The excess of ammonia is removed by diffusion at atmospheric pressure.[255]

In another disclosure a 15% solution of bis(β-chloroethyl)phosphite in ether was used in the full-cell method to impregnate wood. The wood samples were placed in a treatment cylinder. Vacuum of 27 in. Hg was drawn and the solution admitted. A pressure of 150 psi was applied for 1.25 hr. The solvent was then evaporated under reduced pressure, leaving 15% of the compound in the wood.[256]

Another solvent application, $POCl_3$, PCl_3, or $PSCl_3$ in CH_2Cl_2 along with a binder (preferably vinyl chloride–vinyl acetate copolymer), is impregnated in the wood by the vacuum-pressure technique and dried at 65°C. At a take-up of 7.4% $POCl_3$, the wood was stated to be highly flame retardant and stable to prolonged alternating leaching and solar exposures.[257] A trimethyl borate–methyl alcohol–azeotrope mixture impregnated in water-saturated wood at 25 psig for 24 hr is said to impart flame retardancy, decay-resistance, as well as dimensional stability to wood.[258]

Monobasic acid solutions of titanium and antimony dissolved in isopropyl alcohol were already suggested for the impregnation of wood,[259] while the use of chlorinated paraffin, in which antimony trioxide[260] or triphenyl phosphate[261] are suspended, were disclosed for the impregnation of hardboard.

7.4.4. Treatment of Particle Board

The production of particle board involves spraying wood chips with a urea–formaldehyde, melamine–formaldehyde, or phenol–formaldehyde adhesive, and forming and pressing sheets at elevated temperatures (120–220°C) and pressures (10–20 atm). Due to the low cost of the products, only relatively inexpensive processes and flame-retardant chemicals are applied, mainly inorganic salts. Particle board can be impregnated by vacuum-pressure procedures similar to those used for wood. The high density of the boards limits the penetration and retention of the chemicals. Furthermore, most of the adhesives used for the production of particle board are swollen under the impregnation conditions, and consequently the physical properties of the boards are severely impaired. Recent developments concentrated therefore on several possibilities of treating particle board in the production line: one-stage procedures involving (1) dispersion or dissolution of the chemicals in the adhesive before spraying the wood chips, and (2) addition of the chemicals to the chips before spraying the adhesive; and two-stage procedures in which the chips are sprayed with a solution of the chemical, dried, and later sprayed with adhesive, then formed and pressed. Several recent disclosures of these approaches are briefly summarized below.

In a one-stage procedure a mixture of crystals of $(NH_4)_2HPO_4$, NH_4Br, and $(NH_4)_3PO_4$ of less than 20-mesh size is embedded in the resin and uniformly distributed between the chips and pressed. The final laminate contains 20–30% of salts.[262]

In another procedure, the chemical in dry powdered form is added to the wood chips in the drying–mixing vessel before or after the drying of the chips. One or more chemicals, mainly bromides, phosphates, and borates, have been applied. The advantage of this procedure as compared to the addition of the chemical to the chips in the form of a solution or to the addition of the chemical in the adhesive is in the saving of the energy required to evaporate the solvent of the chemical, in reducing the amount of adhesive needed, and in improvements in the physical properties of the final board.[391] When using ammonium bromide and boric acid powders, a combined chemical content of 13.5% on the basis of dry chips was sufficient for the board to pass the Class 1 flame rating according to the French standard.[344]

A mixture of 7:1 powdered boric acid and antimony oxide is added either to the liquid adhesive or to the wood chips before pressing to produce a flame-resistant particle board of the desired density and size.[263] Another patent discloses the application of a mixture of boric acid, phosphoric acid, monoethanolamine and paraffin emulsion, and ammonium chloride, together with the urea–formaldehyde or phenol–formaldehyde adhesive to wood chips and molding under pressure.[264]

A binder slurry of MgO, $MgCl_2 \cdot 6H_2O$, and water in the ratio $1:6:21$, together with 3–12% stearic acid, was sprayed on wood flakes and tumbled in a cement-type mixer; the oven-dry weight of the binder solids is 20% of the flakes. The material was then molded at 120°C at 220 psi for 15 min. To enhance flame resistance, magnesium bromide was suggested. The function of the stearic acid is to decrease swelling and water absorption of the board.[265]

In the one-stage in-process treatments in which the chemicals are added in the form of powders, a "blooming" or efflorescence of the chemical onto the surface of the board is often encountered. In addition, when the chemical is added to the adhesive, a change in the physicochemical properties of the adhesive occurs resulting in a weakening of the bonding of the particles and a subsequent decrease in the physical properties of the final particle board. It has recently been disclosed, however, that in the case of the in-process use of inorganic bromides, borax, boric acid, and ammonium phosphates as additives to the adhesive, the addition of formaldehyde-forming compounds (e.g., paraformaldehyde, hexamethylene tetramine, trioxane, or polyoxymethylene), in amounts of up to 5% by weight of the dry wood improves the physical properties of particle board significantly and restores it to commercially acceptable values. The formaldehyde-forming compound can be added in powdered form to the wood chips or dissolved in the adhesive.[392]

Since inorganic bromides have been shown to be highly soluble in the urea–formaldehyde adhesives,[270] it was suggested that such solutions be applied in the main adhesive spraying stage.

Defined mixtures of sodium and ammonium bromide are preferred in order to preserve the original viscosity and surface-tension characteristics of the adhesive solution.

Some recent disclosures of the two-stage procedure are: prespraying of the chips with mixtures of strong acids, such as HCl, H_3PO_4, or HNO_3

with borax;[266] chlorinated paraffin in which antimony oxide is dispersed;[267] a solution of phosphoric acid and dicyandiamide;[268] a solution of mono- and diammonium phosphate, pentaerythritol, dicyandiamide, and paraformaldehyde.[269] In the last two cases, a partial phosphorylation of the chips probably takes place during the hot pressing.

Several treatments which may be classified as two-stage procedures have lately been developed in which only the outer layer of the wood product, whether plywood or particle board, is pretreated with the flame-retardant chemicals and afterward laminated by the use of adhesives, pressure, and elevated temperature to the untreated core of the composite panel.[393] Paper sheets, plastic films, or thin veneers may serve as these outer layers. Paper sheets impregnated with ammonium polyphosphate were used in such processes.[394] The extent of the flame protection obtained by these procedures is naturally limited. It depends on the nature of the flame retardant, on its amount, on the thickness of the protecting layers, and on the flammability specification applying to the product. The action of the treated outer layer is similar to that of a coating. A formulation by which a flame-retardant plywood is produced by laminating it with aluminum foil and subsequently coating it with a foaming composition containing ammonium polyphosphate was also described.[395] Asbestos fibers mixed with wood chips and boric acid dispersed in a urea–formaldehyde binder are sprayed onto a pressed fiberboard core, and thereby a molded wood composite with an outer flame-retardant layer is formed. The outer layer contained over 10% boric acid and 40–60% asbestos fiber based on the oven-dry weight.[396]

Flame-retardant modifications of the surface of particle board were also obtained by hot pressing the board surface treated with $NH_4H_2PO_4$ powder at 250°C or liquid ammonium polyphosphate at 160–200°C. It has been stated that the strength and surface properties of the resulting particle board are superior to those treated in-process.[397]

8. Testing Methods

In the following, several testing procedures used for the evaluation of the flammability behavior of cellulosic textiles, paper, and wood-base materials are briefly described and discussed. Special emphasis is given to the recent developments, but no attempt is made to fully cover all existing methods or to include complete details of procedures.

8.1. Textiles

The tests used for the evaluation of the flaming properties of textiles are numerous and have been designed and introduced by a considerable number of governmental and safety agencies as well as by technical societies and industrial research departments in several countries. In most cases, the various tests are similar or differ only in minor details, as shown in a recent listing[271] of all tests applied to textiles. The test methods can be differentiated into two groups: tests on fabrics expected to burn (i.e., comparing the flammability of clothing textiles) and tests for fabrics not expected to burn (i.e., measuring the degree of fire resistance of flame-retardant fabrics). In addition, several tests of a more general nature have recently been proposed. Special testing methods have been introduced for specific textile items, such as carpets and rugs. Most of the tests deal with ignition, rate of spread of flame, afterglow, and char length upon burning of relatively small strips of fabric. Lately, attention has been drawn to the possibility of testing whole garments and to the evaluation of the smoke and of the toxicity of the gases produced in the combustion of textiles.

The fire testing of textiles is a rapidly growing field, and further developments, both in the technology of the testing methods as well as in their scientific basis, are to be expected.

8.1.1. Flammability Tests

The 45° Test:[272] The test is designed to indicate textiles, mainly used for apparel, "which ignite easily and, once ignited, burn with sufficient intensity and rapidity to be hazardous when worn." The test was first developed by AATCC. Following the adoption of the Flammable Fabrics Act in 1954,[273] it was accepted as a commercial standard in the United States.[274] Tests on similar lines were accepted by the British Standard Institute,[275] the Swiss Standard Association,[276] and by the Canadian Government.[277]

Oven-dry specimens of 2 × 6 in. of fabric are used in the test. The specimen is held at 45° in a rack placed in a ventilated chamber containing an automatic timing device and a standardized ignition medium. The flame applied is of 5/8 in. length. The distance of the fuel nozzle from the fabric surface is 5/16 in. The distance from the central point of flame impingement on the face of the specimen to a stop-cord made of mercerized cotton sewing thread is 5 in. The time of ignition is adjusted to 1 ± 0.05 sec. The time of flame spread, including ignition from the impingement point to the stop-cord (5 in.) is measured. Flaming times higher than 3.5 sec for textiles without a raised fiber surface or 7 sec for textiles with raised fiber surfaces, are considered of "normal flammability." Textiles of intermediate flammability with raised fiber surfaces have a time of flame spread between 4 and 7 sec. Lower times of flame spread characterize dangerously flammable textiles.

The 45° test has lately been severely criticized as it became known that it is passed by 99% of the commercially available fabrics and that serious burn cases occurred with fabrics meeting this standard which is required by United States law. Both the time of ignition of 1 sec as well as the upper limits of the times of flame spread were found to be too mild for the commercial fabrics. Recent reports indicate that work is now being done on the differentiation and interrelation between time of ignition and rate of flame spread and on their precise measurements. Attention is being given to the behavior of materials which ignite slowly but burn intensely and to the size of the sample.[278,279]

The 45° test has recently been accepted in a modified form in Japan.[280] This test is essentially similar to the AATCC-33 test. but afterflame and afterglow times and charred area are measured. The standards for lightweight fabrics are 3 sec. 5 sec. and 30 cm^2. respectively. The standards for heavy fabrics are 5 sec. 20 sec. and 40 cm.^{2}

Horizontal Spread of Flame Tests: Several tests have been designed for measuring the spread of flame on horizontally placed fabric specimens. In Federal Specification Method 5906[281] specimen strips of 12.5 × 4.5-in. oven-dry fabric are used. Ignition is effected at one end of the sample. and the rate of burning is measured. In Method 5900. the ignition is effected in the center of the horizontal specimen and the largest dimension of the charred area is reported.

A German method similar to 5906 was also announced. Specimens of 35.5 × 10 cm. conditioned at 20°C and 65% relative humidity, are ignited for 15 sec by a propane gas burner of specified dimensions applied to the edge of the specimen from below. The time required for the lower edge of the flame front to travel a specified distance is measured.[398]

Although the horizontal position enables obtaining the greatest difference in the spread of flame on fabrics. these methods are not widely used since it is believed that the usual position of clothing fabric is the vertical.

Vertical Spread of Flame Tests: In the vertical test, introduced in the United Kingdom,[282] the specimen dimensions are 1.5 × 72 in.. conditioned at 70°F and 65% relative humidity. The specimen is ignited with a gas burner at its lower edge. and the rate of flame spread is recorded.

In a similar test developed in Canada.[283,284] the tested specimen is oven dry and its dimensions are 2 × 30 in. The ignition is effected by a microburner. This testing method was recently used by Nielsen and Richards[285] to investigate the influence of several parameters on the rate of burning. It was found that. while the rate of burning decreases linearly with the increase in the weight of the fabric (being lower for polyester–cotton fabrics than for cotton). it was not influenced by the thickness. density. and ratio of air-to-fiber volume of the fabric. The reproducibility of the results is higher in the case of frame-supported specimens than in free-hanging specimens. Resin-treated cotton and cotton–polyester fabrics burn appreciably faster and leave more char than untreated fabrics. The minimum width of the specimen for reproducible results appears to be 2 in.

Ease of Ignition:[286] Oven-dry specimens of 7 × 7 in. are used. The specimens are placed horizontally and a microburner is applied to one edge. The time needed to ignite the specimen is measured. A close linear correlation between the time of ignition and the fabric weight was found on untreated and resin-treated cotton fabrics.[284,285] A method for measuring time of ignition on fabrics conditioned at 65% relative humidity and 20°C has been proposed in Germany. The specimens are mounted horizontally and the propane burner is applied from below to the edge of the specimen.[400] An apparatus comprising an ignition–flame time selector and meter, burning-time meter, and fire- and flame-front detectors was recently described. The apparatus was suggested for the separate determination of ignition time and flame propagation.[401]

8.1.2. Fire Resistance Tests

The AATCC Test Method 34:[287,288,402] This method was first adopted as standard in 1952 for the evaluation of "industrial fabrics" for resistance to fire and was revised in 1966 to include all flame retardant-treated cotton fabrics and fabrics made from inherently flame-retardant fibers.

"Fire resistance" is defined as resistance to flaming, glowing, and smoldering; "afterglow" is defined as the time the specimen continues to glow after flaming ceases. Specimens of $2\frac{3}{4} \times 10$ in. cut in the warp and in the filling direction, conditioned at 21°C and 65% relatively humidity, are used in the test. The specimen is suspended in a specimen holder which keeps it in a vertical position by means of a suitable frame which prevents curling. The holder is placed in a combustion cabinet provided with a vertical movable glass front. The fuel specified for the test is Matheson Gas B with a Btu content of $559/\text{ft}^3$ at 21°C. The ignition is effected at the lower end of the specimen using a Bunsen burner so that the flame covers 3/4 in. of the fabric. The flame is applied for 12 sec. The duration of flaming after removing the burner flame, the afterglow time, and the char length are measured. The char length is defined as the distance from the edge of the specimen exposed to the flame to the upper end of the charred area.

In a recent modification[288] to this test, the procedure is repeated for a 3-sec ignition time since it has been observed that in certain cases, presumably in cases of a marginal level of application of a flame retardant, fabric specimens which show a 5-in. char length when exposed to flame 12 sec burn the entire length when exposed only for 3 sec. Hofman and coworkers suggested[289] that in the case of the short ignition time, the temperature of the fabric in the flame front is higher since the flame is richer in oxygen than at the 12-sec ignition time. Such a higher temperature might induce afterflaming of the fabric. This phenomenon was observed only in fabrics treated with flame retardants in an imperfect way or in insufficient amounts.[403,409] It is to be expected that a similar variation in the char length may be introduced by changing the air flow, i.e., the ease of convection and the mixing of the chemicals in the flame through the combustion chamber, as well as by the size of the ignition flame. A large flame would dilute flammable gases by creating more turbulence compared to a small flame.[290]

The NFPA 701 Test[291] is similar to the AATCC-34 test, but is conducted on oven-dry specimens, thus enlarging the range of conditions of validity of the test results to very low relative humidity environments and to large fires.

High-Heat-Flux Flame Contact[405] is mainly useful in predicting the flame resistance of materials which are not ignited by a low-heat-flux ignition source, but form flammable decomposition products when subjected to a higher heat flux. The specimens used are rectangles of material 7×30.5 cm with the long dimension parallel to either the warp or filling direction of the cloth. A Fisher, high-temperature butane-type burner is used. The flame height is approximately 7.6 cm. The specimen is suspended vertically in the flame for 12 sec, while its lower edge is held 3.8 cm above the center of the burner. After flaming is determined and after it ceases, the specimen is again ignited for 12 sec. Ignition, propagation of flame, melting, shrinkage, and flaming drops and pieces are reported.

8.1.3. The Limiting Oxygen Index (LOI) Test

This test, suggested in 1966 for testing the flammability of polymers,[292-294] has been recently adapted to the testing of fabrics.[189,191b] The LOI is defined as the minimal volume fraction of oxygen in a rising gaseous atmosphere that will sustain candlelike burning of a sample of solid polymer or fabric. The apparatus (see Fig. 4) consists of a gas-metering system and a Pyrex glass chimney as a flame holder. The chimney is 15 in. high and 3.5 in. in diameter. The samples tested are in the form of 2.5×6-in. strips placed into a hinged U-shaped holder, while the free end of the holder is held with clamps. The holder is mounted vertically in the chimney on the chimney axis. A series of known mixtures of O_2 and N_2 in varying ratios are passed upward through the chimney at a velocity of 3–11 cm/sec for 1–2 minutes each. The sample is ignited with a laboratory gas burner at its top end during the flow of the gas mixture. The oxygen concentration below which the flame will not spread downward the entire length of the sample is found, and its volume fraction in the gas mixture is designated as the LOI.

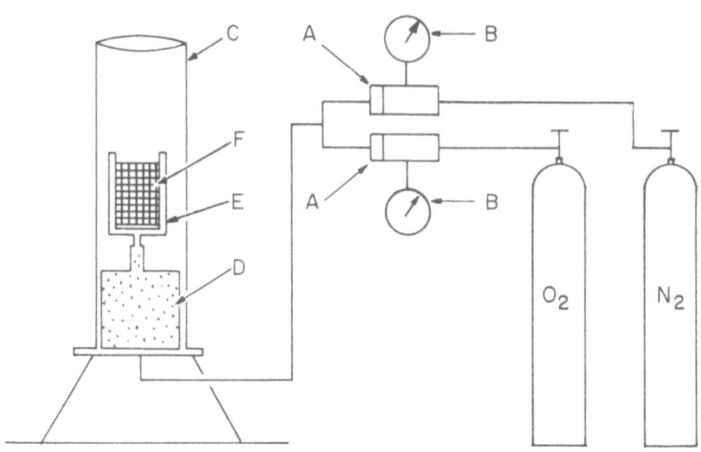

A - METERING ORIFICE   D - BED OF GLASS BEADS
B - BOURDON TEST GAUGE, 0-100 psig   E - SAMPLE HOLDER
C - PYREX CYLINDER   F - TEST FABRIC OR FELT

FIGURE 4. Schematic diagram of oxygen index apparatus.[191b] (From: *Proc. Second Ann. Meeting, Information Council on Fabric Flammability*, 1968, p. 178, Fig. 1.)

Although a limited amount of work has been published to date on this method, it seems to offer a new approach to the testing of flammability. The results have been stated to be highly sensitive, reproducible, and independent of the dimensions and physical form of the sample over a broad range.[191b] The method appears to reflect the effect of varying the chemical composition of fibers and fibrous assemblies. Its relative simplicity enables it to serve as a convenient tool for studying the effect of varying concentrations and ratios of flame-retarding agents on a variety of fibers and blends.

It is evident, however, that the method has several limitations. It does not provide information on the rate of the spread of flame. The conditions of the testing, and in particular the composition of the air, are remote from the actual conditions under which flaming occurs in practice. Furthermore, since the measurements are being carried out under a flow of gas, it is to be expected that in the case of volatile flame retardants the volatility will have a considerable influence on the results. This is the case with halogen-based flame retardants, where the inhibiting activity is based on the concentration of the halogen radicals in the flame. The upward flow of the gas mixture, which is in the direction opposite to the spread of the flame, will tend to dilute the concentration of the active radicals in the flaming zone and thus decrease the accuracy and reliability of the results.

In a recent study,[406] a bottom ignition oxygen index (BIOI) test is suggested and compared with the candlelike LOI test. The use of both of these tests appears to yield a better characterization of a flame resistant material than the LOI by itself. It is stated that a LOI value of at least 27 in addition to a BIOI of 22 is needed for flame retardancy.

The LOI values were found to be sensitive to the moisture content of the sample. A moisture content of less than 6% has very little effect. At higher moisture contents, the LOI values increase, especially in the case of flame-retarded fabrics.[407] External heat sources influence the LOI values. At controlled environmental temperatures of -50 to $250°C$, thermal activities within the specimens or interactions between the specimens and oxidizing atmospheres decrease the LOI values at higher temperatures.[408,409] It was also found that fabric weight had a large effect on LOI values of cotton duck and bleached cotton fabrics, and that construction parameters, such as fabric thickness, thread count, yarn number, and weave pattern had a smaller but still significant effect.[410]

Another complication arises in the case of materials exhibiting an afterglow. The distinction between the afterglow and a weak slow flame is difficult, and erroneous results may be obtained.

In order to draw conclusions from LOI measurements on the flaming properties of an untreated or flame-retardant material, the results obtained have to be checked against other methods. Such comparisons have been published and are reproduced in Fig. 5 for cotton sheeting treated with increasing

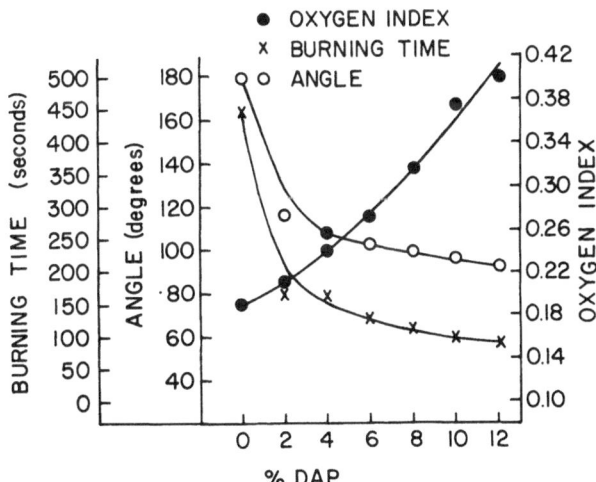

FIGURE 5. Comparison of oxygen index and semicircular flammability tests for rayon challis treated with diammonium phosphate.[191b] (From: *Proc. Second Ann. Meeting, Information Council on Fabric Flammability*, 1968, p. 181, Fig. 7.)

amounts of diammonium phosphate in which a clear correlation can be seen between LOI values and char lengths obtained according to the vertical strip test (AATCC-34-1966) on one hand, and between LOI and burning angles obtained according to the semicircular test.[191b] A low burning angle ($< 100°$) and a low char length are reached at about 7% diammonium phosphate, corresponding to an LOI value of ca. 0.280. A corresponding LOI value of 0.265–0.270 was obtained by other investigators.[189]

8.1.4. The Semicircular Flammability Tester

In the semicircular test,[284,295,296] a fabric sample of 4×10 in. is positioned on semicircular plates placed in a test cabinet and ignited by a gas burner or an alcohol flame on the leading end. In the case of gas ignition, the length of the impinging flame is 1.25 in., and the time of ignition is about 4 sec. After ignition, the sample burns initially in a position approaching that used in the vertical strip test. With the progress of burning, the angle increases, a nearly horizontal position of the flame is reached, and later the flame spreads downwards in a candlelike way. Measurements of the flame front angle are made every 15 sec until the flame stops or it reaches the limit of the test, i.e., 180°.

Three parameters are determined in this test: the angle at which the flame extinguishes itself, the total time of flaming, and the rate of the spread of flame. The test is sensitive to small changes in conditions like air inlet into the test cabinet, smoothing and straightening of the sample strip, etc. Provided the conditions are rigidly standardized, highly reproducible results can be obtained on flammable cellulosic fabrics. Difficulties are, however, encountered when testing fabrics treated with flame retardants which do not continue burning for a time long enough to determine a significant angle.

The proposed German DIN 54331 standard is based on the semicircular test. The ignition time is 15 secs. The specimens are conditioned in standard atmosphere. Duration of flaming, afterglow, distance of flame spread in degrees of arc, time of flame spread to an angle of 90°, and melting behavior are observed.[411]

8.1.5. The TRI Flammability Analyzer

This method[412,413] is based on the principle of keeping the flame stationary by adjusting the feed of the fabric into the flame. The feed rate is governed by the burning properties of the fabric and is equivalent to the rate of the progress of the flame through the fabric. Burning rates at any spatial angle and in any direction relative to the fabric can be studied. The fabric specimen is mounted on the rims of a pair of linked wheels rotating on a horizontal axis. The ignition is done by momentary impingement of a pilot flame on the specimen. The pilot flame is then removed, and the speed of rotation of the wheels is adjusted until the flame remains steady at some point of the circular path. Based on measurements with this analyzer, a *mass burning rate* has been defined. It is obtained by multiplying the flame propagation rate by the area density of the fabric. The mass burning rate appears to be more adequate for comparisons of burning fabrics than the vectorial flame velocity.

The Flammability Analyzer has been used to determine burning rates at various levels of oxygen concentrations in the immediate atmosphere. A linear relationship has been obtained. A characteristic "minimum burning" condition in terms of an oxygen concentration just high enough to support steady-state flame propagation has been suggested to replace the LOI values. By extrapolating the resultant linear relation between burning rate and oxygen concentration to zero burning rate, an intrinsic oxygen index value is obtained. The upward burning oxygen indexes were found to be appreciably lower than the downward indexes.

8.1.6. Flammability of Children's Sleepwear

The recent developments in legislation, standardization, and testing of children's sleepwear constitute at present the most important advances in the introduction of flame retardancy. These developments were a result of a comprehensive collection and analysis of data pertaining to fabric flame accidents which enabled the establishment of candidate priorities for flammability standards by end-use item.[414,415]

8.1.6.1. DOC FF 3-71. This standard[416-418] is limited to sleepwear and fabrics of related materials which are intended or promoted for use in sleepwear of children up to and including size 6X and which have been determined to present the "unreasonable risk" of the "occurrence of fire leading to death, personal injury, or significant property damage."

Specimens of 8.9 × 25.4 cm are fixed between two U-shaped, 0.2-cm-thick steel plates and mounted vertically in a steel cabinet having a door with a glass insert. They are ignited for 3 ± 0.2 sec by a specified gas burner inclined at an angle of 25° from the vertical and connected to a methane gas supply.

Before testing, the specimens have to be dried at 105°C for 30 min, cooled in a desiccator for 30 min and subjected to ignition in the cabinet within 15 sec of removal from the desiccator. Flaming drips or fragments are noted. The residual flame time is measured to the nearest 0.1 sec. The char length is measured by the standard tearing procedure after the afterglow ceases completely. This implies that afterglow is restricted only to its contribution to the ultimate char length, but is per se not prohibited even if it is prolonged.

The standard requires a sample composed of five specimens, two of which are in one fabric direction and three in the opposite direction. The average char length of all five specimens should not exceed 17.8 cm. No individual specimen may have a char length greater than 25.4 cm or a residual flame time greater than 10 sec. The testing is being carried out on finished items (as produced or after one washing and drying) and after they have been washed and dried 50 times according to AATCC Test Method 124-1969.[419]

The Children's Sleepwear Standard is supplemented with a sampling plan covering fabrics and garments. This plan went into effect with the standard on July 29, 1973. The sampling plan provides fabric production acceptance testing, garment prototype qualifications, and garment production acceptance testing. In the case of fabrics, 3 types of sampling are possible: normal, reduced, and tightened. Five thousand yards of fabric constitute a normal sampling unit; for reduced sampling, a unit size may be as much as 10,000 yards. Tightened sampling is required when a unit is rejected under normal sampling. Garment prototype qualifications require testing of seams and trims. A unit for garment production acceptance testing consists of up to 500 dozen finished garments which have a specific identity.

8.1.6.2. DOC PFF 5-73. This standard[420] deals with flammability of children's sleepwear from size 7 to 14, corresponding to ages 7–12 years. The details of the testing and the sampling program are analogous to DOC FF 3-71. Afterflaming time is not reported.

8.1.7. Testing Flammability of Interior Furnishings

Carpets and Rugs: The Standard DOC FF 1-70 recently introduced by the U.S. Department of Commerce[297,415,421] constitutes a modification of the Federal Specification DDD-C-95 test. It involves the exposure of each of eight dry replicate specimens of a given carpet or rug to a standard igniting source (a methenamine timed burning tablet), in a draft-protected environment and measurement of the proximity of the charred portion to the edge of the hole in a prescribed flattening frame.

The size of the specimens is 22.86 × 22.86 cm. Before testing, the specimens are cleaned with a vacuum cleaner until free from all loose ends and any other material left in the pile during manufacturing. If the carpet or rug or its fibers have had a fire-retardant treatment, the selected sample or oversized specimens thereof have to be washed, dry-cleaned, or shampooed 10 times before cutting of the test specimens. The specimens are dried in a forced circulation drying-oven for 2 hr until bone dry and cooled in a desiccator to room temperature for at least 1 hr. If the charred area does not extend to within 2.54 cm of the edge of the hole in the flattening frame at any point for at least seven of the eight specimens, the carpet or rug meets the acceptance criterion. A sampling plan was published and may be included in this as well as in the similar DOC FF 2-70 standard for the flammability of small carpets and rugs.[415] A similar test was introduced in Canada.[422]

The test is designed to check the flammability of the pile fibers in a horizontal position under mild conditions of ignition and does not take into consideration the flammability of the carpet backing. It was found that 26 out of 43 commercial samples passed the test. The measure of protection afforded by carpets passing this test has lately been questioned as a result of a fire to which they were found to contribute.[298] It is questionable whether this test indicates the fire hazard of rugs and carpets used as wall coverings.

More severe requirements have been outlined[299] in the "General Standards of Construction and Equipment for Hospital and Medical Facilities" (PHS No. 930-A-7). These requirements specify a maximum flammability rating of 75 as determined by ASTM E-84-68, the "tunnel test." (See Wood-base Products, Section 8.3.)

In a British test, a hot metal nut is used instead of a pill.[423] The stainless steel hexagonal nut is heated to $900 \pm 20°C$ and placed centrally on the use surface of the material to be tested for 30 ± 2 sec. The times of flaming and afterglow and the greatest radius of the effects of ignition from the points of application of the nut are measured. The specimen size is 30×30 cm. They are conditioned before testing in standard atmosphere.

A similar procedure is used in the British test for non-washable floor rugs.[424] A steel cylinder of 2.5 ± 0.1-cm diameter is heated to $800 \pm 10°C$ and placed in contact for 12 sec with 30×30-cm specimens which were preconditioned at 65% relative humidity and $20°C$. Interlaboratory trials showed that the tests in which the hexagonal stainless steel nuts were used gave more consistent results than the methenamine-pill tests.[425] On the other hand, the tests carried out on the bone-dry samples were the most severe, while the conditioned pill test was the least severe.[426]

A vertical test for floor coverings was adopted in Germany.[430] Ten conditioned samples, 24×10.5 cm cut in the lengthwise direction, are mounted vertically on asbestos cement plates. A propane gas burner is applied at a $45°$ angle to the specimen. The ignition time is 5, 15, or 30 sec, depending upon result of initial tests. Duration of flaming, afterglow, extent of damaged area, and melting behavior are measured.

An interesting test for the determination of surface flash in pile fabrics was introduced in Britain.[427] Two specimens 20×13 cm, with the shorter dimension parallel to the lay of the pile and conditioned in the standard atmosphere, are mounted vertically against an asbestos backplate with the pile pointing downward. The specimens are exposed to a propane or butane gas-jet burner, the flame of which is moved from 5 cm beyond one side of the specimen to 5 cm beyond the other side in 2 ± 0.2 sec at constant speed. It is recorded whether surface flash of the pile occurs.

Mattress Test: A standard for the combustion resistance of mattresses (cigarette test) was issued in 1968 in Canada.[300] The standard uses a 15 × 15-in. mattress specimen, the ticking of which has been leached in water, oven dried, and subjected to ignition from three cigarettes, after which it is examined for evidence of active combustion.

The American Flammability Standard for Mattresses, DOC FF 4-72,[415] which went into effect June 7, 1973, requires that all mattresses on sale after this date comply with it. It is essentially a cigarette test, carried out on a mattress conditioned for at least 48 hr at a temperature of 18–27°C and less than 55% relative humidity. Nine lighted cigarettes are placed 6 in. apart on the upper part of the bare mattress—three each on the smooth surface, tape edge, and quilted or tufted location, if all three exist. Nine additional cigarettes are burned on similar locations on the mattress with each cigarette between two 100% cotton percale sheets. If the matress does not have tufts or quilts, four cigarettes are burned on the smooth surface and five at the tape edge. If the char length on the mattress is smaller than 5.1 cm in any direction from the nearest point of the cigarette, it passes the test. This flammability standard includes sampling plans.

Upholstered Furniture: There is a basic similarity between the flammability hazard of upholstery and mattresses. In both cases, burning cigarettes are considered responsible for development of smoldering fires and consequently lethal atmospheres.[428] The presence of horizontal and vertical fabric areas in upholstered furniture complicates the development of simple testing methods.[415] In a recent study, the use of small samples of upholstery for fire testing has been investigated.[429] The effects of sample size, support, edge condition, and the nature of the materials were studied. The fire propagation rate was determined by measuring the weight loss with a strain-gauge weighing system.

Blankets: A special test for flammability of blankets has been recently developed.[415] It involves subjecting 7 × 7-cm specimens to the impingement of a small flame for 1 sec and determining whether ignition occurs. The specimen in its holder is oven dried at 105°C and cooled in a desiccator for 30 min before testing. An interlaboratory study with the participation of 13 laboratories, each testing 28 blankets both on the face and back, showed that the test method is reproducible and that both face and back of the blanket fabric must be tested.

8.1.8. Testing Flammability of Whole Garments—Heat Transfer and Heat Release

The methods of testing discussed above were concerned with ingition, rate of flame spread, and char lengths produced on specimen strips of fabric in combustion chambers. No characterization was made in these tests of the intensity of burning, the temperatures of the flame and the burning fabric, or of the burn-destruction potential, based on the quantity and rate of heat production and heat transfer to the skin.[301-303,431] A new procedure for igniting garments on fireproof torso forms and mannequins by an open flame using electrical control was recently described.[432] In recent studies heat transfer from burning cellulosic fabrics was measured on circumferential specimens mounted over an iron pipe filled with water in which thermocouples were inserted;[304] on vertical fabric specimens attached to an asbestos board in which a copper disk holding a thermocouple was mounted;[305] on vertical fabric specimens hung from a brass rod alongside an insulating porcelain jacket provided with thermocouples, the tips of which were projecting freely 5/8 in. from the jacket into several locations on the top and bottom of the burning sample;[306] on fiberglass mannequins covered with asbestos and instrumented with heat sensors distributed to give approximately equal weight to equal surface areas;[433] and on dresses made to fit lifesize mannequins of the department store-type coated with paint and with a thin asbestos-cement layer.[306] Thermocouples were inserted in a number of positions of the dummy, both inside and outside the dress and the underwear. It was found that in the case of burning cellulosics, the amount of heat transferred to a nearby surface was proportional to the weight of the fabric and was in the range of 1200 cal/g fabric. All fabrics over 0.3 oz/yd^2 (1.1 mg/cm^2) that burn will transfer sufficient heat to an adjacent body to cause severe burns.[304] Heat flux values of 0.51 and 0.47 cal/cm^2/sec for burning cotton strips and dresses, respectively, were obtained. It was also pointed out that condensable decomposition tar products from burning cotton can contribute to the skin injury.[307] The peak outside temperatures over burning strips of cotton fabric weighing 10.53 mg/cm^2 and over the burning dress of the same fabric on the dummy were 1069°C and 1370°C, respectively. The corresponding peak inside temperatures were 812°C and 1057°C. The temperatures were decreased to 150–250°C when underwear made of glass, modacrylic, or polyvinyl chloride served to insulate the dummy from the burning cotton dress.[306]

In a recent study which included burning of garment assemblies on full-size mannequins as well as *in vivo* experiments on anesthesized rats, garment geometry was found to affect the burn injury potential as much as fabric parameters. Garments made from flame-retardant cotton fabric and 100% thermoplastic fibers did not ignite when used as single layers. When these cotton fabrics were combined with polyester–cotton blends in outerwear–underwear assemblies, they caused less area of the mannequins to be raised to high temperatures than the blend fabrics alone.[436]

The above studies do not include data on flame-retardant cellulosics and the degree of protection from burn which they would impart. They point, however, to a new approach of testing flammability based on whole-garment testing. It has been pointed out that only testing of the whole garment can represent a realistic system of testing under conditions encountered in practice.[305,308,309]

A new and interesting approach to this problem is focusing attention on the rate and total heat evolved when fabrics are burned in air.[434,435] The rate of heat release for steady-state burning at an angle of 45° was found to be independent of fabric weight for cotton fabrics of different weights. The rate of heat release was found to correlate with the flame-propagation rate measured at 45°. Complete heat balances were obtained by correlation of the calorimetric data and the amounts of combustible gases evolved from the burning fabric. The complexity of the phenomenon involved is illustrated by the fact that the total amount of heat released from a 90/10 cellulose triacetate–nylon blend was considerably higher than that expected by assuming additivity of heat released from the triacetate and nylon separately.

8.1.9. Smoke Evaluation

The measurement of smoke produced upon combustion of textiles has recently received some attention and, although it does not yet constitute part of the official accepted standards, it is realized that the reduction in visibility caused by smoke makes escape from a fire and rescue operations difficult. The reduction in visibility depends upon the particle size and the composition, concentration, and distribution of the smoke, as well as on the nature of the illumination.[310] It is to be expected that smoke and toxic products which usually occur together will be taken into consideration, both in the flammability evaluation of fibers and fiber blends as well as of flame-retardant chemicals.

Two methods for smoke measurement were recently applied to textiles:[311]

The Optical Density Measurement: The smoke evolved upon igniting a vertical, freely suspended strip of fabric is measured in terms of the optical density of the surrounding atmosphere. The optical density is the product of the smoke path length, the smoke concentration, and the particle size. The apparatus used is shown in Fig. 6.[311]

The apparatus consists of a cold light source capable of producing a parallel beam of light, a combustion chamber (bell jar), a light-sensing device, and a recording circuitry. A 6 × 2-in. strip of fabric, freely suspended, is ignited, and the maximum optical density is recorded.

The use of photography has also been suggested for the quantitative measurement of visibility in smoke containing atmospheres. A photographic visibility scale has been prepared with ratings from 1 to 10 and with equal increments between the rating. Rating of smoky atmospheres obtained by this method compared favorably with those obtained by the photoelectric cell approach. Both measurements ranked the fabrics in the same order of smoke production.[437]

The Weight Method: The exhaust gases from the combustion of the fabric are passed through a filter and the increase in weight of the filter before and after the filtration is determined. The use of a cigarette filter tip was recommended. A weighed 6 × 2-in. strip of sample is suspended in the bell jar (Fig. 6) and ignited. The air is continually sucked from the top until the chamber is free from smoke. The "smoke factor" is defined as the weight in milligrams of smoke in the filter.

The results obtained by the two methods are not parallel, and cases are reported in which low optical densities are accompanied by high smoke factors and vice versa, depending on the mode of combustion of the fibers. The results obtained with both methods depend on experimental details and work is needed for the adaptation of the testing conditions to the actual behavior of various textiles in a fire.

Experiments in the small scale tunnel (XP-1 Smoke Density Apparatus[312]), in which the two methods were compared for plastics, have shown that the reproducibility of the optical density measurements was poor owing to the large influence of minor changes in air velocity. It was also concluded that the weight and particle size of smoke are not sufficient to relate the amount of smoke generated under test conditions to the hazard presented by smoke. The subsequently designed XP-2 chamber[312,313] concentrated only on the optical density of the smoke produced by a small sample combusted under continuous application of the burner flame in a closed chamber

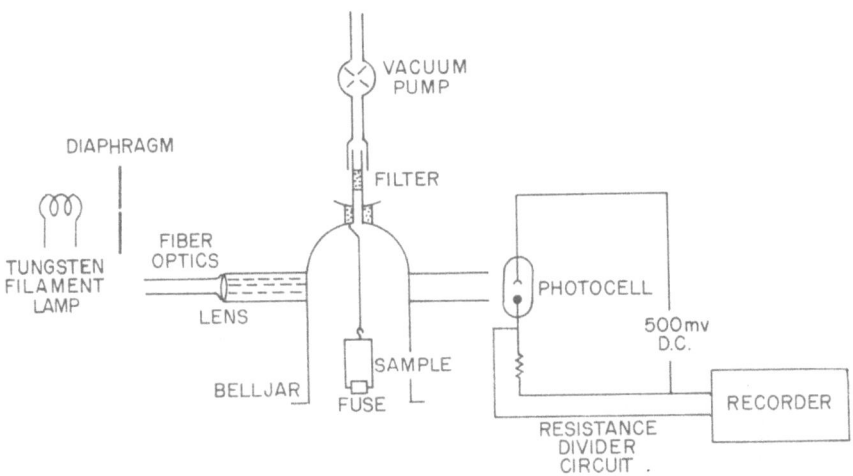

FIGURE 6. Smoke measurement apparatus.[311] (From: *Proc. Second Ann. Meeting, Information Council on Fabric Flammability*, 1968, p. 198.)

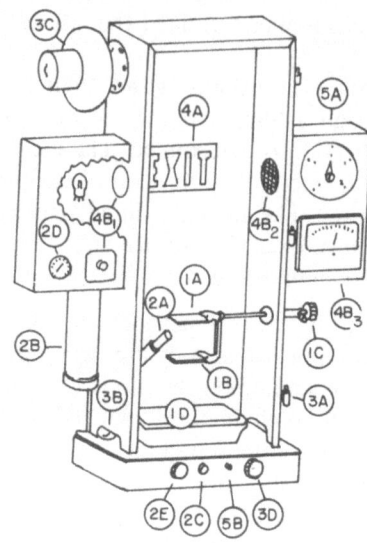

FIGURE 7. XP2 Smoke Density Chamber, schematic diagram.[312]
1. Specimen holder
   A. Stainless steel screen, 2 1/2 in.², 4-mesh, 0.035-in. gage
   B. Asbestos sheet, 1/32 in. thickness
   C. Adjusting knob
   D. Quench pan
2. Ignition
   A. Burner (Bernz-O-Matic, TX-1)
   B. Propane tank (Bernz-O-Matic, TX-9)
   C. Control knob for propane pressure
   D. Pressure indicator, psi
   E. Burner positioning knob
3. Cabinet
   A. Hinges (door, gasketed 3 sides)
   B. Vents (1 in. high opening, 4 sides)
   C. Blower (damper on mounting side)
   D. Control knob (blower on when damper open)
4. Photometer
   A. Visual system (exit sign)
   B. Measuring system
      1. Light source and adjusting transformer
      2. Photronic cell and grid to block stray light
      3. Meter indicating per cent of light intensity
5. Timer
   A. Indicator, 0 to 5 min (friction reset)
   B. ON-OFF switch

(From: *Special Technical Publication No. 422*, American Society for Testing and Materials, 1967, p. 110.)

under 40 psi pressure. Measurements of smoke density are made by means of a barrier-layer photoelectric cell. In addition, with the aid of a timing device, the rate of smoke production in percent per minute and the smoke density rating in percent are determined. The rating is defined as the area under the curve of light absorption versus time divided by the total area times 100. It thus yields absolute values.[312]

In spite of the high reproducibility of smoke-density ratings obtained with the XP-2 chamber, the authors[312] conclude that smoke is a "gross phenomenon" and it is "not practical to make fine distinctions between materials on the basis of the precise quantity of smoke generated under one set of conditions." The chamber should, in their opinion, be used to identify materials which produce large quantities of smoke and, by putting upper limits of smoke-density rating and smoke density, restrict the use of high smoke generating materials.

NBS Smoke Chamber:[438] A relatively simple method for measuring smoke generation of materials and assemblies in thicknesses up to 25.4 mm was developed by the National Bureau of Standards. The attenuation of the light beam by smoke accumulating within a closed chamber due to nonflaming pyrolytic decomposition and flaming combustion is determined. The specific optical density, D_s, is determined from a geometrical factor, $V/LA = G$, and the measured optical density or reduction in light transmittance, T,

$$D_s = \frac{V}{LA} \left[\log_{10} \left(\frac{100}{T} \right) \right] = G \left[\log_{10} \left(\frac{100}{T} \right) \right]$$

where A is the exposed specimen area, V is the volume of the closed chamber, and L the light path.

The method employs an electrically heated radiant-energy source mounted within an insulated ceramic tube and positioned so as to produce an irradiance level of 2.2 Btu/sec/ft^2 or 2.5 W/cm^2 averaged over the central 38.1-mm diameter area of a vertically mounted specimen facing the radiant heater. The nominal 76.2 × 76.2-mm specimen is mounted in a holder which exposes an area of 65.1 × 65.1 mm. In addition to this nonflaming radiation exposure, the lower edge of the specimen area is ignited with a 6-tube burner yielding a row of equidistant premixed air–propane flamelets. The dimensions of the closed test chamber are 914 × 610 × 914 mm. The continuous decrease in light transmission as smoke accumulates is measured by a photometric system with a 914-mm vertical light path.

From plots of specific optical density against time, information on the time to reach maximum specific density, D_m, maximum rate of smoke accumulation, and other parameters for the characterization of smoke emissions of burning materials is obtained.

The NBS smoke chamber has also been used by Gross *et al.* as a source of samples for analysis of the combustion gases obtained in the evaluation of over 140 aircraft interior materials.[439] The use of calorimetric gas detector tubes, gas chromatography methods, and ion-selective electrodes for the determination of the presence and concentrations of toxic and irritating gaseous products has been suggested.

An interlaboratory study, with the participation of 22 laboratories, of the smoke density chamber test method showed that reproducible test results were attainable for a variety of materials tested under flaming and nonflaming exposure conditions.[440] To improve reproducibility and to reduce systematic errors, minor adjustments, such as the location of the pilot burner, the calibrations of flow meters, or controlled ventilation rates were suggested.[441] The results of the interlaboratory study and the fact that the chamber is commercially produced indicate the possibility of its being accepted as a standard method by appropriate organizations.

Smoke Evaluation by LOI Apparatus: The LOI apparatus was modified by mounting a smoke level meter on an extension of the tube. A 150 × 60-mm specimen was placed on a holder and burned in a concentration of oxygen corresponding to LOI and LOI ± 0.5% for the particular sample. For the determination of the smoke level, a system utilizing the attenuation of a beam of light from a tungsten lamp and a selenium photocell for reception is used. The length of the light path was 500 mm.[442]

8.1.10. Toxicity of Burn Gases

The aspect of toxicity which, like smoke density, is not included in the requirements of the accepted flame-retardant standards has lately received some attention and several studies were published. The importance of the toxicity of gases evolved in a fire is evident from the point of view of human survival in a fire, and particularly for the protection of the fire-fighting personnel. It is known that gases and smoke spread much faster in building fires than the flames. It has been shown that mattress fires often cause lethal effects, mainly due to the effect of carbon monoxide.[314-316] An empirical relationship between the maximum time available for escape, T, defined as half of the lethal exposure time in minutes, and the CO concentration (C_{CO}) is given as[314,317]

$$\int_0^T \frac{C_{CO} dT}{4.5} = 1$$

A similar relationship was suggested for the effect of temperature (θ) of dry still air during the exposure:

$$\int_0^T \frac{(\theta - 45)\, dT}{1600} = 1 \qquad 45°C < \theta < 120°C$$

Experiments in simulated mattress fires have shown, however, that the rise in temperature is slow and its contribution is therefore smaller than that of the carbon monoxide. Along with carbon monoxide, a considerable number of other toxic gases are generated and acetic acid has been found in the smoke of cellulosic materials in the early stages of a fire.[318] Formic acid, acrolein, carbon dioxide, and aldehydes are also present. Resin-treated cellulosic textiles may bring about the production of ammonia, hydrogen cyanide, and nitrogen oxides. Little is known on the concentrations of the above compounds produced upon combustion under varying conditions of ventilation. A diagram relating the "maximum allowable concentration for prolonged exposure"[315] of 20 toxic substances to their "dangerous on short (30 min) exposure concentrations" is given by Rasbash.[310] The dangerous concentrations for free halogens, acrolein, and phosphorous trichloride are in the range of 40–60 ppm, those of nitrogen dioxide, sulfur dioxide, and hydrocyanic acid are 100–1000 ppm, while for hydrochloric acid, ammonia, and carbon monoxide the concentrations are 1000–10,000 ppm. Such data for individual compounds are, however, of limited value since a mixture of toxic gases usually is produced in a fire and synergistic effects may be expected. A synergism has in fact been found between CO and CO_2. The presence of CO_2 by increasing respiration will accelerate the uptake of any substance having a high affinity for hemoglobin, such as CO, that may be present.[314]

The apparatus shown in Fig. 6 was used for the determination of the concentrations of 15 toxic gases in the combustion flue. After combustion of the specimen, a known amount of flue is sucked through calibrated gas analysis tubes and the concentrations are obtained from the calibration scales printed on the tubes.[311]

Little is known about the influence of flame retardants on the production of toxic gases in the combustion of cellulosics. While the flame-retardant product may not ignite or the flame may not spread on its surface, it will pyrolyze under the influence of an outside fire. The flame retardants may influence the general combustion process in such a way as to reduce the rate at which carbon monoxide is produced and thus reduce its concentration,[310] but at the same time other toxic substances may be released.

The increasing use of halogen-containing polymers and flame retardants stimulated studies on the degree of toxicity of the gases obtained from their pyrolysis.[319-323] No significant amounts of phosgene and only traces of free halogen were obtained. It was concluded that halogen-containing compositions do not present any enhanced toxic hazard as compared to untreated cellulosic materials and in all cases carbon monoxide was the poisonous gas in the highest concentration found.

8.1.11. Limitations of Tests

The environmental conditions under which a textile may ignite and burn vary widely and would have to be described by a large number of variables. Many variables and the interactions between them are not yet fully understood and defined. It is thus still difficult to construct testing methods which would correspond to actual practical cases. The number of textile fibers and fiber blends, the diversity of clothing assemblies, of finishes, and of end uses, complicate the problem even further. The results obtained with existing testing methods are therefore applicable to particular systems and conditions defined in the tests, but can hardly be considered of general validity. It is to be hoped that in the future a growing number of parameters influencing flammability will be investigated and translated into suitable tests so as to enable the reconstruction of practical fire cases by a series of tests.

8.2. Testing of Paper and Paper Laminates

The test requirements for the flammability of papers are essentially similar to those of textiles and nonwovens. The standard flame-retardancy test[324] most widely used is a vertical strip test applicable to flame-retardant-treated paper and paperboards of less than 1/16-in. thickness. The specimens of 2.75 × 8.25 in. are mounted in a metal frame with the long axis in the vertical position. The test flame is supplied by a bunsen gas burner, with a height of 1.5 in., half of which covers the center of the lower edge of the test specimen. The time of ignition is 12 sec. The char length and the afterglow time are recorded. Papers with char lengths of less than 4.5 in. are considered flame resistant. The Underwriters' Tunnel Test[325] and the Factory Mutual Calorimeter Test[326] are also used for paper and paper laminates (see Wood-Base Materials, below).

8.3. Wood-Base Materials

The effort devoted to the development of testing methods for wood-base materials was much greater than in the case of textiles. The number of methods is considerable. Several methods concentrate on specific aspects of flammability. No single method deals with all factors affecting fires, and, in order to get a more complete picture of the behavior of a given material or of a structural element, several testing methods have to be used.

Some of the methods have been adopted as standards for official testing while others were used as research tools.

The methods are classified into two general groups: tests to evaluate the spread of flame and tests to evaluate the penetration of fire. Additionally, some other criteria are considered: fuel contributed, smoke density, duration of afterglowing, and toxicity of gases produced.

The test methods discussed below represent most of those currently used. The specimen sizes range from 3×0.5 in. to 25 ft $\times$ 20 in.; the exposures range from light to severe with both direct-flame impingement and radiant sources. The portion of the specimen directly exposed to the flame varies from 100% to a small percentage.

8.3.1. Flame-Spread Tests

8.3.1.1. Simple Match Test. In Switzerland and in Israel a preliminary match test is used.[327,328] A specimen of at least 10×10 cm and of original thickness is mounted with one corner pointed downward. The flame of a lighted match is brought in contact with this corner of the specimen for 30 sec. The specimen is classed in group I–III if it burns out completely after removal of the flame; it is classed in group IV–VI if it does not ignite or does not burn out completely.

8.3.1.2. Rod Test. The exact classification of the material is determined in the above standard after performing the rod test. The dimensions of the specimens are $120 \times 12 \times 6$ mm. The specimen is mounted horizontally[327] or at an angle of $45°$,[328] and the free corner is brought in contact for 30 sec with a gas flame of 30 mm height, 5 mm diameter, and a temperature of about $600°$C.

The material is rated as class I if the rod burns out very rapidly, class II if it burns out within 60 sec, class III if it burns out in more than 60 sec, class IV if it continues to burn for more than 5 sec after removal of the gas flame, class V if it continues to burn for less than 5 sec, or class VI if it does not ignite at all.

8.3.1.3. Fire Tube Test. This test[329] measures the progressive loss of weight suffered by specimens exposed to a standard fire. The apparatus consists of a beam balance with a sheet metal tube suspended from one arm. A bunsen burner flame adjusted to furnish a blue flame, 11 in. in height, is introduced at the bottom of the tube to envelop the lower end of a specimen suspended within the tube. Such a flame produces a temperature of $180 \pm 5°$C at the top of the empty fire tube. The standard specimen used is $3/8 \times 3/4$ in. in cross-section by 40 in. long. The gas burner is removed after 4 min.

Before starting a test, the weighing apparatus is adjusted so that the shadow cast on the chart by the pointer will lie on the 100% mark when no specimen is suspended within the tube, and on the 0.0% mark when the specimen is in place.

The percentage loss of weight is recorded at $1/2$ min intervals from the start of the test until no further loss of weight is indicated. The final value indicates the combustibility, and the rate of weight loss indicates the flammability or rapidity of burning.

Various other observations may be made during the test: maximum temperature at the top of the tube, time when maximum temperature is reached, rate of temperature rise, time when flame emerges from the top of the tube, time when flaming ceases, and time when glowing ceases.

8.3.1.4. Modified Schlyter Test. A simple test method described by R. Schlyter of Sweden has been modified by the Forest Products Laboratory, Madison.[330] It is designed to measure the vertical flame-spread characteristics of a material. This method has proved particularly useful in evaluating fire-retarding coatings. Its distinctive feature is a flue between parallel vertical panels for promoting the spread of flame.

The equipment consists of a steel frame arranged to hold two specimens 12 in. wide by 31 in. high in a vertical position with their faces 2 in. apart and with free access for air at the bottom and sides. The igniting flame is supplied by a Bunsen burner with wing tip burning gas at the rate of 37 Btu/min for the "mild" test or by a No. 4 Meker burner with a special T-head burning gas at 291 Btu/min for the "severe" test. The burner is placed for 3 min at the base of the specimens, equidistant from faces and edges. The flame height is observed at 15-sec intervals on a vertical scale at one edge of the frame. Owing to their proximity, the specimen faces are subjected not only to the exposure flame, but also to reradiation from any burning surfaces. Exposure conditions can be easily changed by leaving the burner in place for different time periods and by varying the distance between the faces.

8.3.1.5. Crib Test.[331] The specimen consists of a crib assembled in a special wire frame and composed of 24 weighed pieces of material sawed to $1/2 \times 1/2 \times 3$ in. They are placed 1 in. apart in the wire frame in tiers of two pieces each, with the pieces of each tier placed at right angles to those in the tier below. A Meker burner adjusted to produce a blue flame 10 in. in height and a temperature of $600 \pm 15°C$ is inserted under the ring stand 15 in. below the bottom of the lowest tier of test pieces for 3 min. After all flaming and glowing have ceased, the specimen, including any fallen pieces, is reweighed. The loss in weight is expressed as a percentage of the original weight. Duration of flaming and glowing is recorded.

8.3.1.6. Inclined Panel Test.[332] This is another simple test. The equipment consists of four vertical pointed steel rods on which the 12×12-in. specimen rests at an angle of $45°$. Between the rods is a vertical stand topped by a cork on which a small flat-bottomed metal cup rests. The center of its base is 1 in. below the specimen, 3 in. from the specimen's lower horizontal edge. After pipetting 1 cm^3 of absolute ethyl alcohol into the cup, the alcohol is ignited.

The time for the flame to reach the upper horizontal edge is noted. One minute after the alcohol fuel has been exhausted, any flame and glow in the specimen is noted. The area of char is measured.

8.3.1.7. Radiant Panel Test.[333] A 12 × 18-in. radiant-heat panel is mounted in a vertical position and supplied with a premixed gas–air mixture, the energy output of which is maintained constant. A holder supports a 6 × 18-in. specimen in an inclined position so that the top 6-in. edge of the specimen is 45 in. from the radiant panel and the bottom edge is 9 5/8 in. from it. A gas pilot is located near the top edge of the specimen. A stack, located above the specimen, collects heat and combustion products and supports thermocouples for measuring the gas temperatures and a smoke-sampling device including a glass filter disk to collect smoke deposits. The exposure time is 15 min. When sufficiently heated, the specimen is ignited by the gas pilot.

A flame-spread index is calculated from the time intervals of the flame front at 3-in. positions along the length of the specimen, the maximum observed rise in stack temperature above that observed when testing asbestos-cement board, and a constant arbitrarily chosen to yield a flame-spread index of 100 for red oak. The index computed in this manner recognizes both the ignition and heat-evolution characteristics of the material being tested.

In addition, the weight in milligrams of smoke deposit for the volume flow rate during the test period is reported.

8.3.1.8. SS-A-118b Fire Test.[334] A 36 × 36-in. specimen is supported in a horizontal position on a 2 × 2 × 1/8-in. angle frame assembled to give a clear opening of 30 × 30 in. A flame from a gas burner, with its top 28 3/4 in. below the test specimen, is directed against the center of the lower surface of the specimen. The flame is regulated in accordance with the ASTM standard time–temperature curve.

Degree of fire resistance is classified as: noncombustible (A), fire retardant (B), slow burning (C), or combustible (D). To be class A, no flame issues from the specimen and glow does not progress beyond the fire-exposed area in a 40-min. test. The material is classified as fire retardant (class B), if in a 40-min. test sustained flaming from the specimen does not exceed 10 sec duration; it must occur within a period not exceeding 5 min from the time first observed and the flaming does not reach the angle frame at any point. For class C, in a test lasting 20 min, there can be flaming provided it does not reach the angle frame, and all flaming must cease within 5 min after the gas flame is discontinued.

Flame-spread observations are difficult owing to the limited surface available for flame travel after impingement of the exposure flame on the specimen.

8.3.1.9. FM Calorimeter Test. The Factory Mutual Calorimeter Test[326] is based on the assumption that the fire-spread capability of a material is dependent upon its heat-producing rate. A 4.5 × 5-ft specimen, fitting in the opening in the top of a combustion chamber 4 × 3 3/4 × 17.5 ft, is subjected to fire from gas burners (27,500 Btu/min) for 30 min. The resulting recorded time–temperature curve represents the fuel contribution from both the exposure and the specimen. The specimen is then replaced with a standard concrete cover and the test is rerun under exactly the same conditions except that evaluating burners provide the heat originally produced by the specimen. Metered fuel is introduced through the evaluating burners at the proper rates to follow the curve produced in the first test. With the metered fuel rates recorded in the evaluation test, the total fuel contributed from the specimen and rates of burning for various time periods are computed.

8.3.1.10. 25-ft Tunnel Test. The tunnel test[325] is the best known of the fire tests used in the Unites States and Canada. It correlates the burning characteristics of the material directly to its behavior under conditions of use. The test apparatus consists of a tunnel-like chamber 25 ft in length and 17.5 × 12 in. in cross-section, one end of which contains two gas burners located symmetrically 7.5 in. below the specimen. A 25-ft × 20-in. test sample mounted on the underside of the cover of the tunnel is exposed to the gas flame for 10 min, while the maximum extent of flame spread is observed through glass ports and temperature and smoke density measurements are made down the tunnel. Draft conditions, initial tunnel temperature, and preheat conditions are required to be the same for all tests.

Test results are reported in terms of numerical ratings for flame spread, fuel contributed, and smoke density as compared with select red oak and asbestos-cement board having ratings of 100 and 0, respectively, for each of the mentioned items.

8.3.1.11. 8-ft Tunnel Furnace Test. The smaller tunnel test[335] is designed to provide a relatively simple and economical means for comparing the flame spread of various building materials, and their production of smoke and heat on burning. A 8-ft × 14-in. test sample is placed within the angle-iron frame of the furnace modeled in part after the larger tunnel furnace. This positions the specimen so that the exposed face slopes upward lengthwise from the horizontal at 6°, and slopes upward from the horizontal at 30° across its short axis. An asbestos-faced cover is placed over the back of the specimen. Radiant heat is applied from a stainless steel partitioning plate within the furnace which is heated by a gas burner producing 3400 Btu/min. A small pilot flame at the lower end of the furnace near the specimen surface starts the initial flaming.

The test is conducted for 18.4 min, the average time required for the flames to spread the length of a standard red-oak lumber specimen. The progress of the flame along the face of a given specimen is measured, and expressed as a flame index value relative to the flame-spread index for red-oak lumber, which is arbitrarily assigned a value of 100.

Smoke-density and heat-contribution measurements are also taken, using a light source and photoelectric smoke meter for the former, thermocouples embedded in copper rods within the stack of the furnace for the latter. The respective indexes are computed relative to the red-oak standard.

8.3.1.12. 4-ft Tunnel Test. The 4-ft tunnel[443] is suggested as a screening tool for materials intended for the ASTM E-84-25-ft tunnel and offers substantial reductions in cost of sample preparation and testing. The specimen surface is horizontal as in the 25-ft tunnel. The reference materials used, asbestos-cement board and Kode 25 rigid foam, a self-extinguishing combustible material, exhibit better reproducibility than the red oak prescribed for the ASTM E-84 test. Flame spread along the surface of this material appears to be a linear function of the heat flux applied.

8.3.1.13. Ignitability Test for Materials.[336] A specimen, 228×228 mm and of the original thickness, held in a vertical position, is exposed for 10 sec to a gas flame liberating 1060 cal/min. The copper tube of the gas orifice is inclined at an angle of 45° to the vertical, and the center of the orifice is 3 mm from the center of the face of the specimen.

If any of three specimens flame for more than 10 sec after the removal of the test flame or if burning extends to the edge within this period, the material is classified as "easily ignitable." If no specimen flames for more than 10 sec after the removal of the test flame and burning does not extend to the edge within this period, the material is classified as "not easily ignitable."

8.3.1.14. Fire Propagation Test for Materials. The test[337] makes use of an insulated combustion chamber measuring internally $190 \times 190 \times 90$ mm. It has one open face arranged to accommodate a 228×228-mm specimen of a thickness not exceeding 50 mm. The chamber has a small chimney on the top within which two thermocouples are located. The exposure is initially from a gas jet liberating 7560 cal/min, and after 2 min 45 sec radiant heat is added from two electrical heating elements (1800 W). At 5 min it is reduced to 1500 W. The test duration is 20 min, during which the temperature difference between the ambient conditions and the thermocouples inside the cowl is recorded at 1/2-min intervals, up to 3 min, at 1-min intervals from 4–10 min, and at 2-min intervals from 12–20 min. This curve is compared with a corresponding calibration curve for an asbestos board. The index of performance for the material is determined as follows:

$$1 = \sum_{1/2}^{3} \frac{\theta_m - \theta_0}{10t} + \sum_{4}^{10} \frac{\theta_m - \theta_0}{10t} + \sum_{12}^{20} \frac{\theta_m - \theta_0}{10t}$$

where 1 = index of performance
  t = time in minutes from the origin at which readings are taken
  θ_m = temperature in $^\circ$C for the material at time t
  θ_0 = temperature in $^\circ$C of the calibration curve at time t

8.3.1.15. Surface Spread of Flame Test for Materials. A revision of this British test[338] is in preparation following the publication of the two tests discussed before (ignitability and fire propagation). In the following, a description of the test which is still extensively used is given.

A gas-fired radiant panel is mounted in a vertical position. Perpendicular to one side of the radiant panel at its midpoint is a frame arranged to hold the 9 × 36-in. specimen with its long axis horizontal. A small gas flame is located at the intersection of the specimen and the radiant panel. With this arrangement, the intensity of the heat on the specimen decreases as its distance from the radiant panel increases. Immediately after the specimen is placed in position with the radiant panel functioning, a 7-in.-long vertical gas flame is applied to the hot end of the specimen for 1 min. Observations are made of the time of spread of the flame front for measured distances along the specimen, until the flames have died out or for 10 min, whichever is longer. Ratings are based on the rate and extent of the flame spread:

Class 1—Surfaces of very low flame spread
Class 2—Surfaces of low flame spread
Class 3—Surfaces of medium flame spread
Class 4—Surfaces of rapid flame spread

The method is not sufficiently severe, with too many materials receiving the class 1 rating. In Australia the pilot ignition test was developed as a substitute.[339]

8.3.1.16. Preliminary Surface Spread of Flame Test. This test[338] is essentially the same as the standard test described above except that the dimensions of the apparatus and of the specimens are much smaller. (The size of the test sample is 3 3/4 × 12 in.) This room- and material-saving method proved to be useful for manufacturers and research laboratories. It has the same disadvantage as the big test, being not sufficiently severe.

8.3.1.17. Pilot Ignition Test.[339] The apparatus consists of a 12 × 12-in. gas-fired radiant panel and a wooden frame mounted on a movable carrier. A 24 × 18-in. specimen is secured to the frame with the long dimension vertical and facing the radiant panel. A gas pilot flame is located 2 in. above the center of the specimen and 1/2 in. from its surface. The specimen is moved toward the radiant panel at a predetermined rate. Observations are made of time for ignition and of the flame intensity (by radiation pyrometer) for 2 min after ignition of the sample. Test duration is 6–15 min.

Recorded data are converted to ignitability, spread-of-flame, and heat-evolved index numbers which are combined for a final numerical rating. It is assumed that ignition time, spread of flame, and heat evolved are of little value individually as performance criteria, but when combined represent a truer measure of the fire hazard of a material.

8.3.1.18. Swedish Box Method.[340] The material to be tested should be 10–13 mm thick, its dimensions being 25 × 30 cm. The specimens are placed in a combustion box along its rear wall, its side walls, and its lid. Air is supplied to the combustion box at a rate of 175 liter/min. The electric circuit of the ignition filament above the burner head is closed. Propane is admitted to the combustion box at a proper rate. The temperature of the combustion gases is measured at the cyclonic outlet of the combustion box. The degree of transparency of the smoke is recorded by means of a photocell. Test duration is 5 min.

Ratings are based on the temperature of the combustion gases and on smoke density:

Class 1—Flameproof surface finishes

Class 2—Flame-resistant surface finishes

Class 3—Surface finishes tending to contribute to rapid flame spread and to heavy smoke development

8.3.1.19. German Test Methods. The German standard[341] classifies building materials as nonflammable (class A), difficultly inflammable (class B1), normally inflammable (class B2), and easily inflammable (class B3). The material is considered nonflammable if it passes without flaming (apart from a short glowing or flaming within the first 20 sec) the following test:

In an oven, electrically heated to 750°C, an oven-dry specimen, 50 × 40 × 40 mm, is suspended for 15 min. Six specimens are tested.

The difficulty inflammable rating is determined by another procedure using four specimens, 190 × 1000 mm. These are exposed to a flame for 10 min. No specimen should burn out completely, the average length of the unattacked surface should be at least 15 cm, and the smoke temperature should never exceed 250°C.

A material meeting the following requirements is classified as normally inflammable: (1) None of 10 specimens, 60 × 20 mm, suspended one after the other in the above mentioned oven at 200°C for 15 min should ignite. (2) Ten other specimens of the same dimensions, in contact with the flame of a match, should yield 10 sec after the start of the test a flame height of not more than 5 cm counted from the lower edge of the sample.

8.3.2. Roof-Deck Tests

Roof-deck tests are outlined in Underwriters' Laboratories Class C Requirement Procedure 790, "Test Methods for Fire Resistance of Roof Covering Materials."[342] A total of four tests are conducted: (1) spread-of-flame test; (2) flame-exposure test; (3) resistance-to-burning-brand test; (4) flying-brand test. These fire performance tests are repeated after the "permance of treatment" test. In this test, the samples are exposed to 12 weeks of alternate cycles of 96 hours of rain and 72 hours of drying at 140°F, simulating 800 in. of rain and severe drying over a 10-yr period.

In all the tests, the samples consist of shingles or shakes applied to a slat-type deck made of 1 × 4-in. boards spaced 1 5/8 in. apart. A layer of 15-lb asphalt-saturated organic felt is attached to the decks for waterproofing underlayment. The exposed surfaces of shingles and shakes are 5 and 10 in. respectively, and a 1/4-in. space is left between adjacent shingle and shake courses.

8.3.2.1. The Flame-Exposure Test.[342] The 4 1/3-ft-long by 3 1/3-ft-wide test deck is mounted so that the long dimension inclines 5 in. to the horizontal foot. A 12-mph wind blows up the deck and a gas burner having the width of the test deck is located behind the lower edge of the deck and is adjusted so that the flame covers the exposed surface of the deck and extends from 1 to 2 ft beyond the upper edge of the sample. The temperature of the gas flame is 1300 ± 50°F (700°C). The flame is applied for three 1-min periods, with 2-min off periods between applications. To pass the test, no portion of the shingle or shake covering may blow up or off the deck as flaming or glowing brands; the deck may not become exposed; portions of the deck may not fall away; and the roof covering must protect the combustible supporting deck from sustained flaming.

8.3.2.2. The Spread-of-Flame Test. Here[342] the test deck is 13 ft long and the exposing gas flame is applied continuously for 4 min. The requirements are the same as for the flame exposure, but in addition, the longitudinal spread of flame over the surface must not extend to the top of the sample and there should be no significant lateral burning on the surface.

8.3.2.3. The Resistance-to-Burning-Brand Test. This test[342] has the same requirements as the flame-exposure test, but instead of a gas flame, 25 Class C burning brands* are placed over the joints between the shingles or shakes at 1-min intervals. A 12-mph wind is also applied. To pass the test, no more than 5 of 25 brands may cause sustained flaming of the dry white-pine under-deck slats.

8.3.2.4. The Flying-Brand Test.[342] A deck similar to the flame-exposure deck is constructed. A flame is applied for 4 min with a 12-mph wind for shingles and an 18-mph wind for shakes. Here the wind is allowed to continue after the gas flame is extinguished. The test is discontinued when surface flames go out or when the surface of the supporting deck ignites by flying brands. If brands are produced before the surface flames go out or supporting members ignite, the test indicates a failure.

---

* A Class C brand is kiln-dried white pine, 1 1/2 × 1 1/2 × 25/32 in. thick. One saw kerf 1/8 in. wide and half the thickness of the brand is made across the center of the top, and another is made at right angles across the center of the bottom. Dry weight is 9.5 ± 1.25 g. There is a 2-min preburn before the burning brand is placed on the deck.

8.3.3. Flame-Penetration Tests

The resistance of test-panel assemblies to the penetration of flames or heat is determined in the United States in furnaces of several sizes. One surface of the specimen is subjected to progressively increasing flame temperatures in accordance with ASTM E119-58.[343]

8.3.3.1. Fire Tests of Door Assemblies. This ASTM standard[444] is applicable to door assemblies of various materials. The fire exposure is controlled to conform to the standard time–temperature curve. The points that determine its character are: 538°C at 5 min, 704°C at 10 min, 843°C at 30 min, 926°C at 1 hr, 1010°C at 2 hr, and 1093°C at 4 hr.

The temperatures of not less than nine thermocouples shall be read at intervals not exceeding 5 min during the first 2 hr, and thereafter, the intervals may be increased to not more than 10 mins. Unexposed surface temperatures shall be read at the same intervals. The exposure period is 20 min, 30 min, 45 min, 1 hr, 1.5 hr, or 3 hr. Immediately following the fire-endurance test, the test assembly is subjected to the impact, erosion, and cooling effects of a hose stream directed first at the middle and then at all parts of the exposed surface.

A door assembly is considered as meeting the requirements for acceptable performance when it remains in the opening of the furnace chamber during the fire-endurance and hose-stream tests within certain limitations.

8.3.3.2. Vertical Panel Test. This small-scale test yields preliminary information. A furnace, constructed of fire brick, equipped with pipe outlets for discharging fuel gas to various parts of the interior, has an opening to accommodate the 20×20-in. test specimen. The fuel supply is regulated so that the furnace temperatures will follow the standard time–temperature curve. The endpoint of the test of failure of the sample is reached when fire has burnt through some part of the specimen or when a thermocouple in contact with the outside of the sample indicates a temperature 250°F above room temperature. The period of exposure up to the time of failure is reported as the fire resistance of the material.[343]

8.3.3.3. Vertical Panel Test (10×10 ft): This test is identical in principle with the one just described put permits the use of larger specimens or assemblies possessing practical construction features.[343]

8.3.3.4. Fire-Resistance Tests for Elements of Building Construction:[445] This British standard is in many ways an improvement on the 1953 version.[338] It contains stability, integrity, and insulation criteria, and a limited state of deflection is implied. The time–temperature curve used for tests is now based on a mathematical function instead of being defined by a number of arbitrary points; although not very different from previous standards, this aids computation. In addition, the practice of according specified grading periods to specimens has been abandoned in favor of quoting time to failure under all the relevant criteria. which is less restrictive and more meaningful. A considerable improvement is the requirement that not only should realistic loadings be applied during tests, but also that the methods adopted for supporting or restraining the ends or sides of a specimen should be as far as possible similar to those which would be applied to the element in service. In the past, tests had usually been carried out on simply supported elements.[446]

Similar fire-penetration tests are carried out in other countries where smaller differences in the time–temperature curve and in the requirements can be found.

8.3.4. Combination of Flame-Spread and Flame-Penetration Test

The SS-A-1186 Fire Test described above may be considered as a combined combustibility and surface flame-spread test in which the results depend both on the surface finish and the combustibility of the materials in the assembly.

The dual nature is still more pronounced in the official French Fire Reaction Test.[344] In this test about 25% of the exposed surface of a specimen measuring 30×40 cm is subjected for 20 min to the constant heat given off by an electric radiator consisting of a flat, translucent quartz plate, 0.1 cm thick and 10 cm in diameter, while a collector device forces any gases given off to flow past incandescent rhodium–platinum wires brought to a temperature of 1050°C and thus be ignited if flammable. The exposed surface of the sample is 3 cm from the radiant surface, the whole assembly being inclined at 45°C. Specimens are tested under their normal use thickness, if that is at least 5 mm.

Classification is based on the numerical values of four criteria:

1. Time required for inflammation of the exposed and nonexposed surfaces (inflammability index)
2. Height of the flames during the test (spread index)
3. Maximum-flame-height index
4. Rise of temperature at the top opening of the test chamber (combustibility index)

The flame-spread index and the maximum-flame-height index provide information on the trend of the spread phenomenon, although they do not give a spread rate. The other two indexes give some information on the fire penetration.

Materials are classified as nonflammable, difficulty inflammable, moderately inflammable, and easily inflammable.

The criteria selected for the classification seem satisfactory, but the "thresholds" retained for this classification seem too severe.

In cases where a leach-resistant fire-retardant treatment is required,

TABLE 2[a]

Approximate Relation of Classifications by Various Methods

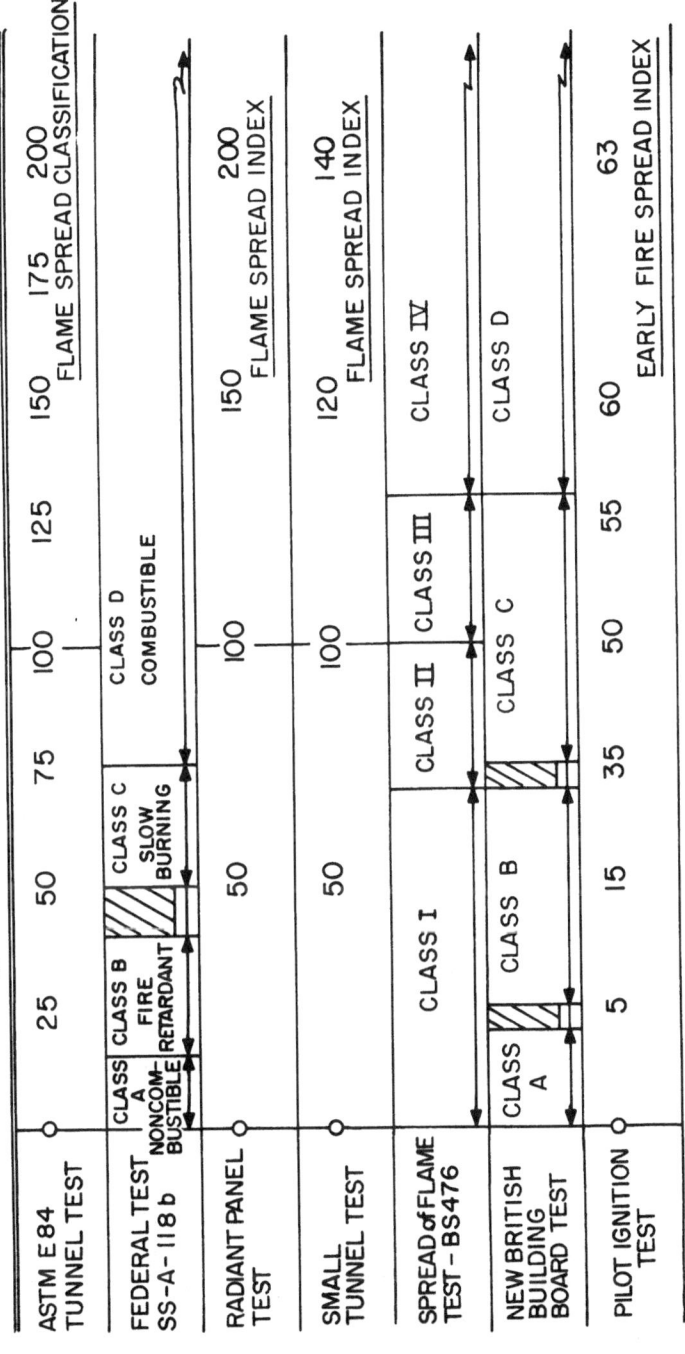

[a] F. J. Rarig and A. J. Bartosic, "Evaluation of the XP2 Smoke Density Chamber." *Symposium on Fire Test Methods — Restraint & Smoke 1966,* ASTM STP-422. p. 106 (1967).

the material is retested after several leaching–drying cycles. Specimens are immersed in water for 64 hr, then dried at 50°C for 8 hr. Three cycles of immersion in water for 16 hr and drying at 50°C for 8 hr follow. Finally, the samples are conditioned and tested.

8.3.5. Correlation Between Tests

The considerable number of testing methods for wood-base materials, differing widely in apparatus, specimen size and position, time of exposure to flame or radiation, and in the flammability characteristics tested, makes it difficult to evaluate the effectiveness of new flame-retardant chemicals and processes on wood-base materials. The need to facilitate research into new systems stimulated some work on the correlation between various testing methods.[345,346] An approximate relation between the classifications of spread-of-flame test methods[347] is shown in Table 2, which is a combination of limited correlations from several sources. More extensive correlations are still lacking, and it has been pointed out that scarcity in identifying data and test results, and even uncertainty in the consistency of test results, preclude the preparation of more precise comparison tables.

8.4. Need for Flammability Testing of Environment

Flammability of textile, paper-base, and wood-base materials are in essence closely related to each other; all these materials are part of the environment in which we live. The distinctions between textile materials (such as curtains and carpets) from building materials (like wall plywood panels or wood floor coverings) as to the flammability test requirements, cannot be too far reaching. The existence and further development of separate testing methods for textiles and building materials for the determination of spread of flame seem not to take adequately into consideration the actual conditions of a fire, when all materials are similarly exposed. It appears that a coordination between the testing methods under which they could all be brought to a common denominator is mandatory. Work in this direction would entail testing of textiles by the suitably modified methods applied for wood-base materials and vice versa. Studies on the influence of thermal radiation, such as occurs in a fire, on the flammability of textiles before and after ignition has taken place, on the contribution of textiles to the flashover times in building fires, and on smoke and toxic fumes developed in a fire would be of considerable interest. Such studies could provide a basis for the development of flammability testing methods for a whole environment rather than of its specific and separate elements. They could further lead to the planning and design of safer materials and environments.

9. References

1. J. E. Ramsbottom, *The Fireproofing of Fabrics*, Department of Scientific and Industrial Research, H.M.S.O., London (1947 and 1930).
1a. J. Wyld, Brit. Pat. 551 (1735).
2. J. L. Gay-Lussac, *Ann. Chim. et Phys.* **18**, (2) 211–216 (1821).
3. W. H. Perkin, *Original Commission 8th International Congress Appl. Chem.: Met. Chem. Eng.* **10**, 636–637 (1912).
3a. A. Kling and D. Florentine, *Genie Civil* **80**, 180 (1920).
4. E. W. Akin, L. H. Spencer, and A. R. Macormac, *Am. Dyest. Rep.* **29**, 418–420, 445–451 (1940).
5. R. W. Little, *Flameproofing Textile Fabrics*, Reinhold, New York (1947).
6. L. J. C. Van de Zande, Fr. Pat. 1,099,874 (1955); U.S. Pat. 2,769,729 (1956).
7. H. P. Knight, Fr. Pat. 1,169,338 (1958).
8. N. D. Clare and A. J. Deyrup, U.S. Pat. 2,823,145 (1958).
9. A. K. Hoffman, U.S. Pat. 2,868,840 (1959).
10. Brit. Pat. 750,262 (1956) (to Chicopee Manufacturing Corp.).
11. W. Gutman and F. Herbst, Ger. Pat. 904,524 (1954).
12. J. R. W. Perfect, *J. Soc. Dyers Col.* **74**, 829–834 (1958).
13. X. Bilger and G. Mangeney, *Melliand Textilber.* **46**, 294–300 (1965).
14. W. Michelitsch, Ger. Pat. 878,792 (1953).
15. R. Aarons, W. H. Baumgartner, and D. R. English, U.S. Pat. 2,935,471 (1960).
16. Swiss Pat. 294,657 (1954) (to Ciba Ltd.).
17. J. L. Kurlychek, U.S. Pat. 2,406,779 (1946).
18. J. S. Vallernaud-Barnier-Bauer, Fr. Pat. 1,086,841 (1955).
19. Brit. Pat. 575,903 (1946) (to British Celanese Ltd.).
20. A. J. Wesson and H. C. Olpin, U.S. Pat. 2,464,360 (1949).
21. R. J. Lincoln and C. H. Campbell, Belg. Pat. 581,368 (1959).
22. J. E. Ramsbottom and A. W. Snoad, *Fireproofing of Fabrics*, Dept. Scientific & Industrial Research—Second Report of the Fabrics Coordinating Research Committee (1930).
23. W. H. Perkin, Jr., U.S. Pat. 844,042 (1907).
24. H. Hopkinson, U.S. Pat. 2,250, 483 (1941); U.S. Pat. 2,343,186 (1944).
25. H. R. Frisch, U.S. Pat. 2,530,458 (1950).
26. I. M. Panik, W. F. Sullivan, and A. E. Jacobsen, *Am. Dyest. Rep.* **39**, 509–516 (1950).
27. A. E. Jacobsen, W. F. Sullivan, and I. M. Panik, *Am. Dyest. Rep.* **40**. P439–443 (1951).
28. I. M. Panik and W. F. Sullivan, U.S. Pat. 2,668,780 (1954).
29. J. P. Wadington, U.S. Pat. 2,691,594 (1954).
30. D. Duane, U.S. Pat. 2,728,680 (1955).
31. H. H. Beacham and I. M. Panik, U.S. Pat. 2,728,691 (1955).
32. W. L. Dills, U.S. Pat. 2,668,784 (1954).
33. W. W. Riches, U.S. Pat. 2,785,041 (1957).
34. L. A. Jordan and L. A. O'Neil, Brit. Pat. 579,328 (1946).
35. C. B. White, U.S. Pat. 2,634,218 (1953).
36. N. J. Read and E. G. Heighway-Bury, *J. Soc. Dyers Col.* **74**, 823–829 (1958).
37. S. A. Rulon, M. J. Sostmann, and I. L. Phillips, *Am. Dyest. Rep.* **35**, P489–492, P501–502 (1946).
38. K. Quehl, Ger. Pat. 1,023,744 (1958); Brit. Pat. 772,364 (1957); Brit. Pat. 784,171 (1957); Ger. Pat. 1,021,326 (1957); Ger. Pat. 1,028,528 (1958); Ger. Pat. 1,029,795 (1958); Fr. Pat. 1,083,963 (1955).
39. K. Quehl, *Melliand Textilber.* **33**, 1115–1116 (1952); **34**, 69–71, 143–145 (1953); **35**, 434–439 (1954).

40. H. J. Reese, *Textil Praxis* **10**, 375–378 (1955).
41. H. J. Reese, *Melliand Textilber.* **37**, 324–326, 457–461 (1956).
42. J. D. Broatch, Brit. Pat. 785,610 (1957).
43. F. G. LaPiana, U.S. Pat. 2,262,770 (1941); U.S. Pat. 2,262,771 (1941).
44. O. C. Bacon, U.S. Pat. 2,413,163 (1946).
45. W. D. Timmons, U.S. Pat. 2,395,922 (1946).
46. D. Lurie, Fr. Pat. 1,138,514 (1957).
47. A. Giordano and W. J. Straka, U.S. Pat. 2,881,097 (1959).
48. P. A. Koenig, *The Cotton Gin and Oil Mill Press* **70**, 19 (April 19, 1969).
49. N. B. Knoepfler, P. A. Koenig, and H. L. E. Vix, *Oil Mill Gaz.* **73**(8), 9–13 (1969).
50. P. A. Koenig and N. B. Knoepfler, *Am. Dyest. Rep.* **58**(17), 30–34, 36 (1969).
51. J. D. Reid and L. W. Mazzeno, Jr., *Ind. Eng. Chem.* **41**, 2828–2831 (1949).
52. D. M. Gallagher, *Am. Dyest. Rep.* **53**, P361–363 (1964).
53. A. C. Nuessle, F. M. Ford, W. P. Hall, and A. L. Lippert, *Text. Res. J.* **26**, 32–39 (1956).
54. Brit. Pat. 604,197 (1948) (to Joseph Bancroft & Sons Co.).
55. F. M. Ford and W. P. Hall, U.S. Pat. 2,482,755 (1949).
56. F. V. Davis, J. Findlay, and E. Rogers, *J. Text. Inst.* **40**, T839–854 (1949).
57. Brit. Pat. 747,014 (1956) (to Ciba Ltd.).
58. A. Berger, U.S. Pat. 2,781,281 (1957).
59. A. M. Loukomsky, U.S. Pat. 2,779,691 (1957).
60. L. A. Fluck and L. J. Moretti, U.S. Pat. 2,784,159 (1957).
61. S. J. O'Brien, *Text. Res. J.* **38**, 256–266 (1968).
62. Brit. Pat. 1,110,116 (1968) (to Courtaulds Ltd.).
63. Brit. Pat. 1,069,946 (1967) (to Asahi Chemical Industry Co., Ltd.).
64. H. Braconnot, *Ann. Chim. et Phys.* **12**(2), 172–195 (1819).
65. J. A. Moede and C. Curran, *J. Am. Chem. Soc.* **71**, 852–858 (1949).
66. T. H. Morton, Brit. Pat. 634,690 (1950) (to Courtaulds Ltd.).
67. R. F. Schwenker, Jr. and E. Pacsu, *Ind. Eng. Chem.* **50**, 91–96 (1958).
68. E. Pacsu and R. F. Schwenker, Jr., *Text. Res. J.* **27**, 173–175 (1957); U.S. Pat. 2,990,232 (1961); U.S. Pat. 2,990,233 (1961).
69. A. McLean and S. F. Marrian, U.S. Pat. 2,470,042 (1949).
70. A. McLean and S. F. Marrian, Brit. Pat. 596,306 (1948) and 633,441 (1949) (to Imperial Chemical Industries Ltd.).
71. J. W. Weaver, J. G. Frick, Jr., and J. D. Reid, U.S. Pat. 2,711,998 (1955).
72. T. D. Miles and A. C. Delasanta, *Text. Res. J.* **38**, 273–279 (1968).
73. C. A. Redfarn, Belg. Pat. 603,640 (1961).
74. Can. Pat. 769,630 (1967) (to Courtaulds (Canada) Ltd.).
75. Fr. Pat. 1,495,909 (1967) (to Courtaulds Ltd.).
76. J. G. Frick, Jr., J. W. Weaver, and J. D. Reid, *Text. Res. J.* **25**, 100–105 (1955).
77. J. W. Weaver, U.S. Pat. 2,778,747 (1957).
78. D. A. Predvoditelev, E. E. Nifantév, and Z. A. Rogovin, *Vysokomolekul. Soedin.* **7**, 1005–1009 (1965).
79. A. Yuldashev, U. M. Muratova, and M. A. Askarov, *Vysokomolekul. Soedin.* **7**, 1923–1926 (1965).
80. J. Truhlar and A. A. Pantsios, U.S. Pat. 2,480,790 (1949).
81. G. C. Tesoro, S. B. Sello, and J. J. Willard, *Text. Res. J.* **38**, 245–255 (1968).
82. G. C. Tesoro and S. B. Sello, U.S. Pat. 3,507,610 (1970).
83. Brit. Pat. 1,082,880 (1968) (to J. P. Stevens & Co., Inc.).
84. D. M. Jones and T. M. Noone, *J. Appl. Chem. (London)* **12**, 397–405 (1962).
85. R. Klement and O. Koch, *Ber.* **87**, 333–340 (1954).
86. J. E. Malowan, U.S. Pat. 2,661,264 (1953).
87. M. L. Nielsen, *Text. Res. J.* **27**, 603–610 (1957).

88. Brit. Pat. 714,214 (1954) (to Monsanto Chemical Co.).
89. J. G. Frick, Jr., R. L. Arceneaux, E. K. Leonard, and J. D. Reid, Development of durable flame retardant finishes for cotton, *Wright Air Development Center Technical Report 58-130*, ASTIA Document No. 206894 (1958).
90. A. Maeder, U.S. Pat. 2,894,019 (1959).
91. Brit. Pat. 835,581 (1960) and 858,582 (1961) (to Courtaulds Ltd.).
92. N. J. Glade, U.S. Pat. 2,971,929 (1961); U.S. Pat. 2,971,930 (1961); U.S. Pat. 2,971,931 (1961).
93. G. L. Drake, Jr., and J. D. Guthrie, *Text. Res. J* **29**, 155–164 (1959).
94. T. D. Miles, F. A. Hoffman, and A. Meiola, *Am Dyest Rep* **49**, P596–599 (1960).
95. R. B. LeBlanc and A. P. Ingram, Jr., *Text. Res. J.* **32**, 284–291 (1962).
96. W. A. Reeves, L. H. Chance, and G. L. Drake, Jr., U.S. Pat. 2,889,289 (1959).
97. L. H. Chance, G. L. Drake, Jr., and W. A. Reeves, U.S. Pat. 2,870,042 (1959).
98. L. H. Chance, G. L. Drake, Jr., and W. A. Reeves, U.S. Pat. 2,886,538 (1959).
99. W. A. Reeves, J. D. Guthrie, and G. L. Drake, Jr., U.S. Pat. 2,906,592 (1959).
100. W. A. Reeves, G. L. Drake, Jr., and J. D. Guthrie, U.S. Pat. 2,933,367 (1960).
101. R. B. LeBlanc, *Text. Res. J.* **35**, 341–346 (1965).
102. Brit. Pat. 1,063,273 (1967) (to Dow Chemical Co.).
103. B. C. M. Dorset, *Text. Manuf.* **95**(3), 112–118 (1969).
104. R. B. LeBlanc and R. H. Symm, U.S. Pat. 3,409,463 (1968).
105. F. Kassack, H. Malz, and F. Lober, Fr. Pat. 1,197,507 (1959).
106. R. C. Steinhauer, U.S. Pat. 3,034,919 (1962).
107. R. Aennshänslin, *Text. Ind.* **133**(11), 99–104 (1969).
108. P. H. P. Valette, Ger. Pat. 1,067,770 (1959).
109. P.H.P. Valette, Brit. Pat. 774,694 (1957).
110. X. Bilger, Brit. Pat. 947,634 (1964).
111. C. Hamalainen, U.S. Pat. 2,681,295 (1954).
112. C. Hamalainen and J. D. Guthrie, *Text. Res. J.* **26**, 141–144 (1956).
113. C. Hamalainen, U.S. Pat. 2,825,718 (1958).
114. J. G. Frick, Jr., and J. W. Weaver, U.S. Pat. 2,686,768 (1954); U.S. Pat. 2,686,769 (1954).
115. Brit. Pat. 961,912 (1964) (to W. R. Grace & Co.).
116. L. E. A. Godfrey, U.S. Pat. 3,455,713 (1969).
117. Belg. Pat. 564,200 (1958) (to Farbwerke Hoechst A.G.).
118. H. Coates and P. A. T. Hoye, Ger. Pat. 1,096,905 (1961).
119. Ger. Pat. 1,040,549 (1959) (to Farbwerke Hoechst A.G.).
120. M. M. Rauhut and A. M. Semsel, U.S. Pat. 3,067,251 (1962).
121. M. Reuter and F. Jakob, Ger. Pat. 1,056,125 (1959).
122. A. J. Hall, *Text. World* **116**(3), 98–102 (1966).
123. A. Hoffman, *J. Am. Chem. Soc.* **43**, 1684–1688 (1921); **52**, 2995–2998 (1930).
124. W. A. Reeves, F. F. Flynn, and J. D. Guthrie, *J. Am. Chem. Soc.* **77**, 3923–3924 (1955).
125. L. M. Kindley, H. E. Podall, and N. Filipescu, *SPE Trans.* **2**(4), 122–125 (1962).
126. W. A. Reeves, G. L. Drake, Jr., J. V. Beninate, and R. M. Perkins, *Text. Chem. Color.* **1**, 365–369 (1969).
127. W. A. Reeves, O. J. McMillan, Jr., and J. D. Guthrie, *Text. Res. J.* **23**, 527–532 (1953).
128. W. A. Reeves and J. D. Guthrie, *Ind. Eng. Chem.* **48**, 64–67 (1956).
129. G. L. Drake, Jr., in *Kirk-Othmer Encyclopedia of Chemical Technology*, 2nd ed., Vol. 9, pp. 300–315, Interscience, New York (1966).
130. L. Diserens, *Neue Verfahren in der Technik der Chemischen Veredlung der Textilfasern*, Vol. 4, pp. 112–113, Birkhäuser Verlag, Basel (1965).
131. W. A. Reeves and J. D. Guthrie, *Dyer*, **3**, 567–570 (1954).
132. W. A. Reeves and J. D. Guthrie, *Text. Weekly* **53**, 1618–1622, 1640 (1954).
133. W. A. Reeves and J. D. Guthrie, *Text. World* **104**, 101, 176, 178, 180, 182 (1954).

134. J. D. Guthrie, G. L. Drake, Jr., and W. A. Reeves, *Am. Dyest. Rep.* **44**, P328–332 (1955).
135. M. F. Stansbury and C. L. Hoffpauir, *Am. Dyest. Rep.* **44**, P645–647, 676 (1955).
136. W. A. Reeves and V. R. Bourdette, *Text. Ind.* **128**, 105–106, 109 (1964).
137. K. M. Decossas, B. H. Wojcik, A. deB. Kleppinger, W. A. Reeves, and H. L. E. Vix, *Text. Ind.* **130**(7), 128–136, 141–143, 156 (1966).
138. W. A. Reeves and J. D. Guthrie, U.S. Pat. 2,809,941 (1957).
139. T. D. Miles, F. A. Hoffman, and A. C. Delasanta, U.S. Pat. 2,993,746 (1961).
140. Brit. Pat. 1,007,580 (1965) and 1,065,547 (1967) (to Hooker Chemical Corp.).
141. G. M. Wagner, U.S. Pat. 3,310,419 (1967); U.S. Pat. 3,310,420 (1967).
142. Brit. Pat. 975,462 (1964) and 975,463 (1964) (to Hooker Chemical Corp.).
143. G. M. Wagner, U.S. Pat. 3,243,391 (1966); U.S. Pat. 3,054,698 (1962).
144. G. M. Wagner and R. A. Schad, U.S. Pat. 3,428,480 (1969).
145. J. D. Reid, J. G. Frick, and R. L. Arceneaux, *Text. Res. J.* **26**, 137–140 (1956).
146. C. Hamalainen, W. A. Reeves, and J. D. Guthrie, *Text. Res. J.* **26**, 145–149 (1956).
147. W. A. Reeves and J. D. Guthrie, U.S. Pat. 2,772,188 (1956).
148. Brit. Pat. 882,993 (1961) (to Albright and Wilson Ltd.).
149. H. Coates, U.S. Pat. 2,983,623 (1961).
150. H. Coates, Brit. Pat. 906,314 (1962).
151. J. R. Perfect, H. Coates, B. Chalkley, and R. Cole, Brit. Pat. 1,075,033 (1967).
152. Brit. Pat. 1,061,134 (1967) (to Ciba Ltd.).
153. W. A. Reeves, G. L. Drake, Jr., L. H. Chance, and J. D. Guthrie, *Text. Res. J.* **27**, 260–266 (1957).
154. G. L. Drake, Jr., W. A. Reeves, and L. H. Chance, U.S. Pat. 2,911,325 (1959).
155. J. V. Beninate, G. L. Drake, Jr., and W. A. Reeves, U.S. Pat. 3,268,360 (1966).
156. L. H. Chance, G. L. Drake, Jr., and W. A. Reeves, U.S. Pat. 3,404,022 (1968).
157. J. V. Beninate, E. K. Boylston, G. L. Drake, Jr., and W. A. Reeves, *Text. Ind.* **131**(11), 110—118 (1967).
158. N. Filipescu, L. M. Kindley, H. E. Podall, and F. A. Serafin, *Can. J. Chem.* **41**, 820–825 (1963).
159. W. J. Vullo, *J. Org. Chem.* **33**, 3665–3667 (1968).
160. J. V. Beninate, E. K. Boylston, G. L. Drake, Jr., and W. A. Reeves, *Text. Res. J.* **38**, 267–272 (1968).
161. J. V. Beninate, R. M. Perkins, G. L. Drake, Jr., and W. A. Reeves, *Text. Res. J.* **39**, 368–374 (1969).
162. J. V. Beninate, E. K. Boylston, G. L. Drake, Jr., and W. A. Reeves, *Am. Dyest. Rep.* **57**, P981–985 (1968).
163. D. J. Donaldson and D. J. Daigle, *Text. Res. J.* **39**, 363–367 (1969).
164. M. Reuter, E. Wolf, L. Orthner, and F. Jakob, Ger. Pat. 1,042,583 (1958).
165. M. Reuter, L. Orthner, F. Jakob, and E. Wolf, U.S. Pat. 2,937,207 (1960).
166. Brit. Pat. 1,131,899 (1968) (to Farbwerke Hoechst A.G.).
167. K. A. Petrov, I. I. Sopikova, and E. E. Nifantév, *Vysokomolekul. Soedin.* **7**, 1667–1669 (1965).
168. A. D. Kiselev, N. A. Aksenova, and L. I. Kutsenko, *Zh. Prikl. Khim.* **38**(6), 1355–1360 (1965).
169. G. L. Drake, Jr., W. A. Reeves, and J. D. Guthrie, U.S. Pat. 2,979,374 (1961).
170. J. R. Costello, U.S. Pat. 2,811,469 (1957).
171. J. R. Costello, Jr., and T. P. Traise, U.S. Pat. 2,841,507 (1958).
172. A. D. F. Toy, R. S. Cooper, and T. P. Traise, U.S. Pat. 2,867,548 (1959).
173. A. D. F. Toy and K. H. Rattenbury, U.S. Pat. 2,735,789 (1956).
174. J. R. Costello, Jr., and T. P. Traise, U.S. Pat. 2,867,597 (1959).
175. M. Reuter and E. Wolf, Ger. Pat. 1,046,047 (1958).
176. Fr. Pat. 1,168,015 (1958) and 1,174,013 (1959) (to Farbwerke Hoechst A.G.).

177. A. J. Erbel and D. L. Kenaga, U.S. Pat. 2,803,562 (1957).
178. R. Schiffner and G. Lange, *Faserforsch. Textiltech.* 9(10), 417–424 (1958).
179. G. C. Tesoro, *Textilveredlung* 2, 435–440 (1967).
180. S. Abdul-Cader Zahir, U.S. Pat. 3,374,292 (1968).
181. R. Aenishänslin, C. Guth, P. Hofmann, A. Maeder, and H. Nachbur, *Text. Res. J.* 39, 375–381 (1969).
182. G. C. Tesoro, S. B. Sello, and J. J. Willard, *Text. Res. J.* 39, 180–190 (1969).
183. R. Aenishänslin, N. Bigler, C. Guth, P. Hoffman, and H. Nachbur, Symposium International de la Recherche Textile Cottonnier, Institut Textile de France, *Proc. 1st Intl. Symp. Cotton Res.*, pp. 703–718 (1969).
184. F. Johnston, U.S. Pat. 2,754,319 (1956); U.S. Pat. 2,754,320 (1956).
185. R. G. Harvey, *Tetrahedron* 22, 2561–2573 (1966).
186. H. Coates and J. J. Lawless, U.S. Pat. 3,050,522 (1962).
187. W. A. Sanderson, W. A. Jueller, and R. Swidler, *Text. Res. J.* 40, 217–222 (1970).
188. H. Petersen, *Synthesis of New Urea–Phosphorus Compounds*, Badısche Anilın- & Soda-Fabrik A.G. (1969).
189. J. J. Willard and R. E. Wondra, *Text. Res. J.* 40, 203–210 (1970).
190. W. A. Reeves, R. M. Perkins, B. Piccolo, and G. L. Drake, Jr., *Text. Res. J.* 40, 223–231 (1970).
191a. G. C. Tesoro, *Text. Chem. Color.* 1, 307–310 (1969).
191b. G. C. Tesoro and C. H. Meiser, Jr., *Text. Res. J.* 40, 430–436 (1970).
192. J. P. Moreau and L. H. Chance, *Am. Dyest. Rep.* 59(5), 37, 38, 64, 65 (1970).
193. M. Reuter, L. Orthner, E. Wolf, and F. Jakob, Ger. Pat. 1,067,812 (1959).
194. Fr. Pat. 1,466,744 (1967) (to Ciba Ltd.).
195. W. H. Schuller, U.S. Pat. 2,822,364 (1958).
196. S. B. Sello, G. C. Tesoro, and R. Wurster, Textilveredlung 5, 391–399 (1970).
197. G. Lask and H. G. Wagner, *8th Symposium (International) on Combustion*, p. 432, Williams & Wilkins Co., Baltimore (1962).
198. W. A. Rosser, H. Wise, and J. Miller, *7th Symposium (International) on Combustion*, p. 175, Butterworth's, London (1959).
199. W. A. Rosser et al., Mechanism of Flame Inhibition, Final Report, Contract No. DA-44-009-ENG-2863 (1958), Department of Agriculture.
200. C. T. Pumpelly, in *Bromine and its Compounds*, (Z. E. Jolles, ed.) Ernest Benn, Ltd., London (1966).
201. P. G. Pape, R. Y. Nulph, and R. C. Nametz, 23rd Ann. Tech. Conference 1968, Reinforced Plastics/Composites Div., The Society of the Plastics Industry, Inc., Section 19-A, page 1.
202. J. A. Schneider, R. G. Pews, and J. D. Herring, Paper presented at the ACS 158th National Meeting, Div. of Organic Coating and Plastics Chemistry, September 1969, New York.
203. M. Goldman, Testing Compounds for Comparative Flame Retardancy—A Simple Paper Strip Flame Test, Institute for Fibers and Forest Products Research, Jerusalem, Israel, Internal Report WC-1-1969.
204. E. Hilscher, Ger. Pat. 1,150,044 (1963).
205. G. A. Pogany and P. W. Croft, Brit. Pat. 927,872 (1963).
206. P. A. Medvedeva et al., *Plasti. Massy* 9, 17 (1963).
207. V. Zvoner, *Chem. Prumyst.* 12, 321 (1962).
208. H. Peters et al., Ger. Pat. 1,139,636 (1962); H. Hahn, Ger. Pat. 1,131,006 (1962).
209. J. Eichhorn, U.S. Pat. 3,058,926 (1962). J. Eichhorn, *ACS Div. Org. Coatings Plast. Chem.* 23, 37 (1963).
210. A. R. Ingram, *ACS Div. Org. Coatings Plast. Chem.* 23, 95 (1963).
211. R. M. Gooch, D L. Kanaga, and H. M. Tobey, *Forest Prod. J.* 9, 325 (1959).

212. W. A. Reeves, R. M. Perkins, B. Piccolo, and G. L. Drake, *Text. Res. J.* **40**, 223 (1970).
213. M. Lewin, U.S. Pat. 3,150,919 (1964).
214. M. Lewin, T. Bernstein-Wasserman, and E. Krakauer, *Israel J. Chem.* **2**, 332 (1964).
215. M. Lewin, A. Lengyel, and B. Toker, *Israel J. Chem.* **3**, 137 (1965).
216. M. Lewin and E. Krakauer, The Bromination of Wood, Internal Report, Institute for Fibers and Forest Products Research, Jerusalem, Israel, 1968.
217. H. W. Eickner and E. L. Shaffer, *Fire Technol.* **3**, 90 (1967).
218a. A. Basch, Y. Halpern, T. Wasserman and M. Lewin, *Cell. Chem. Techn.* **5**, 353–360 (1971).
218b. A. Basch, B. Hirschmann and M. Lewin, *Cell. Chem. Techn.* **7**, 255–265 (1973).
219. Wood Handbook, U.S. Dept of Agriculture Handbook 72, pp. 337–349 (1955).
220. S. H. Clarke, *Proc. 2nd Ann. Conv. Wood Preservers Assoc.* pp. 91–98 (1952).
221. N. D. Mitchel, *Proc. Am. Wood Preservers Assoc.*, pp. 353–371 (1947).
222. Structural Lumber, Fire-Retardant Treatment by Pressure Processes, Standard C20-63, Am. Wood Preservers Assoc. (1963).
223. Plywood, Fire-Retardant Treatment by Pressure Processes, Standard C27-63, Am. Wood Preservers Assoc. (1963).
224. H. W. Eickner, *J. Mater.* **1**, 625–644 (1966).
225. H. W. Eickner and E. L. Schaffer, *Fire Technol.* **3**, 90–104 (1967).
226. F. Kollman, *Technologie des Holzes und der Holzwerkstoffe*, Vol. 2, pp. 109–143, Springer Verlag, Berlin, (1955).
227. Standards for Fire-Retardant Formulations, Standard P10-58, Am. Wood Preservers Assoc. (1958).
228. Qualified Products List of Products Qualified under MIL-L-19140, QPL-19140-27 1965, U.S. Department of Defense.
229. T. S. McKnight, The Hygroscopicity of Wood Treated with Fire Retardant Compounds, Report 190, Forest Research Branch, Canada Dept. of Forestry (1962).
230. F. E. Brink, Naval Civil Eng. Lab., Port Hueneme, Calif., 70 pp. U.S. Govt. Res. Develop. Rept. 41 (23), 104–105 (1966).
231. E. G. Bergin, The Gluability of Fire-Retardant Treated Birch Veneer, Report 191, Forest Products Res. Branch, Canada Dept. of Forestry (1962).
232. R. H. Bescher, W. T. Henry, and W. A. Dreher, *Proc. Am. Wood Preservers Assoc.* **44**, 369–377 (1948).
233. E. E. Pershall, U.S. Pat. 2,637,691 (1953).
234. H. A. Delpew, P. D. Quinn, and G. R. Waitkins, U.S. Pat. 2,868,673 (1959).
235. G. Koeller, Span. Pat. 342,679 (1967).
236. E. R. DuFresne and D. L. Campbell, U.S. Pat. 3,306,765 (1967).
237. Badische Anilin & Soda Fabrik A.G., Ger. Pat. 1,228,797 (1966).
238. Neth. Appl. 6,611,208 (1967).
239. L. Quehl, Ger. Pat. 1,222,785 (1966).
240. L. J. Goldbeck, U.S. Pat. 2,985,540 (1961).
241. *Chem. Process.* **27**, 33 (1964).
241a. W. S. Shisko, U.S. Pat. 3,220,918 (1965).
242. I. S. Goldstein and W. A. Dreher, *Forest Prod. J.* **11**, 235–237 (1961).
243. K. Stumpf and H. Metschke, Ger. Pat. 969,419 (1958).
244. Brit. Pat. 904,954 (1962) (to Chemische Werke Albert).
245. A. S. H. Bruck, Brit. Pat. 869,535 (1961).
246. Fr. Pat. 1,304,492 (1962) (to Statens Skogindustrier, A.B.).
247. Brit. Pat. 922,674 (1963) (to Statens Skogindustrier, A.B.).
248. W. A. Reeves and J. D. Guthrie, Can. Pat. 597,244 (1960); U.S. Pat. 2,927,050 (1950).
249. D. L. Kenaga, U.S. Pat. 3,428,484 (1969).
250. H. P. Wohnsiedler, U.S. Pat. 3,305,436 (1967).

251. R. A. V. Raff, J. W. Herrick, and M. F. Adams, *Forest Prod. J.* **16**, 43 (1966).

252. M. Lewin, Mild Oxidation of Cotton, Project FG-Is-169, Final Report, submitted to ARS, U.S. Dept. of Agric., 1968.

253. M. Lewin and A. Ben-Bassat, *Sirtec, Proc. 1st Intl. Symp. Cotton Res.*, pp. 535-556 (1969).

253a. M. Lewin, Swed. Pat. 302,199 (1968); Brit. Pat 1,101,249 (1968); Israel Pat. 32782 (1967); U.S. Pat. 3,547,687 (1970).

253b. R. A. Davis, U.S. Pat. 3,459,588 (1969).

254 J. B. Doughty, U.S. Pat. 3,081,293 (1963).

254a. K. Sarnecki, *Przegland Papier* **16**, 245–247 (1960).

255. I. S. Goldstein and W. J. Oberley, U.S. Pat. 3,160,515 (1964), Brit. Pat. 1,023,640 (1966).

256. I. S. Goldstein and J. W. Oberley, U.S. Pat. 3,285,774 (1966).

257. Stauffer Chemical Co., Neth. Appl. 6,607,304 (1966).

258. G. A. Martin, U.S. Pat. 3,342,629 (1967).

259. W. L. Dills, Can. Pat. 577,602 (1959).

260. S. Ishihara and T. Maku, *Mokuzai Kenkyu* **37**, 16–31 (1966); *Chem. Abs.* **68**, 4028 (1968).

261. J. E. Dereich and A. Riihimaki, Can. Pat. 565,850 (1958).

262. D. W. Craig, U.S. Pat. 383,274 (1968).

263. M. Eder and R. Groessner, Ger. Pat. 1,209,274 (1966).

264. Chemische Werbe Albert, Belg. Pat. 665,047 (1965).

265. J. A. Clarke, U.S. Pat. 3,245,867 (1966).

266. G. O. Orth, C. V. Pevey, and E. Reichman, U.S. Pat. 3,245,870 (1966).

267. M & I Chemicals, Inc., Brit. Pat. 1,055,759 (1967).

268. R. D. Arsenault, *Forest Prod. J.* **14**, 33–39 (1964).

269. Formwood Ltd., Brit. Pat. 1,089,836 (1967).

270. M. Lewin, Israel Pat. 23679 (1969); U.S. Pat. 3,484,340 (1969).

271. *Proc., 3rd Ann. Meeting*, Information Council on Fabric Flammability, pp. 71–75 (1969).

272. AATCC Technical Manual, Test Method 33-1962; pp. B124–127 (1965); editorially revised 1969; same as ASTM D-1230-61.

273. Flammable Fabrics Act, Public Law 88, 83d Congress, Chapter 164, H.R. 5069, July 1, 1954; amended July 1, 1958; amended and revised December 14, 1967, Public Law 90-189.

274. Commercial Standard 191-53 (Revised), U.S. Department of Commerce, Superintendent of Documents, U.S. Govt. Printing Office, Washington, D.C.

275. British Standard, Method B, 2963-1958.

276. Swiss Standard SNV 98896, 98899.

277. Canadian Government Specification Board 4-GP-2, Method 27-2 (1968).

278. J. V. Ryan, *Proc. 2d Ann. Meeting*, Information Council on Fabric Flammability, p. 230 (1968).

269. L. M. Kushner, *Proc., 3d Ann. Meeting*, Information Council on Fabric Flammability, p. 156 (1969).

280. *Proc., 3d Ann. Meeting*, Information Council on Fabric Flammability, p. 65 (1969).

281. Federal Specification CCC-T191, Method 5906, Federal Supply Service Administration (1951).

282. British Standard 2963-1958, Method A.

283. Canadian Government Specification Board, 4-GP-2, Method 27-3 (1962).

284. H. R. Richards, *Can. Text. J.* **80**, 41–51 (1963).

285. E. B. Nielsen and H. R. Richards, *J. AATCC* **1**, 270 (1969).

286. Canadian Government Specification Board, Schedule 4-GP-2, Method 27-4 (1962).

287. AATCC Test Method 34-1969, *J. AATCC* **2**, 49–50 (1970).

288. R. B. LeBlanc, *J. Am. Assn. Text. Chem. Color.* **2**, 47 (1970).

289. F. Hofman, J. Merz, and F. Raschdorf, Symposium "Hochveredlung" Zürich, (1970), p.31.

290. R. R. Hindersinn and G. M. Wagner, *Encyclopedia of Polymer Science and Technology*, Vol. 1, p. 1 (1967).

291. Standard for Flameproof Textiles, NFPA 701-31, National Fire Protection Association (1951), reprinted 1964.
292. C. P. Fenimore and F. J. Martin, *Mod. Plast.* 141–192 (1966).
293. C. P. Fenimore and F. J. Martin, *Combust. Flame* **10**, 135 (1966).
294. C. P. Fenimore and G. W. Jones, *Combust. Flame* **10**, 296 (1966).
295. G. Schön, *Melliand Textilber.* **48**, 215 (1967).
296. P. St. Hilaire, K. Knoettner, and L. E. Rossiter, *Am. Dyest. Rep.* **61**, P942 (1968).
297. *J. Am. Assn. Text. Chem. Color.* **2**, 13 (1970).
298. *Am. Dyest. Rep.* **46** (1970).
299. W. M. Segall, *Proc., 2d Ann. Meeting,* Information Council on Fabric Flammability, p. 80 (1968).
300. Canadian Government Specification Board, Schedule 35-GP-1 (1968).
301. G. L. Drake, Jr., Fire Resistant Textiles, in *Kirk-Othmer Encyclopedia of Chemical Technology,* pp. 300–315 Interscience, New York (1966).
302. H. A. Freedman, *Bull. N. Y. Acad. Med.* **43**, 663 (1967).
303. A. F. Robertson, *Bull. N.Y. Acad. Med.* **43**, 706 (1967).
304. C. T. Webster, H. G. H. Wraight, and P. H. Thomas, *J. Text. Inst.* **53**, T29 (1962).
305. L. W. Sayers, *Text. Inst. Ind.* **3**, 168 (1965).
306. F. J. Agate, G. P. Crikelair, R. N. Ollstein, J. H. Newman, and F. C. Symonds, *Proc., 2d Ann. Meeting,* Information Council on Fabric Flammability, pp. 7–29 (1968).
307. J. M. Davies, B. McQue, and T. B. Hoover, *Text. Res. J.* **35**, 757 (1965).
308. R. S. B. Holmes, *Bull. N.Y. Acad. Med.* **43**, 716 (1967).
309. R. E. Seaman, *Bull. N.Y. Acad. Med.* **43**, 649 (1967).
310. D. J. Rasbash, *Plast. Inst. Trans. J.* Suppl. No. **2**, 55–60 (1967).
311. M. J. Koroskys, *Proc., 2d Ann. Meeting,* Information Council on Fabric Flammability, p. 182 (1968).
312. F. J. Rarig and A. J. Bartosic, Evaluation of the XP2 Smoke Density Chamber, Symposium on Fire Test Methods—Restraint & Smoke, 1966, ASTM STP422, p. 106 (1967).
313. A method for measuring smoke density, *Quart. Natl. Fire Prot. Assoc.* **57**, 276 (1964).
314. K. Sumi and G. Williams-Leir, *Proc., 2nd Ann. Meeting.* Information Council on Fabric Flammability, p. 207 (1968).
315. Y. Henderson and H. W. Haggard, *Noxious Gases and the Principles of Respiration Influencing Their Action,* 2d ed., p. 168, Reinhold, New York (1943).
316. E. E. Jwillerat, *Bull. N.Y. Acad. Sci.* **43**, 646 (1967).
317. L. T. Minchin, *Ind. Chem.* **30**, 381 (1954).
318. W. H. Easton, *J. Ind. Hyg. Toxicol.* **27**, 211 (1945).
319. A. Fish, N. H. Franklin, and R. T. Pollard, *J. Appl. Chem. (London)* **13**, 506 (1963).
320. R. I. Thrune, *ACS Div. Org. Coatings Plast. Chem.,* Preprints **23**(1), 15 (1963).
321. R. E. Dufour, *Bull. Res.* No. 53, Underwriters' Laboratories, Inc., July 1963.
322. P. Robitschek and C. T. Bean, *Ind. Eng. Chem.* **46**(8), 1628 (1954).
323. M. Goldman, Unpublished Results. Israel Fiber Institute (1968), Jerusalem.
324. TAPPI Standard T461 m-48; ASTM Standard D-777-948 (1946).
325. Method of Test for Surface Burning Characteristics of Building Materials (E84-59T), 1959 Supplement to Book of ASTM Standards, Part 5.
326. N. J. Thompson and E. W. Cousins, The FM construction materials calorimeter. *Natl Fire Prot. Assoc. Quart.* **52**, 186 (1959).
327. Brandschutztechnische Richtlinien fur die Verwendung von brennbaren Baustoffen in Gebäuden, Bern, Juli 1962.
328. Building materials classification according to their behavior during fire, Israel Standard 755, October 1969.
329. A new test for measuring the fire resistance of wood, ASTM-E-69, *ASTM Proc.* Vol. **29**, (Part II), 1929.

330. Fire Test Methods Used in Research at the Forest Products Laboratory, Revised Sept. 1959, Forest Products Lab. Publication No. 1443, p. 4, U.S. Dept. Agric.
331. ASTM-E160-50 (1950).
332. Commercial Standard CS 42-49. Structural Fiber Insulating Board. U.S.
333. Method of Test for Surface Flammability of Materials Using a Radiant Heat Energy Source (E162-60T). 1960 Supplement to Book of ASTM Standards, Part 5.
334. Acoustical Units: Prefabricated, Federal Specification SS-A-1186, August 1954, U.S.
335. Proposed Method of Test for Surface Flammability of Building Materials Using an 8-ft. Tunnel Furnace. ASTM-E286-65T (1965).
336. Fire Tests on Building Materials and Structures. British Standard 476: Part 5: 1968.
337. Fire Tests on Building Materials and Structures. British Standard 476: Part 6: 1968.
338. Fire Tests on Building Materials and Structures. British Standard 476: Part 1: 1953.
339. J. E. Ferris, Fire Hazards of Combustible Wallboards. Commonwealth Experimental Building Station, Special Report No. 18, 1955. Australia.
340. Tendency of Surface Finishes to Contribute to Rapid Flame Spread and to Heavy Smoke Development. Test Method Br 4. Bulletin (Meddelande) No. 123, p. 24, Swedish National Institute for Materials Testing.
341. Brandverhalten von Baustoffen und Bauteilen. Ergänzende Bestimmungen. DIN 4102, Blatt 10, Germany (October 1966).
342. Test Methods for Fire Resistance of Roof Covering Materials. Class "C" Requirement, Procedure 790. *Bull. Res.* No. 23. Underwriters' Laboratories, Inc., 1941. UL790 (1958).
343. Standard Methods of Fire Tests of Building Construction and Materials, ASTM Standards, 1958, Part 5, p. 969.
344. Methode Française Officielle d'Essais d'Inflammabilité, Based on the Decree of the Minister of Interior No. 57-1161 of 17.X.1957, published in the *J. Offi. Republique Française* (Jan. 16, 1958), pp. 611-618.
345. Round Robin Tests on Tunnel Type Flame Spread Furnaces, Final and Supplementary Reports, Southwest Research Institute, Project No. 1-811-2 (1959).
346. A. M. Bisset, A Comparison of Results from Three Fire Propagation Tests, Joint Fire Research Org., F. R. Note No. 368/1958 (Great Britain).
347. J. A. Wilson, ASTM Special Tech. Bulletin No. 301 (1961).
348a. M. Lewin, P. Isaacs, C. Stevens, S. B. Sello, *Textilveredlung* **8**(3) 158-161 (1973).
348b. M. Lewin and P. Isaacs, Ger. Pat. 2,127,188 (1972) (to State of Israel).
349a. S. B. Sello, U.S. Pat. 3,632,297 (1972) (to J. P. Stevens & Co., Inc.).
349b. S. B. Sello, U.S. Pat. 3,681,060 (1972) (to J. P. Stevens & Co., Inc.).
350. R. H. Barker. *Textilveredlung* **8**(3) 180-186 (1973).
351a. D. J. Daigle, A. B. Pepperman, Jr., G. L. Drake, Jr., and W. A. Reeves, *Text. Res. J.* **42**(6) 347-353 (1972).
351b. D. J. Daigle, A. B. Pepperman, G. L. Drake, Jr., and W. A. Reeves, *Am. Dyest. Rep.* **62**(6) 58-60, 80 (1973).
352. C. E. Morris, L. H. Chance, G. L. Drake, Jr., and W. A. Reeves, *Text. Chem. Color* **3**, 136-139 (1971).
353. S. B. Sello, B. J. Gaj, and C. V. Stevens, *Text. Res. J.* **42**, 241-248 (1972).
354. J. Rivlin, Ger. Offen. 2.223,426 (1973) (to Burlington Industries).
355. J. P. Moreau, L. H. Chance, and G. L. Drake, Jr., *Am. Dyest. Rep.* **62**(1) 31-34, 67 (1973).
356. D. J. Donaldson, F. L. Normand, G. L. Drake, Jr., and W. A. Reeves, Paper presented at the 13th Cotton Utilization Research Conference, Southern Regional Research Center, U.S. Department of Agriculture, New Orleans, La., May 1973.
357. H. Petersen: *Textilveredlung* **5** 570-588 (1970).
358. M. Reuter, C. Bermann, and F. Linke, Ger. Offen. 1,930,308 (1970) (to Farbwerke Hoechst A.G.).
359. H. Nachbur and A. Maeder, Ger. Offen. 1,962,644 (1970) (to Ciba, Ltd.).

360. E. D. Weil, U.S. Pat. 3,695,925 (1972) (to Stauffer Chemical Co.).
361. E. D. Weil, Ger. Offen. 2,242,939 (1973) (to Stauffer Chemical Co.).
362. G. V. D. Blunt, Brit. Pat. 1,171,475 (1969).
363. H. W. Eickner, J. M. Stinson, and J. E. Jordan, *Proc. Am. Wood Preserv. Assoc.* **65**, 260–271 (1969).
364. J. Kolodziej, J. Ganszczyk, and A. Kloska, Pol. Pat. 60,535 (1970).
365. D. F. McCarthy, W. G. Seaman, E. W. W. DaCosta, and L. D. Bezemer, *J. Inst. Wood Sci.* **6**(1), 24–31 (1972).
366. Fr. Pat. 2,117,512 (1972) (to Becker und van Hüllen Niederrheinische Maschinenfabrik K.- G.).
367. H. Kobayashi, Jap. Pat. 26,883 (1971).
368. M. Goulet, U.S. Pat. 3,624,233 (1971).
369. A. A. Shamshurin and Y. S. Feldman, U.S.S.R. Pat. 326,062 (1972).
370. B. P. Osanov, B. V. Zubkovskii, A. N. Obaturov, V. T. Gerasimova, B.-I. Shashin, G. I. Brynskikh, and V. F. Bogomolov, U.S.S.R. Pat. 326,194 (1972).
371. W. T. Nearn and R. A. McGraw, *Forest Prod. J.* **22**(7), 47–52 (1972).
372. Y. Okawa, A. Nagai, and H. Kimoto, Jap. Pat. 72-18,079 (1972).
373. C. E. Myers, U.S. Pat. 3,682,675 (1972).
374. C. F. Perirrolo, U.S. Pat. 3,497,454 (1970).
375. T. Murayama and T. Furutani, Jap. Pat. 70-14,436 (1970).
376. E. Proksch, *Proc. Symp. Util. Large Radiat. Sources Accel. Ind. Process,* pp. 467–473 (1969).
377. S. C. Juncja, *Forest Prod. J.* **22**(6), 17–23 (1972).
378. S. Ishihara and T. Maku, *Wood Res.* **52**, 72–89 (1972).
379. R. Peteri, Fr. Pat. 2,119,831 (1972).
380. W. C. Mayer and F. R. Stoveken, Ger. Pat. 2,016,823 (1970).
381. E. D. Hornbaker, B. Sparks, and H. D. Orloff, Ger. Offen. 2,222,756 (1972).
382. M. Lewis, Ger. Offen. 1,958,710 (1971).
383. I. S. Goldstein, U.S. Pat. 3,479,211 (1969).
384. K. Ueno, Japan. Pat. 72-06,595 (1972).
385. K. Hiyoshi and K. Katsuura, Japan. Pat. 72-18,883 (1972).
386. D. N. Demott, U.S. Pat. 3,558,596 (1971).
387. A. E. Gurgiolo, U.S. Pat. 3,501,339 (1970).
388. D. C. Spaulding, Fr. Pat. 1,580,005 (1969).
389. C. A. Lynch, Jr. and E. F. Orwoll, U.S. Pat. 3,678,086 (1972).
390. F. Vulpe, D. Mihaita, and V. Paransanu, Romania Pat. 53,462 (1972).
391. M. Lewin and E. Krakauer, Israel Pat. 27,503 (1970).
392. M. Lewin, E. Krakauer, and P. Smith, Israel Pat. 30,472 (1972).
393. L. O. McMinimy, U.S. Pat. 3,674,596 (1972).
394. T. Okazaki, K. Tashiro, and K. Yanagida, U.S. Pat. 3,671,376 (1972).
395. T. Okazaki, K. Asahara, and T. Kometani, Japan. Pat. 72-30,785 (1972).
396. A. Moralt, Brit. Pat. 1,289,465 (1972).
397. K. C. Shen and D. P. I. Fung, *Forest Prod. J.* **22**(8), 46–52 (1972).
398. Draft for DIN Standard 54 333, Report of ISO/TC 38, Preparatory Working Group on Flammability of Textiles, p. 51 (1973).
399. Australian Standard 1176, Parts 1 and 2, Report of ISO/TC 38, Preparatory Working Group on Flammability of Textiles, p. 37 (1971).
400. Draft of DIN Standard 54 334, Report of ISO/TC 38 Preparatory Working Group on Flammability of Textiles, p. 52 (1973).
401. A. Bernskiold, *18th Hung. Text. Conf.,* 1970, **2**, 409–430 (1971).
402. U.S. Federal Test Method Standard No. 191, Method 5903-2, Change Notice 2 (1971)

similar to AATCC 34-1969; Military Specification Mil-C-43627 A (1973); Mil-C-43122 D (1969) with Admendment-1 (1971).

403. R. J. McCarter, *Text. Chem. Col.* **4**(4), 21–23 (1972).
404. J. W. Weaver, *Text. Chem. Col.* **4**, 116 (1972).
405. U.S. Federal Test Method. Standard No. 191. Method 5905. Change Notice 2 (1971); Military Specification Mil-C-12095 (1971).
406. C. G. Arcand, Jr., and W. Y. Vullo, *Text. Res. J.* **42**, 328–330 (1972).
407. J. E. Hendrix, G. L. Drake, Jr., and W. A. Reeves, *Text. Res. J.* **41**, 854 (1971).
408. J. E. Hendrix *et al.*, *Text. Res. J.* **41**, 360 (1971).
409. J. E. Hendrix, J. V. Beninate, G. L. Drake, Jr., and W. A. Reeves, *J. Fire Flam.* **3**, 2–18 (1972).
410. J. E. Hendrix, G. L. Drake, Jr., and W. A. Reeves, *J. Fire Flam.* **3**, 38–45 (1972).
411. Draft for DIN 54 331; G. Stamm, *Textilveredlung* **6**, 103 (1971).
412. B. Miller and C. H. Meiser, Jr., *Text. Chem. Color.* **3**, 118–122 (1972); B. Miller, H. Lambert and C. H. Meiser, Jr., U.S. Pat. 3.667.227 (1972).
413. B. Miller, B. C. Goswami, and R. Turner, *Text. Res. J.* **43**, 61–67 (1973).
414. J. E. Clark and H. Tovey, *Proc., 5th Ann. Meeting,* Information Council on Fabric Flammability, pp. 208–214 (1971).
415. L. J. Sherman, H. Tovey, and A. K. Vickers, *Proc., 6th Ann. Meeting,* Information Council on Fabric Flammability, pp. 264–306 (1972).
416. Anon., ASTM Studies, DOC Standard FF- 3-71 on Flammability of Children's Sleepwear. *Mat. Res. Stand.* 38–39 (1972).
417. Anon., *Am. Dyest. Rep.* **61**, 64–65 (1972).
418. U.S. Dept. of Commerce, *Text. Chem. Color.* **4**, 71–76 (1972).
419. AATCC Test Method 124-1969. Technical Manual AATCC **46** (1970).
420. DOC PFF 5-73 (Proposed). *Text. Chem. Color.* **5**, 70–75 (1973); Federal Register **38**, 6700–6704 (1973).
421. Federal Register **35**, 38 (1970).
422. Canadian Government Specification Board, 4-GP-2, Method 27.6P (1971).
423. British Standards Institution, B.S. 4790 (1972).
424. British Standards Institution, B.S. 4111 (1967).
425. J. W. Barnes, *Text. Inst. Ind.* **9**, 91–94 (1971).
426. J. W. Barnes, *Text. Inst. Ind.* **10**, 198–199 (1972).
427. British Standards Institution, B.S. 4569 (1970).
428. C. A. Hafer and C. H. Twill, Characterization of Bedding and Upholstery Fires, Southwest Research Institute, March 31, 1970.
429. B. T. Lee and L. W. Wiltshire, *J. Fire Flam.* **3**, 164–175 (1972).
430. Draft for DIN 54 332; Report of ISO/TC 150, Preparatory Working Group on Flammability of Textiles, 50–51 (1973).
431. W. R. Brown and F. A. Vassallo, Cornell Aeronautical Laboratory, Inc., Buffalo, N. Y., CAL Report No. VH-2865-Z-1 (1970). U.S. Department of Commerce, PB 194614.
432. E. L. Finley and W. H. Carter, *J. Fire Flam.* **1**, 166–174 (1970).
433. M. P. Chouinard, D. C. Knodel, and H. W. Arnold, *Text. Res. J.* **43**, 166–175 (1973).
434. M. M. Birky and K. N. Yeh, *J. Appl. Pol. Sci.* **17**, 239–253 (1973).
435. K. N. Yeh, M. M. Birky, and C. Huggett, *J. Appl. Pol. Sci.* **17**, 255–268 (1973).
436. J. F. Krasny and A. L. Fisher, *Proc., 6th Ann. Meeting,* Information Council on Fabric Flammability, p. 79 (1972).
437. H. R. Richards, *Bull. Inst. Text. France* **25**, 675–687 (1971).
438. D. Gross, J. J. Loftus, T. G. Lee, and V. E. Gray, *N.B.S. (U.S.) Bldg. Sci. Ser.* **18**, 1–27 (1969).
439. D. Gross, J. J. Loftus, and A. F. Robertson, ASTM STP 422, 166–206 (1967).

440. T. G. Lee, N.B.S. Tech. Note 708 (1971).
441. J. R. Gaskill and C. R. Veith, ACS Symposium on the Flammability Characteristics of Polymers (1968).
442. Y. Uehara and E. Yanai, *J. Fire Flam.* **4**, 23–41 (1973).
443. C. J. Hilado and P. E. Burgess, *J. Fire Flam.* **3**, 154–163 (1972).
444. Standard Methods of Fire Tests of Door Assemblies, ASTM E-152-72 (1972).
445. British Standard 476, Part 8 (1972).
446. The effect of fire, *BRE News* **23**, 11 (1973).

3

Flame Retardance of Protein Fibers

L. Benisek

1. The Structure of Protein Fibers

Proteins are a class of naturally occurring compounds of high molecular weight. They are extremely widespread in nature, being one of the essential constituents of the tissues of plants and animals. In general, proteins fall into two groups—fibrous and globular. In the fiber field proteins such as wool, silk, mohair, etc., are of great value, while those found in milk or in groundnuts are capable of being transformed into fibers.

Substantial quantities of animal protein fibers come from the hairlike covering of animals; wool from sheep is by far the most important and can be regarded as typical of this class of fibers. Other hair fibers, in general, come from goats (mohair, cashmere), camels, and llamas.

Raw wool may contain large quantities of sand, dirt, suint, vegetable matter, and especially wool wax. These are usually removed before the fiber is converted into textile goods, but the presence of up to 1% residual wax (a mixture of esters of high fatty acids and higher alcohols, including sterols) must be borne in mind since high concentrations of wool wax, usually above 1% may interfere with some flame-retardant finishes. Wool also contains about 8–16% moisture under ordinary atmospheric condi-

L. Benisek · International Wool Secretariat, Technical Center, Valley Drive, Ilkley, Yorkshire, England.

tions, and this favors flame retardancy as oven-dried wool burns much more easily than conditioned wool.

The simplest picture of the wool fiber is one of a polypeptide composed of at least 25 α-amino acid residues. Attached to the α carbon atom are two hydrogen atoms or one hydrogen and a side group, R. These side groups vary markedly in size and chemical nature; some may be hydrophobic, some hydrophilic, some acidic, and some basic. In the special case of cystine, which contains a disulfide bond, the R groups form a cross-link joining chains. A relatively high content of cystine in wool (12%) distinguishes keratin from other proteins, and this amino acid may play an important part in the flammability properties of wool, since it is the most reactive and the degradation of wool with heat begins with the decomposition of cystine and evolution of hydrogen sulfide.[1]

A typical elemental composition of wool is C, 48–50%; O, 23–25%; N, 15–16%; H, 6–7%; S, 3–4%.[2] The relatively high nitrogen content may be also responsible for the relatively high natural flame retardancy of wool fibers when compared with other material or man-made non-nitrogen-containing fibers.

Wool protein, since it contains a high proportion of acidic and basic side chains in addition to end groups, is amphoteric in nature and will therefore adsorb both acids and alkalis. These groups behave as though they were in an ionized state. In acid conditions wool is positively charged owing to protonization of the amino groups, while in alkaline conditions it is negatively charged owing to dissociation of the carboxyl group. This is important with flame-resist treatments based on mordanting with metal complexes which are usually negatively charged and therefore exhausted on the positively charged amino groups.

The amino groups will undergo the normal reactions such as acetylation or deamination with nitrous acid. The carboxyl end groups can be esterified by the usual reagents, e.g., alkyl halides, dimethyl sulfate, epichlorhydrin, etc.

The disulfide grope in cystine is very reactive chemically, and it can be reduced to give thiol groups by thiogycollic acid or sodium bisulfite. The thiol groups produced are very reactive and cross-linkages between the two thiol groups may be formed by reaction with a bifunctional reagent such as ethylene dibromide:

$$2\overset{|}{\underset{|}{C}}HCH_2SH + BrCH_2CH_2Br \longrightarrow \overset{|}{C}HCH_2SCH_2CH_2SCH_2\overset{|}{C}H$$

This may be a useful way of incorporating permanent flame retardants into the wool fibers.

Oxidation of the disulfide group with performic or peracetic acid can yield cysteic acid.

$$\begin{array}{ccc} | & | \\ CO & NH \\ | & | \\ CHCH_2S-SH_2CHC & \xrightarrow{O} 2 \\ | & | \\ NH & CO \\ | & | \end{array} \left[\begin{array}{c} | \\ CO \\ | \\ CHCH_2SO_3H \\ | \\ NH \\ | \end{array}\right]$$

If at least 60% of the original cystine is converted into cysteic acid, wool has a high degree of flame resistance. However, wet strength is significantly reduced because of rupture of cross-links.[3]

Silk is produced in continuous filament form by the larvae of caterpillars when the cocoons are formed. Most silk is produced from the species *Bombyx mori*, an insect which feeds on mulberry leaves. Silk differs from wool in that it contains no cystine and hence no disulfide cross-linkages. The most notable features are the large quantities of the simple amino acids glycine, alanine, and serine. The number of basic and acid side chains in silk is comparatively small when compared with wool; thus the quantity of acid or alkali adsorbed by silk will be less than with wool.

2. Flammability of Wool

Wool is regarded as a safe fiber from the point of view of flammability.[4,6,9-13] It may be ignited if subjected to a sufficiently powerful heat source, but will not usually support flame and continues to burn or smolder for only a short time after the heat source is removed.

The natural flame-resistant properties of the wool fiber are connected with its relatively high nitrogen and moisture content, high ignition temperature, low heat of combustion, low flame temperature, and high limiting oxygen index (LOI) (Table 1).

Another important property of wool fibers is that when ignited it does not melt and drip—a common problem with synthetic fibers and responsible in many cases for skin injuries or the propagation of actual fires (Table 1). The resultant ash or foam from ignited wool can generally be removed fairly easily from the skin. If a burning cigarette is left on a wool carpet, charring of the pile surface can easily be removed by brushing off the ash. With synthetic-fiber carpets, melting can occur, and the damage, in the form of a small black hole, cannot be repaired.

In tests for flammability, the apparent flame properties of wool textiles depend on the test method specified and the fabric construction. A flammabil-

TABLE 1
Flammability Properties of Various Fibers

Fiber	LOI[4-7]	Heat of Combustion,[8-10] kcal/g	Ignition temperature,[8-11] °C	Melting point, °C
Acrilan	18.2	7.6	465–530	235–320
Cotton	18.4	3.9	255	does not melt
Tracetate	18.4		450–520	293
Diacetate	18.6		450–540	255
Polypropylene	18.6	11.1	570	164–170
Rayon	19.7	3.9	420	does not melt
Polyvinyl alcohol	19.7			
Nylon	20.1	7.9	485–575	160–260
Polyester	20.6	5.7	485–560	252–292
Wool	25.2	4.9	570–600	does not melt
PFR Rayon	26.4			does not melt
Modacrylic	26.8			160–190
Nomex Nylon T-450	30.0		800	316
Wool Ti or Zr treated[a]	31.8			does not melt
Polyvinyl chloride	37.1	5.1		100–160

[a] Wool Ti or Zr treated — Wool flame-resist treated with titanium or zirconium complexes.

ity test with the fabric in a horizontal position is much less severe than a 45° or a vertical test. All wool fabrics will pass a horizontal test without any special treatment, but may not pass some 45° tests. Very few wool fabrics will pass a vertical flame test, unless they are of a very dense construction and heavy, i.e., wool felts.

The influence of fabric construction is also very important; the denser and heavier the fabric, the lower the flammability. For example, a conventional wool carpet will pass the Americal "tablet test"[14] without special treatment, while an open shag pile carpet will not pass the same test.

The natural moisture content of wool also significantly increases the flame retardance of wool. Therefore, bone-dry wool burns much more readily than conditioned wool.

When considering flammability it must be remembered that virtually all materials are flammable in certain atmospheres and conditions. Even metals and ceramics flame when exposed to extreme conditions. The best flammability test to compare wool with other fibers is the limiting oxygen index (LOI), which is the minimum volume-percent of oxygen in an oxygen–nitrogen mixture that will just permit a sample to burn in a downward direction.[7,15] This is a vertical flame test which is more or less independent of the physical form and dimensions of the sample over a broad range, and it is the only flammability test which is suitable for isolating the effects of chemical composition of the fabric on flammability from the complicating variables of fabric construction.[4] The LOI value for wool is higher than for most common natural or man-made fibers (Table 1). Only the special "inherently flame-resistant" synthetic fibers, such as modacrylics—copolymers of acrylonitrile with vinyl or vinylidene chloride Nomex—aromatic Nylon, polyvinyl chloride, and others have a higher LOI than wool. However, if wool is flame-resist treated with titanium or zirconium complexes,[16-21] the LOI is significantly increased and is slightly higher than for Nomex.

Based on the values of the LOI, wool can be regarded as a naturally flame-resistant fiber and, if higher degrees of flame resistance are required, this can be accomplished by an effective flame-resist treatment.

Ease of ignition can be measured very well by a horizontal flame test such as Federal Specification CCC-T-191b Method 5906. The results of this test on blankets made from different fibers or blends are summarized in Table 2.[12,22] The index of merit is defined as the reciprocal of the rate of burning and is the ratio of burning time (sec) to char length (in.). In this case wool has the highest index of merit, and it is the most difficult to ignite.

When wool is tested by a vertical freely hanging strip which is ignited with a bunsen burner for 12 sec, it performs very well when compared with the other fibers and blends, but when it is tested by framed vertical strip test—Federal Specification CCC-T-191b Method 5902—it fails to meet the specification (maximum burning time 15 sec, maximum char length 8 in.) and performs similarly as the other fibers or blends (Table 3).[12,22] This is very probably a result of the frame structure in which the sample is enclosed and which prevents convolution and shrinkage of the fabric sample from the flame during the test.

However, when the identical blankets are tested by the Apex test, which consists of attaching the specimen to be tested to a glass-fiber fabric

TABLE 2
Horizontal Flammability Test CCC-T-191b Method 5906

Sample	Char length, in.	Burning time, sec	Index of merit
100% Wool	0.13	9.3	71.5
50% wool 25% rayon 25% cotton	10.0	176.6	17.7
100% polyester	1.0	24.5	24.5
100% acrylic	10.0	97.8	9.8
65% rayon 35% acrylic	10.0	52.0	5.2

TABLE 3
Ease of Ignition and Rate of Burning—Vertical Tests

Sample	Method A[a]			Method B[b]		
	Char length, in.	Burning time, sec	Index of merit	Char length, in.	Burning time, sec	Index of merit
100% wool	0.75	12	16	12	55.9	4.7
50% wool 25% rayon 25% cotton	10.0	28	2.8	12	51.2	4.3
100% polyester	1.25	12	9.6	4.1	24.6	6.0
100% acrylic	10.0	84 93[c]	8.4	12	55.7	4.6
65% rayon 35% acrylic	10.0	48	4.8	12	30.8	2.6

[a] Freely suspended strip.
[b] Method 5902 CCC-T-191b
[c] Burning time of molten drips.

and testing vertically in a cabinet (AATCC Specification 34-1966) and igniting with a paper fuse, wool performs outstandingly, that is similarly to the horizontal flame test and freely hanging vertical strip test (Table 4).[12,22] Although the Apex test is very useful for testing fusible synthetic fibers, as it significantly avoids dripping of the fibers from the flame, it still allows a certain degree of shrinkage of the tested specimen; this explains the favorable results when wool is tested by this method as compared with the framed vertical test.

It follows that wool is a relatively safe fiber which is difficult to ignite

TABLE 4
Ease of Ignition and Rate of Burning—Apex Test

Sample	As obtained			After washing (3 washes)			After dry cleaning (3 cycles)		
	Char length, in.	Burning time, sec	Index of merit	Char length, in.	Burning time, sec	Index of merit	Char length, in.	Burning time, sec	Index of merit
100% wool	0.2	17	85.0	0.2	15	75.0	0.25	17	68.0
50% wool 25% rayon 25% cotton	10	106	10.6	10	114	10	10	127	12.7
100% polyester	1.7	58	34.1	9.6	168	17.5	9.3	146	15.7
100% acrylic	10	169	16.9	10	155	15.5	10	156	15.6
65% rayon 35% acrylic	10	64	6.4	10	62	6.2	10	58	5.8

when compared with common natural or synthetic fibers, but in some cases it needs a flame-resist treatment in order to pass a particular flammability test method and specification.

A few practical observations relating to the behavior of wool have been made. A wool blanket is not ignited by a cigarette end,[23] but it may smolder if an adjacent sheet is first ignited. When an airliner was damaged by a bomb placed in the toilet, but did not disintegrate, a wool curtain which had received a flame-retardant treatment was charred but did not propagate flame. In a crop-spraying accident, a pilot received severe burns to his legs when his nylon trousers melted, but his body was protected by a wool sweater.[24] The Civil Aviation Department of New Zealand noted that flight crews wearing nylon or Orlon next to skin increase their chances of severe burns in the event of an aircraft fire, because of fusing of these synthetic fibers. An inner layer of cotton or wool clothing will act as a barrier to fused nylon materials and thus lessen skin injuries in burning accidents.[22]

The intumescence of wool has been used to advantage by blending the fiber with Kynol.[25] The afterglow normally associated with this otherwise highly flame-retardant fiber may be smothered by the cokelike charred wool, as wool fibers do not afterglow. This is one of the reasons why furnacemen's overalls are made from an all-wool felt.

Blends of wool with asbestos fiber have been suggested for clothing in high-risk occupations.[26]

A blend of 50–50 wool/PFR rayon (permanently flame-resistant rayon) has a higher LOI value than either of its components.[5] However, when wool is blended with common natural or synthetic fibers its natural flame resistance is more or less decreased.

Wool burns more slowly than untreated cotton in compressed air, and its ignition temperature stays high and practically constant as the pressure is increased.[27,28]

Silk is said to burn and fuse rather like wool, but without creating such an unpleasant smell.[29] The difference is very probably due to the absence of sulfur in the fiber. Little quantitative information is available, but silk decomposes rapidly at 170°C, i.e., at a lower temperature than wool because of the absence of cross-links.

3. Mechanism of Thermal Degradation and Combustion

The nature of fire or flame propagation has been the subject of many investigations. A recent review,[30] containing over 100 references, gives a good understanding of the rather complex, physicochemical aspects of flame propagation.

The generally accepted view of polymer combustion can be represented schematically for wool as follows:

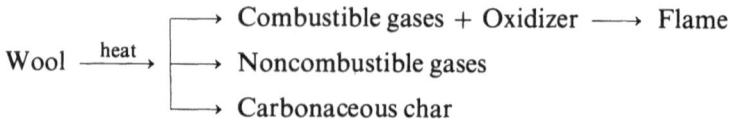

It can be seen that the major components are primarily three: Under noncombustible gases fall CO_2, H_2O vapor, etc. Under combustible gases are fragments of polymers, monomers, etc. Under carbonaceous char are residues which are structurally alike and for all practical purposes independent of their source. Char production hinders the rate of heat transferred from the flame front to the underlying polymer.

The path from combustible gases to flame includes an oxidizer. Burning, therefore, would be hindered if the transport rate of either combustible gases or oxidizer (O_2) is reduced.

There are several distinct stages in the burning process: heating, degradation and decomposition, volatilization, and oxidation. Relatively few flames feed directly on solid or liquid phases: rather the flaming feeds on the gas phase. Flames are either diffusion or premixed, depending on whether the air mixes with the gaseous fuel by diffusion, as in spontaneous fires, or via a mixing apparatus, as in a burner.[31]

Oxidation in the vapor phase is a free-radical process involving H, OH, O, and HO_2 radicals, to name but a few.

Very little is known about the actual mechanism of thermal degradation and combustion of wool.

Wool and silk burn preferentially by charring, while thermoplastic synthetic polyamides such as nylon burn by melting.

Thermogravimetric analysis (TGA) studies show that wool has a high residue (30%) at its decomposition temperature (244°C), when compared with cotton's 10% residue at a decomposition temperature of 366°C.[32]

Recent studies based on differential thermal analysis (DTA), TGA, and mass spectrometry[33] show that wool produces an endothermic peak at a low temperature associated with the loss of water. An endothermic region around 247°C is associated with the release of sulfur compounds owing to the breaking of the sulfur bonds in the wool protein, CS_2 and COS being the main products at this temperature. SO_2 is formed at 315°C, and CO_2 release also starts in this endothermic region.

Early work of Preston[34] indicated that wool undergoes a sort of melting at 240°C. Levau[35] showed that the orthocortex of wool begins to undergo decomposition in air at around 240°C, but the paracortex shows no enhanced degradation at this temperature. Menefee and Yee[36] showed, based on DTA, that at 235°C melting of the major ordered part of the wool protein occurred. This was preceded by a slow formation of amide cross-links at about 160°C, followed by disulfide bond cleavage at 230–250°C and general pyrolytic decomposition above 250°C. DTA studies by Schwenker and Dusenbury[37] as well as by Felix and coworkers[38] ascribed a broad endotherm at about 230°C to cystine cleavage and a small but sharper endotherm at about 245°C to the melting of α-keratin. Simpson[3] suggests that oxidation of cystine may be the initial exothermic reaction when wool burns.

Pinte. Teyssier, and Rochas.[39] while investigating the yellowing of silk. found that oxygen was an important factor in that degradation process. Zerdek and Raffaelli.[40] reported on the possible self-ignition of soaked raw silk and its prevention by the addition of oxidation inhibitors such as β-naphthol to the soaking solution.

From the above-mentioned studies it follows that the disulfide bonds (cystine) in the wool fiber play a major role in thermal degradation and combustion of this fiber and that oxidation and thermal degradation reactions precede and accompany the burning of wool and silk.

4. Smoke Emission and Toxic Fumes

Smoke and toxic products of combustion, such as carbon monoxide, appear to be the cause of a large proportion of fire fatalities, especially in places where large numbers of people gather and in bedding fires: the majority of bedding-fire deaths in the U.S. result from asphyxiation rather than burns.[41,42] With other fire-death statistics a large proportion of the fatalities are the result of smoke and toxic products, but it is difficult to determine the part played by fabrics because other building materials produce appreciable quantities of smoke. For clothing, emission of smoke and toxic fumes is not a primary hazard, but for furnishings and building materials the situation is quite different. The decomposition of the fiber and/or the flame retardant by heat originating from other burning materials may generate smoke and toxic fumes.

It is difficult to separate the effects of smoke and toxic nonparticulate matter as both usually occur together. However, smoke is a double hazard: not only is it toxic and easily able to cause asphyxiation but also, and possibly more important, it reduces visibility. There seems to be general agreement that the danger from smoke precedes that from toxic gases,[43,44] as far as textiles are concerned.

During the past few years, several methods have been developed for measuring the smoke produced from burning plastics and textiles. They use a photoelectric cell to measure the optical density of the smoke. This appears to bear a linear relationship with the "visibility."[43]

Unfortunately. there is little strictly comparable data available on

TABLE 5
Smoke and Gases Produced by Burning Aircraft Interior
Materials under Flaming Conditions

Material	Maximum specific optical density	Gas concentration, ppm		
		CO	HCl	HCN
Nylon fabric	16	30	0	0
75% wool–25% cotton fabric	14	50	0	6
Modacrylic fabric	39	220	110	30
Aromatic polyamide fabric	32	130	0	3
Wool carpet	123	190	0	15
Modacrylic carpet	410	400	1000	70
Nylon plastic	162	500	0	65
PVC plastic	229	550	1200	0
Urethane foam	229	700	150	30

the propensity of different materials to evolve smoke and toxic fumes, but various ad hoc collections of products have been examined and, in particular, a large number of aircraft interior materials has been tested.[45] The results in Table 5 show that wool fibers are not among the most hazardous from the aspect of smoke emission and toxic gases.

The results of smoke emission tests of fabrics of very similar construction and weight made from different fibers or blends, which were measured in a special apparatus devised by Richards,[46] are summarized in Table 6.

The value for wool is better than for most of the fibers compared. A blend of polyester–wool, 55–45, produces significantly more smoke than the individual fibers, and it is important to remember that the flammability or smoke emission of blends can not be predicted from the behavior of

TABLE 6
Smoke Emission from Fabrics Made from Different Fibers
or Blends

Sample	Decrease in visibility, %	Optical density
Acrylic	97	1.5
Cotton	4	0.02
Cotton flame-resist treated	98	1.7
Rayon	4	0.02
Wool	18	0.09
Nylon	6	0.03
Polyester	28	0.14
65% Polyester–35% Cotton	99	2.00
55% Polyester–45% Wool	98	1.70
Polyvinyl chloride	34	0.18

the individual fiber components. In many cases the unpredictable results obtained are due to the fact that different thermodegradation and combustion reactions occur with blends because of interaction of the combustible products.[5,33,47,48] Although cotton shows the lowest smoke emission of the fibers compared (Table 6), when it is flame-resist treated it gives one of the highest smoke emissions, comparable with acrylic fibers. This can be explained either by the less effective combustion owing to the flame-retardant, or by the additional smoke emission of flame-retardants based on organic compounds or both.

For these reasons flame retardants in general more or less increase smoke emission, and a proper choice of a flame retardant when smoke emission has to be low is very important. The titanium and zirconium flame retardants for wool[16-21] are very useful from the point of view of smoke emission, as they are based on inorganic compounds and have only a small effect on smoke emission.

The above discussion only accentuates the complexity of flammability evaluation; in other words, smoke emission can not be considered without evaluating the flame resistance and vice versa.

So far, only carpets for hospitals and other medical facilities receiving aid under the Hill-Burton Program in the U.S. have to pass the tunnel test[49] with a smoke rating of not more than 200. Wool carpets with an adequate backing will pass this specification without any special treatment.[50] The U.S. Department of Housing and Urban Development is proposing a maximum specific optical density of 450 under flaming conditions for carpets when tested by an AMINCO-NBS 4-5800 Smoke Density Chamber. Wool carpets will pass this specification without any special treatment. The British Aircraft Corporation is proposing a specific optical density of less than 90 after 90 sec exposure and less than 200 after 5 min exposure under both flaming and smoldering conditions for aircraft interiors, which includes textile furnishings. Wool seat covers, curtains, and carpets meet this proposed specification.

When wool burns in an excess of air the following gases are formed: carbon dioxide, unsaturated hydrocarbons, saturated hydrocarbons, hydrogen sulfide, sulfur dioxide, methane, ammonia, hydrogen, nitrogen oxides, carbon monoxide, and hydrocyanic acid.[12,51] The last two are the most toxic and are present in measurable quantities. The results of the NBS smoke chamber test under flaming and smoldering conditions with wool carpets show that the concentration of carbon monoxide is 200–400 ppm and of hydrocyanic acid 30–70 ppm,[52] while from the aspect of immediate danger to life the limit for carbon monoxide is 5000 ppm and for hydrocyanic acid 200–300 ppm.[53] Wool fabrics will produce significantly less toxic gases as they are much lighter than carpets.

5. Mechanism of Flame Retardancy

From what was said about the thermal degradation and combustion of wool fibers, it follows that there are a variety of ways to extinguish a flame: cooling the solid, altering the degradation and decomposition process to produce nonflammable volatiles, quenching the volatilization process, adding too much air or eliminating air from the combustion zone, or interfering with the oxidation reactions in the gas phase.[31] Most techniques involve more than one of these. The simple act of blowing out a match involves cooling the substrate and the vapor space and increasing the volume fraction of air in the combustion zone. Chemical fire retardants operate both in the solid phase (altering thermal degradation processes or forming barriers at the surface) and in the vapor phase (interfering with oxidation).

Many theories have been advanced to explain how a textile is made flame retardant and the following are the most important:[54-56]

Physical theories
    1. Glasslike coating—barrier—entrapment
    2. Evolution of noncombustible gases—dilution
    3. Absorption of heat—physical change in retardant
Chemical theories
    1. Modification of thermal decomposition
    2. Reducing the production of combustible gases
    3. Increasing char formation

The first of the physical theories suggests that the flame retardant is decomposed by heat to form a glasslike coating on the individual fibers, which in turn serves as a barrier between the fiber, the source of flame, and the oxygen of the air. The highly flammable tars of decomposition become entrapped in this solid foam and are not available for further combustion reactions. Certain inorganic compounds, in particular borax, boric acid, and diammonium hydrogen phosphate, as well as their mixtures, function in this manner. Those compounds which melt with the evolution of gases to form stable forms are also included in this theory.

A second physical theory implies that selected products decompose at elevated temperatures yielding inert or difficultly oxidizable gases. Flammability is decreased by either diluting the flammable gases during combustion or by blanketing the material with an inert atmosphere. In either case the oxidizing atmosphere is either reduced or eliminated, thus causing major changes in the fuel-to-air ratio, which in turn requires an appreciably higher temperature for ignition. Retardants of this type are the inorganic carbonates and halides, the ammonium salts, and the hydrated salts. This theory has been widely accepted[57] despite the fact that none of the known flame retardants produce inert gases in quantities substantially greater than those of the volatile gases produced by cellulose.[58]

A third physical theory proposes that effective retardants are capable of maintaining fabric temperatures below the minimum combustion temperature. The retardant must be able either to dissipate large quantities of energy internally or to conduct this energy away from the flame front at a rate equal to or greater than that at which it is supplied. It has been shown that numerous compounds which are completely ineffective undergo energy changes during pyrolysis that are comparable to those of known flame retardants.[57]

Chemical theories which are currently used most widely to explain flame resistance state that the retardant substance acts as a catalyst by promoting degradation of the substrate in a direction other than that taken by the untreated fiber, reducing the flammable gas production and/or increasing char formation. Ideally an effective flame retardant should modify thermal degradation and combustion in all the three directions.

From the point of view of reducing the flammable gas production (interfering with oxidation), the action is often attributed to the ability of the added substance to trap radicals which are formed during oxidation in the vapor phase such as H, OH, O. and HO_2 radicals.[31] Bromine compounds are a good example of this:

$$RBr + H \cdot \longrightarrow HBr + R \cdot$$

The generated $R \cdot$ must be a less active radical than the $H \cdot$ which is removed in order for the overall result to be flame inhibition. From the extensive review by Friedman and Levy,[59] which includes discussion of both organic and inorganic compounds of the halogens, it follows that the effectiveness of the halogen atoms decreases in the following order: $I > Br > Cl > F$.

The two important halogens are chlorine and bromine and the latter is about twice as efficient on a weight basis as the former, or four times as efficient on a mole basis. Iodine compounds are too unstable to be of use as additives, while fluorine binds so tightly to carbon that it does not serve as a radical trap.

Volatile antimony compounds are effective free-radical traps and will snuff the flame. While antimony oxide is not effective by itself, an antimony halide SbX_3 is a good retardant. The mechanism here is believed to be formation, on heating, of a volatile halide.[60,61]

$$Sb_4O_6 + 4RCl \longrightarrow 2SbOCl + 2SbCl_3$$

Both antimony and chlorine compounds also have an effect on the degradation reaction in the solid phase. Antimony promotes the charring reactions. It appears that its role is the same as that of phosphorus, and the latter has been studied extensively.

It is well known that most organophosphorus compounds are effective, and only phosphorus salts of nonvolatile metals are not. The phosphorus compounds tend to create a great deal more char and less flammable volatiles. Weight loss on combustion is much reduced, and smoke densities may be increased. Similar results are found with antimony compounds, with nitrogen, arsenic, bismuth compositions, sulfuric acid, or organic sulfates. The key seems to be that an acid is required that is not too volatile. Substances that give rise to acid fragments on mild heating appear to have the desired effect. Phosphoric and sulfuric acids have high boiling points and therefore remain in the solid or liquid phases long enough to act as flame retardants. In fact, after burning a substance treated with a phosphorus compound, a sticky viscous liquid is usually found which is a polymeric acid containing phosphorus.[31]

The described chemical theories of flame resistance are in good agreement with practical results on flame-resist treatments of wool.

Simpson[3] compared the effectiveness of various acids and their sodium salts. Only strong and not too volatile acids (phosphoric, sulfuric, sulfamic) were effective (Table 7). Strong acids which are volatile (nitric, hydrochloric) are ineffective. The sodium salts of the effective acids are much less efficient, as the salts are much less volatile than the acids.

Application of 15% phosphoric acid to wool increases the char residue from 29.8% (untreated wool) to 39.3%, while the decomposition temperature of wool is not affected.[32] This is significantly different from flame-resist-treated cotton, where the decomposition temperature of cotton is decreased by as much as 105°C.

TABLE 7
Flame-Resistance of Wool Impregnated with
Acids and Salts[a]

Acid	Free acid	Sodium salt	Ammonium salt
o-Phosphoric	FP	B	FP
Sulfamic	FP	B	FP
Sulfuric	FP	R	R
Boric	B	R	FP
Hydrochloric	B	B	B
Nitric	B	B	B
m-Phosphoric	FP	G	G

[a] 6% additive in each case.
   Ranking: FP—Flameproof, average char length 2 in.
            G—Good, average char length 2.5–3.5 in.
            R—Some retardant effect, average char length 3.5–5 in.
            B—Burns.

Wool treated with antimony oxide dissolved in hydrochloric acid is flame resistant, but not fast to washing.[3] However, washed fabrics become flameproof again after a brief soaking in diluted HCl, even though this acid alone has no appreciable flameproofing action (Table 7). This emphasizes the effectiveness of volatile antimony halides.

When the disulfide bonds in wool are oxidized to cysteic acid wool is flame-resistant, which can be explained by the strong acid properties of the latter acid which act as a flame-retardant during combustion.[3] Wool treated with relatively unstable polymers derived from reactions of THPC or APO is flame resistant,[3,10] and the mode of action is probably similar to phosphoric acid, that is, increasing char formation and decreasing flammable volatiles.

The action of titanium or zirconium complexes[16-21] on wool can be explained by increased char formation and by the formation of acid fragments on mild heating which trap the flammable volatiles as the effective flame retardant is in this case probably MOX_2 (M is titanium or zirconium, X is halide).[18]

It is well known that certain flame-retardant treatments, when applied to wool fabrics, will yield satisfactory results in the vertical flame test method of evaluation, yet specimens so treated will be completely consumed when ignited by a match. The reason for this deficiency is quite simple.[54] With the vertical test the flame is applied for 12 sec, while with the match the flame application can not be longer than 5 sec. Visual observations, made during the testing indicate that the gases coming from the thermally decomposed samples are actually burning during the first 5 sec of the 12-sec flame-application period. However, by the end of that time the gases have been completely dissipated by oxidation or removed vertically by the hot convection currents. If the time of flame application in vertical test is therefore reduced from 12 to 5 sec, or less, the burning gases will be present, become ignited, and feed back to the test specimen to destroy it completely.

To remedy this situation, such chemical groups as bromine, chlorine, cyano, carboxy, and the haloalkanes[54] should be structurally incorporated into the molecule of the retardant so that upon the application of heat they will be liberated gradually, act as radical traps, and thus reduce the flammability of the gaseous phase.

As the titanium and zirconium treatments at sufficient levels do not show this deficiency, the formation of radical traps during heating is very probable.

Another important but little understood aspect of flame retardancy is glowproofing. The glowing reaction proceeds probably in the following direction:[59]

$$C + \tfrac{1}{2}O_2 \longrightarrow CO \text{ (surface)}$$

$$CO + \tfrac{1}{2}O_2 \longrightarrow CO_2 \text{ (gas reaction)}$$

$$CO_2 + C \longrightarrow 2CO \text{ (surface)}$$

Glowing occurs by reactions in the surface and in the gas phase just above the surface. The reaction to CO_2 is much more exothermic than that to CO, so that prevention of the CO_2 reaction substantially reduces the heat generation and may itself cause glow elimination. It is thought that condensed or polymeric acids will form a film over the surface, thereby blocking the diffusion of oxygen to the burning site. The acid layer also will absorb some of the heat necessary for combustion.

Untreated or flame-resist-treated wool does not show afterglow. When wool is dyed with chrome dyes and a high concentration of bichromate is used for chroming, some wool fabrics can show an afterglow. Chrome trioxide, which is present in chrome-dyed wool, probably acts as a catalyst in the reaction to CO_2. However, this afterglow is significantly shorter than with cotton.

6. Flame-Retardant Treatments

Until recently very little systematic research has been done relating to the flammability of wool fabrics, mainly because wool is fairly regarded as slower burning than most other textiles. With the advent of permanently flame-resistant cotton and nonflammable synthetics, particularly Nomex (aromatic nylon) introduced recently by Du Pont, the advantage once held by wool has vanished in the field of textiles where a high degree of permanent flame resistance is required. The appearance of these flame-resistant textiles has occurred at a time when the general issue of consumer protection has received more attention, and part of the concern has been directed towards the safety of clothing and textiles generally. For instance the U.S. Flammable Fabrics Act of 1964 was strengthened by new legislation enacted in December 1967,[62] which gave the U.S. Secretary of Commerce authority to "promulgate standards wherever necessary to protect the public interest," and extended the act to cover all fabrics and related products.

The introduction of flammability standards for jet aircraft immediately required a permanent flame-resist treatment for wool furnishings and seat covers. But the introduction of the American "tablet test"[14] under the U.S. Flammable Fabrics Act, requiring from April 1971 that all carpets pass this test, created an immediate demand for a permanent treatment for about 15 million pounds of carpet wool. Although traditional high-density wool carpets are acceptable without treatment, fashionable long-pile structures containing a higher ratio of air to fiber are classed as hazardous unless specially treated. More recent regulations covering children's sleepwear[63] will extend the use of treatments in this area.

Thus, three distinctive stages in the research and application of flame-resist treatments for wool can be seen:

1. Use of nondurable treatments mainly based on inorganic borates and/or phosphates for specialized purposes such as theater curtains and aircraft furnishings.

2. Development and application of a durable flame-resist treatment based on a modification of the tetrakis(hydroxymethyl)phosphonium chloride (THPC) polymer treatment for cotton[10] during 1969 as an answer to the revised flammability specification FAR 25-853[64] for jet aircraft.

3. Development and use of cheap and simple durable treatments based on titanium and zirconium complexes[16-21] for carpet wool in order to pass the American tablet test[14] during 1970.

These treatments were further improved to meet the more stringent regulations and tests for fabrics, to supersede the previous treatments, and are the most widely commercially applied for wool textiles.

It is to be expected that there will be increasing use of flammability standards with time, and it is important that wool processors should be capable of meeting the specifications. In view of the latest developments in the flame-resist treatments, wool is in an advantageous position and may be expected, unusually, to compete effectively in certain markets with cotton (for which the permanent flame-retardant treatments are expensive and have adverse effects on fabric properties) and also with inherently flame-resistant synthetic fibers which are rather expensive and in some cases have poor textile properties. The metal complexes are simple and cheap to apply without affecting fabric hand and physical properties, so that wool as a relatively expensive fiber, cheaply protected, may be compared with a cheaper fiber like cotton containing an expensive finish.

However, much basic and development research remains to be done in the field of fire retardation, reducing smoke and toxic gases, the nature of fires and combustion, and last but not least developing realistic test methods that are accurate reflections under which a textile will burn.

An ideal flame-resist treatment should have the following properties.[13]

1. Formulated from efficient, economical chemicals
2. Applied in commercial equipment without unusual requirements in processing
3. Will give reproducible results and be applicable to a broad spectrum of fabric constructions
4. No effect on other processing steps and formulations used in dyeing and finishing
5. Durable under all conditions encountered in use

The properties of an ideal flame-resist-treated fabric should be the following:[13]

1. Will not support combustion (self-extinguishing)
2. No change in flammability in use, laundering, or dry cleaning
3. Does not differ in appearance or performance properties from fabrics accepted for the specified end use
4. Free of toxic, allergenic, or irritating effects
5. Modest cost increase as compared to fabrics accepted for the specific end use

From the durable flame-resist treatments so far available for wool, the titanium and zirconium treatments best fulfil the above conditions.

6.1. Nondurable Treatments

6.1.1. Borates

A systematic investigation in fire-retardant treatments for cotton was carried out by Gay-Lussac in 1820,[65] and borate treatments were the first to be widely applied to cotton. These treatments were also applied to wool for special products such as electric blankets and aircraft furnishings. In fact as recently as 1970, borate treatment was the most common method of achieving flame-resistant wool.

One of the important mechanisms of flame-retardancy with borates is believed to be the formation of glassy films which is supported by the low melting points of boric acid (171°C) and borax (75°C).[66] These are the most important boron compounds for nondurable treatments.

Borax and boric acid together are far more effective than the same weight of either chemical alone,[67] and in most treatments for wool a ratio of 2:1 borax–boric acid is employed. The chemicals are dissolved in water with a wetting agent, and the wool fabric is impregnated in a pad-mangle, winch, or dolly to deposit 5–8 % of solids, depending on fabric weight and construction.

A glassy dustless layer is left on the fiber surface, which is removed during the first washing by dissolving or by mechanical action during the first dry cleaning and the flame-resist effect is lost.

6.1.2. Ammonium Salts

Simpson[3] has compared the effects of various anions in the forms of acids, sodium salts, and ammonium salts by a vertical flame test (Table 7). Only strong and not too volatile acids (phosphoric, sulfuric, sulfamic) and their ammonium salts were effective.

These ammonium salts or acids, into which the former salts are presumably decomposed on heating, alter the route of thermal decomposition as explained in a previous chapter. The low melting point of ammonium sulfamate (125°C), ammonium dihydrogen orthophosphate (190°C), diammonium orthophosphate (155°C), and ammonium hydrogen sulfate (147°C), may also play an important role in the mechanism of flame-retardancy.

The order of effectiveness is free acid > ammonium salt > sodium salt for the strong acids, while the weak boric acid is less effective than its salts. Studies on various polyphosphate systems indicate quite clearly that the more highly polymerized the phosphate, the less effective it is as a flame retardant.[68]

The ammonium salts are preferred as they leave the pH of wool at an acceptable value (4–5), while the acids could significantly decrease the pH value and be responsible for chemical damage and loss in mechanical properties during prolonged wearing.

All the ammonium salts mentioned are effective when the add-on is 10–15% of the weight of wool, best results being obtained with sulfamates and phosphates. The chemicals are dissolved in water and the fabric is impregnated with the solution containing a wetting agent in a pad-mangle, winch, or dolly. The treated fabric is slightly dusty from surface deposits of salts, but this is entirely removed by the mechanical agitation received in the first dry cleaning. The remaining salt is only slowly removed in subsequent dry cleanings, and fastness to at least 10 dry cleanings may be obtained, slightly less with "charge" cleanings. The sulfamate is slightly more resistant to dry cleaning, probably on account of its higher affinity for wool. The ammonium salt treatments are not fast to one washing cycle.

The treatments with ammonium sulfamate or phosphate were used commercially for seat covers of National Airways Corporation of New Zealand[68] but were replaced by the durable titanium and zirconium treatments.

Zinc or cadmium orthophosphates deposited in wool[69] by impregnation with zinc or cadmium chloride followed by phosphates are washfast but have negligible flame retardancy, presumably because these water-soluble compounds are too stable to heat.

Two Japanese references deal with organic nitrogen compounds[70] or a variety of phosphates and borates[71] to improve further silk's flame resistance.

6.2. Phosphorus-Based Treatments

The chemistry of fire-retardants centers around eight elements: phosphorus, antimony, titanium, zirconium, chlorine, bromine, boron, and nitrogen.

Of the eight, phosphorus has the most complex and most fully developed chemistry. In all cases phosphorus is the central element in a compound and there may be an almost infinite variety of substituents in several oxidation states. The chemistry of antimony, titanium, and zirconium is in fact but little developed when compared with phosphorus. Nitrogen has a much simpler chemistry, and for fire retardance the number of nitrogen compounds of interest is limited. In contrast to P, Sb, Ti, Zr, and N, the halogens are not found as central atoms in fire-retardant compounds. Rather, they are substituents on inorganic and organic compounds.

Almost all fire retardants that contain phosphorus are in odd oxidation states, compounds having P in the 5 + oxidation state being the most important and effective. Presence of nitrogen or halides in a flame retardant based on phosphorus increases its effectiveness by synergism and often reduces the need for phosphorus. In many cases this results from chemical interaction which alters behavior in a way not possible with either component. A true synergism occurs in phosphorus–halogen systems, probably by formation of some volatile phosphorus halides which alter combustion by trapping free radicals in the vapor and which facilitate dehydration to char in the solid. The results are of prime importance in lowering the total amount of fire retardant required. With phosphorus–nitrogen systems the nitrogen acts more as an additive flame-retardant, rather than with a synergistic effect. Although nitrogen is less effective than phosphorus, the addition of larger amounts of nitrogen reduce the necessary phosphorus level noticeably.[31,72,73]

6.2.1. Phosphorylation of Wool

It is theoretically possible to incorporate about 3% P in wool by complete phosphorylation of serine, and another 2% by reaction with the hydroxyl groups of threonine. A number of phosphorylating media have been used, including phosphorus pentoxide dissolved in phosphoric acid,[74] phosphorus oxychloride in dimethyl sulfoxide, and phosphoric acid alone,[75] or dissolved in dimethyl formamide, but the phosphorus content of wool was only 0.1–0.3%, which is insufficient to provide adequate flame retardancy.[3,68]

6.2.2. Phosphate Esters

Wool treated with lower trialkyl phosphates from wool-swelling solvents (triethylamine or formic acid) is flame resistant, but the treatment is not fast to washing. The higher esters such as tritolyl and triphenyl phosphate are fast to washing, but the flame-resist effect is not sufficient. Similar results were obtained with monophosphate esters.[3,68]

6.2.3. Phosphorus Polymers Containing Nitrogen or Halogen

Many attempts to produce flame-retardant finishes on wool that are fast to both dry cleaning and washing were carried out by treatments already proven on cellulosic fibers and based on organophosphorus compounds.

A semidurable finish involves forming a basic polymer by polymerizing, for example. 2-methyl-5-vinyl pyridine inside the fiber and using this to bind phosphoric acid.[76] Laundering under alkaline conditions removes the phosphate ion, although the polymer is quite fast.

A similar treatment uses bis(β-chloroethyl)vinyl phosphonate, which is polymerized in wool or in reduced wool from aqueous propanol using ceric catalyst in a nitrogen atmosphere to give an add-on of about 10%.[77]

Cyanamide used in combination with phosphoric acid as a flame retardant for fabrics of cotton, rayon, and wool is an old process.[78,79] This treatment. using an add-on of 6.7% cyanamide 50% and 2.5% phosphoric acid on the weight of wool, is recommended for wool shag-pile carpets in order to pass the American tablet test.[80] The solution is applied to the carpet by padding or spraying, followed by drying and curing at 300°F for 10 min. The mixture does not exhaust on wool, and the curing step is a disadvantage because of technical and economic problems. The treatment is fast to shampooing but only partially resistant to laundering. the latter being required for flame-resist-treated carpets in order to pass the tablet test.

A slight improvement in the washing fastness of this treatment was observed when curing a mixture of orthophosphoric acid, dicyandiamide, and urea (1.0:0.6:2.0 parts by weight) which is applied by a pad–dry–cure (140°C, 5 min) technique. The treatment requires a good control of curing conditions and very high initial solids add-on of about 40%,[3.68] which significantly impairs the fabric handle.

Tris(1-aziridinyl)phosphorus oxide (APO) is another flame-resistant agent for cellulose[81,82] which has been applied to wool. When a boron trifluoride–methanol complex is used as a catalyst,[83] self-polymerization on the wool fiber occurs as APO has three reactive groups and is the reaction product of ethylenimine and phosphoryl trichloride:

$$3\begin{bmatrix} CH_2 \\ | \quad \diagdown \\ | \quad\quad NH \\ | \quad \diagup \\ CH_2 \end{bmatrix} + POCl_3 \xrightarrow{\text{base}} \begin{bmatrix} CH_2 \\ | \quad \diagdown \\ | \quad\quad N \\ | \quad \diagup \\ CH_2 \end{bmatrix}_3 PO$$

A solution of 8:50:2 parts by volume of 80% APO, water, and 35% BF$_3$–methanol is applied to a wool fabric at 100% wet pick-up. dried at 100°C. cured for 5–10 min at 130°C and gives a very good degree of flame resistance which is fast to washing, but there is some discoloration of the fabric.[3,68] Treatments of wool with APO solutions containing various catalysts showed considerable differences between wool and cotton, presumably because the number of reactive hydroxyl groups in wool is much lower than in cotton. APO has also been applied to wool with a resin by a pad, dry, heat cure, and afterwash sequence.[84] So far no commercial use of this formulation on wool has been reported mainly because of toxicity of APO and high cost per unit of phosphorus. This is also the reason why this phosphorus compound is not being applied to cotton fabrics.

As in the latter case, more often, phosphorus compounds have been bound with aminoplast resins to improve wet fastness and also the effectiveness of the treatment.

Aenishanslin et al.[85] describe phosphonates of the general formula

$$
\begin{array}{c}
R_1\!-\!O \quad\ O \\
\diagdown \diagup \\
P \\
\diagup \diagdown \\
R_2\!-\!O \quad CH_2\!-\!CH\!-\!CO\!-\!NH_2 \\
| \\
R_3
\end{array}
$$

for the flame-resist treatment of cellulosic fibers, and the same products, applied with resins, are being developed for use on wool. The treatment has to be applied by a pad–dry–cure technique, curing being carried out at a temperature not very convenient for wool—160–165°C for 4–5 min. After curing, the fabric has to be washed off in an alkaline solution. This treatment has not been commercially applied on wool.

The only organophosphorus treatment which has been commercially applied to wool is based on THPC [tetrakis(hydroxymethyl)phosphonium chloride] which is also the most commercially applied flame-resist treatment for cotton:

$$
\left[
\begin{array}{c}
H \\
O \\
CH_2 \\
| \\
HOCH_2\!-\!P\!-\!CH_2OH \\
| \\
CH_2 \\
O \\
H
\end{array}
\right]^{+} \ Cl^{-}
$$

More has been written about THPC for flame-retardant cotton than about any other flame-retardant topic. Not only are there a large number of papers and individual reports, but there is also a sizable patent literature.[31]

Two general techniques for fixing THPC to cellulosic fabrics are currently practiced. Probably the most common is to impregnate with THPC, dry the fabrics, expose to ammonia vapor to form a nitrogen-containing polymer, and finally neutralize in a winch.[86] Alternatively, the THPC is combined in a thermosetting resin precondensate, which is then applied to the fabric, dried, and heat cured before neutralizing as before.[87]

For wool the treatment involves THPC, urea, and a melamine–formaldehyde resin which is applied by a pad–dry–cure–wash–dry technique.[3,10,68,88] Dyes not fast to reducing treatments must be avoided as THPC is a reducing agent which can reduce disulfide bonds in wool to thiols, [89, 90], and formaldehyde from the thermosetting resin can also have an adverse effect on some dyes.

Residual oil content of wool has to be less than 0.5% methylene chlorine-extractable matter as residual oil acts as a barrier to diffusion of the resin. For the same reason the treatment is not compatible with oil- and/ or water-repellent treatments based on silicones or fluorocarbons. After padding the fabric has to be over dried so as to avoid evaporative cooling during the heat-curing step, which is carried out at 150°C for 4 min.

As the cured fabric contains incompletely polymerized resin, mineral acids, and reducing agents, it has to be after-finished by a rinse with water, oxidizing scour with hydrogen peroxide, and finally rinsed with water or sodium bicarbonate. This sequence leaves the fabric suitably reoxidized and neutralized. Very satisfactory results and wash- and dry-cleaning fastness is obtained when all the necessary conditions during the treatment are carefully controlled. The handle of the treated fabric can be crisp to harsh depending on the storage time of the resin. The relatively short shelf time of THPC, which is 3–9 months depending on temperature, the long and costly procedure, and the incompatibility of the treatment with some dyes are the main reasons why this effective treatment was commercially superseded by the titanium and zirconium treatments. Nevertheless, the THPC treatment played an important role during the period when durable treatments for aircraft furnishings and seat covers[91] were required and other cheaper and simpler treatments were not available.

The THPC treatment is applied to wool fabrics utilizing a THPC–urea condensate (Proban 420A) and in conjunction with a melamine type resin (Proban 420B) in a weight ration 2:1.[88] The total resin solids deposited on the fabric should be about 15%, depending on fabric weight and construction. The bath containing a nonionic wetting agent (0.1%) is padded on the fabric with a wet pick-up of 60–80% and dried at 210–240°F, overdrying being recommended. Since the fabric is hygroscopic once dried, it is suggested that the dried fabric either be cured immediately after drying or be sealed in polyethylene to avoid uneven moisture pick-up which could result in uneven curing. The dried fabric is cured in any conventional (noninfrared) oven at 300°F for 4–6 min. In order to remove excess reactants, the cured fabric is afterwashed with water and surfactant at 80–100°F followed by an oxidizing scour with hydrogen peroxide at 100–120°F for 20–30 min, rinsed with water for 15 min at 80–100°F, hydroextracted, dried, and finished. The use of phosphorus-containing resins as finishing agents has also been claimed to improve the flame retardancy of silk fabrics.[92,93]

6.3. Metal Compounds

6.3.1. Antimony

Wool treated with antimony oxide dissolved in formic and hydrochloric acid to a level of 8% of the oxide is flameproof, but it burns after brief washing.[3,68] The loss in flame retardancy is probably associated with the washing off of chloride ions, which presence is necessary to the formation of volatile antimony halides during burning, the latter trapping the combustible radicals in the vapor phase. In fact, when the washed fabric is soaked in diluted hydrochloric acid it becomes flame resistant again, even though this acid alone has no appreciable flameproofing action (Table 7).

Attempts to incorporate halogen by organic chlorine in a more permanent manner were not successful.[3,68]

Antimony compounds in combination with titanium compounds (e.g., titanium tetrachloride) are effective and fast to washing,[18] but the toxicity and high cost of antimony compounds does not permit a commercial application of this treatment when cheaper treatments utilizing titanium or zirconium complexes are available.

6.3.2. Chrome

Chromium compounds are sufficiently effective to pass the tablet test for carpets when at least 1.5% sodium bichromate reacts with wool by a redox reaction and a minimum of 0.7% chrome trioxide is left inside the wool fiber.[94] This cheap treatment is fast to washing or dry cleaning but the color change is significant (grey-green); therefore, this treatment can be applied only with very dark shades, which severely restricts its application for carpets. Mordanting with chrome is not sufficiently effective to pass a vertical flame test for fabrics. In some cases an afterglow is noticed, particularly with heavier and denser fabrics, probably owing to the catalytic action of chrome trioxide on the glowing reaction. If no afterglow is permitted by a particular flammability specification, other dyes than chrome have to be used.

The formation of coordination bonds between the peptide groups of the wool and the chromium atom is very probably responsible for the wet fastness of this mordanting treatment.[95-97]

Sodium bichromate can be exhausted on wool by any of four standard techniques, i.e., reduced chrome, bottom chrome, meta chrome, or top chrome. Reduced chrome is recommended for the relative ease of shade matching it provides. Sodium bichromate (1.5%) and formic acid (3%) on the weight of wool (oww) are used, the wool is boiled for 45 min, followed by a reductive treatment in the same bath with 1% sodium bisulfite oww for 15 min at the boil. After rinsing, chromed wool can be dyed with chrome, milling, premetalized, or selective acid dyes to obtain dark shades, since the base color is grey-green.

6.3.3. Titanium and Zirconium

The permanent flame-resistant agents which are already available commercially are mainly based on resinous organophosphorus, organochlorine, organobromine compounds, or their mixtures. Their main disadvantage is that they are relatively expensive in terms of chemical cost; they must be applied by an entirely separate operation (pad–dry–cure or spray–nip–dry–cure techniques), which further increases the total treatment cost; and about 15–30% of the product has to be applied on the weight of the fabric to obtain proper fixation of the effective flame-resistant element—phosphorus, chlorine, or bromine.

Such high add-ons influence the original handle of the fabric, which can become firmer, harsh, or stiff, depending on the fabric weight and construction. Unfortunately the lighter the fabric, the more flame-resistant agent is necessary, since light fabrics burn more readily than heavier fabrics. At the same time it is the handle of light fabrics which is most influenced by resin applications.

The disadvantages already listed for commercially available flame-resist agents became more apparent when the new American Federal Specifications DOC FF 1-70 and 2-70 for carpet[14] came into effect on April 15, 1971, for all carpets sold in the United States, whether home-produced or imported.

The specifications require that carpet samples (9 × 9 in.) shall be dried at 105°C for 2 hr in an oven, cooled for 30 min in a desiccator, and subsequently tested within 2 min by methenamine tablet.

The hexamethylenetetramine tablet is positioned in the center of the carpet sample and ignited by a match. The methenamine tablet has a burning time of approximately 2.5 min, and the flame should not spread more than 3 in. from the tablet in any direction if the carpet is to pass this specification.

Conventional wool carpets pass without any special treatment, but shag-pile carpets fail because their very open construction has a higher air-to-wool ratio. Even in these very open constructions the natural moisture content of wool (10–14 %) would be sufficient to prevent flame spread, but the specification requires bone-dry conditions in order to provide safety under the most extreme conditions.

Equipment for applying THPC to carpet was not available, and the total costs were too high for the product. Attention was turned to techniques which could be applied at the dyeing stage.

Several aluminum, zinc, stannous, stannic, chromium, antimonous, titanium, and zirconium compounds have been tested by an exhaustion technique (45 min boil, pH 2–3, liquor ratio 1:25) on wool yarn and a lightweight woven fabric (200 g/m^2). The flame-resist properties were tested by the tablet test[14] after manufacturing a shag-pile carpet (total pile height 3.5 cm, sheared pile weight 768 g/m^2) from the treated yarn or by the vertical flame test on the treated fabric.[101]

Aluminum, zinc, stannous, and stannic compounds are generally not sufficiently effective, whether applied with or without organic complexing acids.[18] A much better improvement in the natural flame resistance of wool was noticed with titanium and zirconium compounds.[16-21,104-109] These compounds readily hydrolyze and polymerize in aqueous solutions which impairs their exhaustion on the wool fiber.[100,102] We were able to avoid these problems by using complexing agents such as some carboxylic acids and fluorides.

6.3.3.1. Titanium and Zirconium Complexes with Carboxylic Acids. Table 8 shows that the acids which are effective are those which contain an α-hydroxy group—citric acid, $C(OH)(COOH)(CH_2COOH)_2$; tartaric acid,

TABLE 8

Influence of Various Organic and Inorganic Acids on the Exhaustion of Titanium on Wool and on Flame-Resist Effect

Application[a]	Exhaustion, %	Vertical flame test	Note
2.5 % $TiCl_4$ oww	2	Fails	
2.5 % $TiCl_4$ + 5 % HCl (37 %) oww	4	Fails	Precipitate
2.5 % $TiCl_4$ + 6 % acetic acid oww	3	Fials	at the
2.5 % $TiCl_4$ + 6 % maleic acid oww	10	Fails	boil
2.5 % $TiCl_4$ + 6 % succinic acid oww	11	Fails	
2.5 % $TiCl_4$ + 6 % oxalic acid oww	94	Passes	
2.5 % $TiCl_4$ + 6 % malic acid oww	95	Passes	
2.5 % $TiCl_4$ + 6 % tartaric acid oww	94	Passes	
2.5 % $TiCl_4$ + 6 % citric acid oww	95	Passes	

[a] All applications by exhaustion—45 min at 100°C, liquor ratio 1:20.

(CHOH·COOH)$_2$; malic acid, CH$_2$COOH·CHOH·COOH—or adjacent carboxyl groups such as oxalic acid, (COOH)$_2$. If this condition is not fulfilled, titanium tetrachloride is hydrolyzed in solution to form TiO$_2$ and the application is ineffective. This is the case with acetic acid, CH$_3$COOH; maleic acid, CH·COOH:CH·COOH; and succinic acid, (CH$_2$COOH)$_2$; where the complex with titanium is not sufficiently stable in solution because of the absence of adjacent hydroxyl or carboxyl groups to the first carboxyl group. The same is applicable for zirconium complexes with carboxylic acids.

Typical exhaustion curves for titanium tetrachloride or potassium titanium oxalate with various concentrations of citric acid are shown in Fig. 1, from which it follows that after 30 min boiling with 4% citric acid, practically all the titanium is exhausted on wool. At lower levels of citric acid, the exhaustion is not complete.[16]

The titanium and zirconium complexes with effective carboxylic acids exhaust on the fiber at the boil, and below this temperature exhaustion is negligible. Another important factor is the pH of the bath. At pH values above 3, the exhaustion of the metal complexes is not complete, and the pH of the bath has to be maintained below 3 throughout to achieve proper

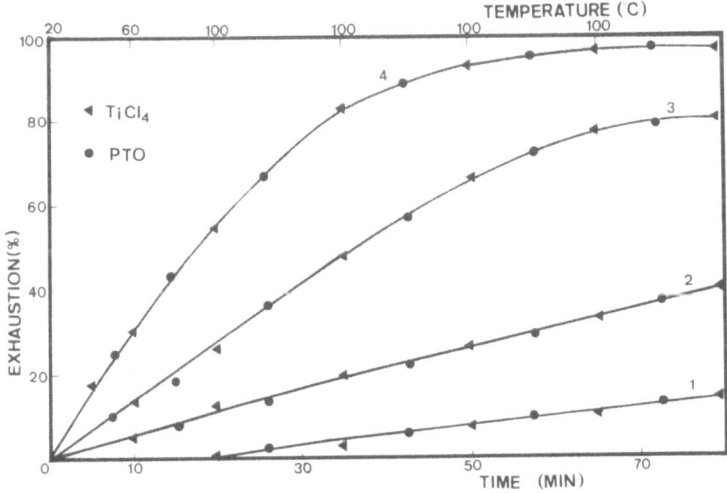

FIGURE 1. Exhaustion curves for TiCl$_4$ (0.8% oww) and potassium titanium oxalate (1.5% oww) with various concentrations of citric acid.

Curve	Citric Acid, % oww
1	1
2	2
3	3
4	4

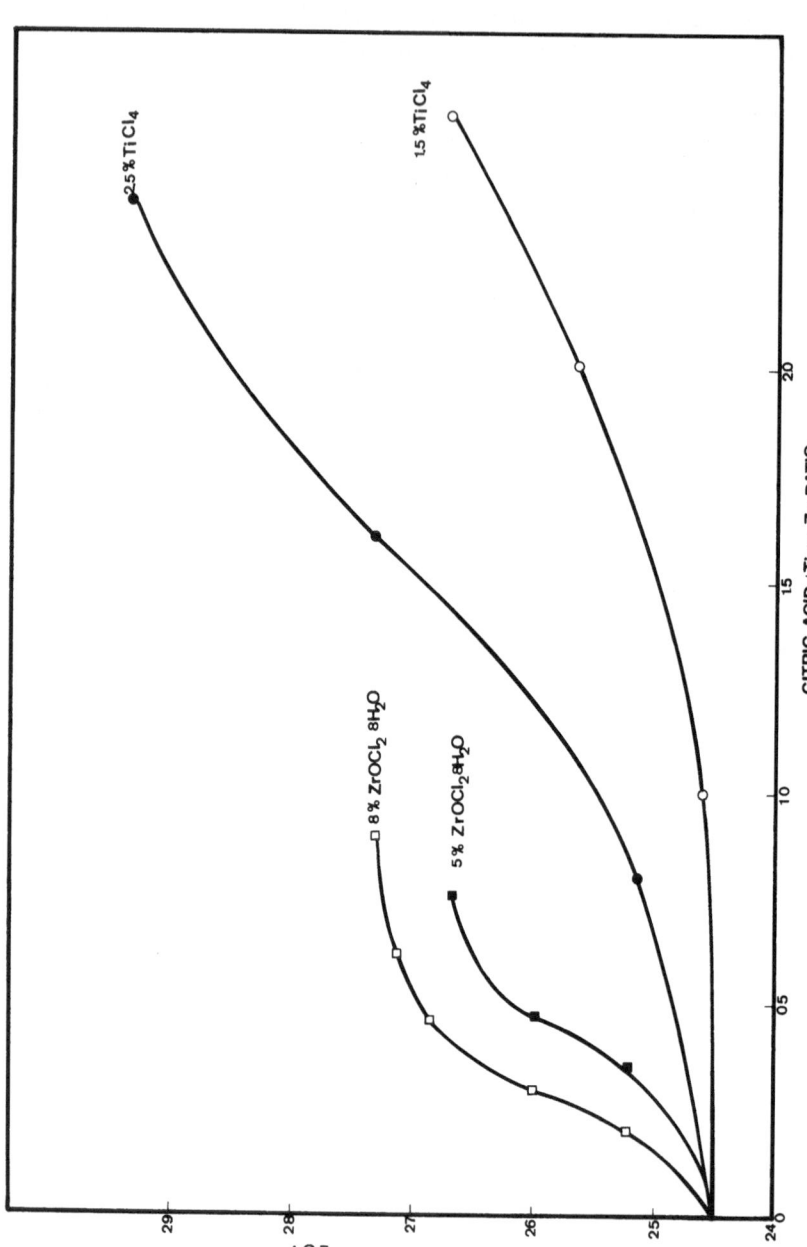

FIGURE 2. Influence of citric acid/Ti or Zr ratio on the LOI of a fabric treated at the boil for 45 min with 5% HCl 37% and the appropriate concentrations of citric acid, titanium tetrachloride, or zirconium oxychloride. Liquor ratio 1:20.

exhaustion and an effective improvement in the natural flame resistance of wool.

The citric acid/Ti or Zr ratio is also important (Fig. 2). For titanium this ratio should be at least 2.5 and for zirconium not less than 0.8. As titanium tetrachloride hydrolyzes very rapidly in solution (Table 8) forming a precipitate, it needs an excess of complexing agent compared with zirconium oxychloride. which gives only a very slight cloudiness in solution in the absence of a complexing agent (Table 9). The smaller atomic radii of titanium when compared with zirconium could also be responsible for this effect.

The increase in the concentration of titanium or zirconium results in a higher degree of flame resistance (Fig. 2): thus it is possible to achieve various degrees of flame resistance according to the end use of the wool textile and flammability standard specified.

The titanium mordanting is generally more effective and occurs at lower metal concentrations than the zirconium application. For example. to achieve an LOI value of 27, which offers a very good degree of flame resistance for common end uses, it is necessary to apply 0.5% Ti to the fiber compared with 1.9% Zr. Although the atomic weight of zirconium is approximately double that of titanium, it appears that the main reason for the higher effectiveness of titanium is connected with its smaller atomic radii, thus achieving better and more even penetration of the fiber.[110]

The fiber damage due to the metal citrate applications is negligible. and in fact it is the same as with ordinary dyeing.[16,18]

Stereoscan evaluation of mordanted fibers shows that the metal complex is located inside the wool fiber.[106] thus the handle and the natural soil-resistance of wool is not affected.

Color fastness to water, washing, and perspiration is not influenced by the metal complex mordant applications which are fast to at least 50 washings at $40°C$ (Woolmark cycle) and/or 50 dry cleanings with Perklone.

TABLE 9

Influence of Citric Acid Concentration on the Effectiveness of the Zirconium Treatment

Application[a]	Vertical flame test	Note
8% ZrOCl$_2$ 8H$_2$O + 5% HCl 37% oww	Fails	Slight Cloudiness at the boil
8% ZrOCl$_2$ 8H$_2$O + 5% HCl 37% + 2% citric acid	Fails	
8% ZrOCl$_2$ 8H$_2$O + 5% HCl 37% + 4% citric acid	Borderline	
8% ZrOCl$_2$ 8H$_2$O + 5% HCl 37% + 6% citric acid	Passes	
8% ZrOCl$_2$ 8H$_2$O + 5% HCl 37% + 8% citric acid	Passes	

[a] All applications by exhaustion—45 min at 100 C. liquor ratio 1 20

The titanium application causes a certain degree of yellowing of the fiber (Table 10) which is not removed by common oxidative and/or reducing treatments. Application of titanium citrates causes less yellowing than with tartarates and oxalates. The yellow color intensifies on exposure to light, and the fastness of this color is graded only 3–4 on the ISO scale. Bleaching with hydrogen peroxide intensifies the yellow color owing to the formation of pertitanic acids.[111] The zirconium citrate treatment gives no yellowing apart from the negligible degree which occurs upon boiling of wool (Table 10). This can be corrected by the addition of various proprietary products recommended for preventing yellowing of wool in the dyebath. Fastness to light of zirconium-treated wool is the same as with untreated wool.

This mordanting technique can be applied simultaneously with acid levelling and 1:1 premetallized dyes. With acid milling and 1:2 premetallized dyes, mordanting should be carried out after dyeing because of the low pH required for metal complex exhaustion. Chrome dyes should always be applied after mordanting, as they form a complex with titanium or zirconium compounds in solution and there could be a color change.

It has been shown that α-hydroxycarboxylic acids readily form chelate rings with the zirconium and titanium atom.[112,113] The negatively charged particle is exhausted on the positively charged protonated amino groups of the wool fiber in strongly acid conditions:

TABLE 10
Influence of Various Titanium and Zirconium Applications on the
Yellowness Index of Wool

Treatment	Yellowness Index
Untreated	7.0
2.5% TiCl$_4$ + 6% oxalic acid oww, boil, 45 min	24.8
2.5% TiCl$_4$ + 6% tartaric acid oww, boil, 45 min	22.5
2.5% TiCl$_4$ + 6% citric acid oww, boil, 45 min	19.7
4% K$_2$TiF$_6$ + 10% HCl 37% oww, 75°C, 30 min	19.0
8% ZrOCl$_2$ 8H$_2$O + 5% HCl 37% + 8% citric acid oww, boil, 45 min	8.5
8% K$_2$ZrF$_6$ + 10% HCl 37% oww, 75°C, 30 min	7.1
10% HCl 37% oww, boil, 45 min	8.4

On the wool the ion may undergo partial hydrolysis which is very probably responsible for the wash fastness.

6.3.3.2. Titanium and Zirconium Complexes with Fluorides. Although the titanium and zirconium citrate complexes are sufficiently effective, their disadvantage is that they have to be applied at the boil: this can cause some problems with fabric structures sensitive to felting and with dyes not sufficiently fast at the boil in acid conditions.

The fluoride complexes of titanium and zirconium are sufficiently stable in acid aqueous solutions and they are in fact one of the most stable metal fluoride complexes known.[114] Figure 3 shows a typical exhaustion of potassium hexafluorozirconate in acid conditions, which is relatively rapid, and 77% exhaustion is achieved at temperatures above 50 C. In order to achieve level results it is necessary to carry out the treatment at 70 C for at least 30 min at a pH below 3 to allow leveling out of the hexafluorozirconate.

With increasing concentration of the hexafluorozirconate or hexafluorotitanate, the LOI value is increased up to the limit of $9-10\%$ K_2ZrF_6 and $6-7\%$ K_2TiF_6 (Fig. 4). As with the metal citrates, the hexafluorotitanate application is more effective and at lower metal concentrations than the hexafluorozirconate treatment. With 4% K_2TiF_6 it is possible to achieve the remarkably high LOI value of 32 when compared with 24.5 for the same untreated fabric. The explanation for the higher effectiveness of the hexafluorotitanate is very probably the same as with the metal citrates. The LOI values obtained with the metal fluoride complexes are higher than with the equivalent metal citrate concentrations, which is probably connected with the smaller molecule radii of the metal fluorides when compared with the metal citrates; thus greater penetration of the former complex into the fiber can be expected.

The F/Ti or Zr ratio of the fluoride complexes has to be at least 5, but preferably 6, to effectively improve the natural flame resistance of wool (Fig. 5). A ratio of 4 has no practical effect on the LOI value of wool. This shows that only a negatively charged penta-, hexa-, or heptafluorozirconate or titanate can be exhausted on the positively charged wool. Zirconium or titanium tetrafluoride which is uncharged has no substantivity to the wool fiber.

The pH during and particularly after the application, as well as the acid applied, has a significant effect on the flame-resist effect and fastness to washing (Table 11). The pH after the application has to be below 3 in order to obtain an effective and wash-fast (50 washings) improvement in the natural flame resistance of wool. The addition of sulfates or the use of sulfuric acid decreases the effectiveness and fastness of the application. This is probably caused by the substantivity of sulfates to the wool fiber which compete with the metal fluoride complex for the available sites in the wool

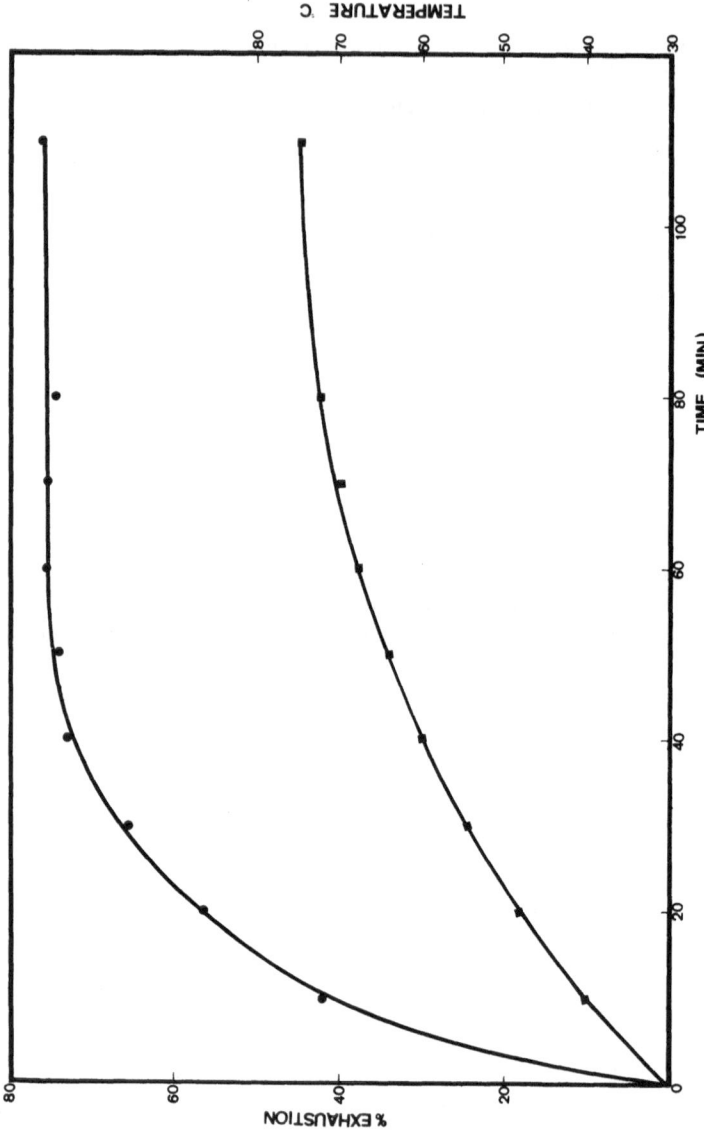

FIGURE 3. Exhaustion of potassium fluorozirconate (8%) in hydrochloric acid solution (10% HCl 37%) on wool. Liquor ratio 1:20.

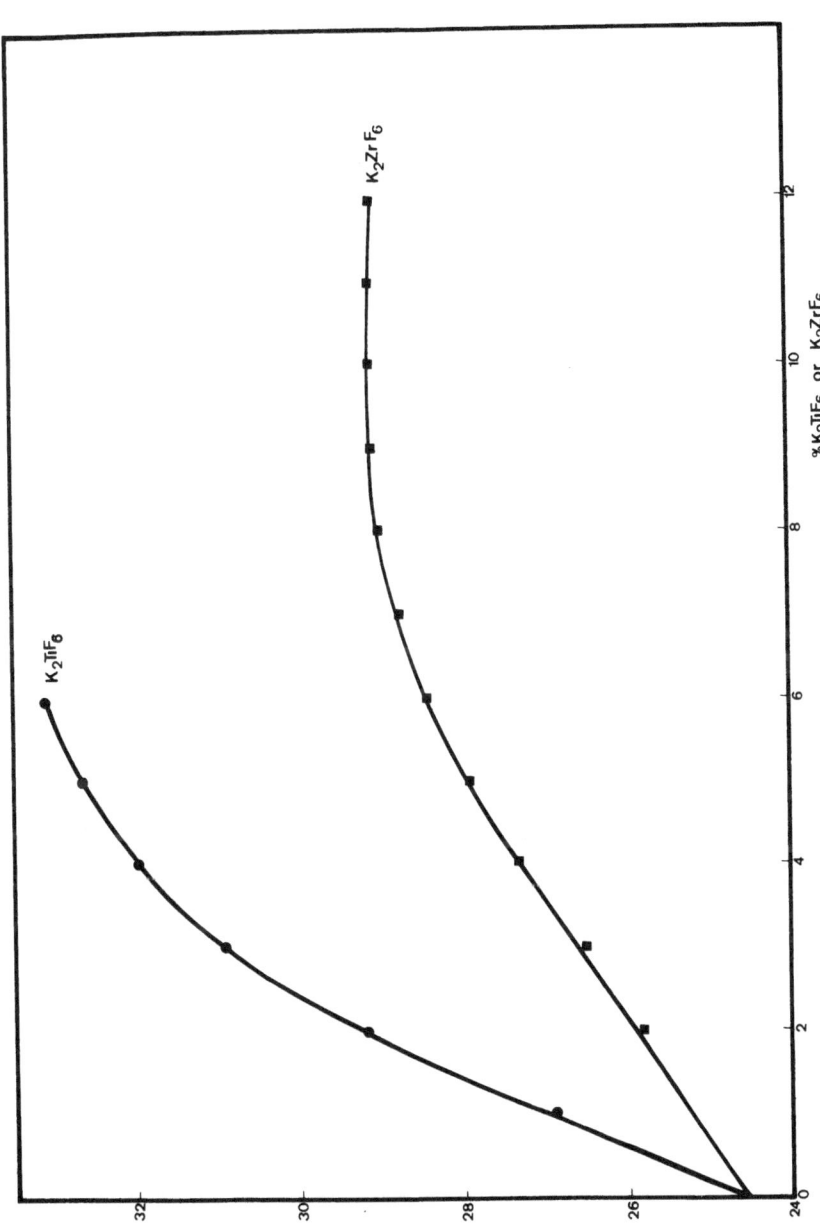

FIGURE 4. Influence of the concentration of potassium hexafluorotitanate or hexafluorozirconate on the LOI of a fabric treated at 75°C for 30 min with 10% HCl 37% and the metal fluorides.

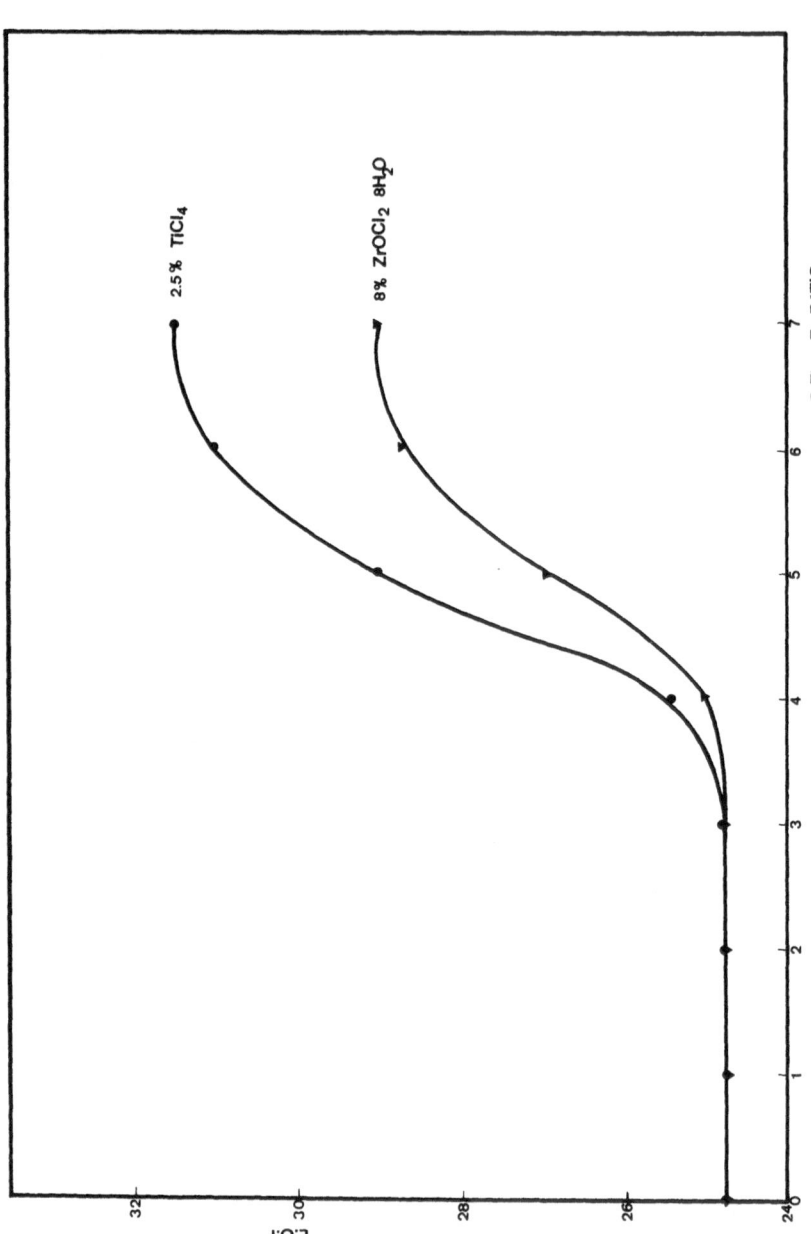

FIGURE 5. Influence of F/Ti or Zr ratio on the LOI of a fabric treated at 75°C for 30 min with 5% HCl 37% and the appropriate concentrations of ammonium bifluoride, titanium tetrachloride, or zirconium oxychloride. Liquor ratio 1:20.

TABLE 11

Effect of pH and Various Acids During the Application of Fluorozirconate on the Flame-Resistance of Wool Before and After Washing

Application[a]	pH		Vertical Flame test					
	Before application	After application	Before washing	After washing				
				10×	20×	30×	40×	50×
$8\% \, K_2ZrF_6 + 6\% \, HCl \, 37\%$ oww	2.5	3.7	Pass	Fail				
$8\% \, K_2ZrF_6 + 10\% \, HCl \, 37\%$ oww	1.9	2.6	Pass	Pass	Pass	Pass	Pass	Pass
$8\% \, K_2ZrF_6 + 10\% \, HCl \, 37\%$ $+ 5\% \, Na_2SO_4$ oww	2.0	2.7	Pass	Pass	Fail			
$8\% \, K_2ZrF_6 + 10\% \, HCl \, 37\%$ $+ 10\% \, Na_2SO_4$ oww	2.1	2.6	Pass	Fail				
$8\% \, K_2ZrF_6 + 3\% \, H_2SO_4$ oww	1.9	3 1	Pass	Fail				
$8\% \, K_2ZrF_6 + 6\% \, H_2SO_4$ oww	1.5	2 4	Pass	Fail				
$8\% \, K_2ZrF_6 + 10\% \, HAc$ oww	3.5	4.7	Fail					
$8\% \, K_2ZrF_6 + 10\% \, HCOOH$ oww	3.6	4.2	Pass	Fail				
$8\% \, K_2ZrF_6 + 20\% \, HCOOH$ oww	2.4	2.7	Pass	Pass	Pass	Pass	Pass	Pass

[a] All applications at 75 C for 30 min. liquor ratio 1 : 20.

fiber, thus decreasing the exhaustion of the latter. Strong acids which have a relatively low substantivity to wool, e.g., hydrochloric and formic acids, are the most suitable ones. Otherwise the application of the low-temperature technique using metal fluoride complexes to wool results in the same properties as with the metal citrate applications, including the yellowing of wool by the fluorotitanate application (Table 10). However, with the zirconium application no yellowing occurs with the low-temperature technique. The latter technique is much more useful as it rarely affects fabric appearance, dye bleeding, or migration when compared with the application at the boil.

Although organophosphorus finishes on textiles are very often degraded by outdoor exposure to light,[115] a wool fabric to which 8% K_2ZrF_6 was applied did not show any loss in flame retardancy and washing fastness after 1000 hr of sunlight exposure in Florida.

The zirconium and titanium application also significantly improves the natural heat resistance of wool. A thermocouple which was exposed to a flame showed an increase of 300°C during 30 sec (Fig. 6). With the untreated fabric there is heat protection of the thermocouple during the first 7 sec, after which the char is destroyed and a significant increase in temperature is noticed, the overall effect being a 190°C rise during 30 sec. The zirconium-mordanted wool forms a thick and dense char on exposure to the flame, and the increase in temperature of the thermocouple is only 47°C during 30 sec. The significantly improved heat resistance of Zr-mordanted wool was exploited for protective clothing for racing drivers,

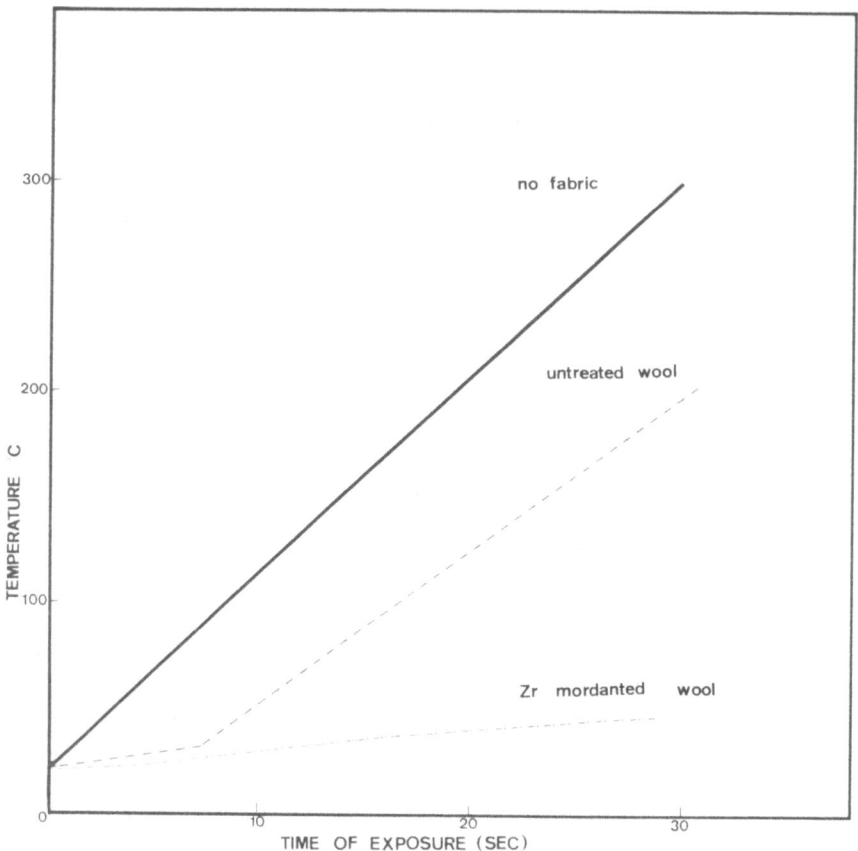

FIGURE 6. Heat protection of untreated and zirconium mordanted wool.

particularly in wool-rich blends with glass fibers, the latter holding the fabric structure during exposure to a flame.

With the low-temperature techniques, no bleeding of dyes occurs under the acid conditions of treatment and fabric felting is negligible. This can be considerable with the metal citrate applications at the boil. The cost of the chemicals for the low-temperature technique is cheaper than for the metal citrates. Because of these technical and economic advantages, the low-temperature technique became the favored one, particularly for fabrics.

Table 12 summarizes the compatibility of the low-temperature exhaustion applications with various dye classes, which depends on the optimal pH for even dye exhaustion.

Because the metal complex reacts with the amino groups inside the wool fiber, there is no change in handle and the treatments do not affect

TABLE 12
Metal Applications by Exhaustion During or After
Dyeing

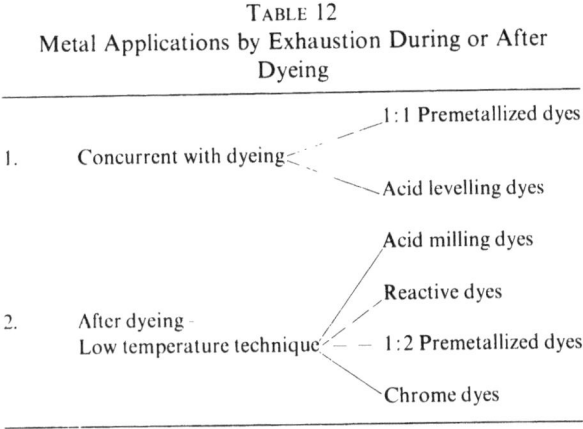

1.	Concurrent with dyeing — 1:1 Premetallized dyes — Acid levelling dyes
2.	After dyeing - Low temperature technique — Acid milling dyes, Reactive dyes, 1:2 Premetallized dyes, Chrome dyes

the natural soil-resistance of wool. Figure 7 shows a stereoscan picture of wool treated with zirconium complexes, Fig. 8 shows wool treated with a commercially available flame-resistant resin for carpets, where surface deposition is noticeable and soiling significantly increased. There is no significant difference in the mechanical properties between the carpets or the fabrics made from the treated yarn compared with the same textiles made from dyes yarn.

Color fastness to water, washing (40°C), and perspiration is not influenced by the flame-resist treatments which are fast to at least 50 washings at 40 C or 50 dry cleanings with Perklone.

It is known that the hexafluorozirconate and hexafluorotitanate ions are stable and largely undissociated, particularly in strongly acid solutions which are used during their application to wool.[116,117] Therefore they are exhausted onto the positively charged wool in acid conditions as the negative ZrF_6^{2-} and TiF_6^{2-} ions according to the equation:

$$MF_6^{2-} + {}^+H_3N-Wool \longrightarrow Wool-NH_3^+MF_6^{2-}$$

The hexafluorozirconate or titanate in the fiber very probably hydrolyzes after several rinses and washes to form $ZrOF_2$ or $TiOF_2$.[118,119] Fluorozirconates gradually lose fluorides during heating, and above 300°C zirconyl fluoride is formed.[120] Therefore it can be assumed that zirconyl or titanyl fluoride is the effective compound in the fiber which is responsible for flame resistance.

6.3.3.3. Pad–Batch and Pad–Dry Techniques. All the previously described applications were by the exhaustion technique during or after dyeing, depending on the dye. For some very sensitive fabric constructions a padding application would be more suitable. Results in Table 13 show that the metal fluorides are effective when applied by a pad–dry or pad–

FIGURE 7. Scanning electron micrograph of wool fibers flame-resist treated by zirconium complexes.

batch treatment, while the metal citrates are not. This is probably connected with the larger molecular radii of the citrate complexes which can not penetrate the fiber unless it is swollen at the boil, while the much smaller molecules of the fluoride complexes can penetrate the fiber at lower temperatures. The addition of formic acid is essential to increase penetration of the fluoride complexes. Potassium hexafluorozirconate and hexafluorotitanate are only slightly soluble at room temperature (1 g/100 ml), therefore the much more soluble ammonium or sodium salts have to be used by mixing the metal chloride with a fluoride.

The titanium fluorides give best results by a pad–dry technique, although the results of the pad–batch technique are sufficiently effective. With the zirconium fluorides, best results are obtained with a pad–batch (1 hr) technique, but if ammonium bifluoride is replaced by sodium fluoroborate as a source of fluorides, the results are significantly improved. This is probably connected with the negative charge of the small molecule of fluoroborate which penetrates the fiber more effectively.

After the pad–dry or pad–batch application the fiber is rinsed in water to remove the excess of chemicals and dried. After this application the improvement in flame resistance is fast to at least 20 washings at 40°C, but it is slightly less effective than the exhaustion technique.

6.3.4. Tungsten Complexes

The titanium and zirconium applications are sufficiently effective to meet the most stringent flammability standards. However, it was noted

FIGURE 8. Scanning electron micrograph of wool fibers flame-resist treated by a commercial resinous flame-retardant.

TABLE 13
Pad–Batch and Pad–Dry Applications

Application	Technique	LOI
2.5% TiCl$_4$ + 6% citric acid + 2% HCOOH	Pad–dry	24.6
2.5% TiCl$_4$ + 6% citric acid + 2% HCOOH	Pad–batch (4 hr)	24.7
7% ZrOCl$_2$ 6H$_2$O + 8% citric acid + 2% HCOOH	Pad–dry	24.5
7% ZrOCl$_2$ 6H$_2$O + 8% citric acid + 2% HCOOH	Pad–batch (4 hr)	24.6
2.5% TiCl$_4$ + 2.5% NH$_4$F · HF	Pad–batch (1 hr)	28.5
2.5% TiCl$_4$ + 2.5% NH$_4$F · HF + 2% HCOOH	Pad–dry	31.9
2.5% TiCl$_4$ + 2.5% NH$_4$F · HF + 2% HCOOH	Pad–batch (1 hr)	31.1
2.5% TiCl$_4$ + 2.5% NH$_4$F · HF + 2% HCOOH	Pad–batch (4 hr)	31.4
2.5% TiCl$_4$ + 2.5% NH$_4$F · HF + 2% HCOOH	Pad–batch (6 hr)	31.5
7% ZrOCl$_2$ 6H$_2$O + 4% NH$_4$F · HF	Pad–batch (1 hr)	25.1
7% ZrOCl$_2$ 6H$_2$O + 4% NH$_4$F · HF + 2% HCOOH	Pad–dry	25.6
7% ZrOCl$_2$ 6H$_2$O + 4% NH$_4$F · HF + 2% HCOOH	Pad–batch (1 hr)	26.4
7% ZrOCl$_2$ 6H$_2$O + 4% NH$_4$F · HF + 2% HCOOH	Pad–batch (4 hr)	25.9
7% ZrOCl$_2$ 6H$_2$O + 4% NH$_4$F · HF + 2% HCOOH	Pad–batch (6 hr)	25.9
7% ZrOCl$_2$ 6H$_2$O + 4.5% NaBF$_4$ + 2% HCOOH	Pad–batch (1 hr)	27.3
Untreated		24.5

that on wool shrink-resist treated by chemical processes, the improvement in the natural flame resistance of wool is not as effective nor as wash fast as on untreated wool. The results in Table 14 indicate that the shrink-resist treatments decrease the flame resistance of wool, and the higher the chemical degradation the lower the LOI value, e.g., the dichloroisocyanuric treatment (DCCA) causes more degradation than permonosulfuric acid. This

TABLE 14
Influence of Chemical Shrink-Resist Treatments on the Effectiveness of the
Mordanting Applications

Treatment	LOI	Vertical flame test					
		Before washing	After washing				
			10×	20×	30×	40×	50×
Untreated	24.5	Fail					
2% DCCA	22.9	Fail					
3% DCCA	22.6	Fail					
4% DCCA	22.5	Fail					
2% H$_2$SO$_5$	23.9	Fail					
3% DCCA + 8% K$_2$ZrF$_6$, 10% HCl 37%	26.9	Pass	Pass	Fail			
2% H$_2$SO$_5$ + 8% K$_2$ZrF$_6$, 10% HCl 37%	27.5	Pass	Pass	Pass	Fail		
3% DCCA + 8% K$_2$ZrF$_6$, 10% HCl 37%, 3% Na$_2$WO$_4$, 37% citric acid	28.1	Pass		Pass	Pass	Pass	Pass Pass
3% DCCA + 8% K$_2$ZrF$_6$, 10% HCl 37%, 5% Na$_2$WO$_4$, 4% citric acid	28.3	Pass		Pass	Pass	Pass	Pass Pass

is probably connected with the partial degradation of the fiber resulting from these chemical shrink-resist treatments, as the wool structure becomes slightly less stable. Further, oxidation of disulfide bonds resulting from these treatments, along with the formation of negative $-SO_3^-$ groups, may decrease the exhaustion of the negatively charged metal complexes and also affect the washing fastness, which is well known in dye chemistry. Several metals were tested including vanadium, molybdenum, and tungsten complexes to improve the flame-resist effect. The former two were effective, but severe discoloration of the fiber occurred. The latter gave an average improvement of flame resistance (LOI 26–27), but discoloration on exposure to light was noted, which was probably due to the formation of tungsten bronzes.[121] Instead of using isopolytungstates which are formed in acid conditions, it was noted that with some heteropolytungstates discoloration on exposure to light did not occur, e.g., phosphotungstates, complexes of tungstates with citric, oxalic, and tartaric acid.[109] The detailed description of these findings will be the subject of a separate paper. With the most effective citratotungstates the effectiveness of the zirconium mordanting treatment on chemically shrink-resist-treated wool is significantly improved and fastness to 50 washes at 40°C is achieved, while without the addition of tungstate complexes the flame-resist effect is lost after 10 or 20 washes, depending on the shrink-resist treatment (Table 14).

The zirconium–tungsten application also has improved fastness to alkaline aftertreatment (Table 15), which is useful when flame-resistant wool is required to have a neutral pH or when improving the natural flame resistance of wool dyed with phthalocyanine dyes. In acid conditions these dyes become yellower, and in order to restore the original shade, neutralization of the wool is necessary. If zirconium-treated wool is immediately neutralized with alkalis before drying, the flame-resist effect is lost; when

TABLE 15

Fastness to Alkaline Aftertreatment of Various Mordanting Applications

	Vertical flame test		
	---	---	---
Application	Before washing	After washing	
		10×	20×
8% K_2ZrF_6, 10% HCl 37%, 30 min, 75°C, $+ 12\%$ $NaHCO_3$, 15 min, 45°C	Fail		
8% K_2ZrF_6, 10% HCl 37%, 30 min, 75°C, dry, $+ 12\%$ $NaHCO_3$, 15 min, 45°C	Pass	Pass	Fail
8% K_2ZrF_6, 10% HCl 37%, 3% Na_2WO_4, 3% citric, 30 min, 75°C, $+ 12\%$ $NaHCO_3$, 15 min, 45°C	Pass	Pass	Pass
8% K_2ZrF_6, 10% HCl 37%, 5% Na_2WO_4, 4% citric, 30 min, 75°C, $+ 12\%$ $NaHCO_3$, 15 min, 45°C	Pass	Pass	Pass

wool is dried before the aftertreatment, the application is effective. The zirconium mordanting very probably requires a slow hydrolysis for the fixation of the metal in the fiber which is not achieved by a rapid alkaline aftertreatment. However, if the zirconium is fixed in the fiber by drying, the effect is fast to alkaline aftertreatment.

6.3.5. Fully Washable and Flame-Resistant Wool

Application techniques which are described in the literature to achieve this very important combination of properties are based on the incorporation of flame-resistant additives into a polymeric shrink-resistant coating[122,123] or application of a polymeric coating which imparts both flame and shrink-resistance.[123] This requires very high add-ons (15–20 %) which impair the handle, and with flame-resistant additives they are not fast to multiple (more than 30) washes. We have tested polyvinylidene chloride latexes containing organophosphorous softeners with an LOI value of 80; at least 15 % of this system on the weight of wool was required to achieve flame resistance of the fiber. Therefore, our approach was to apply a relatively nonflammable resin with a similar or higher LOI value than wool to achieve shrink resistance followed by the flame-resist treatment of the fiber. Hercosett 57, a polyamide epichlorhydrin-type polymer which is being successfully applied for the shrink-resistance of wool,[124,125] has an LOI value of 31.0. That is higher than wool (LOI 24–25), and it was chosen as a suitable resin. The fluorozirconate and citratozirconate treatments on Hercosett-treated wool are effective before washing but not after washing (Table 16). The addition of citratotungstate complexes to the fluorozirconate improves the washing fastness, but not sufficiently. However, the application of fluoro-citratozirconates and -tungstates gives a flame-resist effect which is fast to 50 washings at 40°C. As the wool fiber is chlorinated prior to the Hercosett application in order to increase the surface energy of the fiber and thus achieve efficient adhesion and spreading of small quantities of the polymer[124] on the fiber, the washing fastness of the mordanting application is decreased as already explained. With the hexafluorozirconate and tungsten application, leaching probably occurs before hydrolysis. With fluoro-citratozirconates with an F/Zr ratio of 3 the treatment is fast to 50 washings, while higher F/Zr ratios are not. In the latter case leaching probably occurs before hydrolysis, while in the former case an insoluble fluoride which is fast to washing is formed more rapidly in the fiber. This is also a strong indication of the formation of insoluble $ZrOF_2$ or other low F/Zr-ratio fluorides in the fiber.

TABLE 16
Effectiveness of Various Zirconium and Zirconium-Tungsten Applications
on a Hercosett Shrink-Resist-Treated Knitted Fabric

		Vertical flame test				
Treatment[a]	Before washing	After washing				
		$10 \times$	$20 \times$	$30 \times$	$50 \times$	$50 \times$
8% $ZrOCl_2$ $8H_2O$ + 5% HCl + 8% citric acid oww	Pass	Fail	Fail	Fail	Fail	Fail
8% K_2ZrF_6 + 10% HCl 37%	Pass	Fail	Fail	Fail	Fail	Fail
8% K_2ZrF_6 + 10% HCl 37% + 3% Na_2WO_4 + 3% citric acid	Pass	Pass	Pass	Fail	Fail	Fail
8% K_2ZrF_6 + 10% HCl 37% + 5% Na_2WO_4 + 4% citric acid	Pass	Pass	Pass	Fail	Fail	Fail
8% $ZrOCl_2$ $8H_2O$ + 2% $NH_4F \cdot HF$ + 2% citric acid + 7% HCl + 3% Na_2WO_4	Pass	Pass	Pass	Pass	Fail	Fail
8% $ZrOCl_2$ $8H_2O$ + 2% $NH_4F \cdot HF$ + 2% citric acid + 7% HCl + 5% Na_2WO_4	Pass	Pass	Pass	Pass	Pass	Pass
8% $ZrOCl_2$ $8H_2O$ + 3% $NH_4F \cdot HF$ + 2% citric acid + 7% HCl + 5% Na_2WO_4	Pass	Pass	Pass	Fail	Fail	Fail

[a] All treatments at 75 C for 30 min except the first—30 min at the boil

6.3.6. Multipurpose Finishes

After the titanium and zirconium mordant applications were established in commercial production, further development work was carried out in order to establish the compatibility with other easy-care treatments for wool.

6.3.6.1. Bleaching. If bleaching is required, zirconium mordanting should be applied after the bleaching stage by the low-temperature technique and no discoloration occurs (Table 17). In this case wool can be bleached by reductive bleach—bisulfite or hydrosulfite—or oxidizing bleach—hydrogen peroxide in acid or alkaline conditions.

6.3.6.2. Mothproofing. Commercial mothproofing agents can be applied simultaneously with titanium or zirconium mordanting. The mothproofing agent should be added to the bath before the addition of acid and the metal complex to ensure maximum stability of bath.

6.3.6.3. Heat Resistance. Although most flameproofing treatments lower the decomposition temperature of fiber,[126] it has been found that wool mordanted with titanium or zirconium compounds has enhanced resistance to scorching by radiant heat. A furnaceman's cloth was treated with titanium complexes and tested by the radiant panel test (ASTM E-162-67) without using the pilot flame. With this severe test the sample is exposed for 15 min to a radiant panel with a temperature of 800°C. The untreated samples smoldered in full length (45 cm), while the mordanted ones smoldered for 15 cm. The metal complex applications can be considered not only as flame resistant but also as heat resistant, which is of considerable interest and importance for protective clothing, particularly in heavy industry and for conventional wool carpets in ships, public buildings, and public transport. In many cases a similar radiant panel test is applicable, e.g., Italian Merchant Navy Specification and British Standard 476, Part 7. In order to pass these severe tests it is necessary to use an all jute backing and flame-resistant latex, besides flame-resist treating the wool pile. Polypropylene backing is not suitable because of the low melting temperature and high flammability of this fiber. This also applies to all flammability tests where the flame is in contact with the carpet backing (e.g. vertical-flame test).

6.3.6.4. Shrink-Resist Treatments. Classical chemical processes (chlorination, permonosulfuric acid, Ficlor, Basolan, Orced) are compatible with the metal complex applications. The shrink-resist treatment should be applied before mordanting in order to obtain maximum flame-resist effect and by the low-temperature technique in order to avoid unnecessary damage to wool fibers as described in more detail above.

6.3.6.5. Resinous Shrinkproof Treatments. Various resins are available for the shrinkproofing of wool. In most cases the resin is relatively flammable and it may be difficult to flame-resist resin-treated wool because the resin is mainly deposited on the surface of the fiber and burning at the initial stage is a surface phenomenon. In other words, it is important to select a suitable resin which is compatible with the metal-complex applications.

This multipurpose finish can be achieved by the application of the zirconium–tungsten treatments to Hercosett shrinkproofed wool. Although Hercosett-treated wool burns a little more readily than untreated wool, the advantageous chemical composition of the Hercosett resin from the point of view of flame resistance (nitrogen and chlorine are present), permits a successful flame-resist treatment of Hercosett-treated wool. The flame-resist treatment should be carried out by the low-temperature technique after the dyeing stage and the Hercosett treatment as described previously.

6.3.6.6. Water-Repellent Treatments. Water-repellent agents based on organic derivatives of chromium or wax thermosetting agents are compatible with the metal-complex applications. The water-repellent agent should be applied after the low-temperature flame-resist treatment.

Application of Phobotex FTC (Ciba-Geigy) by the pad–dry–cure technique after a shrink- and flame-resist treatment significantly improves the water-repellent effect without affecting flame resistance (Table 17).

6.3.6.7. Water- and Oil-Repellent Treatments. Fluorocarbons which are used for oil- and water-repellent treatments (e.g., FC.218, 3M Company) are compatible with the metal-complex applications (Table 17). As the

TABLE 17
Multipurpose Finishes

Treatment	Yellowness index	Spray rating	Oil rating	Vertical flame test
Untreated	25.0	50	0	Fail
Bleached[a] 0.5 °₀ H₂O₂ w v	17.2			Fail
Bleached[a] 0.75 °₀ H₂O² w v	16.2			Fail
Bleached[a] 1.0 °₀ H₂O₂ w v	15.2			Fail
Bleached[a] 0.5 °₀ H₂O₂ w v and F.R.T.[b]	17.3			Pass
Bleached[a] 0.75 °₀ H₂O₂ w v and F.R.T.	16.3			Pass
Bleached[a] 1.0 °₀ H₂O₂ w v and F.R.T.	15.4			Pass
Bleached[a] 3g l NaHSO₃. 50 C. 1 hr.	22.1			Fail
Bleached[a] 3g l NaHSO₃. 50 C. 1 hr and F.R.T.	22.2			Pass
Shrink-Resist Treated (3 °₀ DCCA)		0	0	Fail
S.R.T.[c] and 3 °₀ Phobotex FTC (Ciba-Geigy)		100		Fail
S.R.T.. F.R.T.. and 3 °₀ Phobotex FTC (Ciba-Geigy)		100		Pass
S.R.T. and 2 °₀ FC-218 (3M Company)		100	7	Fail
S.R.T.. F.R.T.. and 2 °₀ FC-218 (3M Company)		100	7	Pass
F.R.T. + 3 °₀ Nuva F (Hoechst)		100	7	Pass

[a] Bleaching was carried out with the above concentration of H₂O₂ at pH 8 (phosphate buffer for 1 hr at 55 C)

[b] F R T. Flame-resist treated with 8 °₀ K₂ZrF₆ + 10 °₀ HCl 37 °₀ at 75 C for 30 min

[c] S R T.—Shrink-resist treated with 3 °₀ DCCA

fluorocarbons are not sufficiently effective water-repellent agents, a water-repellent extender in a low concentration has to be added to the fluorocarbons.

The treatment can be carried out in a simultaneous one-step treatment by the pad–dry–cure–rinse–dry technique (curing is necessary for fluorocarbons and the water-repellent extender), successful results being obtained for all the three properties required. It is also possible to apply the treatments separately by the exhaustion or pad–dry techniques. but the fluorocarbons can be applied by the pad–dry–cure technique only. This multipurpose finish is of use on upholstery fabrics for aircraft, hotels, restaurants. bars, and discotheques.

With the newly introduced anionic fluorocarbon Nuva F (Hoechst). which is substantive to the wool fiber, it is possible to achieve water and oil repellency and flame resistance in a simultaneous single-bath application (Table 17).

6.3.6.8. Wool-Rich Blends. The metal complexes are effective with keratinous fibers only. Therefore, the result of the flame-resist treatment will depend on fabric construction. weight, and the flammability properties of the other fibers as well as on the flammability specification.

Generally, the mordanting techniques with a medium level of effectiveness can be applied successfully to blends containing at least 80–90 % wool, while for blends containing only 60–80 % wool the higher level of effectiveness is necessary. Blends containing less than 60 % wool cannot be successfully flame-resist treated with the metal-complex techniques.

6.3.7. Conclusions

There are four basic conditions which must be fulfilled in order to obtain a high exhaustion of an effective metal complex on the wool fiber and an effective flame-resist effect.

1. The metal complex has to be present in solution as the negatively charged (anionic) species. This is achieved by applying an excess of the complexing agent.

2. The wool has to be positively charged in order to ensure exhaustion of the anionic metal complex on the wool fiber. To achieve this condition the pH during treatment has to be below 3.

3. A sufficiently high temperature of the treating bath is required to achieve exhaustion of the metal complex on the fiber. The required temperature varies from 50°C to boil depending on the complex applied.

4. The metal complex in the wool fiber has to be prone to hydrolysis in order to form an insoluble acid residue which is volatile at temperatures below the ignition temperature of wool.

The acid residue is very probably responsible for the increased char formation or acid fragments on mild heating which trap the flammable volatiles in the vapor phase, thus extinguishing the flame.[141]

A better and more even penetration of the metal complex in the fiber results in an improved flame-resist effect. Surface deposition of flame-resistant resins requires over-treatment of the fiber which can impair the handle of textiles. Thus, flame resistance can be compared with dyeing.

The metal complex treatments which are in full commercial production are available in three different levels of flame resistance depending on the amount of metal applied, the chemical composition of the metal and complexing agent, the textile substrate, and the flammability specification which is to be met:

1. Low level of flame-resistance, e.g., for shag-pile carpets to meet the American tablet test.

2. Medium level of flame-resistance, e.g., for wool fabrics to pass the vertical strip test Method 5902.

3. High level of flame-resistance, e.g., for resinous or chemically shrink-resist-treated wool and wool-rich blends to pass the vertical flame test (Method 5902) and for wool furnishings which have to meet

the severe flammability tests for building materials (e.g., BS-476, Part 7).

The various application techniques for the IWS flame-resist treatments are summarized in Table 18. Besides the simplest and cheapest exhaustion technique, the pad–batch–rinse–dry technique may be useful for treating fabrics which are sensitive to the exhaustion treatment at 60°C because of critical construction or because of poor dye fastness. The pad–steam and dip–nip–batch techniques are only modifications of the pad–batch technique.

Because 0.5–2.0% of the metal in the wool fiber is effective, the chemical cost of the mordant applications is low and they can be carried out concurrently with any one of a number of routine processes applied to wool. The effective chemicals are exhausted into the fiber, as a result the handle and the natural soil resistance of wool fibers are in no way impaired. Also the physical and chemical properties of wool are not adversely affected. A very important feature of the new treatments is compatibility with easy-care treatments for wool, including full machine-washability. The mordanting applications exploit the natural flame-resistance of wool and the unique chemical composition and structure of wool without causing any undesirable side effects which are common to other commercial flame retardants.

6.4. Chemical Modification of Wool

Wool has a large number of chemically very different reactive groups and certain compounds can be incorporated into the wool structure. This

TABLE 18

Application Techniques for the Metal-Complex Wool Mordantings

---

1. *Exhaustion Technique*—low temperature or boil
   Loose stock, tops, yarn packages, yarn in hank, fabrics, knitted garments, sheepskins
   Machinery—common dyeing machinery

2. *Pad–Batch–Rinse–Dry Technique*
   Fabrics
   Machinery–pad mangle, tenter, winch

3. *Pad–Steam–Rinse–Dry Technique*
   Tops, fabrics, carpets
   Machinery—for continuous pad–steam dyeing

4. *Dip–Nip–Batch or Dry–Rinse–Dry Technique*
   Sheepskins, loose stock

---

is one of the most promising ways of achieving a permanent flame-resist treatment of wool fibers without significantly impairing the handle and without applying unnecessarily high concentrations of flame retardants and resins for the fixation of the active compounds which otherwise can be expensive and impair the very good textile properties of wool.

The titanium and zirconium flame-resist treatments based on reaction with amino groups are a good example of this approach. The attempts to phosphorylate wool via reaction with the hydroxyl groups of serine and threonine is another example of exploiting the reactivity of the wool fiber, although this attempt has not been successful yet.[3,68]

The oxidation of at least 60% of the original cystine in wool by peracetic acid to cysteic acid, a relatively strong and not too volatile acid which is probably capable of trapping the free radicals formed during burning and results in a flame-resist treatment, is another good example of a flame-resist treatment for wool via chemical modification even though in this case the mechanical properties of wool are significantly decreased because of rupture of cross-links.[3,68]

It should not be forgotten that the disulfide groups in wool can be easily reduced to produce very reactive thiol groups which could be cross-linked with a difunctional reagent containing effective flame-resistant atoms.

These are some of the examples of the possible flame-resist treatments of wool that exploit the reactivity of the wool molecule. This avoids the use of relatively expensive surface coatings which require a high add-on. The necessarily relatively low concentration of the effective flame-retardant atoms has to be fixed by a high concentration of a reactive resin which in many cases results in a change of handle and decrease in mechanical properties.

6.5. Wool/Flame-Resistant Man-Made Fiber Blends

There are several flame-resistant man-made fibers commercially available, but the majority of them are rather expensive and the textile properties are more or less inferior when compared with common textile fibers. A few examples are: Nomex—aromatic polyamide; PBI—poly-benzimidazole; Kynol—cross-linked phenolic; Durette—polydiimide; Kermel—polyamidimide; etc. There are also inherently flame-resistant fibers which may be classified with the bulk-produced synthetic fibers of today in terms of price and textile properties. The majority of these fibers contain combined chlorine, by virtue of which they derive the major part of their flame resistance. The typical types are as follows:

Type	Flame retardant	Trade names
Chlorofibers		
Polyvinyl chloride (PVC)	57 % chlorine	Rhovyl. Movil. Thermovyl. Clevyl. Leavil. etc.
Polyvinylidene chloride (PVDC)	70 % chlorine	Sarans
Modacrylics		
40 PAN 60 PVC	34 % chlorine	Kanekalon. Dynel
50 PAN 50 PVC	36 % chlorine	Teklan. Verel
Flame-resistant rayon	Organophosphorus compound	Darelle. PFR Rayon

Chlorofibers in the unstabilized form (Movil, Rhovyl, Thermovyl) shrink appreciably at temperatures above 70°C, and in the stabilized form (Clevyl T. Leavil), at temperatures above 100°C. Modacrylic fibers — copolymers of acrylonitrile and polyvinyl chloride or polyvinylidene chloride — are manufactured on a comparatively large scale because of their flame resistance and good textile properties.

The flame resistance of knitted fabrics made from wool–modacrylic (Teklan, Courtaulds) and wool-stabilized polyvinyl chloride (Leavil, Montecatini Edison) blends from the point of view of the U.S. Children's Sleepwear Specification DOC FF 3-71 is summarized in Table 19 and Fig. 9.[127] To pass this severe specification, which requires full flame resistance after 50 washes at 60°C, samples are tested in bone-dry conditions by a slightly modified vertical-flame test method with a 3-sec ignition time. Acceptance criteria are: an average char length not more than 7 in., no individual specimen with a char length of 10 in., and no individual specimen with a burning time greater than 10 sec.

Knitted fabrics have to have an LOI value of at least 26.5 in order to pass the children's sleepwear specification (Fig. 9). As modacrylic fibers have an LOI value of 26.7[5] and wool 24.0, wool-rich blends with these fibers cannot achieve this limit (Fig. 3). Polyvinyl chloride fibers have an LOI value of 37.1[5] and wool-rich blends with Leavil easily meet the required specification before, during, or after 50 washings when at least 35 % of the latter fiber is present in the blend. Full washability of this blend at 60°C can be achieved by suitable chemical processes (chlorination) or resinous shrinkproof treatments (Hercosett) of the wool portion.

The recommended 60:40 wool/Leavil blend has significantly improved abrasion resistance, tensile strength, and whiteness when compared with all-wool fabrics (Table 20), and aesthetics are generally satisfactory. This blend exploits the natural flame resistance of wool and the inherent flame resistance of polyvinyl chloride fibers to meet the U.S. children's sleepwear specification.

TABLE 19
LOI Values for Wool/Leavil (polyvinyl
chloride) Blends with and without a Shrink-
Resist Treatment and Wool/Teklan
(modacrylic) Blends

Blend	Shrink-resist treatment	LOI
100% wool		24.0
90/10 wool/Leavil		24.4
80/20 wool/Leavil		24.9
70/30 wool/Leavil		26.0
65/35 wool/Leavil		26.6
60/40 wool/Leavil		27.2
60/40 wool/Leavil	Hercosett	27.0
50/50 wool/Leavil		28.0
50/50 wool/Leavil	Hercosett	27.7
100% PVC		37.1
100% wool		24.0
76/24 wool/Teklan		24.4
70/30 wool/Teklan		24.8
60/40 wool/Teklan		25.5
50/50 wool/Teklan		25.8
100% Modacrylic		26.7

Similar results are obtained with wool/Clevyl T blends, the latter fiber (Rhone-Poulenc Textiles) being slightly less thermally stable than Leavil. Leavil contains 57% chlorine and in the 65:35 wool/Leavil blend approximately 20% chlorine is present, which is the necessary amount of chlorine on the weight of wool to make wool sufficiently flame resistant to pass DOC FF 3-71. This result indicates that to flame resist wool with a highly chlorinated resin, at least 30% of this resin on the weight of wool

TABLE 20
Mechanical Properties of Wool/Leavil Blends

Blend	Yarn		Knitted fabric	
	Tensile strength, g	Elongation, %	Mertindale abrasion resistance, rubs	ICI pilling, test, pills/100 mm^2
90% Wool–10% Leavil	581.3	23.2	13,000	4.5
80% Wool–20% Leavil	636.0	25.1		
70% Wool–30% Leavil	684.8	27.1		
65% Wool–35% Leavil	714.0	26.9		
60% Wool–40% Leavil	744.8	27.3	31,000	5.0
50% Wool–50% Leavil	762.8	27.4		

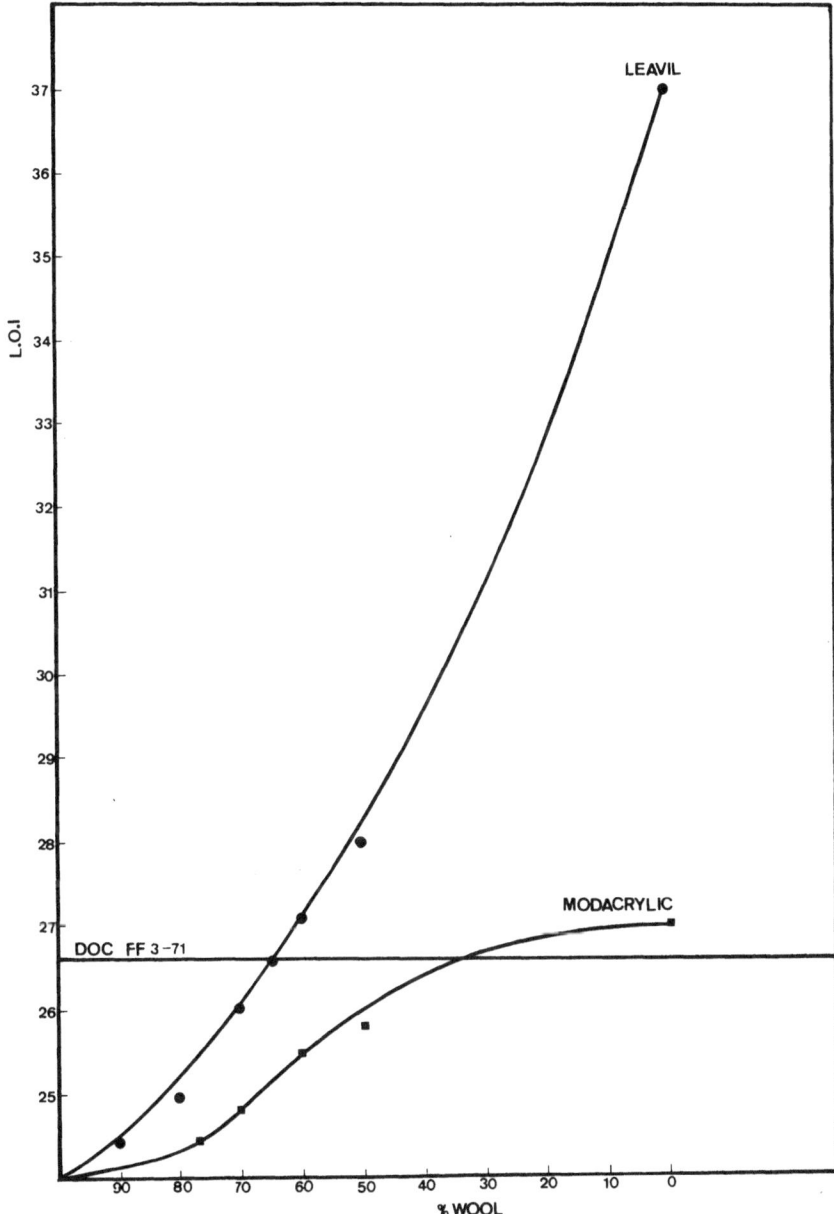

FIGURE 9. Influence of wool content on the LOI value of blends with Leavil (polyvinyl chloride fiber) and Teklan (modacrylic fiber).

or 15 % of a highly brominated resin is necessary (supposing that bromine is twice as effective as chlorine on a weight basis). Such high add-ons could severely impair the textile properties of lightweight wool fabrics for children's sleepwear. These high add-ons are confirmed by other authors when using flame-resistant resins or additives.[122,123]

Therefore, if one considers the several possible approaches to obtain fully washable and flame-resistant wool textiles,[123] the approach based on a flame-resist treatment of the wool fiber and a non-flame-supporting resinous shrinkproof treatment seems the most attractive, since at present, the application of flame-resistant resins, additives, or latexes to achieve flame resistance of wool results in poor esthetic properties of wool textiles.

Another possibility is to use wool blends with flame-resistant manmade fibers. This approach exploits the natural flame resistance of wool fibers, and for wool-rich blends it is necessary to use a flame-resistant man-made fiber with an LOI of at least 35. With these blends it appears that the active flame retardant which is based on phosphorus, chlorine and/or bromine compounds decomposes in the man-made fiber during mild heating to produce acid fragments at the right temperature to trap the free radicals produced by the thermal decomposition of wool.

7. References

1. M. J. Koroskys, *Flammability of Textiles,* U.S. Wool Bureau, New York (1968).
2. P. Alexander, R. F. Hudson, *Wool, its Chemistry and Physics,* Chapman and Hall, London (1963).
3. W. S. Simpson, *J. Appl. Pol. Sci.,* Applied Polymer Symposia No. 18, Proc., 4th International Wool Textile Research Conference, Part II, p. 1177 (1971).
4. G. C. Tesoro and C. H. Meiser, *Text. Res. J.* **40**, 430 (1970).
5. G. C. Tesoro and J. Rislin, *Text. Chem. Color.* 3(7), 156 (1971).
6. P. Perrot, *Text. Chim.* **27**(7/8), 20 (1971).
7. B. Miller and C. H. Meiser, *Text. Chem. Color.* **2**, 205 (1970).
8. M. Biber, *Chemiefasern* **5**, 375 (1969).
9. A. Decorte, *Ind. Text. Belge* 12(12), 33 (1970).
10. G. H. Crawshaw, P. A Duffield, and P. N. Mehta, *J. Appl. Pol. Sci.,* Applied Polymer Symposia No. 18, Proc., 4th International Wool Textile Conference, Part II, p. 1183 (1971).
11. E. R. Kaswell, *Text. Chem. Color.* 4(1), 33 (1972).
12. M. J. Koroskys, *Am. Dyest. Rep.* **58**(6), 15 (1969).
13. G. C. Tesoro, *Text. Chem. Color.* **1**, 307 (1969).
14. *Text. Chem. Color.* 2(9), 2 (1970).
15. J. J. Isaacs, *J. Fire Flam.* **1**, 36 (1970).
16. L. Benisek, *J. Soc. Dyers Color.* **87**, 277 (1971).
17. L. Benisek, *Text. Manuf.* **99**(1164), 36 (1972).
18. L. Benisek, *Melliand Textilber.* **53**, 931 (1972).
19. L. Benisek, Brit. Pat. 1,372,694 (6.11.1974).

20. L. Benisek, Brit. Pat. Appl. 3516/71.
21. L. Benisek, Brit. Pat. 1,379,752 (8.1.1975).
22. L. Straehl, Flammability in Textiles and the Position of Wool in Contrast to Other Fibers. Occasional Publications No. 45, p. 13, Rüschlikon, Zürich, (1969).
23. C. A. Hafer and C. H. Yuill, Characteristics of Bedding and Upholstery Fires, N.B.S. Contract No. CST-792-5-69 (March 31, 1970).
24. *Herald Tribune*, New Zealand, May 23, 1968, p 4.
25. C. Z. Carrol-Porczynski, 2d Shirley International Seminar, Manchester, England (1970).
26. A. Johnson, *J. Text. Inst.* **39**, 561 (1948).
27. E. H. Coleman, *Brit. Welding*, p. 406 (September 1959).
28. G. A. Cook *et al.*, *Text. Res. J.* **37**, 591 (1967).
29. G. Georgavics and C. Salter, *Chemical Technology of Textile Fibers*, Greenwood and Co., London (1902).
30. R. C. Anderson, *J. Chem. Ed.* **44**(5), 248 (1967).
31. J. W. Lyons, *The Chemistry and Uses of Fire Retardants*, Wiley, New York (1970).
32. J. E. Hendrix, T. K. Anderson, T. J. Clayton, E. S. Olson, and R. H. Barker, *J. Fire Flam.* **1**, 107 (1970).
33. C. Z. Carrol-Ponczynski, *Text. Inst. Ind.*, p. 188 (July, 1971).
34. J. M. Preston, *J. Text. Inst.* **40**, T767 (1949).
35. M. Levau, *Bull. Inst. Text. France* **80**, 57 (1959).
36. E. Menefee and G. Yee, *Text. Res. J.* **35**, 801 (1965).
37. R. F. Schwenker and J. H. Dusenbury, *Text. Res. J.* **30**, 800 (1960).
38. W. D. Felix, M. A. McDowall, and H. Eyring, *Text. Res. J.* **33**, 465 (1963).
39. C. J. H. Pinte, Y. Teyssier, and P. Rochas, *Bull. Inst. Text. France* **51**, 7 (1955).
40. M. Zerdik and D. Raffaelli, *Tekstil* **13**, 14 (1964).
41. E. E. Jiullerat, *N.Y. Acad. Med. Bull.* **43**, 646 (1967).
42. International Fire Statistics, 1968. *Fire J.* **63**(6), 59 (1969).
43. D. J. Rasbash, *Trans. J. Plastics Inst. 1967* **55**.
44. International Wool Secretariat, Flammability of Textiles, Technical Information Bulletin No. 22 (February 1970).
45. D. Gross, J. J. Loftus, T. G. Lee, and V. E. Gray, Smoke and Gases Produced by Burning Aircraft Interior Materials, U.S. Dept. of Commerce, Building Science Series 18, (February 1969).
46. H. R. Richards, *Bull. Inst. Text. France* **25**(156), 675 (1971).
47. W. Kruse and K. Filipp, *Melliand Textilber.* **49**, 203 (1968).
48. W. Kruse, *Melliand Textilber.* **50**, 460 (1969).
49. Surface Burning Characteristics of Building Materials, ASTM E-84-68 (1968).
50. C. H. Yuill, *Fire J.* **61**(1), 11 (1967).
51. J. Autian, *J. Fire Flam.* **1**, 239 (1970).
52. N. Wallbank, unpublished results.
53. Hygienic Guide Series, American Industrial Hygiene Association (1961).
54. I. M. Gottlieb, *Text. Res. J.* **26**, 156 (1956).
55. G. A. Greathouse and C. J. Wessel, *Deterioration of Materials*, p. 492, Reinhold, New York, (1954).
56. *Text. Chem. Color.* **1**(25), 540 (1969).
57. R. W. Little, *Flameproofing Textile Fabrics*, Reinhold, New York (1947).
58. H. D. Tyner, *Ind. Eng. Chem.* **33**, 60 (1957).
59. R. Friedman and J. B. Leby, WADC Tech. Rept. 56-568, ASTIA Doc. No. AD 110685, Wright Patterson Air Base, Ohio. See also ref. 31.
60. C. P. Fenimore and F. J. Martin, *Combust. Flame* **10**, 135 (1966).
61. J. J. Pitts, *J. Fire Flam.* **3**, 51 (1972).
62. U.S. Flammable Fabrics Act, 1967 Amendments PL 90-189.

63. *Text. Chem. Color.* **3**(9), 39 (1971).
64. Federal Aviation Administration, Notice of Proposed Rule Making, Federal Register **34**(153), 13036 (1969).
65. J. T. Marsh, *An Introduction to Textile Chemistry*, Chapman and Hall, London (1948).
66. Handbook of Chemistry and Physics, 53rd edition, pp. B-74, B-136, Chemical Rubber Co., Cleveland, 1972-1973.
67. Standard Methods of Fire Tests for Flame-Resistant Textiles and Films, National Fire Protection Association No. 701-1969.
68. W. S. Simpson, The Flameproofing of Wool, Commun. No. 8, Wool Res. Org. of N. Z. (1970).
69. J. W. Bell, C. N. Hutchinson, and C. S. Whewell, *J. Text. Inst.* **57**, T43 (1966).
70. T. Uniski, *Kogyo Kagaku Zasshi* **69**, 2343 (1966).
71. S. Nomura, Jap. Pat. 5550 (July 9, 1956).
72. G. Tesoro, *et al., Text. Res. J.* **38**(3), 245 (1968).
73. G. Tesoro *et al., Text. Res. J.* **39**(2), 180 (1969).
74. R. E. Ferrel *et al., J. Am. Chem. Soc.* **70**, 2102 (1948).
75. K. A. Petrov, *Vysokomolekul. Soedin.* **1963**, 90; *Chem. Abstr.* **60**, 10913c (1964).
76. Z. A. Rogovin, Brit. Pat. 1,022,083 (1966).
77. M. Friedman and S. Tillin, *Text. Res. J.* **40**, 1045 (1970).
78. Brit. Pat. 634,690 (1950) (to Cortaulds Ltd.).
79. U.S. Pat. 2,530,261 (1950) (to Cortaulds Ltd.).
80. S. J. O'Brien and R. G. Wegher, *Text. Chem. Color.* **3**(8), 185 (1971).
81. G. L. Drake and J. D. Guthrie, *Text. Res. J.* **29**, 155 (1959).
82. W. A. Reeves *et al., Text. Res. J.* **27**, 260 (1957).
83. T. D. Miles *et al., Am. Dyest. Rep.* **49**, 596 (1960).
84. R. B. Le Blanc, *Text. Res. J.* **25**, 341 (1956).
85. R. Aeniskanshin *et al., Text. Res. J.* **39**, 375 (1969).
86. H. Coates and B. Chalkley, Brit. Pat. 938,989 (1960).
87. Brit. Pat. 740,269 (1953) (to Albright and Wilson Ltd.).
88. C. A. Duprez, *Am. Dyest. Rep.* **60**(10), 54 (1971).
89. A. D. Jenkins and L. J. Wolfram, *J. Soc. Dyers Color.* **79**, 55 (1963).
90. M. J. Williams, *Text. Chem. Color.* **2**(7), 41 (1970).
91. *Text. Manuf.* **119**(11), 62 (1969).
92. Brit. Pat. 1,069,946 (1967) (to Asaki Chemical Industry Co. Ltd.).
93. Ger. Pat. 1,148,968 (1963) (to B.A.S.F.).
94. M. J. Koroskys, *Am. Dyest. Rep.* **60**(5), 48 (1971).
95. J. F. Gaunt, *J. Soc. Dyers Color.* **70**, 46 (1954).
96. F. R. Hartley, *Austral. J. Chem.* **23**, 275 (1970).
97. F. R. Hartley, *J. Soc. Dyers Color.* **86**, 209 (1970).
98. A. E. Jacobsen *et al., Am. Dyest. Rep.* **40**, P439 (1951).
99. M. I. Panik *et al., Am. Dyest. Rep.* **39**, 509 (1950).
100. S. H. Laurie, *Text. Res. J.* **38**, 1140 (1968).
101. Federal Specification CCC-T-191b, Method 5902 (1951).
102. S. H. Laurie, *Text. Res. J.* **36**, 476 (1966).
103. G. Stamm, *Textilveredlung* **6**(2), 103 (1971).
104. L. Benisek, *Int. Dyer* **147**, 414 (1972).
105. L. Benisek, *Text. J. Austral.* **47**, 32 (1972).
106. L. Benisek, *Textilveredlung* **8**, 318 (1973).
107. L. Benisek, Brit. Pat. Appl. 8412/71, (April 1, 1971).
108. L. Benisek, Brit. Pat. 1,385,399 (26.2.1975).
109. L. Benisek, Brit. Pat. Appl. 23119/73, (May 15, 1973).
110. P. G. Gordon and L. J. Stephens, *J. Soc. Dyers Color.* **90**, 239 (1974).

111. F. D. Snell and C. T. Snell, *Colorimetric Methods of Analysis,* Vol. II, p. 438, Van Nostrand, Princeton (1967).

112. W. B. Blumenthal, *The Chemical Behavior of Zirconium,* p. 331, Van Nostrand, New York (1958).

113. R. Field and P. L. Cowe, *The Organic Chemistry of Titanium,* p. 81, Butterworths, (1965).

114. C. L. Wilson and D. W. Wilson, *Comprehensive Analytical Chemistry,* Vol. 1B, p. 162, Elsevier, Amsterdam (1960).

115. L. W. Mazzeno, H. M. Robinson, E. R. McCall, N. M. Morris, and B. J. Trash, *Text. Chem. Color.* **5**(3), 43 (1973).

116. R. M. Schmitt, E. L. Grove, and R. D. Brown, *J. Am. Chem. Soc.* **82**, 5292 (1960).

117. P. A. W. Dean and D. F. Evans, *J. Chem. Soc.* **A1967**, 698.

118. J. W. Mellor, *Inorganic and Theoretical Chemistry,* Vol. VII, p. 138, Longmans, London (1963).

119. Ref. 46, p. 67.

120. Ref. 40, p. 137.

121. G. D. Rieck, *Tungsten and its Compounds,* p. 107, Pergamon Press, New York (1967).

122. S. Tillin, C. E. Pardo, W. Fong, and M. Friedman. *Text. Res. J.* **42**, 135 (1972).

123. K. W. Fincher, G. B. Guise, and M. A. White, *Text. Res. J.* **43**, 623 (1973).

124. H. D. Feldtman and J. R. McPhee, *Text. Res. J.* **34**, 634 (1964).

125. R. H. Earle and R. H. Saunders, *Appl. Pol. Symposia* No. 18, 707 (1971).

126. J. E. Hendrix, T. K. Anderson, T. J. Clayton, E. S. Olson, and R. H. Barker, *J. Fire Flam.* **1**, 107 (1970).

127. L. Benisek, *Text. Chem. Color.* **6**(2), 23 (1974).

4

Flame-Retardant Polyethylene Terephthalate Fibers

E. L. Lawton and C. J. Setzer

1. Introduction

Polyethylene terephthalate (PET) fiber production reached 3 billion pounds in 1973 in the United States. Because of its outstanding physical properties, ease of care, and low cost, polyester fiber has captured 40% of the total U.S. wearing-apparel fiber market; it accounts for 68% of the synthetic fibers used in apparel. Most of this fiber is used in outerwear as 100% polyester fabric or blended with cotton where flammability of the garment must be considered as an important factor in the overall fabric performance. Owing to increased public awareness and concern and the adoption of more severe fabric flammability standards, the need for the development of fire-retarding polyester fiber or fabric has become critical.

Polyester is manufactured by direct melt polycondensation of ethylene glycol and terephthalic acid:

$$n\text{HOCH}_2\text{CH}_2\text{OH} + n\text{HOOC}\!-\!\langle\bigcirc\rangle\!-\!\text{COOH} \longrightarrow$$

$$\left[\text{CH}_2\text{CH}_2\text{O}\!-\!\overset{\overset{\text{O}}{\|}}{\text{C}}\!-\!\langle\bigcirc\rangle\!-\!\overset{\overset{\text{O}}{\|}}{\text{C}}\!-\!\text{O}\right]_n + 2n\text{H}_2\text{O}\uparrow$$

E. L. Lawton and C. J. Setzer · Monsanto Textile Co., Monsanto Triangle Park Development Center, Research Triangle Park, North Carolina, 27709.

The polymer is also manufactured by melt polycondensation utilizing ethylene glycol and dimethyl terephthalate as raw materials:

$$2HOCH_2CH_2OH + CH_3OOC\!\!-\!\!\langle\bigcirc\rangle\!\!-\!\!COOCH_3 \longrightarrow$$

$$HOCH_2CH_2O\!\!-\!\!\overset{\overset{\displaystyle O}{\|}}{C}\!\!-\!\!\langle\bigcirc\rangle\!\!-\!\!\overset{\overset{\displaystyle O}{\|}}{C}\!\!-\!\!OCH_2CH_2OH + 2CH_3OH \uparrow$$

$$nHOCH_2CH_2O\overset{\overset{\displaystyle O}{\|}}{C}\!\!-\!\!\langle\bigcirc\rangle\!\!-\!\!\overset{\overset{\displaystyle O}{\|}}{C}OCH_2CH_2OH \longrightarrow$$

$$\left[\!-\!CH_2CH_2O\!\!-\!\!\overset{\overset{\displaystyle O}{\|}}{C}\!\!-\!\!\langle\bigcirc\rangle\!\!-\!\!\overset{\overset{\displaystyle O}{\|}}{C}\!\!-\!\!O\!-\!\right]_n + nHOCH_2CH_2OH \uparrow$$

The polycondensation is conducted at temperatures of 265–290°C under vacuum. After reaching sufficient molecular weight, the polymer is either directly spun from the melt into fibers or is extruded as a ribbon and cut into chips. These chips are then fed to a screw extruder for re-melting and spinning as fibers. The spun fibers are subsequently oriented by drawing. The drawing operation induces partial crystallization of the polymer and aligns the polymer molecules along the fiber axis. Drawing develops the essential mechanical properties of the fiber.

Drawn fibers may be crimped, heat-set, and then cut into staple. Polyester staple is often blended with cotton to form the blend yarns used in popular woven wash-and-wear garments. The advent of texturizing processes for polyester continuous filament has resulted in skyrocketing growth in production of knitted outerwear fabrics from 100% polyester.

The scope of the problem of producing flame-retardant fabrics has been reviewed in detail.[1] In the United States, legislation requiring flammability testing of children's sleepwear, mattresses, automobile textiles, aircraft textiles, and carpets is in existence. Legislation covering upholstered furniture, draperies, bedding, children's clothing, and geriatric clothing is anticipated. More stringent standards are also anticipated as commercially feasible solutions to the problem are developed. The fiber producer and fabric producer are interested in modifying their products to meet these legislated standards to protect existing end uses of their products and also to develop products of such salient flammability resistance and end-use properties that they will displace existing fibers in flame-retardant fabrics.

The modification of polyethylene terephthalate textiles to produce a more flame-resistant material may be accomplished at several stages of textile manufacturing. The inclusion of fire-retardant comonomers in the PET melt reaction system can produce ethylene terephthalate copolymers that are melt-spinnable into fibers. Nonreactive fire-retardant additives may be included with monomers in the reaction system or they may be admixed into the polymer prior to melt spinning. The fibers may be modified by surface treatments and chemical grafting at some stage during the drawing or heat-setting operations by the fiber producer. The fabrics produced by the weaver or knitter may be fire retarded during the finishing operation. Finishing covers all the processes which a fabric may undergo after being woven or knitted, including dyeing.

Many criteria must be considered in evaluating the flammability behavior of a fabric. Among these are ignitability, rate of flame spread, heat development, toxic gas production, smoke development, afterglow, and dripping melt. The fiber manufacturers and the fabric manufacturers are concerned with modification of PET to produce a product with improvements in all of these properties. However, the maintenance of the highly desirable textile properties of PET fibers is of paramount concern in this modification effort. Of course, the economics of any modification of the fiber or treatment of fabrics must be compatible with the current low pricing of PET fibers.

Several recurrent problems have emerged in flame retardation of PET fibers. If the polymer is to be retarded by addition of a chemical to the molten polymer prior to spinning, the additive must be able to withstand temperatures of 260–300°C for considerable time without degrading or reacting detrimentally with the polymer. The known additives satisfying this requirement are limited. Besides loss of molecular weight resulting from degradation, many additives are detrimental to the viscous properties of the melt or may not disperse well in the melt. Other problems are common to flame retardants for PET fibers whether introduced before or after spinning. The retardant must not be extracted from the fiber during treatments such as dyeing, home laundering, or dry cleaning. The retardants must not significantly alter the end-use properties of fabrics, such as wash-and-wear properties and dyed lightfastness. The additive must not yield a fiber with undesirable dermatological effects. The fiber should not produce harmful gases or smoke upon accidental burning or in solid-waste incineration. Upon disposal, the fiber must not release materials harmful to the environment.

2. Nonreactive Additives

2.1. Halogen

The flame-inhibiting action of halogenated retardants in polymers is widely documented. The action in polymers has been attributed in some instances to an effect of the additive on the course of decomposition in the solid state.[2-5] However, the action in polymers is more often attributed to thermal decomposition of the flame retardant to produce gaseous products which mix with flammable gases from the decomposing polymer and hinder combustion.[6-10]

The ability of hydrogen halides and other halogenated gases to inhibit flames has been well established.[11,12] For greatest effectiveness, the inhibiting gas should become available in the flame at the time and place where the combustion reactions are most susceptible to its inhibiting effect.[12] The blanketing effect of halogenated gases with their high density and high specific heat has also been cited as hindering combustion of decomposing polymers.[6,7]

Examples of typical halogenated compounds which have been cited in the patent literature for addition to the melt prior to spinning PET fibers are presented in Table 1. This listing is not inclusive but is intended to illustrate the variety of compounds that have been investigated and that these compounds possess certain common characteristics. Brominated compounds are preferred over chlorinated compounds owing to the higher efficiency of bromine as a flame retardant.[13] Bromine is generally incorporated into the additive through bromine atoms bound to phenyl groups rather than through aliphatic bromine to carbon bonds, reflecting the greater thermal stability of the aromatic bond. Bromine is sometimes bound to a neopentyl group where the possibility of thermally induced dehydrobromination is reduced by the lack of a vicinal hydrogen in the group. It is most desirable that the additives contain greater than 50 wt % bromine so that 5–10 wt % bromine will be incorporated in the fiber by the addition of 10 wt % or less of the compound. Antimony–halogen synergism in flame-retarded PET fibers is well recognized. Several patents citing such synergism are noted in Table 1.

Pitts[13] has recently proposed a mechanism for the synergistic effect of Sb_2O_3 on the flame retardation of halogen compounds in polymers. The mechanism was based upon formation in situ of SbOCl upon thermal activation. The SbOCl serves as a reservoir for $SbCl_3$, which is gradually released upon heating of the polymer and constitutes the active retardant in the flame zone. An endothermic reaction at 245°C for SbOCl is also credited with reducing the rate of polymer decomposition.

$$5SbOCl \xrightarrow{245\ 280\ C} Sb_4O_5Cl_3 + SbCl_3 \uparrow$$

$$4Sb_4O_5Cl_2 \xrightarrow{410\ 475\ C} 5Sb_3O_4Cl + SbCl_3 \uparrow$$

$$3Sb_3O_4Cl \xrightarrow{475\ 565\ C} 4Sb_2O_3 + SbCl_3 \uparrow$$

Charring from dehydration occurring in the solid phase is also possible in certain substrates containing halogen and antimony.[13] The thermal decomposition of SbOCl occurs in the temperature range which matches the thermal degradation of many polymers. Thus vaporization of the retardant and combustible gases from the polymer would coincide. Pitts[13] also cites certain examples where antimony (added as triphenyl antimony) in the absence of halogen appears to retard combustion in the gas phase.

In Fig. 1, the thermal gravimetric thermograms of PET fibers containing 6 wt % bromine with and without the addition of 1.5 wt % antimony oxide are compared.[29] The fiber containing antimony exhibited considerably more of its weight loss in the initial decomposition range of 350–400 C than the fiber without antimony. The fiber containing 1.5 wt % antimony oxide exhibited a limiting oxygen index (LOI) of 24.8, an increase of 2.3 units from the value for the fiber without antimony oxide addition. The decomposition range of the fiber with antimony is probably lowered by the thermally activated interaction of bromine and antimony. This lowering of the decomposition range resulting from interaction of antimony and the bromine source has also been observed in thermograms of other antimony–halogen systems.[30,31]

It has been observed that even inorganic compounds added to the

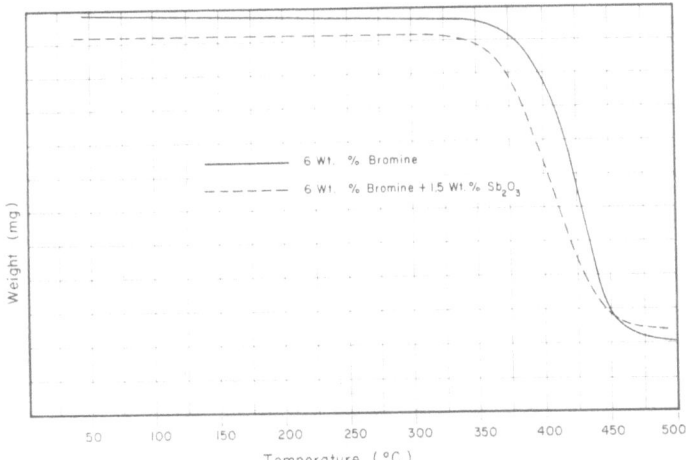

FIGURE 1. Comparison of thermograms in nitrogen atmosphere of fibers with and without antimony synergist (10 C/min).

TABLE 1

Halogenated Additives

Structural characterization	Example	Special features	Reference
Polycarboxylic acid esters of halogen-containing alcohols	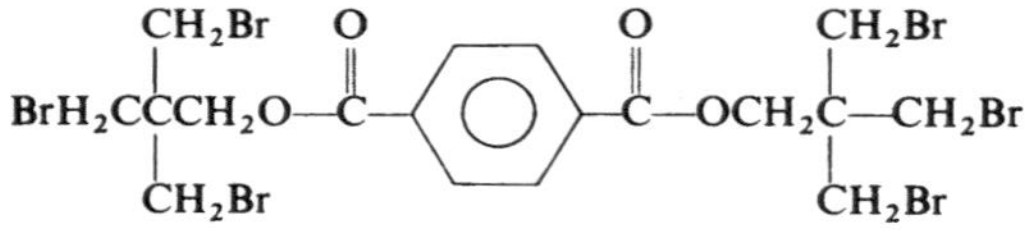 $BrH_2CCCH_2O{-}C(=O){-}[\text{ring}]{-}C(=O){-}OCH_2C{-}CH_2Br$ (with CH_2Br substituents)	Ring may also be halogenated	14
Addition polymer of halogenated neopentylacrylate	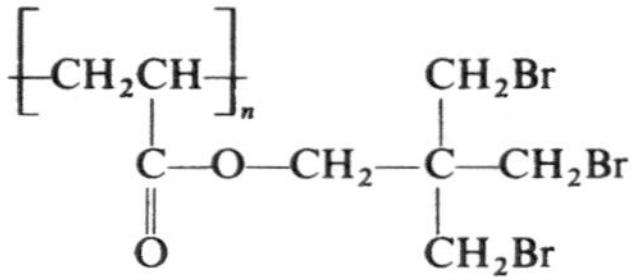 $-[CH_2CH]_n-$; $C(=O){-}O{-}CH_2{-}C({-}CH_2Br)_2{-}CH_2Br$ (with CH_2Br)		15
Polyether of oxetane substituted by $BrCH_2{-}$	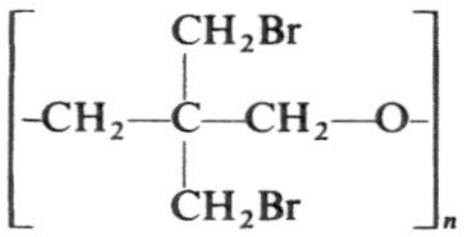 $[-CH_2{-}C(CH_2Br)(CH_2Br){-}CH_2{-}O{-}]_n$	Sb synergist may be added	16
Brominated biphenyl or biphenyl ether	biphenyl with Br_4, Br_4, Br_n, Br_n substituents; $[\text{ring}]Br_4{-}O{-}R{-}O{-}[\text{ring}]Br_n$; $[\text{ring}]Br_3{-}N(H){-}[\text{ring}]Br_3$		17
			18
			19
	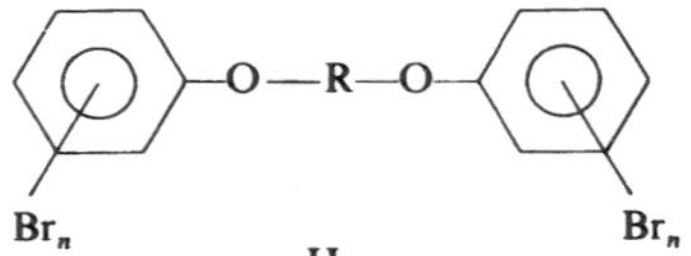		20
	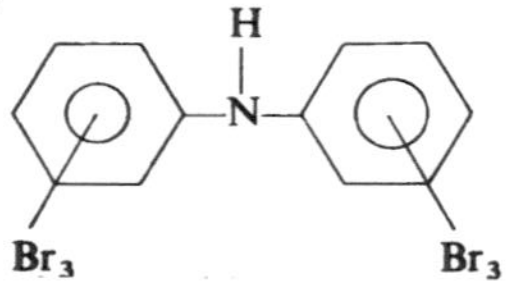	Sb synergist may be added	21

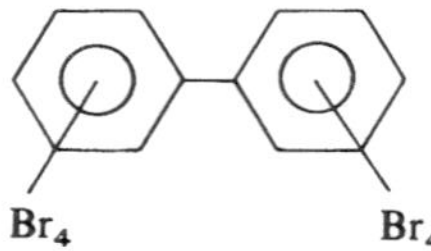

Brominated phenyl structures

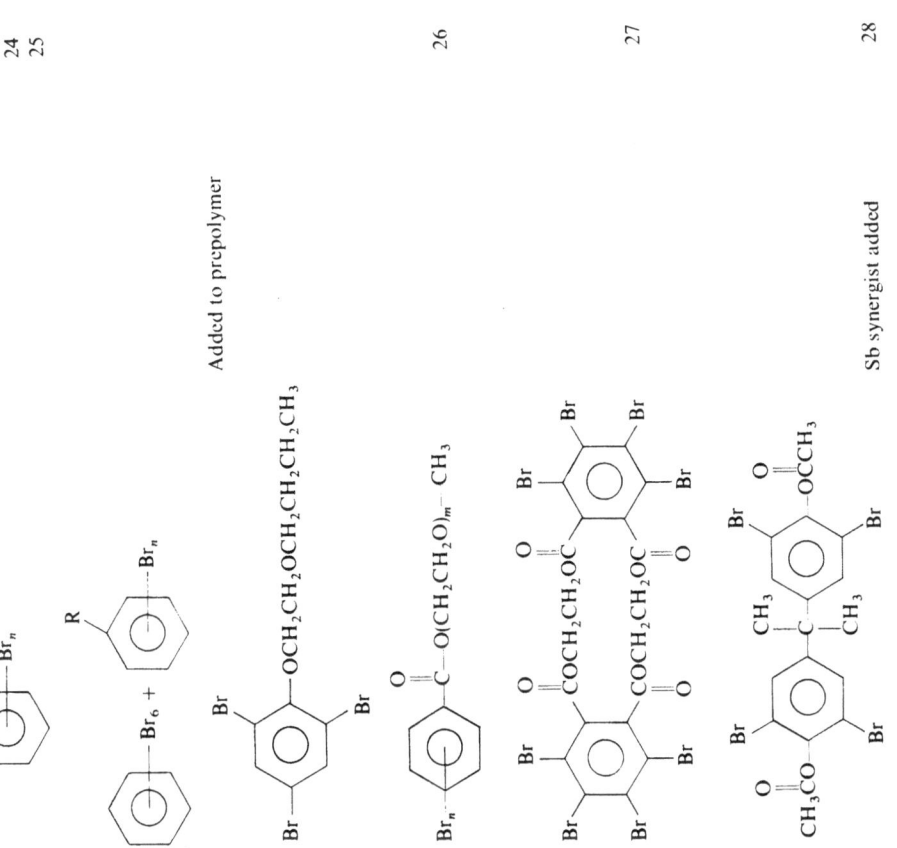

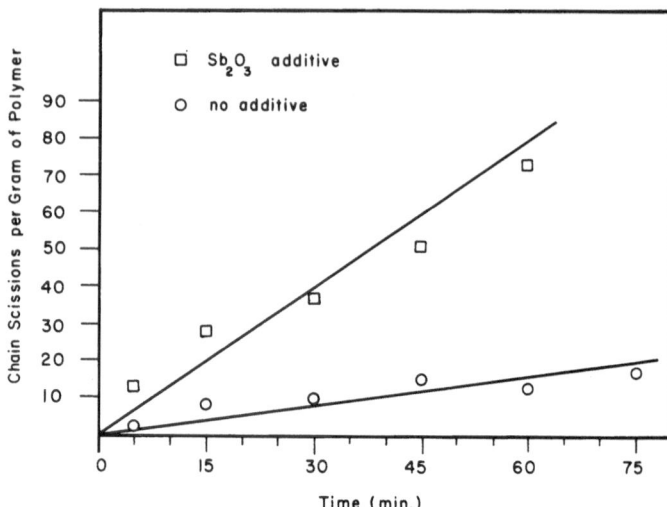

FIGURE 2. Comparison of degradation at 290 C under nitrogen atmosphere of polymer with and without antimony oxide additive.

molten polymer to serve as halogen synergists may have a degrading effect on the polymer. Antimony oxide is widely used at levels of 200–300 ppm as a polycondensation catalyst in the melt polymerization of PET. However, when antimony oxide is added at the 1 wt % level to molten PET under nitrogen, the rate of thermal degradation of the polymer is significantly increased. The rates of loss in molecular weight calculated as chain scissions per gram of polymer are compared in Fig. 2 for polymer at 290°C with and without addition of 1 wt% antimony oxide.[29] The additive definitely accelerated the thermal degradation of the polymer. The use of such compounds as melt additives requires the stringent minimization of contact time between the melt and additive prior to spinning.

While many of the compounds listed in Table 1 impart flame retardancy to PET fiber, this effect cannot be considered permanent owing to possible removal of the additives during consumer use of fabrics made from the fiber. Many of the brominated biphenyls are readily extracted from PET fibers by dry-cleaning fluids such as tetrachloroethylene. Because of limited solubility in water, many of the brominated additives are relatively permanent to home laundering. The dyeing of polyester fibers and fabrics also results in extraction of many of the additives listed in Table 1. To achieve diffusion of dyes into semicrystalline and oriented PET fibers, it is necessary to raise the temperature of the fiber higher than its glass temperature*

---

* The glass temperature of commercial fibers is approximately 120°C, requiring aqueous dyeing under pressure, but this temperature may be lowered to approximately 100°C by the inclusion of organic compounds in the aqueous dyebath which plasticize the fibers.

where the segmental motion of polymer chains in the amorphous regions commences. Such segmental motion allows the dyestuff to diffuse into the fiber and become fixed when the temperature is reduced to below the glass temperature. However, under these dyeing conditions, many additives are partially extracted from the fiber into the dyebath. Many flame-retardant additives plasticize the PET fiber, lowering its glass temperature. While such plasticization might aid dyeability, the effect seldom proves to be of commercial significance due to extraction of the additive during dyeing.

Window glass transmits very little sunlight of wavelengths below 300 nm.[32] However, polyethylene terephthalate begins a strong ultraviolet absorption at 320 nm which continues to below 300 nm. The absorption of sunlight radiation of 320 nm and below results in photodegradation of PET,[32-34] but the degradation is not a significant problem in the end uses of the fiber.

The ultraviolet degradation has been accounted for in terms of photo-chemical reactivity of an electronic (n, π^*) transition localized in the carbonyl group leading to a Norrish type II chain cleavage.[34] Photo-oxidation of PET fibers sensitized by titanium dioxide added to the fibers as a de-lustrant has also been cited[35] as responsible for sunlight degradation. Many of the brominated phenyl compounds listed in Table 1 absorb ultra-violet light of wavelengths above 320 nm. The presence of many of these compounds in PET fibers leads to discoloration of both the undyed and dyed fiber upon exposure to sunlight owing to the photochemical reactivity of the additives.

The ecological compatibility of fibers containing flame retardants must be considered. The recognition[36] that polychlorinated biphenyls are hazardous because of their accumulation in the environment focuses attention on the many halogenated phenyl structures cited as flame retardants for PET fibers in Table 1. Studies by the Haskell Laboratory of the E. I. du Pont de Nemours & Co.[37] have shown that the acute toxicity of polybrominated biphenyls is low, but the compounds are unacceptable as flame retardants for polyester fibers owing to a distinct tendency to concentrate and remain in fat tissue. The compounds cause liver enlarge-ment at low chronic doses and probably would be biomagnified. An environ-mental study of decabromodiphenyl oxide at Dow Chemical Company[38] led to the conclusion that this compound did not behave like the brominated biphenyls in fat or in the liver. It was concluded that decabromodiphenyl oxide was environmentally and toxicologically safe for use as a fire retardant in thermoplastics. Fibers containing fire retardants must be scrutinized for environmental safety by the fiber and fabric manufacturers.

2.2. Phosphorus

The mechanics proposed for retarding combustion of synthetic fibers by a phosphorus-containing additive are rather speculative. Evidence has accumulated indicating that phosphorus may flame-retard synthetic polymers by action in the condensed phase. The phosphorus compounds have been shown to promote char formation, thereby reducing the production of combustible gases during polymer decomposition.[6,39,40] The formation of a glassy coating and a char layer may mask the surface of some polymers containing phosphorus.[7,8] In addition, there is evidence that phosphorus may act as a thermal sink via endothermic reduction of phosphoric acids[7,41] and may reduce the exothermic heat of burning by limiting oxidation of carbon char to the formation of carbon monoxide.[10,40] More recent results from PET fibers containing triphenylphosphine oxide indicate that phosphorus compounds may also be active in the gas phase.[42,43] In some polymeric systems, the oxidation state of the phosphorus in additives appears to affect flame-retarding effectiveness,[44,45] while in other systems nearly any form of phosphorus is equally effective.[46]

Fibers have been prepared containing organophosphates, -phosphonates, -phosphinates, and -phosphine oxides as additives.[29] A plot of the oxygen indexes* of these fibers *vs.* concentration of phosphorus in the fibers is presented in Fig. 3. For a given concentration of phosphorus in these fibers, the LOIs did not differ by more than 2.5 units. While there may be structural features of the phosphorus additives that affect their flame-retarding efficiency in PET fibers, these effects are not pronounced. Characteristics of the fiber, such as melt viscosity[47] (i.e., melt dripping), and the surface-to-volume ratio of the test specimen might also affect the oxygen index. For a given fabric, the LOI value was reproducible to well within the limits of ± 0.5.

Examples of typical organophosphorus compounds cited in the patent literature as melt additives for fire-retarding PET fibers are listed in Table 2. Halogenated additives are often cited for addition in conjunction with these phosphorus compounds, and halogenated organophosphorus compounds are also cited as additives. Generalizations proposed in the previous section concerning thermal stability, extractability, and light stability of brominated compounds apply to these phosphorus additives. The additives

---

* Oxygen indexes were determined on 1 to 3 oz/yd² knit fabrics constructed from a plied yarn composed of a 10-filament bundle of polyester (50 to 100 denier) and a filament bundle of fiberglass (50 denier). The fiberglass tends to minimize flame extinction by dripping of the polyester and allows reproducible measurements. Variations in weight ratio of polyester to fiberglass from fabric to fabric also may affect the measurement through the heat-sink effect of the glass.

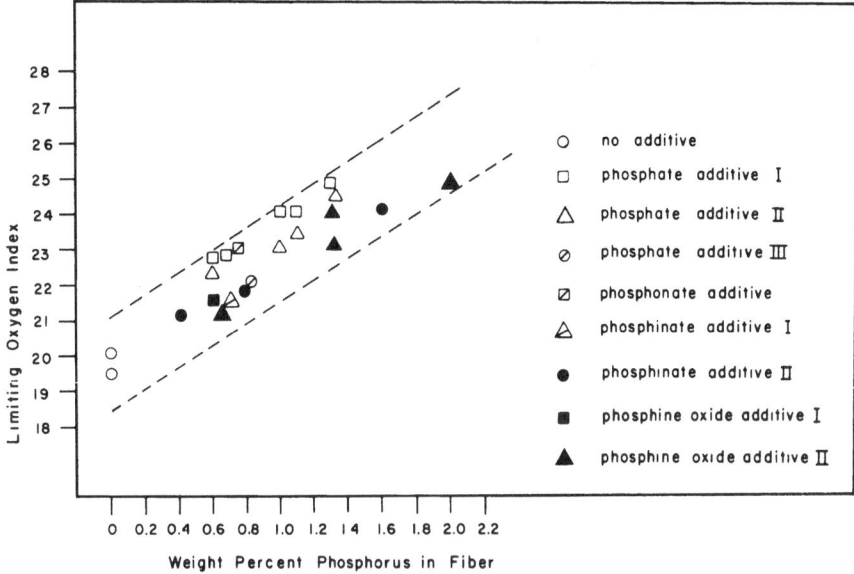

FIGURE 3. LOI of fibers as a function of phosphorus content.

with labile phosphorus acid ester bonds may cause significant degradation of the molten polymer. The effects of two organophosphates on molten PET at 265°C are compared in Fig. 4 where the losses in polymer molecular

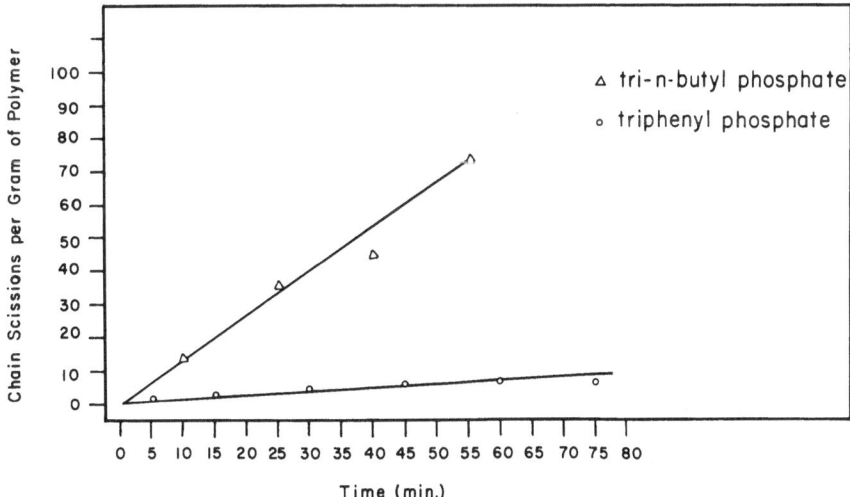

FIGURE 4. Comparison of degradation at 265°C under nitrogen atmosphere of polymer containing 5 wt % phosphate additives.

TABLE 2
Phosphorus-containing Additives

Structural characterization	Example	Special features	Reference
Polyphosphonate			48
			49
			50
		R and R' may be halogenated	51
			52
Phosphate			53

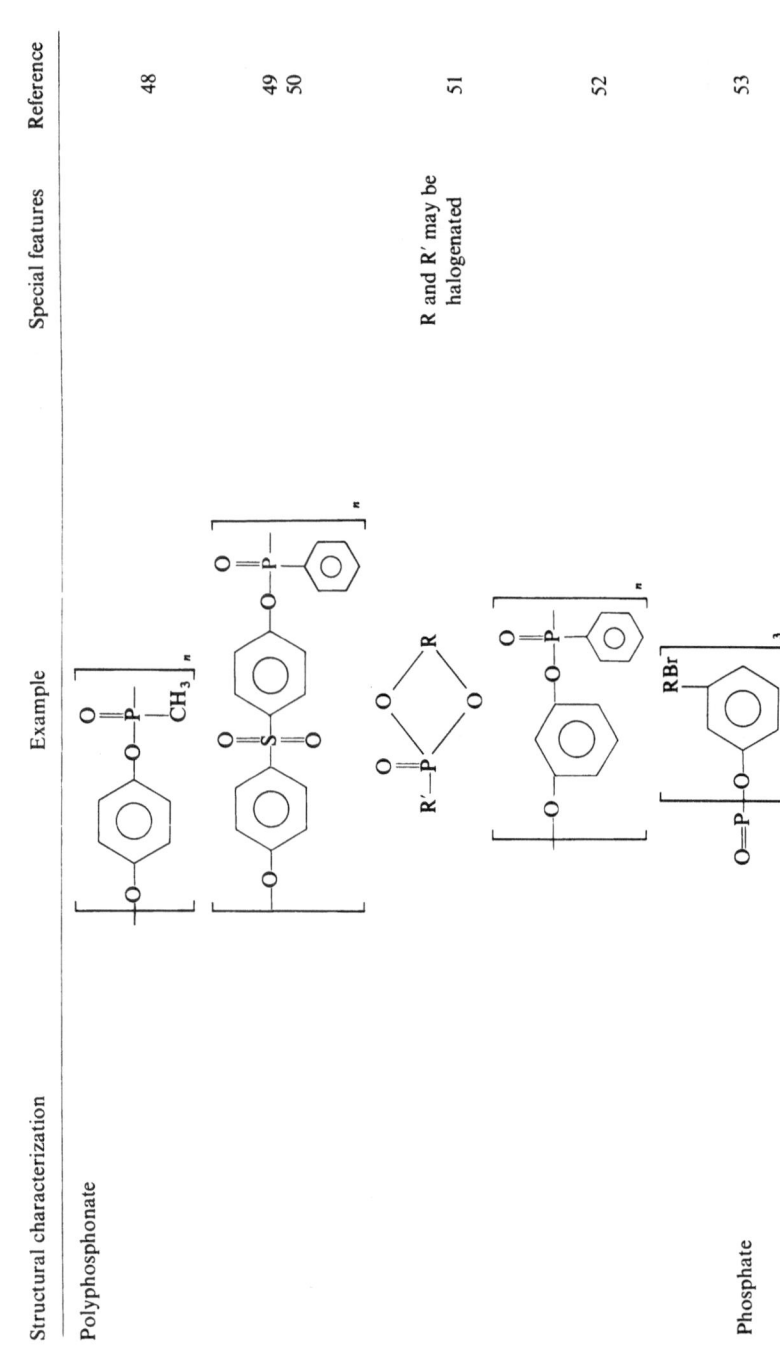

Sb synergist may
be added

54    $O=P(-OCH_2CHClCH_2Cl)_3$

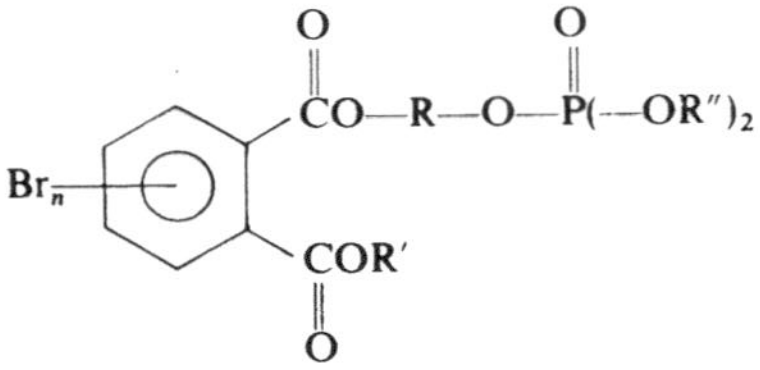

55

56    $Br_6 + (C_6H_5O)_3 \quad P=O$

57

58    $(RO)_3-P=O + ClCH_2CH_2O$

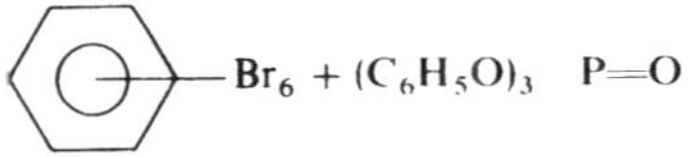

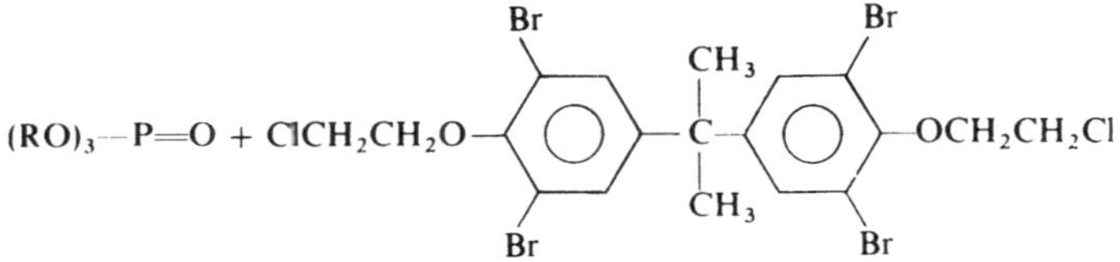

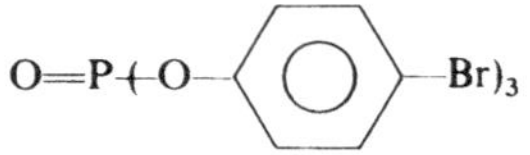

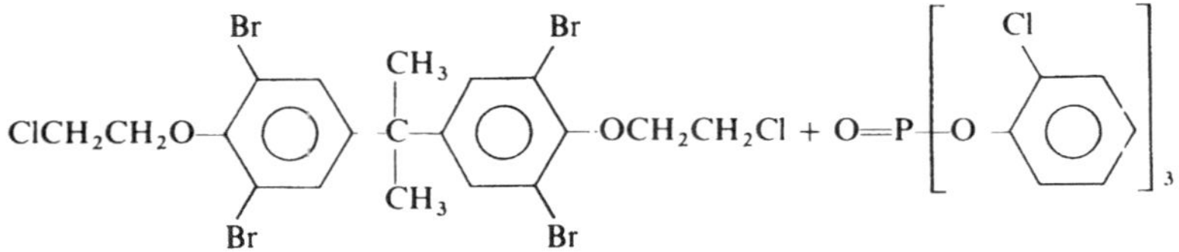

59

60   Phosphite

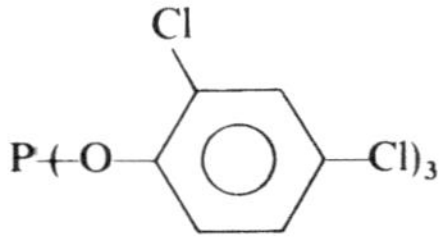

TABLE 2
Phosphorus-containing Additives (Continued)

Structural characterization	Example	Special features	Reference
Phosphine oxide		Synergist mixture	61
		Synergist mixture	62
	nylon 6 + $(C_6H_5)_3-P=O$	Synergist mixture	63
		Aid processing above 200°C	64

weight are plotted against time of mixing additive with polymer.[29] The instability of tri(*n*-butyl phosphate) in the melt results in significant degradation, while triphenyl phosphate is sufficiently stable to not degrade the polymer.

2.3. Phosphorus–Halogen Combinations

Phophorus–halogen synergism in flame-retarding PET fibers has been widely sought, judging from the number of combinations of these elements cited in Table 2. However, there appear to be no substantial examples of such synergism. The generation of phosphorus oxyhalides in the solid phase, which are vaporized to effectively poison flames,[7,44] is often cited as the mechanism for the synergism. Through its action in the solid state, phosphorus might alter the chemical composition of gases from polymer degradation, changing the effectiveness of the halogen in the flame zone. Phosphorus acids formed in the solid state might also affect the production of brominated vapors.[44] The char formation promoted by phosphorus might also cause the halogenated gases to be evolved at a more opportune time during polymer decomposition for flame poisoning. It has also been suggested that phosphorus–bromine–polymer interactions may promote charring.[44]

Sparse reports have appeared which support or discredit bromine–phosphorus synergism in PET fibers. For fibers containing a brominated phenyl additive and a phosphine oxide additive, a plot[65] of LOI *vs.* percent phosphorus in the fiber at several levels of bromine in the fiber is presented in Fig. 5. The lines are approximately linear with increasing phosphorus concentration and are approximately equidistant. This plot implies that the observed oxygen indexes result from the additive effects of the bromine and phosphorus contents of the fibers. Significant synergism was not observed. Phosphorus is approximately seven times as effective in fire retarding PET as bromine, when compared on the basis of weight percent in the fiber.

Although halogen–phosphorus synergism has not been established in PET fibers, a case of an antagonism of phosphorus toward antimony–halogen synergism has been observed.[29] A PET copolymer containing 9 mol % 2,5-dibromoterephthalate groups was spun into fibers with the addition of combinations of antimony oxide and an organophosphate fire retardant to the melt just prior to spinning. The LOI of these fibers is compared in Table 3. The fiber containing bromine and phosphorus appeared to be equivalent in flammability resistance to the fiber containing these two elements plus antimony. The interaction of antimony with bromine would be expected to increase the LOI of the latter fiber by an additional

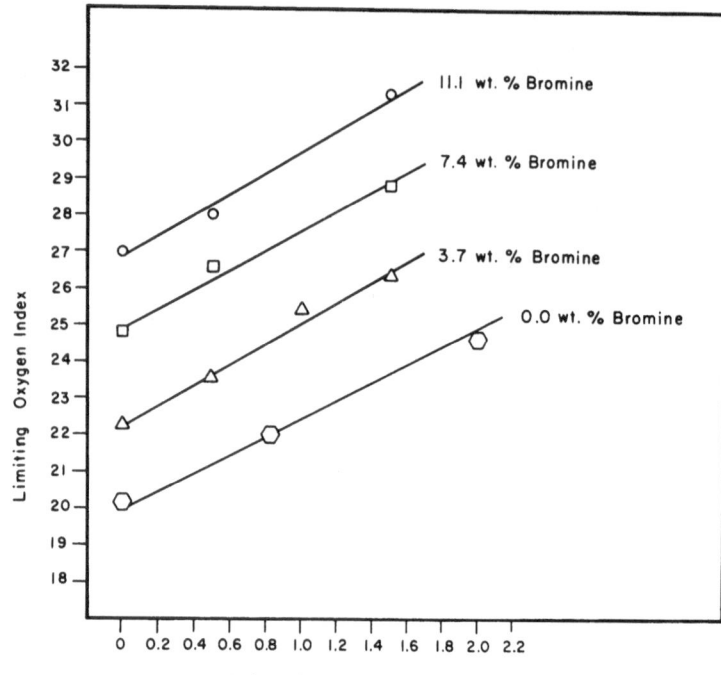

FIGURE 5. LOI of fibers as a function of bromine and phosphorus contents.

two units judging from the second and third entries of Table 3. Apparently the retardant benefit of antimony–bromine synergism or alternately the retardant benefit of phosphorus was lost in the fiber containing all three elements. The complexing of the phosphate with the antimony oxide under thermal activation might hinder the interaction of the antimony with bromine.

The entries in Table 2 under phosphine oxide are noteworthy with reference to synergistic mixtures of triphenylphosphine oxide with other additives. The presence of the nylon 6 in a synergistic mixture with triphenylphosphine oxide has been shown to decrease the rate of volatilization of triphenylphosphine oxide from the fiber and thus make it a more efficient flame inhibitor.[42,43] It appears that a complex of nylon 6 and the phosphine oxide, possibly through hydrogen bonding, is formed in the PET fiber, resulting in the release of the phosphine oxide into the vapor phase at temperatures matching the decomposition of the PET polymer. Triphenylphosphine oxide also appears rather unique in that there is considerable evidence for its predominantly vapor phase activity[42,43] rather than the predominantly condensed phase activity observed for most phosphorus–

TABLE 3
Oxygen Index of Fibers Containing
Bromine, Antimony, and Phosphorus

Elemental components in fiber, wt %			Oxygen index of fiber, %
Br	Sb	P	
0.0	0.0	0.0	20.0
6.1	0.0	0.0	22.5
6.1	1.3	0.0	24.8
6.1	0.0	0.9	25.5
6.1	1.3	0.9	25.5

containing additives in polymers.[6,39,40] The incorporation of triphenyl-phosphine in epoxy resins has been shown to have similar flame-retarding effects in atmospheres of oxygen–nitrogen or nitrous oxide–nitrogen, indicating condensed phase activity where the nature of the oxidative atmosphere was immaterial.[66] However, similar experiments[42] with PET fibers containing triphenylphosphine oxide demonstrated that the compound was an effective retardant in an oxygen atmosphere but not in a nitrous oxide atmosphere indicating vapor phase activity. The final entry of Table 2 illustrates the formation of Lewis acid complexes of triphenyl-phosphine oxide to lower the volatilization of the compound when added to PET polymer prior to melt spinning.

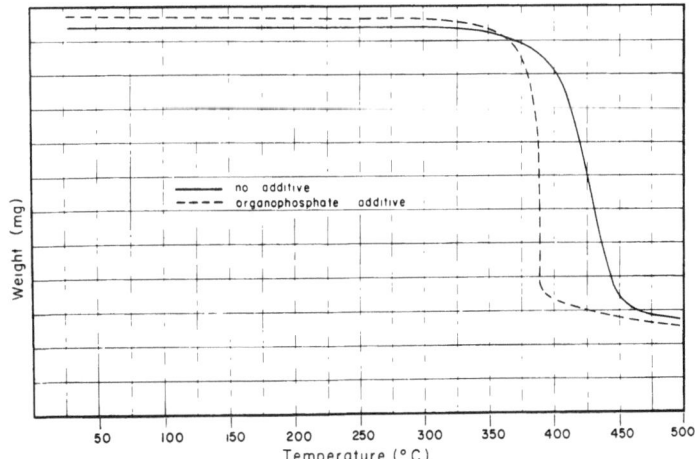

FIGURE 6. Comparison of thermograms in air atmosphere of fibers with and without organophosphate additive (10°C/min).

Bostic and Barker[42] have reported that triphenylphosphine oxide in PET fibers produced only minor changes in the energetics of pyrolysis as measured by thermogravimetric analysis. However, such an observation should not be generalized to encompass other phosphorus-containing additives. For instance, in Fig. 6, the thermograms of a PET fiber without additive and with an organophosphate additive are compared.[29] The major decomposition of the fiber without additive occurred in the 400–450°C interval; the major decomposition of the fiber containing 1.3 wt % phosphorus was shifted to the 350–400°C interval. Volatilization of the phosphate additive in the 350–400°C interval was far too small to account for the weight loss of the fiber in this interval. The mechanisms of flame inhibition by phosphorus in PET has not been established and may be many faceted.

Observations by Bostic and Barker[42] and by Bostic, Yeh, and Barker[67] have indicated that low levels of phosphorus in polyester fibers, i.e., below 0.1 wt %, may increase the flammability of the fibers. At low levels, the phosphorus may enhance char formation to produce a scaffolding that supports the molten polymer to aid burning, and this effect outweighs the flame-inhibiting effect of the phosphorus.

2.4. Special Techniques

Special melt spinning techniques have been utilized to avoid detrimental effects of additives on molten PET. Patents describe the spinning of bicomponent fibers in which the core of the fiber is fire retarded and the sheath is unmodified PET. For instance, a sheath–core fiber has been described[68] in which the core of PET contains a halogenated aromatic additive and a phosphate additive, while the sheath is PET without additive. The outer sheath portion of the fiber is intended to approximate commercial PET fibers in surface properties, color, and physical properties while problems of polymer discoloration and poor physical properties are masked in the core portion of the fiber. The thickness of the sheath of the composite fiber also affects the permanence of the additives to dyeing and home laundering. Of course, the overall properties of such fibers are dependent upon the responses of the fiber as a composite. Another patent[69] cites a similar composite fiber approach in which the core is a brominated copolyamide or copolyester. Additives which are incompatible with or react with polymer melts might be included in the melt after being placed in microcapsules whose skins are decomposed at inflammation temperatures of the fibers but not at spinning temperatures. Of course, the microcapsules must be able to pass through the filtration medium and spinnerette holes of the melt spinning system. This concept of inclusion of microcapsules of fire retardants in fibers has been cited in the patent literature.[70]

A blend of polyester fibers in which one portion of the fibers contains an antimony additive and a second portion contains a brominated additive has been described in a patent.[71] Chemical interaction of the brominated additive with the antimony additive in the polymer melt is avoided by this technique. But the synergistic interaction of the bromine and antimony is apparently obtained when the blend fabric is exposed to inflammation temperatures.

Several elements in addition to phosphorus, bromine, and chlorine have been claimed as fire retardants for PET. Patents describe the use of organoarsenic compounds such as ethylenebis(diphenylarsine)[72] and sulfur.[73]

3. Ethylene Terephthalate Copolymers

The examples of copolymers with the ethylene terephthalate unit plus a halogenated or phosphorus-containing third component that have been reported are far too numerous to be listed here. Fiber manufacturers are interested in such copolymers from economically feasible comonomers for production of a flame-retarded polyester fiber that retains the highly desirable properties and favorable economics of polyethylene terephthalate. Although many other homopolymers and copolymers are known and can be conceived which would produce flame-resistant fibers, it is unlikely that a polymer can be found which would be competitive with PET in fiber properties and manufacturing cost. However, the fiber manufacturers seek to produce a copolyester fiber of sufficient flame resistance to meet end-use requirements and at the same time retain the essential properties of PET homopolymer by copolymerizing less than 10 mol % of a third component with ethylene terephthalate units.

The copolyester fibers with ethylene terephthalate units as the main component have been reviewed by Morimoto.[74] Those monomers containing bromine or phosphorus have been of greatest interest for imparting flame resistance to ethylene terephthalate copolymers. These monomers must be thermally stable to temperatures in excess of 230°C for several hours during the melt copolymerization. Monomers with brominated phenyl structures and certain phosphorus-containing monomers have been utilized as the third component in these copolymerizations. Copolymer fibers containing brominated phenyl structures often exhibit sensitivity to exposure to ultraviolet light as do those with nonreactive brominated phenyl additives. When the flame retardant is incorporated in the polymer chain as a coreactant, the problem of extraction of the retardant from the fiber during dyeing, laundering, or dry cleaning is eliminated. Some examples of the monomers of interest to fiber manufacturers for preparation of ethylene terephthalate copolymers are listed in Table 4. The variety of synthetic methods which can be applied to such copolymerizations has been discussed by Morimoto.[74] The production of a copolymer with properties acceptable for a commercial fiber often depends upon controlling subtle features of the melt polymerization.[74,75]

TABLE 4
Ethylene Terephthalate Copolymers

Third component	Reference
	75
	76
	77
	78
	79
	80

A comparison of fibers prepared from ethylene terephthalate copolymers of the first two comonomers listed in Table 4 illustrates the importance of copolymer structure–property relationships. To achieve a copolymer containing 6 wt % bromine* requires the substitution of 4 mol % 2,2-bis[3,5-dibromo-4-(2-hydroxyethoxy)phenyl]propane for ethylene glycol or the substitution of 9 mol % 2,5-dibromoterephthalic acid for terephthalic acid. Both compounds contain 50 wt % bromine, but the molecular weight of the diol is nearly twice that of the diacid. The weight percentages of brominated diol and brominated diacid in the copolymers containing 6 % bromine are nearly equal. The copolymer from the diol melts at 246°C, the copolymer from the diacid at 233°C, and the homopolymer at 256°C. Random copolyesters have been shown to exhibit a melting point depression from the homopolymer that is dependent upon the mol % of the third component.[81,82] For ethylene terephthalate copolymers, the depression is to 2–3°C/mol % of the third component.

Fibers prepared from either copolymer are very similar in flame resistance, but differ in physical properties because of the differing effects of the comonomers on fiber morphology. The challenge to the fiber producer is to alter the processing (spinning, drawing, heat setting) of such copolymers so as to achieve fiber properties approaching those of the homopolymer fiber for commercial acceptance. Of course, the possibility exists that the copolymer fiber may have a property improvement such as increased dyeability relative to the homopolymer owing to the interruption of fiber crystallinity and orientation by the third component. In the commercial market such an improvement might offset other deficiencies in properties of the copolymer fiber.

The incorporation of blocks of a fire-retardant polymer of molecular weight above 1000 into ethylene terephthalate copolymers also exists. Although block copolyesters of ethylene terephthalate have been investigated[74] widely to achieve property modifications of PET fibers, their utilization in flame-retarding polyester fibers has been limited.

4. Fiber and Fabric Treatments

The thermoplastic fibers such as polyester and nylon have been traditionally considered to be less hazardous than cotton and acrylics because they melt away from flames. When blended with nonmelting fibers such as cotton, the combination definitely burns more easily. The non-

---

* It appears that the presence of approximately 6% bromine in ethylene terephthalate copolymer fibers imparts sufficient flame retardancy for the fiber in lightweight 100% polyester fabrics to pass the U.S. Children's Sleepwear Test (DOC FF 3-71).

melting fiber of the blend forms a carbonaceous gridwork during combustion that supports the molten thermoplastic fiber and prevents dripping and shrinkage of the fabric away from the flame. Kruse termed this behavior of polyester–cotton blends as the "scaffolding effect."[83] The prominence of blend fabrics in the apparel market has attracted attention to the unique and formidable problems of flame retarding such fabrics.

Several investigators[84-86] have demonstrated that major modification of the flammability of both the cotton and polyester components of blends are required to achieve significant flame retardance in fabrics. However, the flammability behavior of blend fabrics has not been found to be predictable from the behavior of the component fibers.[87]

Tesoro[84] has summarized the difficulties that fabric manufacturers have encountered in finding satisfactory flame-retardant finishes for polyester–cotton blends. Several of the more prominent finishes described in the patent literature are listed in Table 5. Hofmann and Loss[88] reported that finishes derived from tetrakishydroxymethylphosphonium chloride are more effective, based on equal weight of phosphorus on the fabric, than finishes derived from N-methylol dimethylphosphonopropionamide in fire-retarding polyester–cotton blends. Tesoro[84] has reported similar behavior for blends in which cellulose hydroxyl groups were reacted with

TABLE 5

Fabric Finishes for Cotton–Polyester Blends

Fabric finish	Reference
Coating fabric with reaction product of dimethyloable nitrogen-containing compound, a phosphonopropionic acid amide, and formaldehyde and with tris(2,3-dibromopropyl) phosphate	89
Various condensates of tetrakis(hydroxymethyl) phosphonium	90
	91
	92
	93
Curable polyepoxides containing phosphorus	94
Coating fabric with (a) tetrakis(hydroxymethyl)phosphonium chloride and urea, (b) reaction product of melamine with formaldehyde, (c) fatty acid ester of polyethylene glycol and forming insoluble reaction products on fabric	95
$(CH_3CH_2O)_2PCH_2OCCH=CH_2$ (with two $\overset{O}{\overset{\|}{}}$ groups)	96
$(CH_3CH_2O)_2PCH_2NHCCH=CH_2$ (with two $\overset{O}{\overset{\|}{}}$ groups)	97
Ammonium salt of bis(hydroxymethyl)phosphinic acid	98

the following two reagents differing only in the oxidation state of the phosphorus atom:

$$\begin{array}{c} CH_3O \\ \diagdown \\ \diagup \\ CH_3O \end{array} \overset{O}{\overset{\|}{P}}-CH_2CH_2CONHCH_2OH$$

$$\begin{array}{c} CH_3 \\ \diagdown \\ \diagup \\ CH_3 \end{array} \overset{O}{\overset{\|}{P}}-CH_2CH_2CONHCH_2OH$$

The phosphine oxide was more effective in increasing the oxygen index of blends than the phosphonate, while the phosphonate was more effective on 100% cotton. Such findings suggest that the mechanism by which phosphorus interacts with cellulose in the solid state to retard combustion may be changed by the presence of polyester. It should be noted that finishing treatments result in the insolubilization on the fabric of the phosphorus-containing reagent by condensation reactions to yield cross-linked polymers. The distribution, molecular weight, penetration, etc., of the deposited polymer, as well as the state of oxidation of phosphorus, may affect the effectiveness on blend fabrics.

Several approaches to the application of fire-retardant finishes to polyester–cotton blend fabrics have been reported,[84,85,87,88,99] but in general these approaches have not been commercially acceptable.[84] It has been suggested that a PET fiber with an oxygen index greater than 28 and a flame-retardant finish for the cotton component would be required to produce a significantly flame-retarded blend fabric.[65]

Many possibilities exist for the utilization of unmodified and fire-retarded polyester fibers in blend fabrics. For instance, blends of 80% fire-retarded cellulose acetate fiber and 20% unmodified polyester fiber in knit fabrics have been reported to pass the U.S. Children's Sleepwear Test (DOC FF 3-71).[100] Blend fabrics of cotton–polyester–cellulose acetate can be flame retarded by application of two finishes, one specifically for the cotton and another for the polyester and acetate.[101] Blend fabrics of polyester–cellulose acetate can be fire retarded by the application of the finishing agent, tris(2,3-dibromopropyl)phosphate.[102]

The use of tris(2,3-dibromopropyl)phosphate as a flame-retarding finishing agent for 100% polyester fabrics has received considerable commercial attention. The compound is padded onto the fabric, the fabric dried at $110°C$ for one minute, and then the fabric is subjected to a thermosol-type dry heat treatment at $200–220°C$ for 1 min. The chemical is absorbed into the fibers and is relatively fast to removal by home laundering. Approximately 6–8 wt$\%$ add-on of the chemical to 100% polyester lightweight woven fabrics allows these fabrics to pass the U.S. Children's Sleepwear Test. The chemical is also being applied to fabrics by exhaustion from conventional dyeing equipment.[103]

Several surface treatments of PET fibers diffuse fire-retardant additives into the fibers. One patent[104] describes the contact of a solution of 2-(2,4,6-tribromophenoxy)phenol with the fiber at temperatures above the glass temperature of the polymer. The additive thus is diffused into the fiber. This brominated phenol may also be used as a carrier in aqueous disperse dyeing where it serves to plasticize the fiber during dyeing and is retained after dyeing to serve as a fire retardant.[105] A surface treatment designed for application by the fiber manufacturer consists of exposure of PET fibers to baths of halogenated organophosphorus compounds followed by a drawing of the fiber resulting in impregnation of the fiber with the additive.[106] Many other compounds have been cited as effective flame retardants for surface application to fibers, such as $(BrCH_2BrCHCH_2O)_3 \cdot P{=}O,$[107,108] $(ClCH_2)_3P{=}O,$[109] and Br_5C_6OH in combination with $(C_6H_5O)_3P{=}O.$[110]

Several chemical modifications of PET fibers which may be classified as graftings are cited in the patent literature. One[111] describes the application of allyl phosphates such as

$$(CH_2{=}CH{-}CH_2O)_2\overset{\displaystyle O}{\overset{\displaystyle \|}{P}}{-}OCH_2CHBrCH_2Br$$

to PET fabrics and then in the presence of a free-radical generator forming an addition polymer of this monomer or grafting the monomer to the PET chains. Another[112] concerns the impregnation of phosphorus-containing vinyl monomers into fibers followed by *in situ* polymerization of the monomer. Surface chlorination of polyester fibers is cited as a modification.[113] Additives may also be included in the fiber which are more susceptible to reaction than PET. For instance, a patent[114] describes the incorporation of polyethylene oxide, polystyrene, or polyethylene into polyester fibers and subsequent treatment of the fiber with PCl_3 in the presence of oxygen. Apparently phosphorus is incorporated into the fiber through reaction of the PCl_3 with the added polymer.

Several patents have been issued[115,116] describing the chlorination and/or bromination of PET while dissolved in mixtures of chlorinated alkanes and trichloroacetic acid. The resulting chlorinated and brominated PET was soluble in chlorinated alkanes such as chloroform and thus probably not of significance as a fiber candidate.

5. Fabric Flammability

Several studies of the burning of fabrics utilizing instrumented man-nequins have been reported. These studies attempt to correlate the heat transfer from burning fabrics with the possible physiological effects of such fires on human subjects. In the studies of Chouinard, Knodel, and Arnold,[117] 100% polyester dresses burned with a melt–drip process which inhibited flame spread, and the garments were extinguished before mea-surable energy reached the mannequin. The melted drops of polyester were shown to contain considerable energy but the melt–drip process inhibited flame spread and heat generation to the extent that less energy was transferred by the burning polyester fabrics than by cotton, polyester–cotton, and acrylic fabrics. Evidence was presented which indicated that burning cotton, nylon, or polyester fabrics could produce burns deeper than moderate second degree. Polyester fabrics were found to produce the least injury. These workers correlated the burn hazard of fabrics to the wearer with the fabric's ability to transfer heat to the skin.

Mannequin experiments by Krasny and Fisher[118] indicated that the garment geometry could affect the burn-injury potential as much as fiber parameters. The effects of fabric layers in outerwear–underwear assemblies were shown to affect the burn potential in garment fires. Polyester–cotton blend fabrics appeared to have more potential for causing injury than 100% cotton or 100% polyester fabrics. The 100% polyester fabrics ap-peared to have the lowest injury potential in this study as in others. Carter, Finley, and Farthing[119] have shown that for burning of polyester–cotton blend fabrics, the mannequin surface temperature decreased as the poly-ester fiber content of the blend increased. The increasing polyester content resulted in increased heat radiation away from the mannequin surface. It was shown that polyester–cotton fabrics burn more rapidly than 100% cotton fabrics. The assessment of the danger of garment fires from manne-quin data is a new area of investigation and open to a variety of inter-pretations.

Several calorimetric studies of the combustion of polyester and poly-ester–cotton fabrics have been reported.[42,67,120] The study of Birky and Yeh[120] indicated that burning PET fabrics in air released only 42% of the standard heat of combustion and that rate release was considerably lower for PET fabrics than for cotton or acrylic fabrics. The heat release from cotton–PET fabrics burning in air increased linearly as the polyester content of the fabric increased.[67] The generation of smoke during com-bustion appeared to increase also as the polyester content of cotton–PET blend fabrics increased.[86]

Of major concern in the United States has been the flame retarding of 100% polyester fabrics to the extent that they will pass the U.S. Children's Sleepwear Test (DOC FF 3-71). As experience with the application of this test has been gained, it has been recognized that besides the inherent tendency of the polyester fiber in a fabric to support combustion, other factors of fabric construction also greatly affect the performance of the fabric in the test. Failures of seamed fabrics to pass the test have even been traced to the type of lubricant used on the seaming thread.[121] Factors such as increasing fabric weight, which hinders the shrinkage of the 100% polyester fabrics from the flame, allow the fabrics to reach their ignition temperature under conditions prescribed for in DOC FF 3-71 tests.[122]

6. References

1. B. M. Baum, *Chem. Tech.* **3**, 167, 311, 416 (1973).
2. C. P. Fenimore and G. W. Jones, *Combust. Flame* **10**, 295 (1966).
3. A. R. Ingram, *J. Appl. Polym. Sci.* **8**, 2485 (1964).
4. C. P. Fenimore and F. J. Martin, *Combust. Flame* **10**, 135 (1966).
5. C. P. Fenimore, *Combust. Flame* **12**, 155 (1968).
6. C. J. Hilado, *Flammability Handbook for Plastics*, p. 82, Technomic Publishing Co., Stamford, Conn. (1969).
7. J. K. Jacques, *Plast. Inst. Trans. J., Conf. Suppl.* **2**, 33–37 (1967).
8. C. J. Hilado, *Ind. Eng. Chem., Prod. Res. Develop.* **7**, 81 (1968); also see *Am. Chem. Soc., Div. Org. Coatings Plast. Chem., Pap.* **28**, **1**, 317–334 (1968).
9. R. C. Nametz, *Ind. Eng. Chem.* **59** (5), 99 (1967).
10. H. Piechota, *J. Cell. Plast.* **1**, 186 (1965).
11. W. G. Schmidt, *Plast. Inst. Trans.* **33**, 247 (1965).
12. E. C. Creitz. *J. Res. Nat. Bur. Stand., A* **74**, 521 (1970).
13. J. J. Pitts, *J. Fire Flam.* **3**, 51 (1972).
14. Can. Pat. 919,693 (1973) (to Hercules Inc.).
15. U.S. Pat. 3,755,498 (1973) (to Eastman Kodak Co.).
16. U.S. Pat. 3,645,962 (1972) (to Hercules Inc.).
17. Swiss Pat. 537,991 (to Inventa AG).
18. Japan. Pat. 4,602,959 (1971) (to Celanese Corp.).
19. Belg. Pat. 773,986 (1972) (to Dow Chemical Co.).
20. Japan. Pat. 72-14,500 (1972) (to Toray Inds. Inc.).
21. Japan. Pat. 72-27,137 (1972) (to Toray Inds. Inc.).
22. Japan. Pat. 71-32,865 (1971) (to Toray Inds. Inc.).
23. Japan. Pat. 72-32,430 (1972) (to Toray Inds. Inc.).
24. Japan. Pat. 72-32,298 (1972) (to Toray Inds. Inc.).
25. Japan. Pat. 72-14,501 (1972) (to Toray Inds. Inc.).
26. Japan. Pat. 72-19,181 (1972) (to Teijin Ltd.).
27. Japan. Pat. 71-29,622 (1971) (to Toray Inds. Inc. and Sanyo Chem. Ind. Co. Ltd.).
28. Brit. Pat. 1,281,937 (1972) (to Imperial Chem. Ind. Ltd.).
29. C. J. Setzer and E. L. Lawton, Monsanto Triangle Park Development Center Inc., unpublished results.
30. I. Touval, *J. Fire Flam.* **3**, 130 (1972).
31. I. N. Einhorn, *J. Macromol. Sci.* **D1**, 113 (1971).

32. A. S. Tweedie, M. T. Mitton, and P. Z. Sturgeon, *Text. Chem. Color.* **3**, 22 (1971).
33. M. J. Wall and G. C. Frank, *Text. Res. J.* **41**, 38 (1971).
34. D. M. Wiles, *Polym. Eng. Sci.* **13**, 74 (1973).
35. L. M. Lock and G. C. Frank, *Text. Res. J.* **43**, 502 (1973).
36. R. Edwards, *Chem. Ind.* **47**, 1340 (1971).
37. J. G. Aftosmis, O. L. Dashiell, F. D. Griffith, C. S. Hornberger, M. E. McDonnell, H. Sherman, F. O. Tayfun, and R. S. Waritz, Toxicology of brominated biphenyls, presented at the Society of Toxicology Meeting in Williamsburg, Virginia, March 5–9, 1972.
38. J. M. Norris, J. W. Ehramantraut, C. L. Gibbons, R. J. Kociba, B. A. Schwetz, J. Q. Rose, C. G. Humiston, G. L. Jewett, W. B. Crummett, P. J. Gehring, J. B. Tirsell, and J. S. Brosier, Toxicological and environmental factors involved in the selection of decabromodiphenyl oxide as a fire retardant chemical, *Appl. Polym. Symp.* **22**, 195 (1973).
39. C. J. Hilado, *J. Cell. Plast.* **4**, 339 (1968).
40. J. M. Church, The role of phosphorus compounds in the fire resistant treatment of textiles, Department of Chemical Engineering Report, Columbia University, New York (1952).
41. E. A. Dickert and G. C. Toone, *Mod. Plast.* **42** (5), 197 (1965).
42. J. E. Bostic, Jr., and R. H. Barker, *J. Appl. Polym. Sci.*, in press (1975).
43. J. W. Hastie and G. D. Blue, *Am. Chem. Soc., Div. Org. Coatings Plast. Chem.* **33**, 484 (1973).
44. J. W. Lyons, *J. Fire Flam.* **1**, 302 (1970).
45. A. J. Papa and W. R. Proops, *J. Appl. Polym. Sci.* **16**, 2361 (1972).
46. J. R. VanWazer, *Phosphorus and Its Compounds.* Vol. II, p. 1955, Interscience Publishers, Inc., New York (1961).
47. H. K. Reimschuessel, S. W. Shalaby, and E. M. Pearce, *J. Fire Flammability* **4**, 299 (1973).
48. Japan. Pat. 4,739,154 (1972) (to Toyobo Co. Ltd.).
49. Japan. Pat. 72-32,299 (1972) (to Toyo Spinning Co. Ltd.).
50. Japan. Pat. 4,743,041 (1972) (to Toyobo Co. Ltd.).
51. Japan. Pat. 70-25,989 (1970) (to Toray Inds. Inc.).
52. Fr. Pat. 2,096,798 (1972) (to Fiber Industries Inc.).
53. Japan. Pat. 70-23,030 (1970) (to Toray Inds. Inc.).
54. Japan. Pat. 69-28,596 (1969) (to Toyo Rayon Co. Ltd.).
55. Japan. Pat. 70-37,667 (1970) (to Teijin Ltd.).
56. W. Ger. Pat. 2,162,437 (1972) (to Allied Chemical Corp.).
57. Japan. Pat. 72-47980 (1972) (to Teijin Ltd.).
58. W. Ger. Pat. 2,253,207 (1973) (to General Electric Co.).
59. Japan. Pat. 72-13,065 (1972) (to Teijin Ltd.).
60. Japan. Pat. 69-28,998 (1969) (to Toyo Rayon Co. Ltd.).
61. W. Ger. 2,139,395 (1973) (to Badische Anilin-und-Soda-Fabrik A.G.).
62. U.S. Pat. 3,681,281 (1972) (to Celanese Corp.).
63. U.S. Pat. 3,629,365 (1971) (to Akzona Inc.).
64. U.S. Pat. 3,660,350 (1972) (to M. T. Chemicals Inc.).
65. E. L. Ringwald, An overview of flame retardant man-made fibers, Flammability Seminar, American Association of Textile Chemists and Colorists Technical Center, Research Triangle Park, N. C., Nov. 3, 1973.
66. F. J. Martin and K. R. Price, *J. Appl. Polym. Sci.* **12**, 143 (1968).
67. J. E. Bostic, Jr., K. N. Yeh, and R. H. Barker, *J. Appl. Polym. Sci.* **17**, 471 (1973).
68. U.S. Pat. 3,658,634 (1972) (to Toray Inds. Inc.).
69. Neth. Pat. 72-02,052 (1972) (to Imperial Chemical Inds. Ltd.).
70. Fr. Pat. 2,119,743 (1971) (to Kane-Gafuchi Boseki KK).

71. U.S. Pat. 3,763,644 (1973) (to Eastman Kodak Co.).
72. U.S. Pat. 3,776,883 (1973) (to Celanese Corp.).
73. Can. Pat. 924,835 (1973) (to Celanese Canada Ltd.).
74. S. Morimoto, Copolymers, in: *Man-Made Fibers* (H. F. Mark, S. M. Atlas, and E. Cernia, eds.) Interscience Publishers, a division of John Wiley & Sons, Inc., Vol. 3, (1968).
75. Belg. Pat. 796,793 (1973) (to Emery Ind. Inc.) and U.S. Pat. 3,794,617 (1974) (to Emery Ind. Inc.).
76. J. P. Nelson, Flame retardant polyester by incorporation of 2,5-dibromoterephthalic acid, 165th American Chemical Society Meeting, Dallas, Tex., April 9–11, 1973.
77. Brit. Pat. 1,248,835 (1971) (to Farbwerke Hoechst A.G.).
78. Neth. Pat. 6,401,071 (1964) (to Gelsenberg Benzin A.G.).
79. Japan. Pat. 7,245 (1963) (to Teijin Ltd.).
80. U.S. Pat. 2,646,420 (1953) (to duPont).
81. D. Coleman, *J. Polym. Sci.* **14**, 15 (1954).
82. O. B. Edgar and R. Hill, *J. Polym. Sci.* **8**, 1 (1952).
83. W. Kruse, *Melliand Textilber.* **50**, 460 (1969).
84. G. C. Tesoro, *Text. Chem. Color.* **5**, 235 (1973).
85. G. C. Tesoro, J. Rivlin, and D. R. Moore, *Ind. Eng. Chem., Prod. Res. Develop.* **11**, 164 (1972).
86. H. Stepniczka and J. Dipietro, *J. Appl. Polym. Sci.* **15**, 2149 (1971).
87. G. C. Tesoro and C. H. Meiser, Jr., *Text. Res. J.* **40**, 430 (1970).
88. P. Hofmann and R. Loss, *Textilveredlung* **8**, 310 (1973).
89. U.S. Pat. 3,669,725 (1972) (to Ciba Ltd.).
90. Fr. Pat. 2,151,136 (1973) (to Ciba-Geigy AG).
91. Belg. Pat. 788,595 (1973) (to Ciba-Geigy AG).
92. Belg. Pat. 788,593 (1973) (to Ciba-Geigy AG).
93. Belg. Pat. 788,594 (1973) (to Ciba-Geigy AG).
94. U.S. Pat. 3,558,668 (1971) (to Sandoz Ltd.).
95. U.S. Pat. 3,681,124 (1972) (to J. P. Stevens and Co.).
96. Belg. Pat. 795,978 (1973) (to Hooker Chemical Corp.).
97. Neth. Pat. 72-04,125 (1972) (to Hooker Chemical Corp.).
98. U.S. Pat. 3,600,219 (1971) (to Farbwerke Hoechst AG).
99. P. Linden, L. G. Roldan, S. B. Sello, and H. S. Skovronek, *Textilveredlung* **6**, 651 (1971).
100. F. Novak, Flame retardant acetate fabrics for children's sleepwear, Proc. 1973 Symp. on Textile Flammability, LeBlanc Research Corporation (1973).
101. U.S. Pat. 3,692,559 (1972) (to Celanese Corp.).
102. U.S. Pat. 3,729,340 (1973) (to Celanese Corp.).
103. E. Baer, Durable fire retardant finishing of synthetic textiles, Proc. 1973 Symp. on Textile Flammability, LeBlanc Research Corporation (1973).
104. Fr. Pat. 2,152,888 (1972) (to Dow Chemical Co.).
105. U.S. Pat. 3,749,600 (1973) (to Dow Chemical Co.).
106. Japan. Pat. 71-19,759 (1971) (to Toray Ind. Inc.).
107. Japan. Pat. 69-25,398 (1969) (to Teijin K.K.).
108. U.S. Pat. T900-021 (1972) (to Eastman Kodak Co.).
109. W. Ger. Pat. 1,282,599 (1968) (to Farbwerke Hoechst AG).
110. Brit. Pat. 1,292,878 (1972) (to Eastman Kodak Co.).
111. U.S. Pat. 3,708,328 (1973) (to Farbwerke Hoechst AG).
112. Japan. Pat. 71-19,758 (1971) (to Toray Inds. Inc.).
113. Fr. Pat. 1,593,827 (1970) (to Eastman Kodak Co.).
114. Japan. Pat. 72-24,253 (1972) (to Toray Inds. Inc.).
115. U.S. Pat. 3,575,927 (1971) (to Eastman Kodak Co.).

116. Brit. Pat. 1,103,774 (1968) (to Eastman Kodak Co.)
117. M. P. Chouinard, D. C. Knodel, and H. W. Arnold, *Text. Res. J.* **43**, 166 (1973).
118. J. F. Krasny and A. L. Fisher, *Text. Res. J.* **43**, 272 (1973).
119. W. H. Carter, E. L. Finley, and B. R. Farthing, *J. Fire Flam.* **4**, 106 (1973).
120. M. M. Birky and K. N. Yeh, *J. Appl. Polym. Sci.* **17**, 239 (1973).
121. R. R. Sanders, *Text. Chem. Color.* **5**, 285 (1973).
122. J. L. Throne, *Text. Res. J.* **43**, 642 (1973).

NOTE ADDED IN PROOF

The attention of the reader is directed to the following publications which have appeared since the preparation of this manuscript and which deal directly with flame-retardant PET:

J. W. Hastie, *J. Res. Nat. Bur. Stand.* **A77**, 733 (1973).
J. R. Bercaw, Some recent developments in man-made flame resistant fiber and testing technology, Proc. 1974 Symp. on Textile Flammability, LeBlanc Research Corporation (1974).
E. Nagai, Toyobo flame retardant polyester fibre—"HEIM", Proc. 1974 Symp. on Textile Flammability, LeBlanc Research Corporation (1974).
H. E. Stepniczka, *J. Fire Flam./Fire Retardant Chem.*, **2**, 30 (1975).
P. J. Koch, E. M. Pearce, J. A. Lapham, and S. Shalaby, *J. Appl. Polym. Sci.* **19**, 227 (1975).

5

Flame Retardance of Rubbers

T. H. Rogers, Jr. and R. E. Fruzzetti

1. Introduction

According to American Society for Testing and Materials (ASTM) definition.[1] rubber is "a material that is capable of recovering from large deformations quickly and forcibly, and can be, or already is, modified to a state in which it is essentially insoluble (but can swell) in boiling solvent, such as benzene, methyl ethyl ketone, and ethanol–toluene azeotrope. A rubber in its modified state, free of diluents, retracts within one minute to less than 1.5 times its original length after being stretched at room temperature (18 to 29°C) to twice its length and held for one minute before release." This definition includes not only the more common types of rubbers such as styrene–butadiene, natural, neoprene, and the like, but flexible polyurethanes and some of the flexible plastics. In this chapter these latter two categories are not included because they are handled in other parts of the book.

To understand the flame retardance of rubbers, we must know something about their chemical makeup. When the average person thinks of rubber, he is somewhat familiar with natural rubber, the grandfather of all rubbers, which he sees in rubber bands and truck tires; and he has heard about synthetic rubber, known to the chemist as styrene–butadiene rubber, which is used in shoe soles and automobile tires. However, there are many more different types of rubber, and within each type there may be

T. H. Rogers, Jr. and R. E. Fruzzetti · Research Division, Goodyear Tire & Rubber Company, Akron, Ohio.

TABLE 1
Commercial Rubbers

Name	ASTM symbol	Chemical configuration	Relative flammability
Natural rubber	NR	$-CH_2-C=CH-CH_2-$ $\quad\quad\quad\vert$ $\quad\quad\quad CH_3$	Very high
High *cis*-polyisoprene rubber	IR	Same	Very high
styrene-butadiene rubber	SBR	$(-CH_2-CH=CH-CH_2)_3-CH_2-CH-$ (phenyl)	Very high
Polybutadiene rubber	BR	$-CH_2=CH=CH-CH_2-$	Very high
Butyl rubber	IIR	$\left[\begin{array}{c} CH_3 \\ \vert \\ -CH_2-C- \\ \vert \\ CH_3 \end{array} \right]_{50} -CH_2-C=CH_2-CH_2- \\ \quad\quad\quad\quad\quad\quad\vert \\ \quad\quad\quad\quad\quad\quad CH_3$	Very high
Ethylene-Propylene-Diene rubber	EPDM	$(-CH_2-CH_2-)_3(-CH_2-CH-)-$ $\quad\quad\quad\quad\quad\quad\quad\quad\vert$ $\quad\quad\quad\quad\quad\quad\quad\quad CH_3$ $(-CH_2=CH-CH_2-CH_2-CH=CH_2-CH_2-)$	Very high

Name	Abbreviation	Structure	Flammability
Butadiene–acrylonitrile rubber	NBR	$-CH_2-CH=CH-CH_2-CH_2-CH-$ with CN	Very high
Chloroprene rubber (neoprene)	CR	$-CH_2-C=CH-CH_2-$ with Cl	Self-extinguishing
Polysulfide rubber (Thiokol)	TR	$-CH_2-CH_2-S-S-$, $S=S$	Very high
Chlorohydrin rubber	ECO	$-CH_2-CH-O-CH_2-CH_2-O-$ with CH_2Cl	Low burning rate
Silicone rubber	MQR	$-O-Si-O-Si-$ with CH_3 groups	Self-extinguishing
Hypalon rubber	CSM	$(-CH_2-CH_2-CH_2-\underset{Cl}{\overset{H}{C}}-CH_2-CH_2-CH_2)_x$, $\underset{Cl}{\overset{H}{C}}-SO_2$	Low burning rate
Fluorocarbon rubber	FKM	$-CH_2-CF_2-CH_2-CF_2-CH_2-CF_2-CH_2-CF-$ with CF_3	Nonflammable

several chemical variations. Table 1 shows some of the more popular types of rubber with their individual molecules which are linked together to give the giant macromolecules of rubber. Also included in this table are the ASTM symbols for each rubber and the relative degree of flammability of each rubber in its original state.

As can be seen by the chemistry, the ones with a high degree of flammability are essentially hydrocarbon in makeup, while those having medium to high resistance in flammability have other atoms than carbon and hydrogen incorporated within their structures. Fluorocarbon rubber is least flammable, followed closely by silicone rubber, with neoprene and Hypalon about on the same level, and chlorohydrin rubber not quite as good as Hypalon. For practically all applications rubber is mixed with other materials and then vulcanized in order to obtain optimum properties. During this compounding step it is possible to add a variety of materials to rubber to make it more flame resistant. Most raw rubbers are elastomeric gumlike materials which, when masticated on a two-roll mill or in a banbury (an internal screw-type mixing machine), become much softer and take on the appearance of a highly viscous liquid. During this breakdown period, chemicals required for vulcanization, such as sulfur, antioxidants, accelerators, pigments, and the like, are added and thoroughly mixed into the viscous mass. During the compounding step it is also possible to add materials such as pigments, liquids, or other rubbers and polymers, to rubber to make it more flame resistant.

2. Nature of Burning of Rubber

Before continuing on to the various ways of treating rubber to make it flame resistant, a discussion of the nature of the burning process and the types of testing required for rubber products are in order. There are several distinct stages in the burning process, including the heating of the rubber, the degrading and decomposition of the compound, the ignition of the gases resulting from this latter step with their subsequent combustion, and finally the propagation of the flame itself by the repetition of these steps. Frequently some low-melting-point rubbers, such as flexible urethanes, will melt just below the thermal degradation temperature, and the section adjacent to the flame will fall away from the main sample giving the impression that the rubber resists burning. However, if the melted and flaming droplets land on a combustible material, additional flaming will take place which indicates that the rubber is far from flame resistant.

As was mentioned earlier, the chemical structure of the rubber molecule is the factor that determines the thermal stability of rubber; it also determines the nature of the decomposition products. As an example, in the burning of raw natural rubber, the heat of combustion[2] is 10,547 cal/g. Thermal degradation of natural rubber occurs from 200 to 475°C,[3] and the temperature is dependent on the state that the rubber is in, i.e., the degree to which it has already aged and its molecular weight. Thermal degradation results mainly in chain scission of the polymer[4] with products[5] such as isoprene, dipentene, and *p*-menthene evolving. Natural rubber has no inherent flame retardancy, and when ignited it burns vigorously with emission of large quantities of black smoke. Reaction products of natural rubber can be made that will not burn, and commercial types include the chlorinated product which is marketed as Alloprene (ICI) and Parlon (Hercules). These products do not possess the elastomeric properties of rubber but are resins which are used in paints and other nonflexible applications.

3. Flammability Testing

At the ISO/TC 45 (International Standards Organization Technical Committee 45 on Rubber) meeting held in the Hague in October 1969, the delegations which represent practically all of the industrial countries of the world were asked to submit test procedures used for determining flammability of rubber. There were 116 test procedures submitted which could be classified into three major categories based on the approximate intensity of the heat applied. These were: low-heat ignition (examples— alcohol cup, match), medium-heat ignition (examples—Bunsen burner, Tirrill burner), and high-heat ignition (examples—electric coil with sparking, silicon carbide rod at 950°C).

The test that is of major importance today and will be for some time to come is known as United States Safety Standard No. 302 for Flammability of Motor Vehicle Interior Materials.[6] This test is quite similar to ASTM D-1692 in that the sample is held in a horizontal position during burning, a Bunsen burner is the source of flame, and a rectangular hood enclosure is used in which the specimen is burned. Although No. 302 is specifically designed for motor vehicle interior materials, including rubber and urethane foams that are used for cushioning and padding, it can be and is applied to other rubber products with slight modification.

Test No. 302 involves clamping a 14 × 4 × 1/2-in. foam sample between matching U-shaped metal frames and applying for 15 sec a 1 1/2-in.-high yellow flame from a Bunsen burner. In order to obtain the same burning temperature each time the test is run, the composition of the gas is specified as one which has a flame temperature equivalent to natural gas. Burning rate is determined by the time in minutes that it takes for the flame to advance along the horizontally held sample. In some cases the sample under test melts and falls away from the main portion of the test piece, giving the appearance of a nonburning or self-extinguishing sample. By placing finely divided cotton batting under the sample, the molten droplets will ignite the cotton which would indicate that the product is indeed flammable.

The simple match test is one of the easiest to run for screening. A wooden match is used and the rapidity of burning, the type of smoke generation, and the odor can tell quite a bit about the compound.

One rubber product that must pass a vigorous flame test specified by the Bureau of Mines is the conveyor belt used in mines.[7] A sample of belt is inserted in a bracket and ignited by a Bunsen burner for 1 min. When the burner is removed a fan is turned on to pull air at a velocity of 300 ft^3/min past the sample, sucking the flame and hot combustion gases back along its length. To pass the test the flame must extinguish within 1 min after the air suction starts, and all glow spots must disappear within 3 min. The air blast simulates conditions that would be experienced in the mine's ventilating system.

The limiting oxygen index (LOI) test, introduced in 1966, is a useful laboratory tool for determining the fire-resistant characteristics of a rubber. The LOI of a material is the percentage concentration of oxygen in a mixture of oxygen and nitrogen which will maintain equilibrium burning conditions. The heat generated during combustion just balances the heat lost to the surroundings. Physically, this is the lowest concentration of oxygen which will support sustained combustion of the test specimen. In the test, a sample held in a vertical position within a test column is

TABLE 2
Oxygen Index Results on Rubbers

Identification of rubber	Oxygen index
EPDM	21.9
Hypalon	25.1
Silicone GE SE9035	25.8
Neoprene	26.3
Silicone GE SE9014	27.9
Silicone GE SE9044A	30.4
Silicone GE SE9090	33.7
Silicone GE SE5537	34.0
Silicone GE Expl 2-2123A (Fire resistant type)	39.2

ignited on top like a candle. Two lines, one oxygen and the other nitrogen, are accurately measured before the streams are mixed and flowed up the column. By adjusting the gases and determining the minimum concentration of oxygen that is required to just support combustion, the LOI is obtained. Table 2 shows some rubbers that have been subjected to this test.[8] As the index increases, more oxygen is required to maintain burning and the rubber is more flame resistant.

Most flammability tests for rubber products are designed to simulate conditions under which the products are used. Modification of ASTM D-1692 is frequently made to adapt it to end-product use.

4. Smoke Generation

Equally important in the burning of rubber is the problem of the quantity and type of smoke generated.[9] The news media are full of stories of nursing homes and public buildings which experienced fire, but human casualties resulted from asphyxiation. Quite often smoke is highly toxic,[10] especially when burning takes place with a paucity of oxygen and large volumes of carbon monoxide[4] are generated. In the case of vulcanized rubber, sulfur- and nitrogen-containing gases may be present in addition to carbon dioxide and carbon monoxide. The presence of certain fire retardants may increase the potential amount of smoke[11] and toxic gases, but, depending on the severity of the ignition source in a real fire, the presence of the fire retardants may prevent the fire from developing to a point where smoke and toxic gases are a serious problem.

Several methods and types of apparatus have been set up for measuring smoke density and determining types and quantities of products produced at fires.[4,12-15] The principle employed in determining smoke density is to burn a standard-size sample with a known heat source under controlled conditions in a chamber and, by optical means, determine the time required to reach a certain density. In the case of the National Bureau of Standards Smoke Chamber, the increase in smoke density may be obtained in several time increments to show the results graphically. A comparison of the XP-2 smoke density chamber (ASTM D-2843) and the NBS chamber shows the NBS chamber is a more versatile and more accurate measure of smoke development resulting from pyrolysis.[16]

Gases produced from burning rubber insulation in excess of air have been analyzed as 13.55% carbon dioxide, 7.55% carbon monoxide, 2.65% unsaturated hydrocarbons, 0.45% saturated hydrocarbons, 7.40% methane, 10.00% hydrogen, 0.12% hydrogen sulfide, 6.55% oxygen, and 51.73% nitrogen.[10]

5. Flame Retardancy of Rubbers

The flammability of rubber or any polymer is essentially a three-stage process beginning with ignition by a flame or other heat source, then decomposition of the polymer with the formation of volatile gases, and finally the ignition of the volatiles when sufficient oxygen is present from the surrounding environment. The flame propagates as more volatiles are generated by the heat that is transferred from the flame to the adjacent rubber. Flame-resistant rubber compounds are made by incorporating materials into the rubber which (1) generate volatiles that inhibit combustion, (2) act as a heat sink and thus decrease the intensity of the flame which becomes insufficient to generate more combustible gases, and (3) melt away so that sufficient flaming rubber drops from the rubber specimen thus decreasing the heat source and preventing additional volatilization of the gases whereby the flame is extinguished. This last phenomenon is common with some urethane rubber foams.

The addition of halogen atoms into the rubber molecules, as in the case of neoprene, or the mixing of a halogen-containing material into rubber are means by which flame retardancy of rubber is obtained.[17] Of the halogens, bromine is more effective than chlorine, and chlorine is more effective than fluorine. However, the way in which the halogens are tied into the rubber and the nature of other materials that may be present are also important. A possible reason for bromine being more effective than chlorine is the faster generation of hydrogen bromide over a smaller range of temperature than hydrogen chloride during the ignition stage. Some free-radical contributors such as dicumyl peroxide, which may be used in certain types of vulcanization systems, act in a synergistic manner with halogen-containing complexes. Antimony-containing materials, particularly antimony trioxide, have a synergistic effect on halogens when used in rubber compounds; this results in a greater quantity of char residue.[18]

Phosphorus compounds such as tributyl phosphate and triphenyl phosphate are used in rubber compounds as flame retardants, and it is thought that their effectiveness is the result of the formation of nonvolatile complexes during burning which inhibits the burning mechanism and results in greater char.[19,20] Phosphorus compounds have less deleterious effects on aging of the rubber than halogens, and they are better as non-stainers. Frequently combinations of phosphorus, halogens, and antimony are used together to develop the right balance of physical properties of the vulcanizate with flame retardancy.

Recently the use of large quantities of aluminum trihydrate with halogen-containing materials, such as chlorinated paraffin and antimony trioxide, have been found to reduce flammability in latex foam and also to have relatively low smoke generation.[11]

6. Commercial Rubbers

6.1. Natural Rubber (NR)

Practically all rubbers, including natural, are used in the vulcanized state. During the compounding stage, when sulfur, accelerators, anti-oxidants, and other vulcanization ingredients are added, combustion-retarding materials such as halogen-containing materials, borates, metallic oxides, and various pigments, may be added to retard flammability in the end product. Most of these materials, when used in sufficient quantities to yield a nonflammable product, decrease other physical properties such as stress/strain values, aging, and rebound. The cost of the product is substantially increased also. A natural rubber may be compounded with combustion-retarding materials, such as a combination of chlorine donor and hydrated alumina, to make it sufficiently flame retardant to pass one of the medium-heat ignition tests.

Flame-resistant halogenated natural rubber is made[21] by combining the rubber chemically with the halogen trichlorobromomethane. Centrifuged natural rubber latex is stabilized with sodium dodecyl sulfate solution to which is added tertiary butyl hydroperoxide and an anionic stabilizer and ammonia. Tetraethylenepentamine is then added as a 10% solution, and the system is reacted for 16 hr at a temperature of 30°C or lower until 95% of the trichlorobromomethane is combined with the rubber. This results in a latex from which vulcanized rubber films are produced that are self-extinguishing in less than 1 sec. The stress/strain values are decreased, and the tension set is increased. It is also possible to react larger quantities of this halogen with natural rubber in the latex form and then blend this, which is a non-film-former, with natural latex to achieve flame resistance. The compounding of these rubbers utilizes antimony trioxide to enhance flame retardancy. Although this product has not become commercial, with the increased emphasis on nonflammability of products, it may become so.

6.2. Synthetic cis-Polyisoprene

Synthetic *cis*-polyisoprene (IR), made by the polymerization of isoprene in solution and then converted to either the raw rubber stage or to a latex by dispersing the polymer in water, has physical properties, including flammability, equivalent to natural rubber. Except for its lower gum strength, it is completely equivalent to natural rubber and may be used wherever natural rubber is used. The decomposition temperature is 295–450°C.[3] Flame retardancy may be built into IR the same as with NR.

6.3. SBR

Styrene–butadiene rubber (SBR) is the most important and widely used rubber in the world today. It is a copolymer of styrene and butadiene, generally in the ratio of 1:3 and is produced chiefly by emulsion polymerization, but increasing quantities are being made by solution polymerization. It has inferior gum tensile strength as compared with NR, but when reinforced with carbon black, it achieves strengths of greater than 3000 psi. It is used in making automobile tire treads and in many industrial products such as conveyor belts, hoses, flooring, and the like. The decomposition temperature[3] of SBR is 378°C, and products of decomposition include butadiene with a mixture of saturated and unsaturated hydrocarbons.[5,22] The heat of combustion[23] is 10,400 cal/g. Although SBR has no fire retardance, it can be compounded with halogens, phosphorus, and antimony complexes similar to those used in natural rubber, and made to pass flammability tests. A typical fire-resistant formulation for an SBR mechanical goods application would contain, based on 100 parts rubber, 20–60 parts chlorinated resin (Chlorez 700) and 10 parts antimony trioxide in addition to the regular vulcanization ingredients. Stocks made with these compositions would be self-extinguishing within 5 sec using a medium-intensity-flame test procedure.[24] Chlorinated paraffin may be substituted for the Chlorez with approximately the same nonflammability properties. SBR latex foam compositions have been made self-extinguishing by incorporating 20 parts chlorinated paraffin, 10 parts antimony trioxide, 8 parts tris(2,3-dibromopropyl phosphate), and 4 parts tetrabromobisphenol, all based on 100 parts rubber; this is in addition to the pigments and chemicals required for vulcanization. Because of the low-density nature of foam rubber, usually from 3.0 to 20 lb/ft^3, this represents one of the more difficult rubber products to make flame resistant.

6.4. Polybutadiene Rubber

Polybutadiene rubber (BR) is used in blends with NR and SBR to increase resilience and in tire treads to improve mileage. The decomposition temperature is 382°C.[3] The heat of combustion is 10,700 cal/g.[25]

This rubber has no basic flame-resistant properties, but like NR and SBR it may be compounded with conventional flame retardants to obtain this property. Brominated[26] and tetrachlorinated[27] reaction products of BR also yield products that will pass conventional flame tests.

6.5. Butyl

Butyl rubber (IIR) is produced by the copolymerization of isobutylene with small quantities (up to about 5%), of isoprene in solvent system. Many types are produced, including small quantities of chlorinated butyl rubber. Because of its low air permeability factor, its major applications are for truck-tire inner tubes and as an inside layer in passenger-car tires. It also is used in wire and cable insulation, hoses, gaskets, and mechanical goods. Thermal degradation occurs by depolymerization;[28] thermal decomposition temperature[3] is 260°C; and heat of combustion[29] is 11,200 cal/g. There is no inherent fire retardance, but nonflammability can be built into the compound using those methods described for NR and SBR.

6.6. EPM and EPDM

Ethylene propylene diene rubber (EPM and EPDM) are polymers of high-molecular-weight olefins produced by solvent polymerization systems. The EPM is a saturated copolymer; the EPDM is a terpolymer with diene used to create vulcanizable sites on the chain. Although chlorination of these rubbers has been accomplished for the purpose of covulcanizing them with NR and SBR, the quantity has been insufficient to make them flame resistant.[30] Flame retardancy can be built into the rubber by addition of halogens, phosphorus complexes, oxides of antimony, or a combination of these.

6.7. Nitrile

Butadiene–acrylonitrile rubber (NBR), usually called nitrile rubber, has a high degree of flammability, but it can be made nonflammable with additives. It is a copolymer of butadiene and acrylonitrile made by emulsion polymerization, and it has excellent oil and chemical resistance. Its major uses are for oil seals, conveyor belts, printers' rolls, and adhesives. The thermal degradation temperature[31] is 380°C. It can be made flame resistant by compounding it with similar materials used for NR and SBR. A compound that will extinguish instantly when the flame is removed consists of 100 parts NBR, 30 parts siliceous filler, and 30 parts tricresyl phosphate, in addition to the regular materials required for vulcanization.

6.8. Neoprene

Chloroprene rubber (CR) is also called neoprene and was the first American-made synthetic. Like the other synthetics, it represents a family of polymers, and its major building block is chloroprene, which contains 40% chlorine by weight. Products made from CR are normally self-extinguishing, but for making compounds with exceptionally good flame resistance, mineral fillers are usually added.[32-34] Hydrated alumina, hard clays, and calcium silicate may be used. Barite (barium sulfate) is undesirable; petroleum plasticizers are not recommended. A model compound for flame resistance would contain a blend of 75 parts Neoprene W and 25 parts Neoprene FB with 30 parts hydrated alumina, 20 parts hard clay, 10 parts zinc borate, and 15 parts antimony trioxide.[35] CR does not use sulfur or the organic accelerators used in NR and SBR for vulcanization but rather metallic oxides such as magnesia or zinc. The blending of the ingredients into CR is by use of regular rubber mixing equipment. CR vulcanizates have high stress/strain values, good abrasion resistance, and outstanding sunlight resistance. They also have resistance to swelling and deterioration by oils and naphtha. Neoprenes are used for molded and extruded goods, hoses, belts, wire and cable, coated fabrics, gaskets, and seals. Neoprene latex is used to make flame-resistant foam-rubber mattresses for the U.S. military. The thermal degradation temperature is 382°C,[3] and the decomposition products include HCl.[36]

A unique system based on neoprene latex has been developed to produce high-quality foam products with fire-retardant characteristics superior to all other known foam-cushioning materials. Additionally, these foams have high load-bearing capacities at lower densities than heretofore possible with neoprene latex foam. The new system of preparing neoprene latex foams is achieved by incorporating a blend of undistilled polyisocyanates in a frothed latex compound. The uncured foams rapidly develop high wet-gel strength, which reduces linear shrinkage and permits the attainment of densities as low as 3 lb/ft^3. The rapid development of wet-gel strength is also sufficient to allow the demolding of 3–5-lb/ft^3 foams from pin core molds after 15–30 min at room temperature without the usual steam vulcanization precure. The cured foams have 2–3 times the load-bearing capacity (per unit density) of conventional neoprene foams.[37]

6.9. Thiokol

Polysulfide rubber (TR), commonly called Thiokol, has good solvent resistance and relatively low stress/strain properties. This family of rubbers is employed in gasoline hoses, putties, gaskets, printers' rolls, and some mechanical goods. Thermal degradation[38] occurs at 320–360°C, with SO_2 and H_2S included in the volatiles. This rubber is not inherently fire resistant, and the applications to which it has been put has not encouraged much work on making it resistant.

6.10. Chlorohydrin Rubbers (ECO)

They have moderate flame resistance and may be made more resistant by additives. HCl is one of the products of combustion. This rubber has good chemical resistance, low gas permeability, and good heat aging and ozone resistance. It is recommended for seals, gaskets, wire and cable jackets, mechanical goods, and adhesives.

6.11. Silicone Rubbers (MQR)

They differ from organic-type rubbers in that the backbone consists of a silicone–oxygen structure. They have practically no initial gum strength, but when reinforced they go as high as 1000 psi. They exhibit inherent flame resistance in practically all applications,[39-42] including foams.[43] Thermal degradation[3] occurs between 350 and 437°C, leaving a silica residue.

6.12. Hypalon Rubber (CSM)

This is moderately flame resistant[44] and may be made more so by adding halogen complexes and antimony oxide. Decomposition temperatures[45] go up to 487°C, and the volatile products include HCl. However, HCl evolution can be minimized by compounding, as in neoprene. Acid acceptors such as fine-particle-size calcium carbonate are blended into the compound to neutralize HCl formed during pyrolysis.

6.13. Fluorocarbon Rubbers (FKM)

This is a family of elastomers noted for chemical and heat resistance and especially flame resistance.[46] They have good gum tensile strength (1800–3000 psi) and are used in seals, gaskets, valve linings, and protective coatings. Maximum service temperature is 205°C, and thermal degradation proceeds by depolymerization to the monomer in addition to 13% HF.[47] Thermal decomposition temperature is 478°C.[3] An important subclass of fluorocarbon rubbers is the nitroso rubbers, which are reaction products of trifluoronitrosomethane and tetrafluoroethylene. They may be produced in bulk, solution, or water-suspension systems. Carboxynitroso rubbers are made by reacting 3-nitrosotetrafluoropropionic acid and 4-nitrosohexafluorobutaric acid. Carboxynitroso rubber is the only rubber to be nonflammable in 100% oxygen atmosphere.[48,49]

7. Conclusion

Most rubbers of commerce consisting of hydrocarbons are used in products that do not normally require flame resistance, e.g., tires, gaskets, and washers. Where fire retardancy is required, burning resistance may be built into the rubber by the addition of halogen-containing materials, phosphorus compounds, oxides of antimony, and combinations of these materials. Some rubbers that contain atoms that inhibit burning (such as halogens and silica), have self-extinguishing properties. The addition of flame retardants to rubber decreases the physical properties of the vulcanizate (such as stress/strain values and permanent set), but make the rubber resistant to burning.

8. References

1. ASTM, 1971 Annual Book of Standards; Rubber, Carbon Black, Gaskets; Part 28: D-1566.
2. T. H. Rogers, *Encyclopedia of Chem. Technology*, Vol. 17, 2d ed., Wiley-Interscience (1968).
3. A. K. Sircar and T. G. Lamond, *Rubber Chem. Tech.* **45**(1) 329–345, 1972.
4. D. J. Rasbash, Smoke and toxic products produced at fires, *Trans. J. Plast. Inst.* # **2**, 55–61, Jan. 1967.
5. S. Straus *et al., J. Res. Nat'l. Bur. Std.* **42**, 499 (1949).
6. C. J. Hilado and W. Patten, An evaluation of safety standard no. 302, *J. Cell. Plast.* (Sept/Oct 1971).
7. Bureau of Mines, Title 30, Chap. 1, Sub E, Part 34.10.
8. J. L. Isaacs, The oxygen flammability test, *J. Fire Flam.* **1**, 43 (1970).
9. The Los Angeles Fire Department, *Operation School Burning,* No. 2, NFPA, Boston (1961).
10. J. Autian, Toxicological aspects of flammability and combustion of polymeric materials, *J. Fire Flam.* **1**, 239 (1970).
11. K. C. Hecker, R. E. Fruzzetti, and E. A. Sinclair, paper delivered at Rubber Division ACS, Boston, April 27, *Rubber Age,* April 1973, p. 25.
12. F. J. Rarig and A. J. Bartosic, Special Technical Presentation, No. 422, ASTM, (1967).
13. A Method of Measuring Smoke Density, National Fire Prevention Association, Boston, Mass. (Jan. 1964).
14. A. A. Loehr and P. F. Levy, Measurement of Smoke Density by TGA/Photometric Analysis, American Laboratory (Jan. 1972).
15. J. R. Gaskill, Smoke development in polymers during pyrolysis or combustion, *J. Fire Flam.* **1**, 183 (July 1970).
16. L. G. Imhof, Evaluation of New Materials for Flame Retardant Applications, Regional Technical Conf SPE Oct. 14, 1971, Newark Section, p. 7.
17. V. A. Pattison and Hindersinn; Kirk-Othmer *Encyclopedia of Chemical Technology,* Vol. 7, p. 1 Wiley-Interscience, New York (1967).
18. J. J. Pitts, Antimony–halogen synergistic reactions in fire retardants, *J. Fire Flam.* **3**, 51 (1972).

19. J. W. Lyons, Mechanisms of fire retardation with Phosphorus compounds, *J. Fire Flam.* **1**, 302 (Oct 1970).
20. J. K. Jacques, *Trans. J. Plast. Inst.* 1–67 (Eng).
21. E. G. Cockbain, T. D. Pendle, E. C. Pole and D. T. Turner, *Proc. Rubber Technol. Conf.,* 4th, London, 1962.
22. S. Straus and S. L. Madorsky, *J. Res. Natl. Bur. Std.* **61**, 77 (1958).
23. L. A. Wood, in *Synthetic Rubber,* (G. S. Whitby, ed.) Wiley, New York (1954).
24. Dover Technical Application Data Bull. 535 (April 1968).
25. R. S. Nelson, R. S. Jessup and D. E. Roberts, *J. Res. Natl. Bur. Std.* **48**, 206 (1952).
26. T. P. Dolezal *et al., Rubber Age* **104** (2), 37 (1972).
27. H. Rosen *et al., J. Appl. Polym. Sci.* **13**(8), 1721–1728 (1968).
28. S. L. Madorsky, *Thermal Decomposition of Organic Polymers,* Wiley-Interscience Publishers, New York (1964).
29. G. S. Parks and J. R. Mosley, *J. Chem. Phys.* **17**, 691 (1949).
30. R. T. Morrisey, *Rubber Chem. Tech.* **44**(4), 1025–1042 (1971).
31. G. S. Skinner and J. H. McNeal, *Ind. Eng. Chem.* **40**, 2303–2308.
32. R. N. Conklin, *Rubber News* **2**(6), 21 (1962).
33. I. T. Gridunov *et al., Khim. I. Khim. Tekhnol.* **5**(3), 821 (1962).
34. M. V Polemkina and L. N. Kireenkova, *Kauch. i Resina* **25**(9), 25–27 (1966).
35. C. E. McCormack, *Rubber Age,* Flame retardant compositions of neoprene and hypalon. pp. 27–36 (June 1972).
36. L. A. Wall, *J. Res. Natl. Bur. Std.* **41**, 315 (1948).
37. F. J. Asti and A. L. O'Meara, paper delivered at Div. Rubber Chemistry, ACS (April 1968).
38. R. E. Fruzzetti, unpublished data.
39. J C. Caprini, Society for the Advancement of Material and Processing Engineering # 7 (1964).
40. Anon., Tough silicones, *Chem. Week* **96**(24), 76 (1965).
41. Germ. Patent 1,221,010 (1966).
42. T. L. Laur and L. B. Guy, *Rubber Age,* (Dec. 1970).
43. Anon., Fire resistant silicone rubber foam, *Mater. Des. Eng.* **62**(1), 5 (1965).
44. H. J. Lanning, *Rubber Plast. Age* **37**, 227–232 (1956).
45. D. A. Smith, *Kautsch. Gummi Kunstst.* **19**, 477 (1966).
46. D. C. Miles, *Rubber Plast. Weekly* **141**, 536 (1961).
47. L. A. Wall and S. Straus, Pyrolysis of fluorocarbon polymers, *J. Res. Natl. Bur. Std.* **65A**, 227 (1961).
48. R. R. Hindersinn and G. Wagner, in *Encyclopedia of Polymer Science and Technology,* Vol. 7, pp. 1–64 Wiley-Interscience, New York (1967).
49. N. B. Levine, *Appl. Polym. Sym. 11,* pp. 135–156 (1959) (Eng).

6

Retardation of Combustion of Polyamides

Eli M. Pearce, S. W. Shalaby, and R. H. Barker

1. Introduction

Polyamides can be defined as polymers with reoccurring units linked by —NR—CO— groups.[1] They can be either synthetic or naturally occurring polymers. This chapter will be limited to discussions pertaining to synthetic polyamides or nylons. Polyamides can be classified into groups according to their base monomer types.[1-10] Polyamides made by the condensation of diacids and diamines (or suitable derivatives) represent the first group and are referred to as **AA-BB** polyamides (e.g., nylon 6-6). Those which are derived from condensed amino-acid structures constitute the second group and are usually referred to as A-B polyamides. They can be prepared from amino acids (e.g., polybenzamide and nylon 11), by lactam ring opening (e.g., nylon 6), or from unsaturated monomers by a hydride transfer mechanism (e.g., nylon 3). Typical examples of the different types of polyamides and general routes for their synthesis are summarized in Table 1.

Polyamides represent a sizable fraction of the total volume of synthetic polymers, which are now being used as engineering thermoplastics, foams, fibers, composites, and other related forms. In many of their applications,

Eli M. Pearce · Polytechnic Institute of New York, Brooklyn, N.Y. *S. W. Shalaby* · Chemical Research Center, Allied Chemical Corporation, Morristown, New Jersey. *R. H. Barker* · Department of Textiles, Clemson University, Clemson, South Carolina.

TABLE 1

Typical Polyamides and General Synthetic Approaches

Type of polyamides	Monomers and synthetic approaches	Ref.
Group I: AA-BB Polyamides	$nR(NH_2)_2 + nR'(COOR'')_2 \xrightarrow[-2nR''OH]{}$	
	$\begin{bmatrix} \overset{H}{\overset{\mid}{N}}-R-\overset{H}{\overset{\mid}{N}}-\overset{O}{\overset{\parallel}{C}}-R'-\overset{O}{\overset{\parallel}{C}} \end{bmatrix}_n$	1
Nylon 6-6	$R = (CH_2)_6;$ $R' = (CH_2)_4$	1
6-10	$R = (CH_2)_6;$ $R' = (CH_2)_8$	1
6-T	$R = (CH_2)_6;$ $R' = $ ⟨O⟩	1
10-2	$R = (CH_2)_{10};$ $R' = $ none	2
MPD-1	$R = R' = $ ⟨O⟩	16
Group II: A-B Polyamides		
a. Polyamino-acids	$nH_2N-R-COOR' \xrightarrow[-nR'OH]{} \begin{bmatrix} \overset{O}{\overset{\parallel}{C}}-R-NH \end{bmatrix}_n$	1
Poly(p-benzamide)	$R = $ ⟨O⟩	3
Nylon 11	$R = (CH_2)_{10}$	4
b. Polylactams	$n(CH_2)_m \overset{CO}{\underset{NR}{\diagup\diagdown}} \longrightarrow \begin{bmatrix} (CH_2)_m - \overset{O}{\overset{\parallel}{C}} - \overset{H}{\overset{\mid}{N}} \end{bmatrix}_n$	1
Nylon 3	$m = 2$	5
4	$= 3$	6, 7
6	$= 5$	1
12	$= 11$	8
c. H-transfer polyamides	$nCH_2{=}C-CONHR' \xrightarrow{Base} \begin{bmatrix} CH_2-CH-\overset{O}{\overset{\parallel}{C}}N \\ \quad R \quad R' \end{bmatrix}_n$ (R below)	
Nylon 3	$R = R' = H$	9

polyamides could be exposed to a direct flame or a high temperature in an oxidizing atmosphere. The possible routes to achieve or approach non-flammability in polyamides are discussed and critically evaluated.

In preparing this chapter, reports[10-30] pertaining to the different aspects of polymer flammability were interrelated and analyzed. This was done in an attempt to present the problem of retarding the combustion of polyamides as a challenge to polymer scientists with varying technical backgrounds. Before discussing different approaches to polyamide flame retardancy. the following should be considered: (1) the effect of certain molecular parameters and bulk properties on the course of the combustion and related processes. (2) the type and extent of the different physico-chemical changes which occur upon degradation. and (3) the different techniques available for assessing most of the changes or evaluating the products of both the degradation and combustion processes. New trends and approaches to achieve flame retardancy in polymers will be discussed in terms of their general applicability to polyamides.

2. Factors Affecting the Combustion and Related Processes of Polyamides

It is generally accepted[10,11] that the performance of any polymer in the presence of a fire is dependent on its behavior in the pre- and post-ignition processes, as well as the ignitability of its degradation products. Any attempt to understand the stabilization of polyamides against heat and fire must be associated with a basic understanding of those processes and their dependence on the polymer structure and bulk properties. Initially the preignition processes—heating, transitions, degradation, decomposition, and oxidation will be discussed. The ignition process (combustion) and the postignition processes (propagation) will be discussed subsequently.

2.1. Heating

When a polymer is exposed to a heat source (direct flame or radiant energy), the efficiency of transmitting the thermal energy (in terms of the heat flux) in the polymer, will determine the rate at which it will undergo physical and mechanical changes due to heating. On a macroscale the thermal conductivity and heat capacity (or specific heat) of the polymeric materials are the most important factors which affect the heat flux and hence the heating process. On a molecular level both the heat capacity and thermal conductivity of a polymer are affected to a great extent by its chemical structure and the manner in which the chains are packed. The effect of the chemical structure on the heat capacity of polyamides can be demonstrated by considering the case of nylon 6 and nylon 7.[31] At 100, 200, and 300°C, nylon 6 shows heat capacities (C_p) of 16.3, 28.0, and 41.0 cal/deg mol, respectively. The corresponding C_p values for nylon 7 were shown to be 19.1, 35.4, and 56.4 cal/mol deg. Specific heats of nylon 6, 6-6, 6-10, and 11 were shown to be 0.38, 0.40, 0.40, and 0.58 cal/g °C, respectively.[32] On the other hand, effect of structure on conductivity is rather complex.[33] However, different nylons show different thermal conductivities. The thermal conductivities of different polyamides (in cal/sec/cm²/1 (°C/cm) × 10⁻⁴) are 5.8, 5.9, 5.2, 7.0, and 5.3 for nylon 6, 6-6, 6-10, 11, and 12, respectively.[34,35]

2.2. Transitions

Prior to thermal degradation, most polyamides soften or melt. Hence, both the second- and first-order transitions measured as T_g (glass transition temperature) and the T_m (melting temperature) of any particular polymer should be considered in assessing the factors affecting its degradation. High T_g and T_m are usually associated with polymers of reasonable or excellent thermal stability and fire resistance. Both the first- and second-order transitions are endothermic processes, but the energy input required to achieve melting of a crystalline polymer is much higher than that needed for the glass transition of the same weight of an amorphous polymeric material (providing that all other conditions are similar).

The T_g of polyamides is affected by chain flexibility, which in turn is determined by chain structure and steric parameters. The T_m is determined by intermolecular forces, chain rigidity, chain regularity, molecular weight branching, and other factors. Structurally, the frequency of the amide groups along the chain, the steric requirement about the chain segments, and the amount of branching are the main determining factors for both the T_g and T_m values of polyamides. The T_g and T_m values of some important polyamides are summarized below.[36]

Nylon	T_m, °C	T_g, °C	Nylon	T_m, °C	T_g, °C
6-6	265	80	8-20	175	48
MXD-6	238	115	10-10	192	49
6-I	210	142	12-16	167	50
6-8	226	51	14-16	158	48
6-10	219	50	18-16	158	50
8-10	200	50			

The percent crystallinity of a polymer is also a contributing factor to the thermal stabilization of a polyamide. A high degree of crystallinity will decrease the rate of oxygen diffusion through the polymer and thus show an improved thermal-oxidative stability.

Other related thermodynamic processes which will affect the transition process are the latent heat of fusion (function of crystallinity) and latent heat of vaporization (function of the concentration of volatile low-molecular-weight species). both of which may accelerate the ignition process.

2.3. Degradation

Polymer degradation may be defined as the process in which the polymer dissociates into smaller fragments. The temperature of initial degradation is usually the temperature at which the least thermally stable bond fails. Degradation may be of two types: thermal degradation in the absence of oxygen, and thermal-oxidative degradation, which is influenced by both heat and oxygen.

In the absence of oxygen, chain degradation will be governed by (1) the decomposition temperature of the different chemical bonds, in particular the relatively weak ones; (2) the fraction of the least stable bonds; (3) the latent heat of dissociation of the least stable linkages; (4) the presence of reactive impurities; and (5) the concentration and position of certain functional groups in the polymer chain.

Examples of different chemical bonds and their characteristics can be summarized as follows:[37,38]

Chemical bond	Average bond energy, kcal/mol	Chemical bond	Average bond energy. kcal/mol
C—H	98.2	N—H	93.4
C—F	102	C—C	77.7 (ethane)
C—Cl	78		79.0 (propane)
C—Br	65		80.1 (isobutane)
H—F	135		84.9 (solid carbon)
H—Cl	102.1	C=C	140.0 (ethylene)
H—Br	86.7	C≡C	193.3 (acetylene)
C—N	69.0	C⋯C	123.8 (benzene)
C—O	84.0		

Considering the bond energy values of the different chemical bonds, one can see that: (1) In most polyamides the $-CH_2N-$ bond is the weakest thermal linkage along the chain. (2) A polyamide in which the bond order of some of its carbon–carbon bonds is more than one, will be more thermally stable than those with a saturated paraffinic structure. Examples of higher-order bonds are those present in alkenes, alkynes, and aromatics. (3) Substituting the hydrogen atom of the paraffinic components of polyamide chain will improve the C–C thermal stability, because of increased bond order about these linkages through hyperconjugation.[38a] This increases the thermal stability of the C–C bonds in the chain. (4) The presence of halogens on the main chain or its side groups, or as part of an organic additive, will lead to a facile formation of stable hydrogen halide. These acids may accelerate certain chemical processes such as hydrolysis, thus leading to noticeable chain degradation. The consideration of bond-energy values alone has a serious drawback in that it does not take into account other degradative mechanisms such as elimination reactions, activation of sites toward oxidation, and contribution of adjacent groups to steric, polar, and resonance effects.

2.4. Decomposition

Decomposition will be used to denote extensive degradation where a large number of the polymer bonds are ruptured. *Degradation* will denote loss of physical properties, while the polymer residue may show only a small weight loss. An example of the former is the complete decomposition of nylon 6-6 at $> 550°C$ (the decomposition temperature of nylon 6 and 6-6 is 310–380°C). An example of polymer degradation is the formation of cross-linked residual char which occurs in the case of certain aromatic polyamides at elevated temperatures. In the absence of molecular rearrangements or structural changes in the polymer, the degradation and decomposition stages can be separated only when the failure temperature of the least stable bonds is significantly lower than the decomposition temperature of the majority of the bonds in the polymer. When a polymer contains a variety of bonds with an almost continuous spectrum of decomposition temperatures, the degradation and decomposition stages merge into a single stage. Therefore, the factors affecting the degradation process can be also pertinent to the decomposition process for some systems. General factors affecting the polyamide decomposition are:

1. Temperature of initial decomposition (T_d): This is the lowest temperature at which decomposition occurs. When two hypothetical polyamides of equal specific heat and thermal conductivity are exposed to heat at their respective surfaces, the extent to which decomposition proceeds into the mass is largely a function of the T_d. Although most aliphatic polyamides have comparable T_d's, one may discriminate between certain polylactams and polyamides made from diacids and diamines, because of their different depolymerizability below the conventional T_d. Polylactams,

especially those made from 4C- to 6C-lactams, undergo equilibration with their monomers between 200 and 300°C.

2. Latent heat of decomposition: This is the heat absorbed or released during decomposition. If decomposition is exothermic, the heat released increases the rate of temperature rise, but an endothermic decomposition will slow the heating process. In the former case a polymer will undergo more facile ignition as compared to the latter.

In polyamides the most common decomposition mechanism is the chain cleavage. However, the polymers can be modified structurally to allow endothermic reactions to occur prior to or simultaneously with the process of chain cleavage. Alternatively an additive which undergoes an endothermic reaction below or at the T_d of the polyamide may also be used to slow the rate of the polymer degradation and render it less vulnerable to burning.

3. Decomposition behavior: The manner in which a polyamide decomposes will determine the relative amounts of resulting combustible and noncombustible gases, liquids, solid residue, and solid particles, as well as the pattern of the observed phase changes. As in most cases, since ignition and combustion occur in the gas phase, the complete elimination of combustible gases would effectively preclude flaming. Aliphatic polyamides cannot be reduced to highly carbonaceous residues without releasing some combustible volatiles. Structural modification of polyamides can lead to lower volatiles formation which render the polymer less ignitable and more flame resistant.

Decomposition reactions leading to graphitization and/or cross-linking are most desirable for improving polymer flame resistance. Both reactions will produce appreciable amounts of solid residues which tend to help in maintaining the structural integrity of the polyamides, protecting the adjacent unit masses from decomposition and impeding the mixing of air with the combustible gases. Certain molecular features are felt necessary to achieve graphitization and/or cross-linking in polyamides. These can be inherent features of the polymer chain or may be introduced by proper structural modifications. Cross-linking may also be introduced by external agents or additives. In some instances, violent decomposition may lead to the fragmentation of the solid residue, and the fragments will be entrained in the moving gases. These fragments appear as smoke particles and, if swept into the flame itself, as incandescence. Most aliphatic polyamides show little or no smoke generation upon burning.

2.5. Oxidation

At a high enough temperature and in the presence of sufficient oxygen, aliphatic polyamides will undergo rapid oxidation to produce heat and flame. On the other hand, at relatively lower temperatures ($\sim 200°C$) different polyamides behave differently, depending on their chemical structure. It is known that polyamides react with oxygen at moderate temperatures to form hydroperoxides at the —CH_2—N— group.[39] Thus the concentration of this group per unit mass of the polymer will determine its susceptibility to hydroperoxidation. Hydroperoxides may lead to the formation of a cross-linked surface upon prolonged aging and may render the polyamide less flammable. If aging is not associated with the hydroperoxide formation, these species will accelerate the polyamide decomposition. When aromatic diamines are used to form the polyamides, no —CH_2—NH— group will be available for hydroperoxidation and this may contribute to their improved thermal-oxidative stability.

2.6. Ignition

The ignition process is dependent on the nature of the decomposition process. All the factors affecting the latter will determine to a great extent the path of the former process. In addition, the ignition process is significantly affected by the following characteristics of the polyamides, which are in turn determined by their chemical structure (in the absence of additives or plasticizers).

1. Flash ignition temperature (FIT): This is the temperature at which the decomposition gases can be ignited by a spark or flame. The FIT values of aliphatic polyamides are higher than their decomposition temperatures. For example, nylon 6 generates an appreciable amount of caprolactam between 300 and 400°C but undergoes flash ignition at 421°C.[40]

2. Self-ignition or autoignition temperature (SIT): This is the temperature at which reactions within the material become self-sustaining to the point of ignition. The SIT of polyamides is affected by their chemical structure; for instance nylon 6 and nylon 6-6 have SIT values of 424 and 532°C, respectively.[40]

3. Limiting oxygen index (LOI): This is the minimum level of oxygen required to sustain ignition and combustion. For normal use, most commercial polyamides cannot continue burning with less than 21% oxygen and hence are considered "self-extinguishing."[10,11] The chemical structure of the different polyamides is expected to affect their LOI values.[41,42]

2.7. Combustion

This process describes a full-scale or fully developed thermal-oxidative process or fire. The important factor in the combustion process is the heat of combustion, that is, the heat released by the combustion of a unit mass. Flame inhibition will depend on the net heat of combustion, that is, the difference between the heat of combustion and the heat required to energize the polymer from its initial state to the combustion stage. A negative net heat of combustion will encourage flame inhibition, and a positive one will support flame propagation. In the case of polyamides, the type of combustible gases will determine the value of the heat of combustion. For instance, this will be high if carbon monoxide predominates in the pyrolyzate prior to ignition.[43] On the other hand, a lowered heat of combustion can result from excessive formation of ammonia.

2.8. Propagation

For the flame to propagate, all the precombustion processes in the condensed and gas phases must proceed in consecutive order and at optimum rates. Generally speaking, there should be (1) a constant supply of fuel precursors, (2) a constant flow of combustible gases, (3) sufficient oxygen, (4) a positive net heat of combustion, and (5) no barrier for the heat flux in all phases. Any disturbance in the sequence of the precombustion processes or these five specific requirements will lead to flame retardation and possibly fire extinguishment. In the case of polyamides, the chemical structure and hence the decomposition mechanism will determine both the fuel and combustible gas supply. Structurally modified polyamides or blends with certain additives may lead to the formation of a barrier (char formation) and hence disturb the heat flux between the gas and condensed phase. Certain additives may produce gases with reduced combustibility which may dilute the oxygen supply in the gas phase or result in an overall decreased heat of combustion.

3. Thermal and Thermal-Oxidative Degradation of Different Polyamides

Most of this section will be dedicated to nylon 6, 6-6, and 6-10, which are heavily investigated, in view of their abundance and relative importance. Studies on other polyamides are relatively few and will be briefly referred to in this review.

3.1. Nylon 6, 6-6, and 6-10

3.1.1. Thermal Degradation

A number of studies on the decomposition of polyamides have been reported, and several degradation products were identified.[44-49] The type and concentration of these products were shown to be functions of (1) polyamide structure, (2) temperature, (3) sample "environment," and (4) gas pressure in the reactor. In 1951, Achhammer[44] pyrolyzed samples of nylon 6, nylon 6-6, and nylon 6-10 at 400°C *in vacuo*. The volatile decomposition products (analyzed by mass spectroscopy) were shown to include carbon dioxide, carbon monoxide, cyclopentanone, water, and a series of low molecular weight hydrocarbons. The reaction scheme shown below was suggested to account for the formation of cyclopentanone and carbon dioxide. However, the measured concentration of CO_2 exceeded that ex-

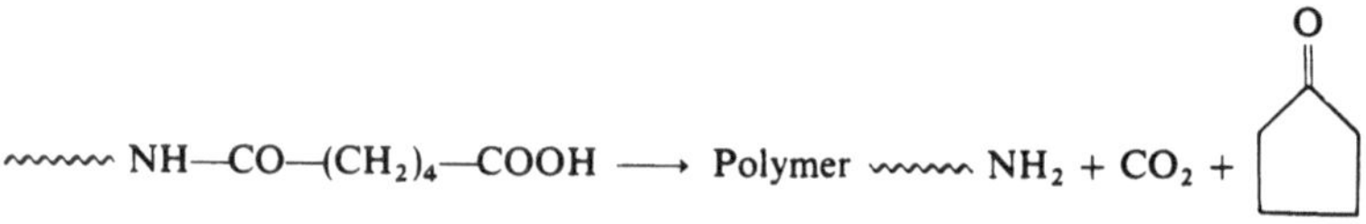

$$\text{NH—CO—(CH}_2)_4\text{—COOH} \longrightarrow \text{Polymer} \quad \text{NH}_2 + CO_2 + \quad \text{[cyclopentanone]} \quad (1)$$

pected on the basis of reaction (1). To account for this, it was suggested that the weak —CH_2—NH— link (66 kcal) undergoes homolytic cleavage (rather than the relatively strong C—C bond with 82.5 kcal) to produce reactive radicals which will form cyclopentanone and carbon dioxide through an unknown mechanism. The presence of traces of water in the decomposing polyamide will induce amide hydrolysis, thus producing more —COOH end groups. This will produce more CO_2 than expected on the basis of initial molecular weight of the nylon, and the following reaction scheme (2) may illustrate this proposal. Water may be present as

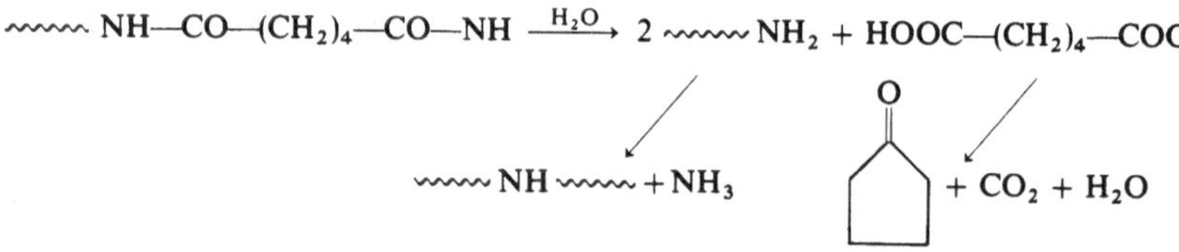

$$\text{NH—CO—(CH}_2)_4\text{—CO—NH} \xrightarrow{H_2O} 2 \quad \text{NH}_2 + \text{HOOC—(CH}_2)_4\text{—COOH}$$

$$\text{NH} + \text{NH}_3 \qquad \text{[cyclopentanone]} + CO_2 + H_2O \quad (2)$$

water of hydration or may result from the decomposition of the residual $^+NH_3{}^-OOC$ group (present in nylon 6-6) which will lose H_2O and form an —NH—CO— upon heating.

$$\overset{+}{\text{NH}}_3\overset{-}{\text{OOC}} \longrightarrow \text{CO—NH} + H_2O$$

The water consumed in the hydrolysis process will be regenerated upon cyclization of the diacid, reaction (2), and the process will thus be repeated.

In his study of the pyrolysis of nylon 6-6, Hasselstrom[45] demonstrated that the volatile products formed upon heating between 280 and 400°C contained appreciable amounts of CO_2 and NH_3. The residues were also analyzed, and the data were used to suggest that nylon 6-6 undergoes appreciable depolymerization at 280–400°C. This was ascribed mostly to a hydrolytic and not a homolytic amide cleavage.

Using dibutyl adipamide as a model compound for nylon 6-6, Goodman[46,47] indicated that under anhydrous conditions at 350°C the major decomposition product was CO_2 with the production of equivalent quantities of hydrocarbon, and less than 1% of the volatiles consisted of carbon monoxide and ammonia. Although these data are in partial disagreement with Achhammer's and Hasselstrom's conclusion, no alternate mechanism was suggested by Goodman.

In their classical work, Straus and Wall[12] studied the kinetics of thermal decomposition of aliphatic polyamides. They suggested that (1) the polyamide chain degrades at random; (2) the energy of decomposition was lower than that encountered in an exclusively free-radical process and is affected by the molecular weight; $E_a = 27$ and 34 kcal/mol for nylon 6 having $\overline{M}_n = 30,000$ and 60,000, respectively; (3) the presence of trace amounts of water and acid accelerate the rate of decomposition; and (4) both the homolytic (free radical at 380°C) and hydrolytic cleavages of the chain were proposed as mechanisms for the chain degradation, in view of the low E_a and the effect of water. In some respects this is consistent with Achhammer's free-radical mechanism and Hasselstrom's ionic mechanism and the previously proposed reaction (2), which were used to explain the degradation behavior of nylons.

Upon heating nylon 6 at temperatures between 257°C and 305°C, Kamerbeek[48] showed that carbon dioxide, ammonia, and water were the major volatile products formed and reactions (3a) and (3b) were proposed. The measured amount of water did exceed that predicted by reaction (3a), but part of this can be accounted for by considering reaction

$$\sim\sim\sim\overset{\text{O}}{\overset{\|}{C}}-NH-(CH_2)_5-COOH + HOOC-(CH_2)_5-NH-CO\sim\sim \longrightarrow$$

$$H_2O + CO_2 + \sim\sim\sim C-NH-(CH_2)_5-\underset{\underset{\text{O}}{\|}}{C}-(CH_2)_5-NH-CO\sim\sim\sim\sim \quad (3a)$$

and

$$\sim\sim\sim NH-CO-(CH_2)_5NH_2 + HN_2-(CH_2)_5-\underset{\underset{\text{O}}{\|}}{C}-NH\sim\sim\sim \longrightarrow$$

$$NH_3 + \sim\sim\sim NH-CO-(CH_2)_5-NH-(CH_2)_5-CONH\sim\sim\sim \quad (3b)$$

(3c), where water may form upon dehydration of amide end group.[12,49] On the other hand, small amounts of water may be present as tightly held water of hydration in the polymer matrix. Reaction (3c) not only accounts

$$\sim CO-NH-CH_2+CH_2\text{)}_4 \overset{\overset{O}{\|}}{C}-\overset{\overset{H}{|}}{N}\sim \longrightarrow \sim CONH_2 + CH_2\text{=}CH-(CH_2)_3-CO-NH\sim$$
$$\downarrow$$
$$\sim\sim C\text{≡}N + H_2O \tag{3c}$$

for the water formation in excess of that predicted by equation (3a), but also explains the presence of nitriles and olefins in the volatile products of pyrolyzed nylon 6. It is also consistent with the suggestions made by Straus and Wall[12] that CH_2—N bonds are quite reactive at elevated temperatures.

Recent work on the degradation of nylon 6-6[50] classified the volatiles into highly volatile materials (CO_2, NH_3, and H_2O), less volatile neutral fractions (cyclic monomer, cyclopentanone, cyclopentylidinecyclopentanone, and cyclopentylcyclopentanone), and basic fractions (hexylamine, hexamethyleneimine, and hexamethylenediamine and some pyridine bases). A number of mechanisms were proposed to account for the presence of most of these volatile products.

Polylactams show the additional feature of thermal equilibration with their monomers, which becomes an important factor at moderate temperatures (200–300°C) as compared to those polyamides made from diacids and diamines. The lactam concentration increases with the increase in temperature and decrease in the steric requirements of the lactam ring.[51-54]

Of all polylactams, nylon 6, nylon 11, and nylon 12 are the most important ones commercially. The equilibrium concentration of lactam in nylon 12[55] is very small in comparison to that of nylon 6. A number of studies were concerned with the caprolactam formation upon heating nylon 6 as a contributary feature to its thermal degradation.[12,50,56-61] Few mechanisms have been suggested for the appreciable depolymerization of nylon 6 into caprolactam. Smith[52] has postulated that caprolactam is formed by a depolymerization mechanism that is based on hydrolysis and both inter- and intramolecular segment interchange involving end groups. This postulation may lead one to expect that nylon 6 and nylon 6-6 have a negligible, if any, difference at all in their thermal stability. However, other data indicate that nylon 6-6 under certain conditions is more thermally stable than nylon 6.[41,42,45] Herman, Keikens, and Smith[53] suggested that the hydrolysis reaction (common to both nylon 6 and 6-6) plays a minor role in the formation of caprolactam. They also reported that the rate of caprolactam generation is a function of both amine and carboxyl end group as follows:

$$\text{Rate of monomer generation} = \frac{dm}{dt} = k'_3[NH_2][COOH]$$

A value of 42 kg/hr mol was reported for k'_3 at 250°C.

Studies of the thermal aging of nylon 6 at temperatures below 300°C has been the subject of few reports. Patemkina *et al.*[60] heated nylon 6 at 200°C in oxygen and were able to detect caprolactam although the effect of oxygen on the nonvolatile fraction of the degrading polymer was not determined. The effect of oxygen was later referred to by Reimchuessel and Dege.[61] These authors have also pyrolyzed nylon between 250 and 290°C in an oxygen-free atmosphere and measured the rate of decarboxylation and deamination. They proposed a new mechanism, reaction (4a), for the formation of ammonia, which results in an aminoether end group.

$$
\left[(CH_2)_5 \bigg\langle{\!\!{}^{NH}_{CO}} \right] \rightleftharpoons \left[(CH_2)_5 \bigg\langle{\!\!{}^{N}_{C-OH}} \right] + \sim\!\!\sim\!\!\sim NH_2 \longrightarrow
$$

$$
\sim\!\!\sim\!\!\sim O-\overset{\displaystyle N}{\underset{\displaystyle \|}{C}}-(CH_2)_5 + NH_3 \quad (4a)
$$

For the formation of carbon dioxide, reaction scheme (4b) was also proposed by those authors.[61]

$$
\sim\!\!\sim\!\!\sim COOH + \text{Caprolactam} \longrightarrow (CH_2)_5 \bigg\langle{\!\!{}^{N=C\sim\sim\sim}_{C}}\,O \longrightarrow
$$

$$
(CH_2)_5 \bigg\langle{}^{N}_{}\!\!\!\!{=}\,C\sim\!\!\sim\!\!\sim + CO_2 \quad (4b)
$$

In recent studies[49] on the thermal degradation of nylon 6, the polymer was decomposed isothermally at 400 and 500°C. It was shown that caprolactam was the main decomposition product. Oligomers of caprolactam were detected in the pyrolyzate, and only small amounts of carbon dioxide and ammonia were detected. Kinetic data suggested an approximately first-order free-radical decomposition with an activation energy of 45 ± 3 kcal/mol.

3.1.2. Thermal-Oxidative Degradation

Studies on the thermal-oxidative degradation of polyamides are not as frequent in the literature as those of the thermal degradation. This is because of the complexity of the reactions encountered in oxidizing media. Recently, attention has been given to the oxidative degradation of nylon in order to understand and also try to avoid some of the side reactions caused by traces of oxygen in polyamide melts. It has been shown by many authors that hydroperoxides are formed through the reaction of oxygen with the CH_2—N— group in polyamides and that the degradation products in air are somewhat different from those formed in nitrogen.[62-65]

Valk, Krüssmann, and Diehl[62] degraded nylon 6 in air at $200 \pm 10°C$ for 3 hr. The analysis of the hydrolyzed residue revealed the following components: formic acid, acetic acid, propionic acid, butyric and n-valeric acid, cyclopentanone, methyl oxalate, malonic acid, methyl succinate, methyl glutarate, adipic acid, $MeNH_2$, $EtNH_2$, $PrNH_2$, $BuNH_2$, and n-pentylamine. Adipic acid and valeric acid were the major components, and their production was thought to indicate a primary attack of oxygen on the N-vicinal methylene groups. The alkylamines were thought to be formed by the decarboxylation of the ω-amino carboxylic acid. Further, Valk and Krüssmann[63] suggested that cyclopentanone is probably produced through the cyclization of adipic acid, which is one of the major degradation products. Model air-oxidation studies of polyamides were made by Rieche and Schön,[64,65] who also concluded that in both nylon 6 and 6-6 degradation, the oxygen attacks the methylene group adjacent to the nitrogen. It was also suggested that caprolactam hydroperoxide (an oxidation product which forms during the polycondensation of caprolactam) changes easily to the imide of adipic acid.[64,65] Further support for the facile peroxidation of nylon 6 was shown later on by Valk et al.,[39,66] who reported that the N-vicinal methylene group in the trans conformation of the amide group, a component of the trans–gauche mixture in the reacting nylon 6, is about 60 times as reactive as N-vicinal in the gauche position (characteristics of caprolactam cyclic dimer, 1,8-diazacyclotetradecane-2,9-dione). The higher reactivity of the trans conformation resulted from the more intense hyperconjugation between the protons of the —CH_2— with the 4π electrons of amide group. This work signifies the contribution of the polymer chain conformation to the thermaloxidative stability of polyamides, which in turn affects the flammability.

In his studies on the thermal-oxidative degradation of polyamides, Neiman et al.[57] reported the presence of H_2O, CO_2, CO, HCHO, CH_3CHO, and CH_3OH in the volatile decomposition products of nylon 6. A mechanism was proposed for the formation of those products and was based on the initial formation of the following peroxide and peroxide radical:

$$\underset{\sim\sim\sim\sim CO-NH-\overset{|}{C}H-CH_2CH_2\sim\sim\sim\sim,}{\overset{O-O\cdot}{\overset{/}{}}} \qquad \underset{\sim\sim\sim\sim CO-NH-\overset{|}{C}H \quad \overset{|}{C}H_2-CH_2\sim\sim\sim\sim}{\overset{O-O}{\overset{|\quad|}{}}}$$

In their studies on the effect of molecular weight of polyamides on their thermal and thermal-oxidative stability, Okuhashi and Kuwahara[67,68] showed that in fractionated high-molecular-weight nylon 6-6 the thermal-oxidative stability decreases with the decrease in the molecular weight.

3.1.3. Effect of Aging on the Properties of Polyamides

3.1.3.1. Effect on Molecular Weight and Cross-Linking. When fused nylon 6-6 was heated in a stream of nitrogen at 270°C for a definite period of time, the molecular weight did not decrease. Similar treatment at 300°C revealed a decrease in the molecular weight (in terms of viscosity) to a definite minimum value, independent of the initial $\overline{M}_n$ of the different polymers (Fig. 1).[69] On the other hand, at 330°C, in addition to a decrease in $\overline{M}_n$ of the soluble portion, heating led to the formation of a cresol-insoluble substance. The amount of the insoluble fraction was shown to increase with time and reached 97% after 6 hr. Kamerbeek, Kroes, and Grolle[28] indicated that part of the gelation occurs from the condensation of two amine end groups, followed by amide formation according to the reaction shown. They have also demonstrated the presence

$$\mathrm{NH_2 + H_2N} \longrightarrow \mathrm{N} + \mathrm{NH_3}$$

of dihexylamine in the hydrolyzate of the gelled nylon.

In a recent study on the thermal degradation of nylon 6-6, Peebles and Huffman[50] reported that (1) the rate of gel formation in pyrolyzing nylon 6-6 was dependent upon the rate of removal of the volatile products of degradation; (2) if a sample was heated above its T_m in a sealed tube, the material showed a gradual increase in intrinsic viscosity and passed through a maximum followed by a minimum and then an abrupt formation of an insoluble material; and (3) if the volatiles were allowed to escape, rapid gelation took place, even in the absence of any oxygen. The molecular weight minimum observed in the closed system was attributed to simultaneous hydrolysis and aminolysis of the amide group, which was accompanied by the formation of multifunctional cross-linking agents. These

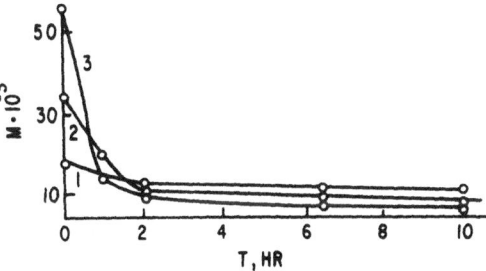

FIGURE 1. Dependence of the molecular weight of polyhexamethyleneadipamide on the duration of heating at 300°C (in a stream of nitrogen) for three samples with initial molecular weights of 18,000 (1), 35,000 (2), and 55,000 (3).[69]

will then participate in gel formation. The residue obtained by pyrolysis was hydrolyzed, and the hydrolyzate was shown to contain many components.

In reference to cross-linking in polyamides, Kamerbeek[28] suggested the following reaction to account for the gel formation upon pyrolysis:

$$\sim\!\!\sim\!\!\sim COOH + HOOC \sim\!\!\sim\!\!\sim \longrightarrow \sim\!\!\sim\!\!\sim C \sim\!\!\sim\!\!\sim + CO_2 + H_2O$$

with the carbonyl group reacting further:

$$\underset{\displaystyle \sim\!\!\sim\!\!\sim N}{\overset{\displaystyle +\sim\!\!\sim\!\!\sim NH_2}{\sim\!\!\sim\!\!\sim C \sim\!\!\sim\!\!\sim}}$$

Another mechanism was suggested by Neiman,[29] in which an ether linkage is formed.

$$2\sim\!\!\sim\!\!\sim CO\!-\!NH \sim\!\!\sim\!\!\sim \; \rightleftharpoons \; 2\sim\!\!\sim\!\!\sim \overset{\displaystyle OH}{\underset{\displaystyle }{C}}\!\!=\!\!N\sim\!\!\sim\!\!\sim \longrightarrow$$

$$\underset{\displaystyle \sim\!\!\sim\!\!\sim N\!\!=\!\!C\sim\!\!\sim\!\!\sim}{\overset{\displaystyle \sim\!\!\sim\!\!\sim C\!\!=\!\!N\sim\!\!\sim\!\!\sim}{|\,O\,|}} \quad + H_2O$$

This reaction also describes one of the routes for water formation during nylon degradation. Generally speaking, at minimum cross-linking only a slight change in the chemical structure of the polyamide was recorded.[70,71] In most of the reported mechanisms of cross-linking in polyamides, the participation of the paraffinic part of the chain was essentially overlooked. In fact, the hydroperoxidation and subsequent cross-linking of the polyamide aliphatic segments may follow mechanisms similar to those known for some polyolefins.

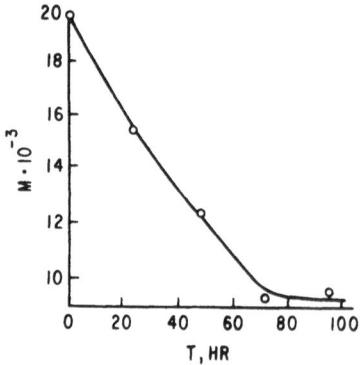

FIGURE 2. Dependence of the molecular weight of polycaproamide on the duration of its heating in sealed ampoules in nitrogen at 250°C.[70]

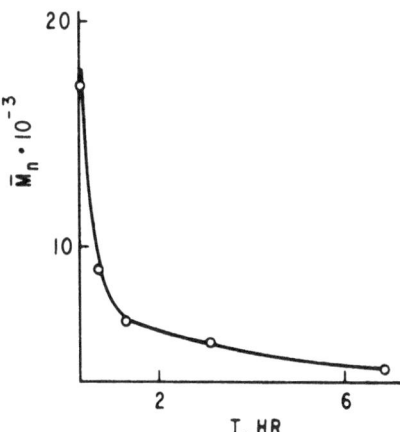

FIGURE 3. Dependence of the molecular weight of polycaproamide, determined according to the viscosity in 85% formic acid, on the duration of heating in air at 200°C.[72]

The molecular weight of polycaproamide was shown to decrease steadily with time, upon heating at 250°C in nitrogen. After 80 hr a definite minimum in nylon 6 $\overline{M}_n$, similar to that observed for nylon 6-6, was reached[70] (Fig. 2). Heating nylon 6 in air at 200°C resulted in an accelerated degradation as indicated by the fast decrease in the $\overline{M}_n$ as a function to time[72] (Fig. 3).

3.1.3.2. Effect on Mechanical Properties. Heating polyamides in the presence of air or oxygen leads to considerable changes in their physico-mechanical properties. Prolonged aging results in a decrease of the ultimate tensile strength and ultimate elongation with increased brittleness in both fibers and films. Nylon 6 fibers lose 0, 14.5, 50.8, and 89.3% of their original strength as a result of 8-hr heating in air at 80, 100, 120, 140, and 160°C, respectively[73] (Fig. 4). The data in Fig. 5 show a substantial reduction of the breaking strength and elongation of polyamide films heated at 140°C

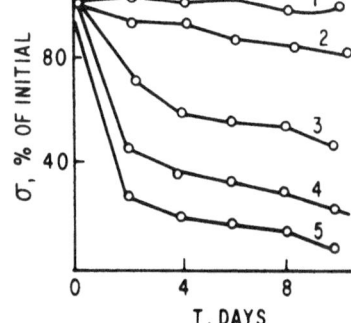

FIGURE 4. Dependence of the breaking strength of polycaproamide fiber on the duration of heating in air at various temperatures: (1) 80°; (2) 100°; (3) 120°; (4) 140°; (5) 160°C.[73]

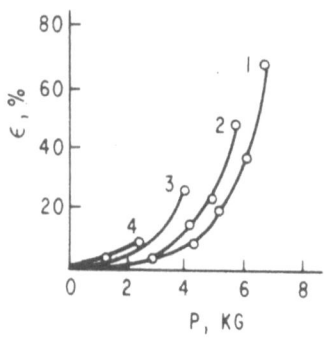

FIGURE 5. Dependence of the stretching on the load for polyamide film heated in air at 140°C: (1) initial film; (2) after 4 weeks of heating; (3) after 8 weeks of heating; (4) after 16 weeks of heating.[74]

in air for different periods of time.[74] Figure 6 illustrates a more significant deterioration in the elongation of a nylon 6 sample which was degraded in air as compared with that degraded in nitrogen. The oxidized film at 140°C (8 hr) loses 40% of the breaking strength in comparison with the initial value and shows no stretchability.[74]

It has been also shown that the introduction of a limited amount of cross-linking by oxidative aging results in improved toughness of polyamides.[75]

3.1.3.3. Effect on Thermal Behavior. A typical example of the effect of thermal degradation on the thermal behavior of polyamides is illustrated in Fig. 7, where two samples undergo degradation in two different atmospheres, namely air and nitrogen, and show noticeably different thermograms.[76] In many cases, changes due to minor differences in degradation can also be shown by DTA (or DSC).

3.2. Less Common Polyamides

The polyamides to be discussed in this section are either produced in small quantities commercially or have not been commercialized yet. Consequently they have not been fully examined for their thermal and thermal-oxidative stability or their flammability characteristics.

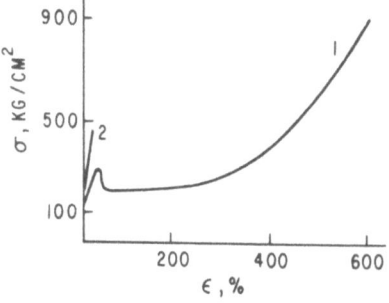

FIGURE 6. Dependence of the stress (σ) on the deformation (ε): (1) initial capron film; (2) film oxidized in air at 140°C for 8 hr.[74]

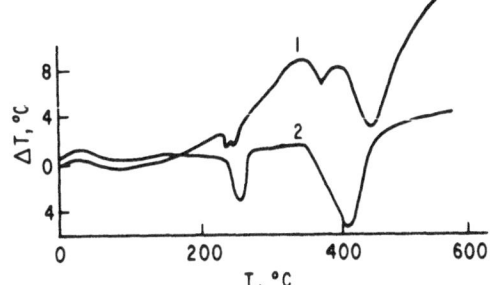

FIGURE 7. Thermogram of polyhexamethyleneadipamide: (1) in air;
(2) in nitrogen.[76]

3.2.1. Aliphatic Polyamides

3.2.1.1. Nylon 3. Poly-β-alanine is a highly crystalline polyamide[77] made by the polymerization of either an energy-rich β-lactam[5] or acrylamide.[9] It has little tendency to revert to its monomer by thermal degradation.[9,78] In addition, the polymer high melting temperature[77] and crystallinity favor its thermal stability. On the other hand, the ratio of the weak —CH_2N— to the C–C bonds in the repeat unit suggests appreciable —CH_2N— chain cleavages at the corresponding decomposition temperatures.

3.2.1.2. Nylon 4. This polyamide is made from a stable lactam and has a very high tendency to form pyrrolidone upon heating.[6] This and the high concentration of the thermally weak H_2C–N bonds per unit mass suggest that nylon 4 will not perform better than nylon 6 and 6-6 as to thermal-oxidative stability, in spite of its high melting,[55,79] crystallinity,[80] and claims as to improved thermal stabilization.[7,81]

3.2.1.3. Nylons 11 and 12. Nylon 12 is known for its reduced tendency for lactam regeneration as compared to nylon 6.[8,55] This and its low concentration of the thermally weak —CH_2N— group per unit mass favor its improved thermal stability. However, its low melting temperature[8] and the increased concentration of paraffinic hydrogens may result in poor flammability characteristics. Nylon 11 is prepared commercially from the corresponding amino acid.[4] It should have comparable thermal stability to nylon 12.

3.2.1.4. Polyoxamides. This class of polyamides is prepared from oxalic acid derivatives and different diamines.[2] The thermal transitions of aliphatic polyoxamides occur at higher temperatures than those of other structurally comparable polyamides,[82] but their thermal stabilities are not significantly different.[83] On the other hand, the tendency of polyoxamides to generate excessive amounts of carbon monoxide upon thermal degradation[83,84] may reflect a decrease in their flame resistance as compared to other polyamides, such as nylon 6-6, in view of the high heat of combustion of carbon monoxide. Some aromatic polyoxamides[85] have been prepared and show some promising thermal stability, but are still a scientific curiosity.

3.2.1.5. *Poly(piperazine amides)*. A number of these polyamides, containing tertiary amide groups, have been prepared and showed some resistance towards thermal aging in inert atmospheres.[86-88] However, their thermal-oxidative stabilities are noticeably poor.[89] The susceptibility of these polymers to oxidation is expected if one considers the high tendency of the piperazine moieties to form hydroperoxides.

3.2.2. Aromatic Polyamides

The synthesis and thermal properties of aromatic polyamides have been discussed in excellent reviews by Black and Preston.[17,90] The thermal properties of both aliphatic–aromatic and wholly aromatic polyamides were described and correlated with their chemical structure. It has been shown that the replacement of some of the aliphatic segments of polyamides with aromatic units resulted in a class of polymers having superior thermal stability as compared with wholly aliphatic ones; e.g., nylon 6-T (from hexamethylene diamine and terephalic acid) is more stable thermally than nylon 6-6. On the other hand, fully aromatic polyamides were shown, or predicted, to have even better thermal stability and to have great potential for use as thermally stable fibers and related products such as papers. One of these polyamides, Nomex,[91] is presumed to be a poly(m-phenyleneisophthalamide) and is now in commercial production. Other polyamides are now being evaluated and expected to be commercialized, such as the polyterephthalamide of N,N'-m-phenylene-bis(m-amino benzamide).[92]

Fiber B is a new fiber being developed by Du Pont. It is believed to be a homopolymer or a copolymer based on p-phenylene diamine and terephthalic acid and was claimed to have excellent mechanical and thermal properties.[93]

HT-4 Fiber is a new polyamide of undisclosed structure and is claimed by Du Pont to supercede Nomex in its flame resistance, but it is inferior to Nomex in prolonged thermal aging.[94]

The excellent thermal stability and flame resistance of partially or completely aromatic polyamides have been attributed[13,90] to (1) their high T_g or softening temperature, (2) their high energy of decomposition, (3) the presence of a minimum amount, or complete absence, of oxidizable paraffinic hydrogen, (4) their low volatility upon degradation, and (5) their tendency to form an infusible solid barrier upon decomposition.

4. Assessment of Physicochemical Changes in Degrading Polyamides

We have referred earlier to a few techniques which can be used to monitor the degradation process of polyamides, namely, viscosity and molecular-weight determination, extent of gel formation, and changes in tensile properties. Unfortunately, there are some limitations for the use of such techniques, e.g., the amount of sample required and the solubility and swellability of the polymer. A number of other techniques are now available and can be used effectively to study degradation of different polyamides. The different techniques and a brief description of their use are discussed below in order of decreasing significance.

4.1. Thermogravimetric Analysis (TGA)

Degradation in nitrogen atmosphere results in a progressive decrease in the weight of the nonvolatile polyamide residue as a function of temperature at a constant heating rate (programmed heating). In the case of isothermal degradation, weight-loss values are recorded as a function of time. In studying polymer degradation in an air or oxygen atmosphere one must not overlook the fact that as the polymer loses weight by degradation, oxidative reactions may increase the weight of the residue by introducing oxygen moieties. Careful handling of the TGA data provides basic information pertaining to degradation mechanisms of any particular polyamide. Many excellent reviews on this subject are now available.[14,95]

4.2. Differential Thermal Analysis (DTA)

DTA is one of the most powerful techniques used to understand the degradation of polymers, including polyamides. Both the position and area of endotherms or exotherms in DTA or differential scanning calorimetry (DSC) thermograms provide useful information regarding the dominant thermochemical processes in the degrading sample. DTA data obtained in both air and nitrogen are equally important. In interpreting the DTA thermograms of polyamides degraded in air, the overlap of the decomposition endotherm, starting at about 180°C, with the melting endotherm around 200°C should be considered. The air data can be intelligently interrelated with the flammability characteristics of the polymer. The theory and application of DTA to polyamides have been the subject of a number of excellent reviews.[14,96,97]

4.3. Infrared Spectroscopy (IR)

The use of IR spectroscopy to study both the volatile and nonvolatile products of degraded polyamides has been demonstrated by many authors.[28,98] The change in chemical structure and creation of new functional groups or the change in the concentration of a particular dipolar group can be followed by recording the IR spectra of the degrading polymer. In some cases, molded films can be heated in a special attachment to the spectrophotometer, and minor degradation may then be followed. A relatively new technique, the attenuated total reflectance (ATR), is now being used to study polymer degradation.

4.4. Mass Spectrometry (MS)

The use of MS to analyze the low-boiling volatile products of degrading nylons, was demonstrated by Straus and Wall as early as 1958.[12] Since then, many improvements have been introduced to this technique, and high-boiling volatiles can now be analyzed.[99] It is worth mentioning that an intelligent interpretation of the fragmentation patterns of the different volatiles may lead to a good understanding of the degradation mechanism of polyamides. Any changes in the degradation mechanism owing to the presence of certain additives or stabilizers in the polymer can be recognized by MS. Needless to say, traces of additive or chemical species which are part of the polymer backbone (or network) may be detected by MS. The use of mass spectrometry in conjunction with a gas chromatograph has been described as a powerful approach for studying polymer degradation.[99]

4.5. Pyrolysis–Gas Chromatography (P–GC)

This technique was used by many early workers to study the degradation products of polymers, including polyamides.[15] This technique is useful for both quantitative and qualitative studies. The different applications of P–GC[100,101] and GC in conjunction with TGA, namely TG–GC, have been described in an excellent review by Bauer.[102]

4.6. Election Spin Resonance (ESR)

In the thermal degradation of polymers, including polyamides, free radicals do form and hence, the use of ESR as a tool for studying free-radical-bearing moieties was invoked.[103] The thermaloxidative degradation of nylon 6 was recently studied by ESR, and a mechanism for the nitroxide radical formation was proposed.[104] This was based on the oxidation of secondary amine groups in the polymer.

4.7. *Nuclear Magnetic Resonance (NMR)*

In the study of thermal and thermal-oxidative degradation of polymers NMR may be used, both for the identification and quantitative determination of low-molecular-weight products and mainly for assessing the changes in any particular polymer during its breakdown.[105]

The wide-line NMR technique was used to study the thermal-oxidative degradation of nylon after heating at 210°C for 2 hr. The change in the intensity of the narrow components of the NMR line of the polyamide film as a function of temperature (40–140°C) was more noticeable in the degraded sample as compared with the intact one.[105,106]

4.8. *Other Techniques*

Many other less important techniques may be used in studying polyamide degradation. These include (1) polarography[107] for estimating hydroperoxide content, (2) light scattering,[99] osmometry,[99] and ultracentrifugation[99] for molecular-weight measurements, (3) melt rheology to study the effect of thermal degradation on the flow of the polymer,[108] and (4) ultraviolet to examine the effect of degradation on the type and concentration of chromophoric compounds present in the degraded polymer.[46]

5. *Stabilization Against Combustion*

The stabilization of polyamides against combustion will also call for the improvement of their thermal and thermal-oxidative stability. This will involve disruption of the preignition, ignition, and postignition processes, and may be achieved by one or more of the following approaches:

1. The use of additive-type stabilizers and flame retardants
2. Modification of existing polyamides
3. Synthesis of structurally modified polyamides
4. Synthesis of high-temperature polyamides

The first approach provides a simple means of providing thermal stabilization and/or flame retardance, particularly at the end-use or compounding stage. This approach is quite suitable for molding resins, but it has its limitations when the polyamides are to be spun into fibers. A number of difficulties may be encountered in the spinning operation due to the presence of these additives.

The second approach was developed in an attempt to incorporate certain chemical entities of similar structure to effective stabilizing and flame-retardant additives, as a permanent part of the polymer structure. Depending on the required level of stabilization or flame resistance, the polyamides may be subjected to minor structural modifications or major chemical reactions on their backbones. This approach has the advantage of providing permanent flame retardancy or thermal stabilization with minor changes in the physical and physicomechanical properties of the polyamides. However, there may exist some limitations to this approach if the modifications are to be made prior to molding or spinning. Such an approach could be quite effective when applied to shaped articles, where the thermal stabilization and flame retardance are imposed on the polyamide in postfabrication treatments.

The third approach was developed on the basis of prior knowledge of effective flame retardants and heat stabilizers. This approach pertains to the synthesis of monomers (diamines or diacids) containing certain groups similar to those present in effective flame retardants or heat stabilizers. The monomers are then polymerized or copolymerized at the desirable concentrations with common monomers. Such an approach can be quite useful for molding resins, but it may have its limitation in spinning resins. In fibers, certain amounts of crystallinity are required to attain an acceptable mechanical performance, and a copolymer may not furnish enough crystallinity for such performance.

The fourth approach consists of synthesizing polyamides which are inherently heat resistant and hence flame resistant in most cases. Most of these polyamides are aromatic in nature, and with the improvement in the solution spinning technology, their use as flame-resistant fibers is rapidly growing. Unfortunately this approach provides little help in imparting flame retardance to existing commercial polymers. The use of the high-temperature polymer as a substitute of the commercial, reasonably priced polyamides is questionable.

5.1. Use of Additive-Type Stabilizers and Flame Retardants

Considering the different factors affecting polyamide heat and flame resistance (Section 2), an effective additive may be capable of (1) redirecting the decomposition or combustion reactions towards the evolution of gases which are noncombustible and heavy enough to interfere with normal interchange of combustion gases and combustion air. This will disrupt the normal decomposition, ignition, combustion, and propagation processes; (2) changing the decomposition and combustion processes so as to lower the heat of combustion. This will perturb the normal combustion and propagation processes; (3) retaining the physical integrity of the polyamide, with reduced access of oxygen and heat. This may interfere with the normal decomposition, ignition, combustion, and propagation processes; (4) increasing the heat capacity and thermal conductivity of a material so as to have an increased heat dissipation and slow temperature rise due to heat-energy absorption. This will lower the amount of thermal energy accessible for propagation, and will slow the normal heating process and in effect may cause favorable changes in the subsequent processes; and (5) interfering with free-radical formation and hence retarding the degradation and oxidation processes in both the gas and condensed phase.

A number of additives have been used to stabilize polyamides against thermal and thermal-oxidative degradation. Quite a few additives were shown to be effective flame retardants for polyamides. Unfortunately, very little is known about the mechanism of action of most of these additives. The different additives tabulated in Tables 2 and 3 will be discussed in

TABLE 2

Additive Stabilizers and Flame Retardants for Spinning and Molding Polyamide Resins

Name and/or structure of additives	Use[a]	Ref.
1. Phosphorus-containing systems, with or without halogen moieties		
Phosphonium halides; $[R_3PCH2R']^+ + X^-$, $[R_3PNHR''] + X^-$	AO	132
Complex compounds of Cu_2I_2 and hydrocarbon phosphines	AO	133
Caprolactam phosphonate	FR	134

Phosphonitrile compounds; $(NPX_2)_3 \cdot (NPX_2)_{3-4}$	TS	135
Copper compounds plus phosphonium halides;		
$Cu(OAc)_2 + [R_3PR']^+X^-$	TS, TOS	136
Organic phosphinate plus manganese hypophosphite	TS	137
Phosphinylhydrocarbyloxy esters of pentavalent phosphorus compounds	FR	138
Phosphonates plus copper salts; $Cu(OAc)_2 + ArP(OEt)_2$	TS	139
Phosphoric acids plus halogenated anhydrides	FR	140

[a] TS = thermal stabilization, TOS = thermal-oxidative stabilization; AO = as antioxidant; FR = flame retardation.

TABLE 2 (continued)

Name and/or structure of additives	Use[a]	Ref.
Organoaminophosphonic acid or anhydrides; $N[CH_2P(OH)_2]_3$ $\parallel$ O	FR	141
Phosphinic and phosphonic acids or esters	TS	142
Phosphinic oxides plus chlorinated anhydrides	FR	143
Thermally stable phosphine oxides	FR	143a
Tertiary phosphinedihalides plus copper salts; $Ph_3PI_2 + Cu(OAc)_2$	TS	144
Phosphorus compounds plus amines; $R\overset{O}{\overset{\parallel}{P}}HOH + Ph_2NH$	TS	145
Trialkylphosphines; $sec\text{-}Bu_3P$	FR	146
Alkyl phosphoramides; $[(R_2N)_2\overset{O}{\overset{\parallel}{P}}]_2O$	TS	146a
Sodium phenylphosphinate plus sodium aminoalkylphosphinate	TS	147
Metal compounds of hexaalkyl phosphoramides	TS	147a
Hexamethylenediammonium phenylphosphinate + copper acetate + potassium iodide	TS	148
Naphthylphenylenediamine plus N,N'-bis(pyrocatechinphosphite)-p-phenylene diamine	TS	148a
Organoaminopolyphosphonate	FR	149
Phosphorthioates	FR	150
Organic halogenated diphosphates	FR	151
Halogenated phosphine pyranones	FR	152
Tris(bromomethyl)phosphine oxide	FR	153
Salts of phosphorus acid plus carboxylic acids	TS	154
Hypophosphates plus cerium or titanium salts	TS	155
Diphosphinium halides plus chlorinated hydrocarbon	FR	156
Phosphoncarbonamides	FR	156a
2. Organohalogen compounds, with or without metal oxides		
Organic halides plus metal oxides; chlorinated biphenyls + oxides of Sn, Cu, Fe, or Sb	FR	157
Alkyl halides plus copper compounds	TS	157a
Highly chlorinated biphenyls	FR	158
Brominated aromatic compounds	FR	158a
Highly chlorinated organic compounds plus antimony oxide or sulfide	FR	159
Perhalocycloalkanes plus antimony trioxide plus color stabilizer (organotin compound)	FR	160
Adducts of hexahalocyclopentadiene and alkadienes	FR	161
Adducts of halogenated cyclopentadiene and polyunsaturated cycloaliphatic compounds plus antimony trioxide	FR	162
3. Salts and complex compounds		
Halides of transition metals (e.g., Zn, Cd, Pb) with or without tin or copper oxides	TS	163
Copper salts; formate, acetate, or borate	TS	164
Copper salts plus iodoform	TS	164a

TABLE 2 (continued)

Name and/or structure of additives	Use[a]	Ref.
Lead borate plus inorganic halides	FR	165
Nickel dibutyldithiocarbamate	TS	166
Copper complexes of diaminotetracarboxylic acids plus potassium iodide	TS	167
Copper complexes of α,α-disubstituted polymethylene-bisiminoacetic acid plus iodides	TS	168
Lead borate plus chlorinated hydrocarbon	FR	168a
Tin salts	TS	169
Copper compounds plus aryl-substituted secondary amines	TS	170
Copper complexes of 8-hydroxyquinoline	TS	171
Sodium tetraborate plus halogenated hydrocarbons	TS	172
Copper phthalate plus inorganic iodides	TS	173
Copper salts plus halolactams	TS	174
4. Miscellaneous additives		
Derivatives of melamine or the melamine condensation product, melam	FR	175
2-Mercaptobenzothiazole or its copper salts	TS	176
2,4,6-Triamino-1,3,5-triamine plus triphenylchloromethane	FR	177
N-Substituted aromatic amines	AO	178
Intramolecular Schiff bases	TS	179
Lead arsinate plus perhalopentacyclodecane	FR	180
2-Hydroxy-1,3-bis(naphthylaminophenoxy)propane	TS	181
Triazine plus trityl compounds	FR	181a
Hydrolyzed cotton lignin	TS	182
Naphthylphenylenediamine plus mercaptobenzothiazole	TS	182a

TABLE 3

Additive-Type Flame Retardants for Reinforced Polyamide Resins

		Ref.
Melamine (or its derivatives) plus organic halogen compounds	Glass fibers or asbestos	184
Pentabromotoluene plus antimony trioxide	Glass mica or asbestos	185
Phosphorus-containing coupling agents, e.g., $(EtO)_2 P\!\!-\!\!(CH_2)_{10}\!\!-\!\!CH_3$, $\parallel$ O	Mineral fillers	186
Polyhalogenated aromatic hydrocarbon plus antimony trioxide	Glass fibers	187
Red phosphorus	Glass fibers	188
Brominated phenyl ethers plus antimony trioxide	Inorganic fillers	189
Dechlorane (dimer of hexachloropentadiene) plus Sb, Fe, or Sn oxide	Glass beads	190
Chlorinated norbornene derivatives plus antimony trioxide	Glass fibers or asbestos	191

groups of common structural features or similar modes of action or methods of application. The mechanism by which the different additives impart heat and fire resistance to polyamides will be discussed in relation to their effect on any of the stages of the burning and preburning processes. In these mechanisms distinctions will be made between the additive performance in the gas and condensed phases.

5.1.1. Phosphorus-Containing Additives

Phosphorus additives in the form of red phosphorus, organophosphorus, or inorganic phosphorus compounds are likely to perform via a general mechanism. However, the efficiency of a particular additive will depend on its degree of dispersion in the polyamide, its thermal stability (in the presence or absence of the polyamide), and its chemical structure.

Phosphorus additives appear to retard burning by redirecting some of the preignition and postignition processes, which occur primarily in the condensed phase, according to the following mechanisms:

1. To promote char formation, and hence reduce the concentration of combustible carbon-containing volatiles in the gas phase.[109,110] This will disrupt the propagation and perhaps the combustion processes. The presence of a char will result in lowering the heat transfer from the flame to the condensed phase, which will interfere with the heating and decomposition processes.

2. To stop the oxidation process of carbon at the carbon monoxide stage, thus decreasing the exothermic heat of combustion and indirectly damping the thermal-transition processes.[111-113]

3. To form phosphoric and related acids which will act as a heat sink as they undergo endothermic reduction.[114-117] This will upset the heating process.

4. To form a thin glassy or liquid protective coating on the condensed phase, thus lowering oxygen diffusion and heat and mass transfer between the gas and the condensed phases. This barrier will disrupt or redirect the heating, oxidation, combustion, and propagation processes.

5. To form anhydrides of phosphoric and related acids, which may act as dehydrating agents and discourage hydrolytic degradation and promote char formation. This may affect the decomposition and degradation processes. In addition to the above mechanism, other authors[118-121] have proposed numerous general explanations for the effectiveness of phosphorus and its compounds as flame retardants.

In a review by Kodolov, Sapogova, and Spaskii[26] on the fire resistance of phosphorus-containing polymers, it was concluded that the most effective retardants of the combustion of polymers are (1) those containing both phosphorus and nitrogen, (2) those bearing both phosphorus and halogen, and (3) oligomers containing phosphorus with phenylene rings in the chain, with functional groups capable of cross-link formation. In their discussion, it was suggested that in order to obtain a nonburning material it is necessary to form on its surface a polymer with comparatively high cross-link density containing sufficiently strong bonds. Using previous data on phosphorus chemistry,[26,120,121] it was suggested that such a three-dimensional surface network can be formed by the polymerization of metaphosphoric acid. This is known to form phosphorus-containing materials upon decomposition.

$$ n\text{P(OH)}_3 \longrightarrow $$

5.1.2. Halogen-Containing Compounds

Halogen fire retardants are believed to retard the burning of polymers by redirecting some of the essential processes for burning, in both the gas and condensed phases.[122] Some of the proposed mechanisms for the action of halogens are summarized below:

1. Generation of effective free-radical chain-terminating agents which will disrupt both the ignition and propagation processes.[122a]
2. Formation of a char through dehydrogenation reactions.[121c,123] The effect of char on the different processes will be similar to that described in the case of phosphorus. The char formation caused by halogen is much more effective than that characteristic of phosphorus additives.
3. Formation of a high-density blanket of hydrogen halides[125] which will act as a gas barrier between the gas and the condensed phases. Because of its high heat capacity, this blanket will act as a heat sink and protect the condensed phase from atmospheric oxygen. This will result in changes in the heating and oxidation processes.

Although halogen additives were described as very effective flame retardants for many synthetic polymers, they are less effective in the case of polyamides. This may be attributed to some chemical degradation which can be initiated or catalyzed by the hydrogen halides. In studies by Reardon[49] on the effect of brominated flame retardants on nylon 6 thermal stability, it was reported that: (1) A control sample of nylon 6 heated isothermally at 400 and 500°C underwent a first-order decomposition with an activation energy of 45 ± 3 kcal. A free-radical decomposition mechanism was suggested. (2) The incorporation of brominated flame retardants (e.g., hexabromobiphenyl) in nylon 6 resulted in lowering both the activation energy and the decomposition temperature. (3) Bromo compounds do not appear to alter the decomposition mechanism of nylon 6, since caprolactam was still the major decomposition product. It was further suggested that the bromo compounds catalyze the decomposition of nylon producing flammable products similar to those characteristic of the control nylon sample. It was concluded that these bromo compounds do not appear to be effective flame retardants for nylon 6.

5.1.3. Synergistic Systems

A synergism may be defined as a case in which the effect of two components taken together is greater than the sum of their effects taken separately. This concept has been tested in flame-retarding polyamides. Of the many synergistic systems examined, those comprised of phosphorus–nitrogen, phosphorus–halogens, and antimony–halogen will be discussed, and examples are shown in Tables 2 and 3.

5.1.3.1. Phosphorus–Nitrogen. The phosphorus–nitrogen synergism was demonstrated in polyesters and polyurethanes.[124] The latter example is quite relevant to potential synergism in polyamides. It has been shown that polyesters require about 5% P to achieve flame retardance, while polyurethanes (containing 10% nitrogen) can be flame retarded with only 1.5% phosphorus.[124a] This was attributed to the synergistic effect of phosphorus with the nitrogen present in the polyurethane chain. Subsequently, it is logical to expect similar behavior in case of polyamide whose nitrogen content usually varies between 5 and 15%. There are a number of conflicting views[109,122,125] regarding the existence of phosphorus–nitrogen synergism.

5.1.3.2. Phosphorus–halogen. Phosphorus–halogen synergism is likely to occur in both the condensed and gaseous phase according to the following general mechanism, which can be also applied to polyamides:

1. Phosphorus will act as a catalyst for halogen-atom formation at the surface. Chlorine will then act as a free-radical scaverger and thus alter the decomposition mechanism.[111]

2. Formation of volatile phosphorus halides and oxyhalides are highly effective in the gas phase as free-radical chain terminators.[109,111,126]

5.1.3.3. Antimony–halogen*. Antimony–halogen synergism has been the subject of many studies, and it was generally accepted that this system

---

* Other metals, such as As, Bi, and Sn, may perform through a similar mechanism to that of P or Sb.

manifests its flame retardance in the gas phase according to the following general mechanisms, which are likely to be valid for polyamides:

1. Formation of antimony halides and oxyhalides which are extremely active free-radical chain terminators. This may alter the decomposition and propagation processes.[127]

2. The formation of very small particles of nonvolatile solid oxides from the volatilized antimony halides, in the flame. At the interface with any of these airborne solids in a combustion reaction, the energy is dissipated on the surface of the solid, and the flame reaction mechanism will be altered. This is called the "wall effect."[128] The formation of HO_2 will be favored over the $\dot{O}H$ radical, and the flame reaction will become less energetic. This may affect both the decomposition and propagation processes.

5.1.4. Less Commonly Used Additives for Polyamides

A number of additives may have been effective in many synthetic polymers, and are of limited use for polyamides. The type and possible mechanism[129] of action of these additives can be summarized as follows:

5.1.4.1. Bismuth Compounds. These are less active than antimony but may follow a similar mechanism.

5.1.4.2. Boron Compounds. These are likely to manifest their flame retardancy in the condensed phase by redirecting the decomposition process in favor of carbon formation rather than CO or CO_2. A second mechanism may involve the formation of a surface layer of protective char which prevents the oxidation of carbon by limiting the accessible oxygen.[129] In both cases a char is formed.

5.1.5. Thermal and Thermal-Oxidative Stabilizers

A number of reviews and patents have described the different types of stabilizers which can be applied to polyamides.[29,130] However, no attempt has been made to study the mechanism of their action.[29] Examples of thermal and thermal-oxidative stabilizers or antioxidants are listed in Table 2. The most common classes of stabilizers are based on copper salts or chelates, manganese salts, aromatic amines, phenols, organosulfur compounds, organophosphorus compounds, and different mixtures of two or more of these systems. The mechanism of their action as stabilizers or antioxidants is most likely as free-radical scavengers interfering with free radicals formed by thermal homolytic cleavage. They may also interfere with peroxide or hydroperoxide formation and the subsequent thermal reactions.

5.1.6. Reinforced Polyamides

The use of engineering thermoplastics, including polyamides in composite forms with inorganic fillers, has recently attracted the attention of a number of molding-resin producers. The use of reinforced thermoplastics in the construction and automotive areas is most promising.[131] Faced with the problem of flammability regulations, interested parties are now introducing flame retardants into their composites. A few patents on flame-retardant composites have appeared in the last three years and are summarized in Table 3. Most of the flame retardants used are similar to those previously applied as additives in unfilled polyamides, namely phosphorus, halogen, and antimony compounds, as well as some melamine derivatives.

5.2. Modification of Existing Polyamides

In this approach existing polyamides are stabilized by minor modification imposed on their main backbone. This may include (1) end capping, (2) grafting with monomer-containing flame-retardant elements, (3) substituting with reactive additives containing flame-retardant elements or capable of rendering the polymer flame retardant through a special mechanism.

These modifications can be imparted to the polymer prior to or after its fabrication, depending on the type of polyamide and modifier in question. The possible operative mechanisms involved in the use of these modifiers can be summarized as follows:

1. Modification of the polymer to redirect the decomposition and combustion processes toward the formation of heavy noncombustible gases which will interfere with the normal diffusion of oxygen into the combustion zone.

2. Modification of the polyamide so as to increase its minimum ignition and decomposition temperatures and hence render it less susceptible to ignition.

3. Modification of the polymer so as to lower its heat of combustion. This is to minimize the amount of thermal energy accessible for other processes, such as phase transitions and decomposition.

4. Modification of the polyamide to increase the amount of solid residue, in order to maintain structural integrity, minimize the heat transfer, and limit the oxygen diffusion to the condensed phase.

Numerous approaches to the modification of polyamides have been examined. Those reported in the last 15 years are summarized in Table 4. The data in Table 4 and other pertinent literature information seem to

TABLE 4

Finishes for Heat- and Flame-Shaped Articles Made of Polyamides

Type of finish	Ref.
1. Methylolated urea, thiourea, melamine and derivatives	
Urea–thiourea–formaldehyde resins	194
Bisulfite-modified urea–thiourea–formaldehyde resins which are partially alkylated (water soluble)	195
Partially alkylated thiourea and urea–formaldehyde mixed resins (water soluble)	196
Alkyleneurea–thiourea condensates (water soluble)	197
Thiourea formaldehyde resins	198
Methylolated urea–formaldehyde resin plus ammonium bromide	199
Blends of urea–formaldehyde and sulfonated thiourea–formaldehyde resins	200
Methylolated phenylmelamine/boron fluoride	201
2. Other methylolated nitrogen compounds	
Aminotriazine-formaldehyde condensate plus ammonium bromide	202
Halogen-containing methylol compounds, e.g., $A-CO-NH-CH_2OH$ (A = halogenated group)	203
Fully methylolated heterocyclic compounds having a saturated ring with at least two nitrogens plus thiourea;	204

$$HO-CH_2N \overset{\displaystyle \overset{O}{\underset{C}{\diagdown}}}{} N-CH_2OH$$

Methylolated dicyandiamide/boron fluoride	205
Methylolated disubstituted guanidine/boron fluoride	206
3. Phosphorus- and halogen-containing compounds	
Halocycloalkenyl acyl halides	207
Chloramides plus chlorinated paraffins	208
Reaction product of phosphonyl chloride and ammonia	209
Triazinylaminoalkyl phosphonate	210
4. Other reactive chemicals and special treatments	
4-Vinyl-1-cyclohexene dioxide	211
Unsaturated organic acids or anhydrides (fumaric, itaconic and maleic)	212
Formaldehyde (for cross-linking amorphous regions in fibers)	213
Formaldehyde plus oxidizing agents	214
Polymers based on mono- and divinyl pyridines	215
Condensation products of sulfamic acid with urea derivatives (ethyleneurea or guanidine)	216
Iodo-s-triazine plus copper complex [Cu(II)acetonyl acetonate] and monosodium phosphate	217
Cyanuric chloride in presence of caustic soda	218
Molybdenum compounds, e.g., molybdenum disulfide	219
Sulfur treatment of aromatic polyamides, followed by heat treatment	220
Oxyhalides and halides such as $COCl_2$, $SOCl_2$, SCl_2, or PCl_5 used for treating aromatic polyamides at 250–500 C	221

indicate that a very effective means of imparting flame retardancy is by
polymer postreaction on the fabricated articles.[192-221] For most fiber-
forming polyamides modifications could be done as part of fabric finishing
processes.

In fabric finishing (or aftertreatment), many of the modifiers consisted
of urea–formaldehyde and similar condensates.[192,193] When a thiourea–
formaldehyde condensate was used, the T_m of nylon was lowered about
40°C.[192] This led Douglas[192] to suggest that flameproofing was accomplished according to the physical principle of removal of the melted, decomposing nylon by dripping. The exact chemical effect of thiourea on
nylon was not reported. A number of other urea or melamine derivatives
are shown in Table 4.[194-200] Serious drawbacks of the aftertreatment
techniques could be (1) unsatisfactory retention of modifier in some cases,
after repeated washing, (2) changes in fiber aesthetics, (3) poor reproducibility, and (4) inadequate incorporation of desirable amounts of flame
retardants.

Some interesting work in this area consisted of surface treatments
to further improve the flammability characteristics of some high-temperature
polyamides. Treatment of certain aromatic polyamides with sulfur improved
their dimensional stability and flame resistance.[220] Gas-phase reaction
of aromatic polyamides with certain inorganic halides improved their
flame resistance.[221] The flame resistance of poly(m-phenylene isophthalamide) was claimed to be improved by treating the fabric with certain
molybdenum compounds.[219]

Reactive halogen, boron, phosphorus, and nitrogen compounds were
effective in improving the flame resistance of the finished polyamides.[207-210]
Unsaturated monomers were also allowed to react with nylon fibers to
improve their heat resistance.[211,212]

5.3. Synthesis of Structurally Modified Polyamides

This approach involves the synthesis of polyamides and copolyamides
from monomers bearing certain elements or groups which were effective
as part of flame-retardant or thermal-stabilizing additives.[222-236] The
efficiency of these groups or elements is likely to be higher when they are
present as an inherent part of the polymer chain, rather than as a component
of a physical mixture. Stabilization by end-group capping has been shown
effective in a few cases.[237-241] A survey of the recent literature in the area
of stabilization and flame retardancy by the synthesis of those modified
structures of polyamides was made, and a summary of this survey is shown
in Table 5.

The most interesting monomers for the synthesis of structurally modified

TABLE 5

Chain and End-Group Modifiers for Improving the Heat and Flame Resistance of Polyamides

Type of modifier	Modification parameter[a]	Use[a]	Ref
1. R—$\overset{\text{O}}{\overset{\|}{\text{P}}}$(C$_6H_4$COOH)$_2$ + diamine	P	FR	222–227
2. C$_6$H$_5$$\overset{\text{O}}{\overset{|}{\text{P}}}$[(CH$_2$)$_n$—NH$_2$]$_2$ + diacid; n = 2–12	P	FR	233
3. R^1—$\overset{\text{O}}{\overset{\|}{\text{P}}}$(C$_6H_4$COOH)$_2$ + R^3—$\overset{\text{O}}{\overset{\|}{\text{P}}}$(C$_6H_4NH_2$)$_2$ *or* R^2—P(CH$_2$CH$_2$COOH)$_2$ *or* R^4—$\overset{}{\underset{\overset{\|}{\text{O}}}{\text{P}}}$(CH$_2CH_2CH_2NH_2$)$_2$	P	FR	222–232
4. C$_6$H$_5$$\overset{\text{O}}{\overset{|}{\text{P}}}$(CH$_2CH_2$COOH)$_2$ + hexamethylene diamine with or without caprolactam	P	FR	234
5. CH$_3$—$\overset{\text{O}}{\overset{\|}{\text{P}}}$(C$_6H_4NH_2$)$_2$ + diacid CH$_3$—$\overset{}{\underset{\overset{\|}{\text{O}}}{\text{P}}}$(C$_6H_4$COOH)$_2$ + diamine	P	FR	235
6. S(—R—NH$_2$)$_2$ + diacid	S	TOS	236
7. Bis(4-aminocyclohexyl), Bis(4-aminocyclohexyl)methane, *or* bis(4-aminocyclohexyl)ethane + diacid	High T_g and T_m	TS	237
8. 1,3-Dihydroxy-5,7-dimethyl adamantane + dinitrile + H$_2$SO$_4$	High T_g	TS	238
9. 2,2-Bis(4-aminocyclohexyl)propane + adipic + caprolactam	High T_g	TS	239
10. Diacylamide + polyamides	End group	TS, TOS	240
11. Carboxylic acids and derivatives + nylon 6	End group	TS	241

[a] Structural or physicochemical feature felt responsible for the performance of the modifiers.
[b] TS = thermal stabilization; TOS = thermal oxidative stabilization; FR = flame retardance.

polyamides are those containing phosphorus. Phosphorus-bearing additives were proven quite effective as flame retardants for polyamides. Hence, it was anticipated that polyamides made from phosphorus-bearing monomers would also be flame-resistant. Polyamides based on bis(p-carboxyphenyl)phosphine oxides were first prepared by Korshak, Frunze, and coworkers[222-225] and others.[226,227] Other phosphorus-bearing diacids and diamines were used successfully to prepare flame-retardant or thermally stable polyamides.[228-235]

A limited number of examples of the thermal-oxidative stabilization of polyamides by structural modification can be found in the literature. This may be associated with the questionable stability of the claimed thermal-oxidative stabilizing moieties at the melting temperature of most polyamides. Polyamides with a divalent sulfur in the main chain were prepared and shown to have good heat and light stability.[236]

Scattered reports described different techniques to achieve modest heat stability in polyamides by structural modification of their chains. These include the use of monomers capable of forming rigid backbones with high glass-transition or melting temperatures as shown below:

1. Bis(4-aminocyclohexyl), bis(4-aminocyclohexyl)methane, and 1,2-bis(4-aminocyclohexyl)ethane were reacted with suitable diacids to give thermally stable polyamides.[237]

2. Isophthalonitrile and 1,3-dihydroxy-5,7-dimethyladamantane in the presence of sulfuric acid form a thermally stable rigid polyamide.[238a] Fibrous polyamides based on adamantane-1,3-dicarboxylic acid, were reported to be thermally stable.[238b]

3. 2,2-Bis(4-aminocyclohexyl)propane was reacted with adipic acid and caprolactam to give thermally stable polymers.[239]

In an attempt to moderately stabilize polyamides against thermal and thermal-oxidative degradation, the polymers were treated with diacylamide to block their amine end groups.[240,241]

5.4. Synthesis of High-Temperature Polyamides

The recognition of thermal and thermal-oxidative stability as an important factor in determining the flame resistance of polyamides directed the attention of many authors to the synthesis and characterization of high-temperature polyamides. The partial or complete replacement of aliphatic chains by aromatic nuclei was shown to improve the thermal stability of polyamides. Fully aromatic polyamides were shown to have higher decomposition temperatures than analogous aliphatic systems.[242-247] Most authors[242-248] tend to attribute the improved thermal stability of polyamides to (1) the higher dissociation energy of the carbon–carbon bonds in the aromatic rings and those directly attached to them, which results in a high decomposition temperature; (2) the high temperatures of transition, e.g., T_g and T_m, associated with high chain rigidity and high heat of fusion, which will require high heat input to mobilize the chain prior to decomposition; (3) the presence of a minimum concentration of oxidizable hydrogen atoms in the polymer chain, which results in a decreased heat of combustion, ΔH_c. The ΔH_c's of nylon 6, nylon 6-6, and Nomex are 7658, 7439, and 6462 cal/g, respectively;[248a] and (4) the predominance of aromatic nuclei in the backbone which may encourage char formation.

Although the degradation mechanism of aliphatic polyamides has been the subject of many studies, similar investigations of aromatic polyamides are relatively scarce. Krasnov, Legunova, and Sokolov[248] studied the degradation of aromatic polyamides, based on isomeric phenylenediamines and benzenedicarboxylic acids, in the 320–550°C range. It was concluded that polyamides based on the para isomers are more thermally stable than those based on the meta isomers. The activation energy (E_a) of the degradation process of aromatic polyamides made from a para-diamine + para-diacid, meta-diamine + para-diacid, para-diamine + meta diacid, and meta diamine + meta diacid were shown to be 53.2, 42.4, 31.4, and 33.8 kcal/mol, respectively. They[248] also reported that the decomposition products of these isomeric polyamides are qualitatively and quantitatively identical. They were shown to consist of carbon dioxide and water at the low temperature range, and as the temperature was raised, carbon monoxide, benzene, toluene, and benzonitrile were detected in the condensable fraction. At the maximum temperature, hydrogen and ammonia were shown to form. Mechanisms to account for the formation of these products were also proposed, but their discussion is beyond the scope of this chapter. On the other hand, an excellent analysis of these mechanisms can be found in a recent review by Conley and Guadiana.[249]

The thermal stability and other related properties of aromatic polyamides suggested their potential use as flame-retardant polymers. Indeed, a number of partially or fully aromatic polyamides are now being produced commercially or are being evaluated as flame-resistant fibers, foams, or films. Unfortunately, large-scale production of aromatic polyamides is limited by their present high cost and the need for less conventional synthetic and processing techniques. A number of reported approaches to simplify and optimize the synthesis and processing of these polymers are summarized below.

1. The synthesis of certain aliphatic–aromatic polyamides was considered in an effort to have polymers with reasonable flexibility and because they are more suitable for solvent processing than fully aromatic ones.[16 18]

However, the introduction of aliphatic moieties into the backbone decreases polyamide thermal stability.

2. Inorganic salts in high-boiling amide solvents were used for the preparation of polyamide solutions suitable for the spinning and casting processes. Poly(p-benzamide) dope for fibers, films, and coatings was made from $\geq 4\%$ polymer in an amide solvent (including lactams) containing LiCl or CaCl$_2$.[250] These salts are likely to coordinate with the amide groups along the chain. This results in decreased effectiveness of inter-molecular polar forces and thus renders the polymer more soluble.[246]

3. The synthesis of partially N-substituted polyamides or copoly-amides results in lowering the level of hydrogen bonding and thus crystal-linity, which will facilitate the polyamide solubilization. N-Methyl-p-phenylenediamine and 4,4'-diaminophenylmethane were reacted with dicarboxylic acid dichlorides in solution to form reasonably soluble polymers.[251]

4. Comonomers, especially isomeric ones, have been used for the synthesis of aromatic copolyamides. Polymers made from isomeric monomer mixtures, which may be denoted as stereo copolyamides, are expected to have lower chain regularity and thus a lower T_g or T_m than any of the cor-responding homopolymers.[19,20] Copolyamides with variable concentra-tion of the different repeat units such as (I) and (II) were made and exhibited satisfactory physical properties, including transition temperatures.[252]

(I)   (II)

R^1 and R^2 = H, alkyl, halogen, etc.
R^3 = difunctional aromatic group

To illustrate the different types of aromatic polyamides and the scope of their application, typical examples of homo- and copolyamides (both partial and fully aromatic) were selected from the recent literature and summarized below.

1. The aliphatic–aromatic polyterephthalamides were the first to be investigated.[253] These were usually prepared from aliphatic diamines (primary or secondary) by interfacial polymerization with terephthaloyl chloride.

2. Wholly aromatic polyamides were prepared by either self-condensa-tion of aminobenzoic acid derivatives[3,254,255] (the HCl salt of the amino-acid chloride in a base) or preferably by the polycondensation of arylene or heterocyclic diamine with diacid chlorides.[257-260]

3. Du Pont's Nomex was the first commercial. thermally stable aromatic polyamide. It was prepared by the solution polymerization of *m*-phenylenediamine and isphthaloyl chloride.[258]

4. The concept of preparing "ordered copolymers" was adopted by Monsanto's workers as a means of tailoring thermally stable polyamides to achieve the desired balance of properties (mainly solubility. mechanical and thermal performance).[247,261-264] Such an approach was conducted by first preparing a multifunctional block possessing structural symmetry. This was condensed with another monomer under conditions such that no bond rearrangements occur. and can be illustrated by the following example:

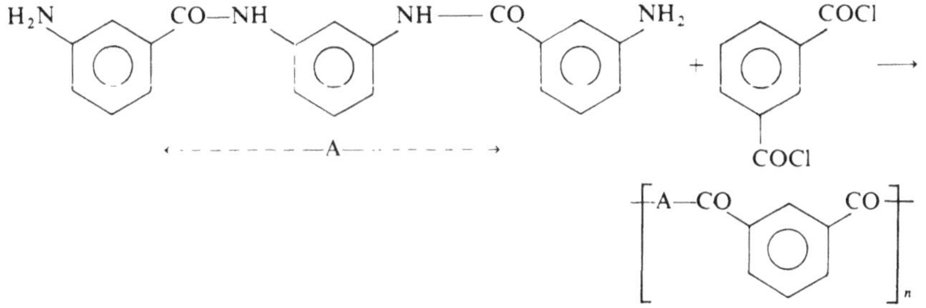

5. Polyoxamides of high molecular weights were prepared by melt condensation of aromatic diamines with diphenyl dithioloxalate in tetramethylene sulfone.[85a]

6. A number of sulfur-containing aromatic polyamides with improved solubility characteristics were prepared by a variety of approaches.[265-267] A recent example[267] of these polymers is shown below:

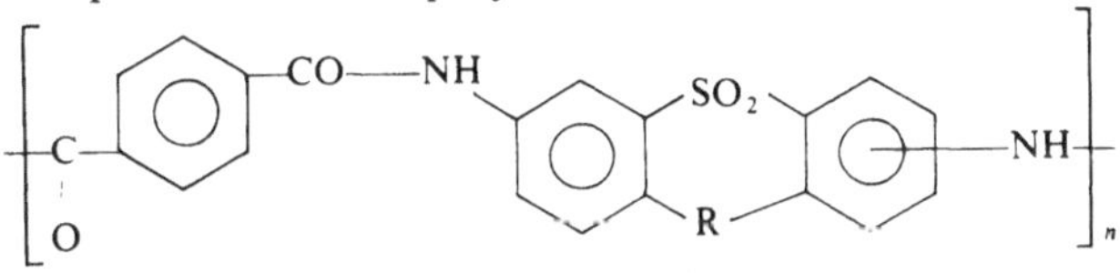

R = CH_2. *i*-Pr. O. S. SO_2. NR. etc.

7. Linear polyphthalamides have been prepared by reacting diimides with a suitable diamine (preferably hexamethylene diamine) to give heat-resistant polymers[268]:

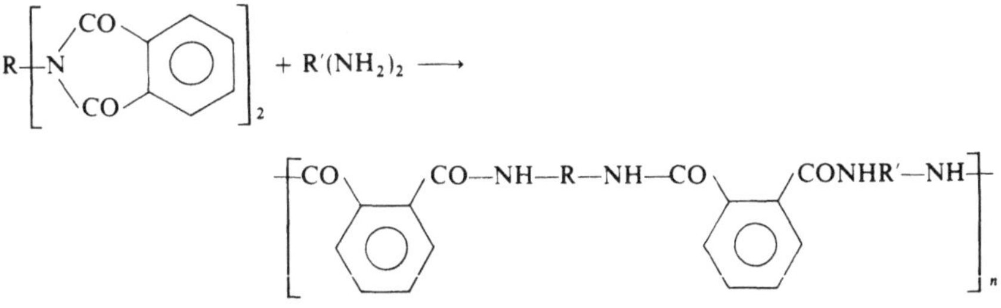

8. Heat-resistant foams were made by heating the polyamide shown below with a certain solvent content at 300°C or higher[269]:

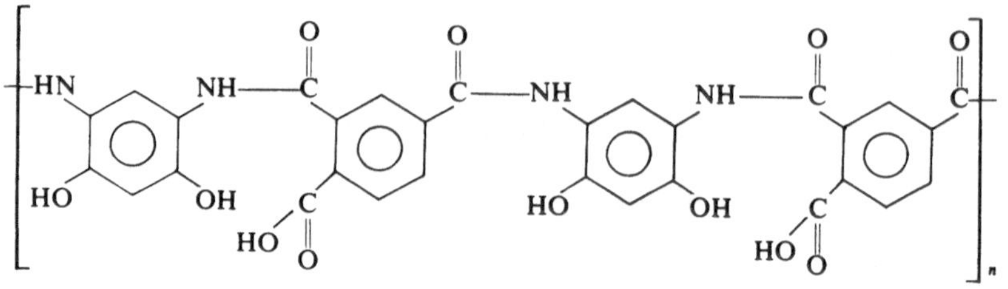

9. Unsaturated diacids, such as $C_6H_4(CH{=}CH{-}COOH)_2$, $R{-}C_6H_3$-$(CH{=}CH{-}COOH)_2$, and $C_6H_4(CH{=}CH{-}(CH_2)_n{-}COOH)_2$, were condensed with diamines to produce high-melting polyamides.[270] These were claimed to be suitable for the production of high-temperature fibers and as vehicles or binders for high-temperature composite articles. The acid chloride of phenylene-bis(acrylic acid) when reacted with hexamethylenediamine gave a polyamide with a $T_m > 400°C$.

10. Claims for increased flame retardancy have been made for aromatic polyamides prepared from halogenated monomers, e.g., polyamides made from isophthaloyl or terephthaloyl dichloride with chloro- or 2,6-dichloro-p-phenylenediamine[271]:

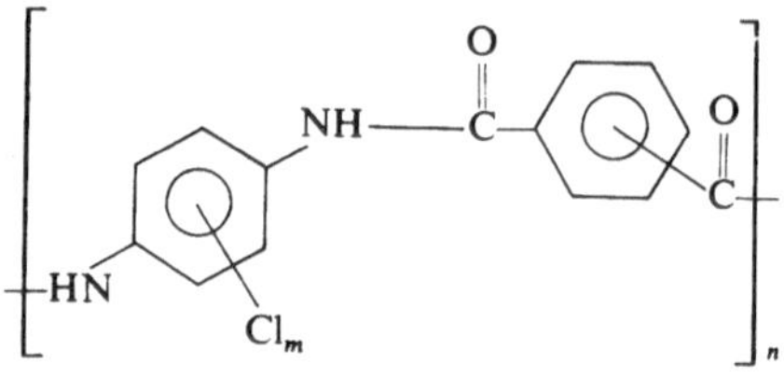

$m = 1, 2$

Other halogenated aromatic polyamides based on the diamine, shown below, were claimed suitable for flame- or heat-resistant fibers and protective coatings[271]:

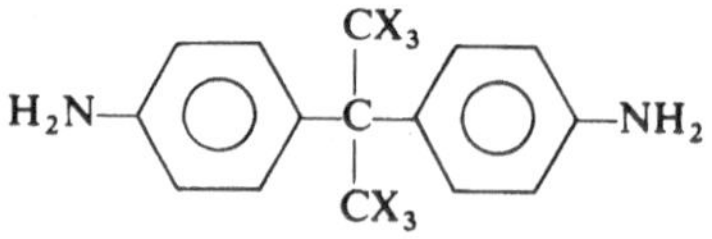

$X = Cl, F$

Aromatic polyamides-imides showed improved thermal stability when compared to aromatic polyamides. This was attributed to the greater inherent thermal stability of the imide over the amide group. When compared to aromatic polyimides, these copolymers have more favorable solubility characteristics and are more flexible.[272] Some of the recent examples of these copolymers are shown on the following page.

1. Trimellitic anhydride and diamines[272-274]: The same polymer can

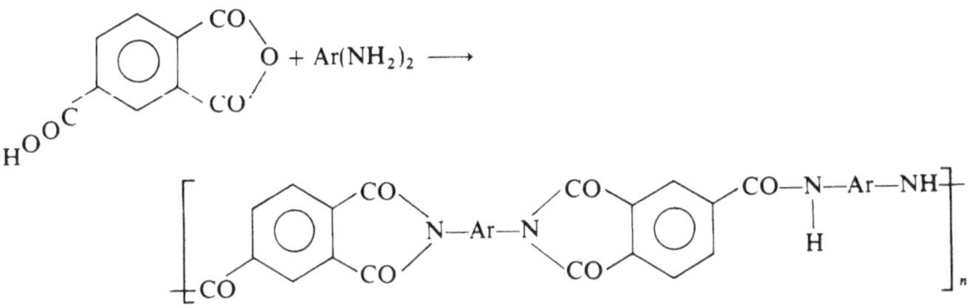

be obtained by reacting trimellitic anhydride with a suitable disocyanate, $Ar(N=C=O)_2$.[272]

2. From diamines with a preformed imide ring,[275] for example:

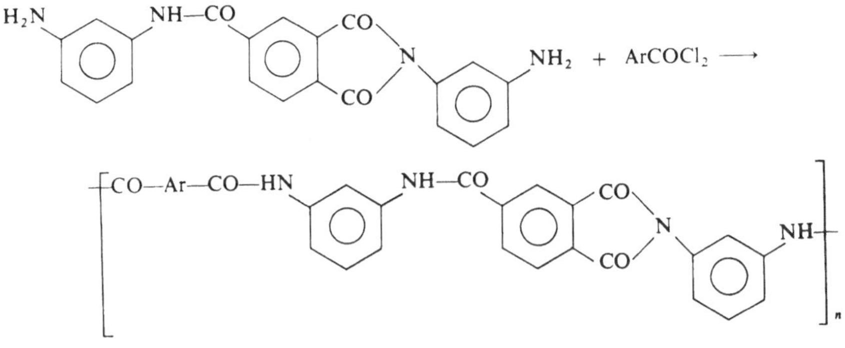

6. New Trends

Prior to the discussion of new trends in the area of thermal stabilization and flame resistance of polymers, one ought to outline the main drawbacks of the present approaches to these problems in order to evaluate any particular trend effectively. Deficiencies of the present approaches are summarized below.

1. Additives may lead to difficulties in melt spinning of polyamide fibers when high concentrations are required. This is less of a problem in molding-resin applications where fabrication requirements are not as stringent.

2. Modifying existing polyamides by reacting the polymer with some reactive flame retardants or stabilizers is sometimes associated with poor reproducibility and poor surface properties. A high surface-to-volume

ratio of the shaped articles is usually required for effective applications, and this limits the scope of this approach.

3. The synthesis of structurally modified polymers requires the preparation of some unconventional and initially costly monomers. Polyamides and copolyamides based on these monomers may suffer loss in crystallinity and mechanical properties, and thus may not become suitable for the desired end use.

4. The synthesis of high-temperature polyamides for use as flame-retardant fibers or protective coatings has been demonstrated as an excellent approach for some specialized applications. These relatively expensive polymers are not economic substitutes for aliphatic polyamides, and they have other property deficiencies.

The search for an effective approach to achieve flame retardancy in polyamides and polymers in general must be recognized as a continuing challenge. A number of new approaches are outlined in the following paragraphs:

Char Forming and/or Cross-Linked Polyamides. The promotion of char formation should be advantageous in rendering polyamides flame retardant by:

1. Producing low heat of combustion pyrolyzate. (The ΔH_c's of CO_2, CO, and solid carbon are 94, 26, and 0.0 kcal/g mol, respectively.[43])
2. Conservation of mechanical integrity.
3. Acting as a physical barrier towards oxygen diffusion.

Char formation has been achieved in many instances by improving the tendency for cross-linking, but the level of charring is still unsatisfactory. Several of these approaches are summarized below:

1. Self-extinguishing foams with uniform pore structure were prepared using alkenyl aromatic polymers, foaming agent, an organic bromine compound, and a small amount of *N,O*-azetidinone peroxyacetyl of the formula[276]

2. Curing a prepolymer, made from diethylenetriamine with adipic acid and halogenated anhydrides resulted in flame-retardant, dimensionally stable cross-linked polyamides.[277] An example of a typical anhydride is 5,6,7,8,9,9-hexachloro-1,2,3,4,4a,5,8,8a-octa-hydro-5,8-methano-2,3-naphthalene.

3. Hirsch[278] reported that when poly[*m*-phenylene-bis(*m*-benz-amide)terephthalimide] was heated in air for 150 min at 425°C, the originally flammable polymer did not burn in air. This was attributed to excessive cross-linking in the partially degraded polymer.
4. Cross-linked polyamides or polyamide–imides were prepared and exhibited good flame resistance, e.g., aromatic polyamide-imides made from aromatic diisocyanate and trimellitic anhydride or trimesic acid,[279] and polyamide made from a polyalkylene-polyamine, epichlorohydrin, and phosphorus-containing compounds.[280]

Chelatable Polymers. A new class of flame-retardant organometallic polymers has been recently developed.[281] These are chelated poly(tereph-thaloyl-oxalic-bisamidrazone), and can be made from amines and acid chlorides as follows:

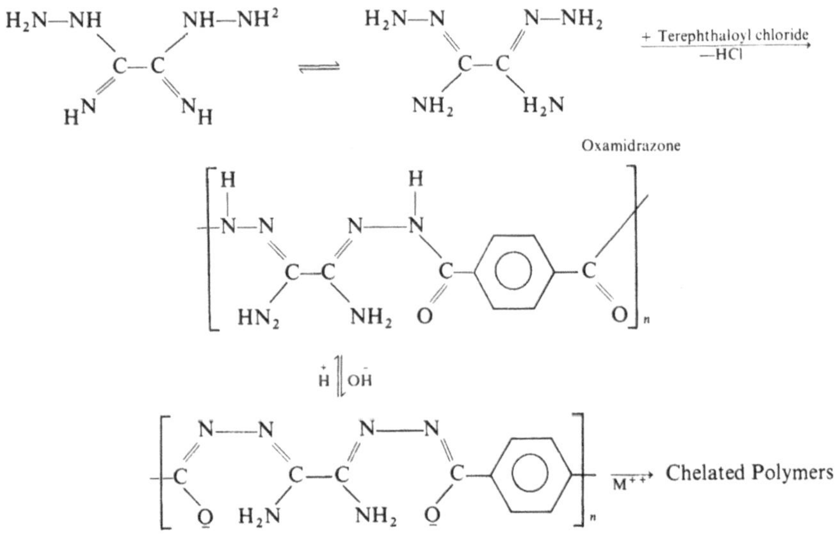

Oxamidrazone

Chelated Polymers

Graphitization. The high cost of graphitization has limited the application of this technique to some interesting polymeric systems. In recent years, the graphitization of polyamides attracted the attention of few workers. Nylon 6-6 and 6-T were graphitized into fire-resistant, smokeless fibers.[282] The graphitized nylon 6-T revealed a 10 million psi modulus and 100,000 psi tenacity with a 1% elongation.

Special Treatment of Polyamide Fibers. Textile fibers or fabrics made of certain polyamides were rendered flame retardant by the following techniques:

1. Semicarbonization of thermally stable aromatic polyamides in a fiber or fabric form by heat treatment in air, under moderate conditions.[283] The resulting fibers or fabrics were essentially nonflammable, thermally stable, chemically inert, and exhibited good dimensional stability at elevated temperatures.

2. Aromatic polyamides were rendered fire resistant and dimensionally stable by heating to 200–500°C and then treating with chlorine in the absence of oxygen for a maximum of 12 hr.[284]

3. Textiles made from aromatic polyamides containing sulfur were made fire resistant upon exposure to hydrocarbon flame for ≥ 10 sec.[285]

4. Shaped articles made of thermally stable polyamides were flame retarded by treating them with halides and oxyhalides of group IV, V, and VI elements.[221,286]

7. Conclusion

In dealing with the problem of combustion retardation of polyamides, one ought to treat these polymers in small subgroups of common structural features, comparable thermal and thermal-oxidative behavior, or similar processability and end use. A thorough understanding of the thermal and thermal-oxidative behaviors is necessary for fruitful studies of the combustion retardation of any particular polyamide.

In spite of the limitations of present approaches for retarding combustion, each of these approaches can be utilized successfully for specific application; for example, additives have been shown to be quite suitable for molding resins and reinforced thermoplastics, while the synthesis of more combustion-resistant, high-temperature aromatic polyamides is a viable approach for specific industrial and textile fiber end uses.

New approaches to the problem of combustion retardation should be investigated bearing in mind (1) the significance of redirecting the thermodynamic and kinetic processes prior to or during combustion; (2) the role of liquid or solid barrier formation between the gas and the condensed phases; (3) the mechanism for char-forming reactions, particularly those resulting in cross-link formation; (4) some of the recent trends in the field of flame retardance, especially those describing high temperature surface treatments; (5) the effect of other materials and polymers in contact with the polyamide; and (6) the cooling effect of the flame by certain noncombustible gases.

8. References

1. For general previews on the chemistry of polyamides, see: R. W. Lenz, *Organic Chemistry of High Polymers*, Chap. 4, Wiley-Interscience, New York (1967): O. E. Snider and R. J. Richardson, *Encyclopedia of Polymer Science and Technology*, Vol. 10, Wiley-Interscience, New York, 1969, p. 347; V. V. Korshak and T. M. Frunze, Synthetic Hetero-Chain Polyamides, Israel Program for Scientific Translations, Jerusalem, Israel (1964); P. W. Morgan, Condensation Polymers, Wiley-Interscience, New York (1965).
2. J. A. Sommers, *Man-made Text.* **32**, 60 (1956).
3. Brit. Pat. 901, 159 (1962) (to E. I. duPont deNemours and Co.).
4. M. Genes, *Angew. Chem.* **74**, 535 (1962); S. R. Gunliffe and J. G. Hawkins, *Brit. Plast.* 682 (April) (1963).
5. R. Graf, G. Lohaus, K. Börner, E. Schmidt, and H. Bestian, *Angew. Chem.* **74**, 523 (1962).
6. U.S. Pat. 2,638,463 (1953) (to Arnold, Hoffman, and Co., Inc.)
7. Japan. Pat. 70-16,716 (1970) (to Asahi Chem. Ind. Co., Ltd.).
8. W. Griehl and D. Ruestem, *Ind. Eng. Chem.* **62**(3), 16 (1970).
9. D. S. Breslow, G. E. Hulse, and S. Matlack, *J. Am. Chem. Soc.* **79**, 3760 (1957).
10. C. J. Hilado, *Flammability Handbook for Plastics*, Technomic Publishing Co., Stamford, Conn. (1969).
11. J. W. Lyons, *The Chemistry and Uses of Fire Retardants*, Wiley, New York (1970).
12. S. Straus and L. Wall, *J. Res. Natl. Bur. Std.* **60**, 39 (1958).
13. R. T. Conley and R. A. Guidiana in *Thermal Stability of Polymers*, (R. T. Conley, ed.) Vol. 1, Marcel Dekker, New York (1970).
14. L. Reich and S. S. Stivala, *Elements of Polymer Degradation*, McGraw-Hill, New York (1971).
15. S. L. Madorsky, *Thermal Degradation of Organic Polymers*, Wiley, New York (1964).
16. (a) A. H. Frazer, High temperature resistant polymers, in *Polymer Reviews*, Vol. 17 (H. Mark and E. H. Immergut, eds.) Wiley-Interscience, New York (1968); (b) C. L. Segal, ed., *Polymers in Space Research*, Marcel Dekker, New York (1970); (c) S. B. Sello, *J. Elastoplast.* **4**, 2 (1972); (d) F. H. Winslow, L. D. Loan, and W. Matreyek, Polymer Structure and Pyrolytic Behaviour, paper presented at the 161st National Meeting ACS, Los Angeles, Calif. (April 1971); (e) V. V. Korshak, Heat Resistant Polymers, Israel Program for Scientific Translations, Jerusalem, Israel (1971).
17. J. Preston (ed.), High temperature resistant fibers from organic polymers, *Appl. Polym. Symp. No. 9*. Wiley-Interscience, New York (1969).
18. B. K. Patnaik, *J. Sci. Ind. Res.* **27**, 417 (1968).
19. J. Preston, *Chem. Tech.* 664 (Nov. 1971).
20. W. B. Black, *Trans. N.Y. Acad. Sci.* **32**(7), 765 (1970).
21. A. D. Delman, *J. Macromol. Sci., Revs. Macromol. Chem.* **C3**(2), 281 (1969).
22. W. Lincoln Hawkins, *Polymer Stabilization*, Wiley, New York (1971).
23. L. I. Chudina, S. I. Litovchenko, T. N. Spirina, and A. M. Chukurov, *Soviet Plast.*, No. 8, 7 (1970).
24. E. A. Dickert and G. C. Toone, *Mod. Plast.* **42**(5), 197 (1965).
25. R. C. Nametz, *Ind. Eng. Chem.* **59**(5), 99 (1967).
26. V. I. Kodolov, L. A. Sapogova, and S. S. Spaskii, *Soviet Plast.*, No. 10, 37 (1969).
27. G. C. Tesoro, S. B. Sello, and J. J. Willard, *Text. Res. J.* **39**, 180 (1969).
28. B. Kamerbeek, G. H. Kroes, and W. Grolle, Thermal Degradation of Some Polyamides, Soc. Chem. Ind. Monograph No. 13, Society of Chemical Industry, London (1961), p. 357.
29. M. B. Nieman, *Aging and Stabilization of Polymers*, Consultant Bur., New York (1965).
30. I. N. Einhorn, *J. Macromol. Sci., Revs. Polym. Technol.* **D1**(2), 113 (1971).

31. V. P. Kolesov, I. E. Paukov, and S. M. Skuratov, *Zh. Fiz. Khim.* **36**, 770 (1962).
32. Ref. 10, p. 35.
33. (a) D. R. Anderson, *Chem. Revs.* **66**, 677 (1966); (b) W. D. Freeston, Jr., *J. Fire Flam.* **2**, 57 (1971).
34. *Modern Plastic Encyclopedia*, **46**, 976 (1969), McGraw-Hill, New York.
35. Ref. 10, p. 37.
36. Eli M. Pearce, *Trans. N.Y. Acad. Sci.* **31**, 629 (1969).
37. J. Hine, *Physical Organic Chemistry*, McGraw-Hill, New York (1960).
38. L. Pauling, *The Nature of the Chemical Bond*, p. 85, Cornell Univ. Press, Ithaca, N.Y. (1960).
38a. N. Muller and R. S. Mulliken, *J. Am. Chem. Soc.* **80**, 3489 (1958).
39. H. Krüssman, G. Valk, G. Heidenmann, and S. Dugal, *Angew. Chem. Internat. Edit.* **8**, 215 (1969).
40. Ref. 10, p. 39.
41. S. W. Shalaby, E. M. Pearce, and H. K. Reimschuessel, unpublished work.
42. K. B. Goldblum, *SPE J.* **25**, 50 (1969).
43. Ref. 10, p. 78.
44. B. G. Achhammer, F. W. Reinhart, and G. M. Kline, *J. Res. Natl. Bur. Std.* **46**(5), 391 (1951).
45. T. Hasselstrom, H. W. Coles, C. E. Balmer, M. Hannigan, M. M. Keeler, and R. J. Brown, *Text. Res. J.* **22**, 742 (1952).
46. I. Goodman, *J. Polym. Sci.* **17**, 587 (1955).
47. I. Goodman, *J. Polym. Sci.* **13**, 175 (1951).
48. Ref. 28, p. 365.
49. T. J. Reardon, M.S. Thesis, Dept. of Textiles, Clemson Univ., Clemson, S.C.
50. L. H. Peebles, Jr., and M. W. Huffman, *J. Polym. Sci., Part A-1* **9**, 1807 (1971).
51. H. K. Reimschuessel in *Kinetics and Mechanism of Polymerization*, Vol. 2, Ring Opening Polymerization (Frisch and Reegen, eds.), Marcel Dekker, New York (1969).
52. S. Smith, *J. Polym. Sci.* **30**, 459 (1958).
53. P. H. Herman, D. Keikens, and S. Smith, *J. Polym. Sci.* **38**, 265 (1959).
54. R. H. Boyd, *J. Polym. Sci., Part A-1* **5**, 1573 (1967).
55. K. Dachs and E. Schwartz, *Angew. Chem., Internat. Edit.* **1**, 430 (1962).
56. G. Valk and H. Krüssmann, *Angew. Chem., Internat. Edit.* **6**, 1008 (1967).
57. L. I. Levantovskaya, B. M. Kovarskaya, G. V. Drayuk, and M. B. Neiman, *Vysokomol. Soedin* **6**(10), 1885 (1964).
58. S. R. Rafikov and R. A. Sorikinia, *Polym. Sci. U.S.S.R.* **1**, 189 (1960).
59. L. A. Wall and S. Straus, *J. Polym. Sci.* **44**, 313 (1960).
60. Z. I. Patemkina, L. G. Tokareva, L. I. Ananeva, and N. V. Mikhailov, *Khim. Volokna* (1), 32 (1969); *Chem. Abstr.* **70**, 97463g (1969).
61. H. K. Reimschuessel and G. J. Dege, *J. Polym. Sci., Part A-1* **8**, 3265 (1970).
62. G. Valk, H. Krüssmann, and P. Diehl, *Makromol. Chem.* **107**, 158 (1967).
63. G. Valk and H. Krüssmann, *Angew. Chem.* **79**, 1021 (1967).
64. (a) A. Reiche, *Kunstoffe* **54**, 428 (1964); (b) A. Reiche and W. Schön, *Chem. Ber.* **99**, 3238 (1966).
65. A. Reiche and W. Schön, *Kunstoffe* **57**, 49 (1967); *Ger. Plast.* **4**, 22 (1967).
66. G. Valk, G. Heidemann, S. Dugal, and H. Krüssmann, *Angew. Makromol. Chem.* **10**, 135 (1970).
67. L. T. Okuhashi and M. Kuwahara, *Kobunshi Kagaku* **27**(305), 628 (1970); *Post—J.* **9**(5), 2470n (1971).
68. T. Okuhashi and M. Kuwahara, *Kobunshi Kagaku* **28**(309), 91 (1971); *Post—J.* **9**(5), 2471p (1971).
69. V. V. Korshak, G. L. Slonimskii, and E. S. Krongauz, *Izvest. Akad. Nauk U.S.S.R.; Oddel. Khim. Nauk*, p. 221 (1958).

70. M. Staudinger and H. Schnell, *Makromol. Chem.* 1, 49 (1947).
71. S. R. Rafikov and R. A. Sorokina, *Vysokomol. Soedin.* 3, 21 (1961); *Chem. Abstr.* 55, 26580*i* (1961).
72. W. Sbrolli and T. Capaccioli, *Chim. Ind.* 42(12), 1325 (1960).
73. A. Agster, *Melliand Textilber.* 37, 1338 (1956).
74. C. Reimer, *Kunstoffe* 45, 367 (1955).
75. U.S. Pat. 2,364,204 (1944) (to Bell Telephone Labs., Inc.).
76. R. F. Schwenker, *Text. Res. J.* 30(8), 624 (1960).
77. J. Masamoto, K. Sasaguri, C. Ohizumi, and H. Kobayashi, *J. Polym. Sci., Part A-2* 8, 1703 (1970).
78. H. Bestian, *Angew. Chem., Internat. Edit.* 7, 278 (1968).
79. N. Ogata, *J. Polym. Sci., Part A* 1, 3151 (1963).
80. R. J. Fredericks, T. H. Doyne, and R. S. Sprague, *J. Polym. Sci., Part A-2* 4, 899 (1966).
81. *Chem. Eng. News*, p. 9 (Nov. 3, 1969).
82. (a) U.S. Pat. 3,247,168 (1966) (to E. I. duPont de Nemours & Co.); (b) U.S. Pat. 3,432,575 (1969) (to E. I. duPont de Nemours & Co.).
83. S. W. Shalaby, E. M. Pearce, R. J. Fredericks, and E. A. Turî, *J. Polymer Sci., Polymer Phys. Ed.* 11, 1 (1973).
84. Ye. P. Krasnov, L. B. Sokolov, and T. A. Polyakova, *Vysokomol Soedin.* 6, 1244 (1964).
85. (a) H. K. Hall, Jr., and J. W. Berge, *J. Polym. Sci., Part B* 1, 277 (1963); (b) L. B. Sokolov, L. V. Turetskii, and L. I. Tugova, *Vysokomol. Soedin.* 4, 1817 (1962).
86. L. Mortillaro, M. Russo, V. Guidotti, and L. Credali, *Makromol. Chem.* 138, 151 (1970).
87. D. C. Allport, *Soc. Chem. Ind. (London), Monogr. No. 26,* 143 (1967).
88. J. Preston, *Polymer Preprints* 11(2), 347 (1970).
89. L. Credali, P. Parrini, L. Mortillaro, L. Russo, and M. Simonazzi, *Angew. Makromol. Chem.* 19, 15 (1972).
90. W. B. Black and J. Preston in *Man-Made Fibers,* Vol. 2, p. 297 (Mark, Atlas and Cernia, eds.), Wiley-Interscience, New York (1968).
91. Properties of Nomex High Temperature Resistant Nylon Fiber, NP-33 Bulletin, E. I. duPont de Nemours and Co., Wilmington, Delaware.
92. (a) J. O. Weiss, H. S. Morgan, and M. R. Lilyquist, *J. Polym. Sci., Part C* 19, 29 (1967); (b) J. Preston, R. W. Smith, and C. J. Stehman, *J. Polym. Sci., Part C* 19, 7 (1967).
93. J. Zimmerman and R. E. Wilfong, Future Organic Tire Fibers, paper # 26, Symposium on Synthetic Fibers and Related Monomers, ACS 163d National Meeting, Boston, Mass. (April 1972).
94. J. C. Shivers, A New High Temperature Fiber, a paper presented at the 42d TRI Annual Research and Technology Conference, New York (April 5, 1972).
95. (a) C. D. Doyle, Thermal Analysis, Vol. 1, Chap. 4, in *Techniques and Methods of Polymer Evaluation* (Slade and Jenkins, eds.), Marcel Dekker, New York (1966); (b) Ref. 14, Chap. 3.
96. L. Reich, in *Macromolecular Reviews* (Peterlin *et al.,* eds.), Vol. 3, p. 49, Wiley-Interscience, New York (1968).
97. B. Ke. in Newer Methods of Polymer Characterization (B. Ke, ed.), Vol. 6, Chap. IX, *Polymer Reviews,* Wiley-Interscience, New York (1964).
98. S. D. Bruck, *Polymer* 5, 435 (1964).
99. (a) J. C. P. Schwarz (ed.), *Physical Methods in Organic Chemistry,* Holden Day, Inc., San Francisco (1964); (b) G. P. Shulman, *J. Makromol. Chem.* A-1, 107 (1967).
100. E. Hagen, *Plaste Kaut.* 15, 711 (1968).
101. E. P. Krasnov *et al., Vysokomol. Soedin.* 8, 380 (1966); *Chem. Abstr.* 64, 19810 (1966).
102. G. M. Bauer, Thermal Characterization Techniques, Vol. 2, Chap. 2, in *Techniques and Methods of Polymer Evaluation* (Slade and Jenkins, eds.), Marcel Dekker, New York (1970).

103. T. L. Squires, *An Introduction to Electron Spin Resonance*, Academic Press, New York (1964).
104. T. C. Chang and J. P. Sibilia, *J. Polym. Sci., Part A-1* **10**, 605 (1972).
105. (a) I. Ya. Slomin and A. N. Lyubinov, *The NMR of Polymers*, Plenum Press, New York (1970), pp. 289–293; (b) M. Modena, C. Garbuglio, and M. Raggazzini, *J. Polym. Sci., Part B* **10**, 145 (1972).
106. O. A. Mochalova, Diploma work, MITKHT (1962) (from Ref. 105).
107. I. M. Kolthoff and J. J. Lingane, Polarography, Vols. 1 and 2, Wiley-Interscience, New York (1952).
108. A. M. Kotliar, Degradation of Condensation Polymers, ACS Middle Atlantic Regional Meeting, Polymer Chemistry paper # 15, Philadelphia, Pa. (Feb. 1972).
109. Ref. 10, pp. 82, 83.
110. C. J. Hilado, *J. Cell. Plast.* **4**, 339 (1968).
111. J. K. Jacques, *Plast. Inst. Trans. J., Conf. Suppl. No. 2*, 33 (1967).
112. C. J. Hilado, *Ind. Eng. Chem., Prod. Res. Develop.* **7**, 81 (1968).
113. C. J. Hilado, *Am. Chem. Soc., Div. Org. Coating Plast. Chem.* **28**(1), 317 (1968).
114. N. E. Boyer, *Plast. Technol.* **8**(11), 33 (1962).
115. J. M. Church, The Role of Phosphorus Compounds in the Fire Resistant Treatment of Textiles, Department of Chemical Engineering Report, Columbia University, New York (1952).
116. E. A. Dickert and G. C. Toone, *Mod. Plast.* **42**, 197, 204, 264, 267 (1964).
117. J. R. Van Wagner, *Phosphorus and its Compounds*, Vol. II, Wiley-Interscience, New York (1961), p. 1955.
118. H. Piechota, *J. Cell. Plast.* **1**, 186 (1965).
119. I. N. Einhorn, *Am. Chem. Soc., Div. Org. Coatings Plast. Chem.*, **28**(1), 291 (1968).
120. (a) C. P. Fenimore and G. W. Jones, *Combust. Flame* **10**, 295 (1966); (b) L. N. Kireenkova *et al.*, *Soviet Rubber Tech.* No. 6 (1968).
121. (a) N. A. Pudovik and V. K. Khairullin, *Russ. Chem. Rev.* **37**(5), 317 (1968); (b) W. A. Rosser, J. H. Miller, S. H. Inami, and H. Wise, Stanford Research Institute Contract D. A. 44-009-Eng. 2863, Phase II, Final Report (June 30, 1958); (c) C. P. Fenimore and F. J. Martin, *Combust. Flame* **10**, 135 (1966); (d) R. I. Thrune, *Am. Chem. Soc., Div. Org. Coatings Plast. Chem.* **23**(1), 15 (1963).
122. G. L. Drake, Jr., in *Kirk-Othmer Encyclopedia of Chemical Technology*, 2nd ed., Vol. 9, p. 300–315, Wiley-Interscience, New York (1966).
122a. Ref. 11, p. 122.
123. (a) C. P. Fenimore and G. W. Jones, *Combust. Flame* **10**, 295 (1966); (b) A. R. Ingram, *J. Appl. Polym. Sci.* **8**, 2485 (1964).
124. (a) N. E. Walsh, E. G. Uhig, and T. M. Beck, *Am. Chem. Soc., Div. Org. Coatings Plast. Chem.* **23**, 1 (1963); (b) Ref. 11, p. 20; (c) R. H. Barker, J. E. Bostic, Jr., T. J. Reardon, and R. A. Strong, Abstracts, 164th ACS National Meeting, New York, 1972.
125. P. Robitschelk, *J. Cell. Plast.* **1**, 395 (1965).
126. D. R. Miller, R. L. Evers, and G. B. Skinner, *Combust. Flame* **7**, 137 (1963).
127. J. J. Pitts, P. H. Scott, and D. G. Powell, *J. Cell. Plast.* **6**(1), 35 (1970).
128. W. G. Schmidt, *Plast. Inst. Trans.* **33**, 247 (1965).
129. Ref. 10, pp. 86, 87.
130. W. L. Hawkins (ed.), *Polymer Stabilization*, Wiley, New York (1971).
131. *Modern Plastics*, p. 152 (April 1972).
132. U.S. Pat. 3,532,667 (1970) (to E. I. duPont de Nemours & Co.).
133. U.S. Pat. 3,428,597 (1969) (to Stamicarbon N. V.).
134. U.S. Pat. 3,583,981 (1971) (to Upjohn Co.).
135. U.S. Pat. 3,448,086 (1969) (to Dart Industries, Inc.).
136. U.S. Pat. 3,477,986 (1969) (to Farbenfabriken Bayer A.G.).

137. U.S. Pat. 3,328,342 (1967) (to E. I. duPont de Nemours & Co.).
138. U.S. Pat. 3,020,306 (1961) (to Monsanto Co.).
139. Ital. Pat. 846,021 (1969) (to Montecatini Edison S.p.H.); *POST-P*, **9**(10), 5731*j* (1971).
140. U.S. Pat. 3,522,204 (1970) (to American Cyanamid Co.).
141. U.S. Pat. 3,043,810 (1962) (to American Cyanamid Co.)
142. U.S. Pat. 3,533,986 (1970) (to Imperial Chem. Ind. Ltd.).
143. U.S. Pat. 3,532,668 (1970) (to American Cyanamid Co.).
143a. U.S. Pat. 3,629,365 (1971) (to Akzona Inc.).
144. U.S. Pat. 3,553,161 (1971) (to Farbenfabriken Bayer A.G.)
145. U.S. Pat. 3,595,829 (1971) (to Imperial Chem. Ind. Ltd.).
146. Japan. Pat. 69-28,598 (1969) (to Toyo Rayon Co.).
146a. Japan. Pat. 72-058,17 (1972) (to Toray Ind. Inc.).
147. U.S. Pat. 3,374,288 (1968) (to E. I. duPont de Nemours and Co.).
147a. Japan. Pat. 72-016,95 (1972) (to Unitika Ltd.).
148. U.S. Pat. 3,377,314 (1968) (to E. I. duPont de Nemours and Co.).
148a. U.S.S.R. Pat. 293,820 (1971) (to N. Mikhailov *et al.*).
149. U.S. Pat. 3,541,046 (1971) (to Monsanto Co.).
150. Ger. Offen. 2,024,392 (1970) (to Mobil Oil Corp.).
151. U.S. Pat. 3,157,613 (1964) (to Monsanto Co.).
152. U.S. Pat. 3,243,406 (1966) (to Hooker Chem. Corp.).
153. U.S. Pat. 3,468,678 (1969) (to Monsanto Co.).
154. U.S. Pat. 3,228,898 (1966) (to Badische Anilin & Soda Fabrik A.G.).
155. U.S. Pat. 3,108,091 (1963) (to Badische Anilin & Soda Fabrik A.G.).
156. U.S. Pat. 3,422,048 (1969) (to American Cyanamid Co.).
156a. Belg. Pat. 771,037 (1972) (to Ciba-Geigy AG).
157. U.S. Pat. 3,418,267 (1968) (to E. I. duPont de Nemours and Co.).
157a. U.S. Pat. 3,629,174 (1971) (to Snia Viscosa S.p.A.).
158. U.S. Pat. 3,440,211 (1969) (to E. I. duPont de Nemours and Co.).
158a. Belg. Pat. 770,299 (1972) (to Imperial Chem. Ind. Ltd.).
159. U.S. Pat. 2,930,838 (1960) (to Organico, S.A.).
160. U.S. Pat. 3,418,263 (1968) (to Hooker Chem. Corp.).
161. U.S. Pat. 3,396,201 (1968) (to Hooker Chem. Corp.).
162. U.S. Pat. 3,403,036 (1968) (to Hooker Chem. Corp.).
163. U.S. Pat. 3,468,843 (1969) (to E. I. duPont de Nemours and Co.).
164. Japan. Pat. 70-36,827 (1970) (to Toray Industries, Inc.).; *Chem. Abstr.* **75**, 118947*r* (1971).
164a. U.S. Pat. 3,629,188 (1971) (to Soc. Dela Viscose Suisse).
165. U.S. Pat. 3,630,988 (1971) (to E. I. duPont de Nemours and Co.).
166. Hung. Pat. 158,565 (1971) (I. Gaspar *et al.*); *POST-P* **9**(6), 3385*a* (1971).
167. U.S. Pat. 3,573,245 (1971) (to Badische Anilin & Soda Fabrik A. G.).
168. U.S. Pat. 3,458,474 (1969) (to Badische Anilin & Soda Fabrik A. G.).
168a. U.S. Pat. 3,630,988 (1971) (to E. I. duPont de Nemours and Co.).
169. U.S. Pat. 3,067,168 (1962) (to E. I. duPont de Nemours and Co.).
170. U.S. Pat. 3,275,594 (1966) (to E. I. duPont de Nemours and Co.).
171. U.S. Pat. 3,272,773 (1966) (to E. I. duPont de Nemours and Co.).
172. U.S. Pat. 3,385,819 (1968) (to Hooker Chem. Corp.).
173. U.S. Pat. 3,457,325 (1969) (to E.I. duPont de Nemours and Co.).
174. U.S. Pat. 3,313,769 (1967) (to Badische Anilin & Soda Fabrik A. G.).
175. (a) U.S. Pat. 3,321,436 (1967) (to Badische Anilin & Soda Fabrik A.G.); (b) U.S. Pat. 3,660,344 (1972) (to Farbenfabriken Bayer A. G.).
176. U.S. Pat. 3,308,091 (1967) (to E. I. duPont de Nemours and Co.).
177. Japan. Pat. 72-017,14 (1972) (to Toray Ind. Inc.).
178. Ref. 29, p. 247.

179. Japan. Pat. 66-17,678 (1966) (to Kanegafuchi Spinning Co.).
180. U.S. Pat. 3,382,204 (1968) (to Hooker Chem. Corp.).
181. U.S.S.R. Pat. 286,228 (1969) (to N. V. Mikhailov *et al.*); *POST-P* **9**(6), 3210*q* (1971).
181a. Japan. Pat. 72,017,14 (1972) (to Toray Ind. Inc.).
182. U.S. Pat. 3,660,344 (1971) (to Farbenfabriken Bayer A. G.).
182a. Japan. Pat. 72,054,24 (1972) (to Firestone Tire & Rubber Co.).
183. V. E. Bronovitskii, *Uzb. Khim. Zh.* **15**(1), 53 (1971); *POST-J* **9**(4), 1975*u* (1971).
184. Brit. Pat. 1,235,813 (1971) (to Farbenfabriken Bayer A. G.).
185. Brit. Pat. 1,192,363 (1970) (to Imperial Chem. Ind., Ltd.).
186. U.S. Pat. 3,344,107 (1967) (to Monsanto Co.).
187. Brit. Pat. 1,253,062 (1971) (to Monsanto Co.).
188. Ger. Offen. 1,931,387 (1970) (to Badische Anilin & Soda Fabrik AG).
189. Ger. Offen. 1,941,189 (1971) (to G. Illing).
190. Dutch Pat. 2,064,169 (1970) (to Wellman, Inc.).
191. Belg. Appl. 770,954 (1970) (to Dart Ind. Inc.).
192. D. O. Douglas, *J. Soc. Dyers Color.* **73**, 258 (1957).
193. M. W. Ranney, *Flame-Retardant Textiles*, Noyes Data Corp., Parkridge, N.J. (1970), p. 321.
194. U.S. Pat. 3,308,098 (1967) (to American Cyanamid Co.).
195. U.S. Pat. 2,854,437 (1958) (to American Cyanamid Co.).
196. U.S. Pat. 2,859,206 (1958) (to American Cyanamid Co.).
197. (a) U.S. Pat. 2,881,152 (1959) (to American Cyanamid Co.); (b) H. E. Bille and H. A. Petersen, *Textilveredlung* **2**(5), 243 (1967); *Chem. Abstr.* **67**, 33686G (1967).
198. U.S. Pat. 2,922,726 (1960) (to American Cyanamid Co.).
199. U.S. Pat. 3,017,292 (1962) (to H. H. Mosher).
200. (a) U.S. Pat. 2,923,644 (1960) (to American Cyanamid Co.); (b) U.S. Pat. 2,999,847 (1961) (to American Cyanamid Co.).
201. Japan. Pat. 72-020,76 (1972) (to Unitika Ltd.).
202. U.S. Pat. 2,953,480 (1960) (to American Cyanamid Co.).
203. U.S. Pat. 3,620,818 (1971) (to Monsanto Co.).
204. U.S. Pat. 3,032,440 (1962) (to General Tire & Rubber).
205. Japan. Pat. 72-020,75 (1972) (to Unitika).
206. (a) Japan. Pat. 72-020,77 (1972) (to Unitika Ltd.); (b) Japan. Pat. 72-020,78 (1972) (to Unitika Ltd.).
207. U.S. Pat. 3,409,386 (1968) (to Universal Oil Products).
208. U.S. Pat. 2,926,097 (1960) (to United States Army).
209. U.S. Pat. 2,795,513 (1957) (to Monsanto Chem. Co.).
210. Neth. Appl. 71,085,83 (1971) (to Stauffer Chem. Co.).
211. Ger. Pat. 1,619,164 (1971) (to Soo Valley Co.); *Chem. Abstr.* **75**, 119159*X* (1971).
212. U.S. Pat. 2,999,056 (1961) (to E. I. duPont de Nemours and Co.).
213. U.S. Pat. 3,294,755 (1966) (to E. I. duPont de Nemours and Co.).
214. Czeck. Pat. 102,383 (1962).
215. U.S. Pat. 2,702,763 (1955) (to Phillips Petroleum Co.).
216. Japan. Pat. 72-025,19 (1972) (to Unitika Ltd.).
217. T. Okuhashi and M. Kuwahara, *Kogyo Kagaku Zasshi* **74**(6), 1214 (1971); *Chem. Abstr.* **75**, 130675*h* (1971).
218. R. W. Montcrieff, *Text. Manuf.* **92** (1093), 58 (1966).
219. Ger. Offen. 2,001,239 (1971) (to Clark, David Co., Inc.); *Chem. Abstr.* **75**, 110978*f* (1971).
220. U.S. Pat. 3,576,590 (1971) (to Monsanto Co.).
221. U.S. Pat. 3,549,307 (1970) (to Monsanto Co.).
222. (a) V. V. Korshak, *J. Polym. Sci.* **31**, 319 (1958); (b) T. M. Frunze, V. V. Korshak, V. V. Kurashev, G. S. Kolesnikov, and B. A. Zhubanov, *Izvest. Akad. Nauk. USSR Otdel. Khim. Nauk* **1958**, 783 (1958).

223. T. M. Frunze, V. V. Korshak, and V. V. Kurashev, *Vysokomol. Soedin.* **1**, 670 (1959).
224. T. M. Frunze, V. V. Korshak, L. V. Kozlov, and V. V. Kurashev, *Vysokomol. Soedin.* **1**, 677 (1959).
225. S. V. Vinogradova, V. V. Korshak, G. S. Kolesnikov, and B. A. Zhubanov, *Vysokomol. Soedin.* **1**, 357 (1959).
226. U.S. Pat. 2,646,420 (1953) (to E. I. duPont de Nemours and Co.).
227. K. A. Petrov and V. V. Parshena, *Zh. Obshchli Khim.* **30**, 1342 (1960).
228. V. V. Korshak, G. S. Kolesnikov, and B. A. Zhubanov, *Izvest. Akad. Nauk. USSR Otdel. Khim. Nauk* **1958**, 618 (1958).
229. Ger. Pat. 1,086,896; 1,086,897 (1960) (to H. Niebergall).
230. U.S. Pat. 2,671,077 (1954) (to E. I. duPont de Nemours and Co.).
231. L. A. Errede and W. A. Pearson, *J. Am. Chem. Soc.* **83**, 954 (1961).
232. J. P. Pellon and W. G. Carpenter, *J. Polym. Sci., Part A* **1**, 863 (1963).
233. (a) Brit. Pat. 897,680 (1962) (to American Cyanamid Co.); (b) U.S. Pat. 3,043,810 (1962) (to American Cyanamid Co.).
234. (a) U.S.S.R. Pat. 765,899 (1969) (to Institute of Organic Chemistry); (b) U.S. Pat. 3,346,543 (1967) (to Institute of Organic Chemistry).
235. T. Ya. Medved, T. M. Frunze, K. Chen-Mei, V. V. Kurashev, V. V. Korshak, and M. I. Kabachnik, *Vysokomol. Soedin.* **5**(9), 1309 (1963).
236. U.S. Pat. 3,308,091 (1967) (to E. I. duPont de Nemours and Co.).
237. U.S. Pat. 3,598,789 (1971) (to E. I. duPont de Nemours and Co.).
238. (a) Fr. Pat. 1,536,229 (1968) (to Sun Oil Co.); (b) E. Dyson, D. E. Montgomery, and K. Tregonning, *Polymer* **13**, 85 (1972).
239. Fr. Pat. 1,541,384 (1968) (to Badische Anilin & Soda Fabrik).
240. Neth. Appl. 69-182,95 (1969).
241. G. Reinisch and K. Dietrich, *Faserforsch. Text. Tech.* **21**, 367 (1970).
242. R. A. Dine-Hart, B. J. C. Moore, and W. W. Wright, *J. Polym. Sci., Part B* **2**, 369 (1964).
243. F. Dobinson and J. Preston, *J. Polym. Sci., Part A* **1**, 2095 (1966).
244. E. P. Krasnov, V. M. Savinov, V. I. Legunova, U. K. Belyakov, and T. A. Polyakova, *Vysokomol. Soedin.* **8**, 380 (1966).
245. J. M. Lancaster, B. A. Wright, and W. W. Wright, *J. Appl. Polym. Sci.* **9**, 1955 (1965).
246. S. Nishizaki and A. Fukami, *J. Polym. Sci., Part A* **1**, 1769 (1968).
247. J. Preston, *J. Polym. Sci., Part A* **1**, 529 (1966).
248. E. P. Krasnov, V. I. Logunova, and L. B. Sokolov, *Vysokomol. Soedin.* **8**, 1970 (1960).
248a. J. E. Bostic, Ph.D. thesis, Dept. of Textiles, Clemson Univ., Clemson, S.C., 1972.
249. Ref. 13, Chap. 10.
250. Brit. Pat. 1,262,002 (1972) (to E. I. duPont de Nemours and Co.).
251. Japan. Pat. 71,039,96 (1971) (to Toyo Rayon Co. Ltd.); *POST-P* **9**(7), 3821q (1971).
252. Brit. Pat. 1,260,362 (1972) (to Farbenfabriken Bayer A.G.).
253. V. E. Shashoua and W. M. Eareckson, *J. Polym. Sci.*, **40**, 243 (1959).
254. Belg. Pat. 637,260 (1964) (to Monsanto Co.); *Chem. Abstr.* **60**, 12164 (1964).
255. (a) Belg. Pat. 622,574 (1963) (to Monsanto Co.); *Chem. Abstr.* **59**, 5285 (1963); (b) Belg. Pat. 620,511 (1963) (to Monsanto Co.); *Chem. Abstr.* **58**, 9254A (1963).
256. U.S. Pat. 3,063,966 (1962) (to E. I. duPont de Nemours and Co.).
257. H. F. Mark, S. M. Atlas, and N. Ogata, *J. Polym. Sci.* **61**, S-49 (1962).
258. Ya. Fedotova, M. L. Kerber, and I. P. Losev, *Vysokomol. Soedin.* **5**, 881 (1963).
259. (a) U.S. Pat. 569,760 (1958) (to E. I. duPont de Nemours and Co.); *Chem. Abstr.* **55**, 3115 (1961); (b) Belg. Pat. 569,760 (1958) (to E. I. duPont de Nemours and Co.).
260. A. P. Terentev, G. G. Rukhadze, I. G. Mochalina, and G. V. Panova, *Vysokomol. Soedin.* **5**, 837 (1963).
261. J. Preston and F. Dobinson, *J. Polym. Sci., Part B* **2**, 1171 (1964).
262. J. Preston, R. W. Smith, and C. J. Stehman, *J. Polym. Sci., Part C* **19**, 7 (1967).

263. J. Preston and W. B. Black, *J. Polym. Sci., Part A-1* **5**, 2429 (1967).
264. J. Preston, W. F. DeWinter, and W. B. Black, *J. Polym. Sci., Part A-1* **7**, 283 (1969).
265. A. P. Terentev, E. V. Rukhadze, G. V. Panova, and I. G. Mochalinir, *Vysokomol. Soedin.* **5**, 842 (1963).
266. W. C. Stephens, *J. Polym. Sci.* **40**, 359 (1959).
267. Brit. Pat. 1,263,963 (1972) (to General Electric Co.).
268. M. A. Abishev, *Vestn. Nauk. Kaz. SSR* **26**(11), 73 (1970); *POST-J* **9**(4), 1830*t* (1971).
269. Japan. Pat. 71,179,904 (1971) (to Toray Inds. Inc.).
270. U.S. Pat. 3,637,602 (1972) (to Celanese Corp.).
271. U.S. Pat. 3,349,062 (1967) (to E. I. duPont de Nemours and Co.).
272. W. Wrasidlo and J. M. Angl, *J. Polym. Sci., Part A* **1**, 321 (1969).
273. R. Pigeon, *Textilia* **47**(7), 35 (1971); *Chem. Abstr.* **75**, 110937*j* (1971).
274. U.S. Pat. 3,573,260 (1971).
275. W. M. Alvino and L. W. Frost, *J. Polym. Sci., Part A* **9**, 2209 (1971).
276. Brit. Pat. 1,229,124 (1971) (to Farbewerke Hoechst A.G.).
277. U.S. Pat. 3,384,626 (1968) (to Universal Oil Products Co.).
278. S. S. Hirsch, Symp. Polym. Flammability, Hoboken, N.J. (April 1970); *Chem. Eng. News* **48**. 34 (1970).
279. Brit. Pat. 1,262,009 (1972) (to Schenectady, Midland Ltd.).
280. U.S. Pat. 3,591,529 (1971) (to National Starch & Chem. Corp.).
281. (a) Brit. Pat. 705,592 (1966); 713,132 (1967); 712,798 (1967); 748,357 (1969); 748,358 (1969) (to ENKA Corp.); (b) F.C.A.A. Van Berkeland and H. Grotjahn, *Am. Chem. Soc., Div. Org. Coatings Plast. Chem.* **32**(2), 42 (1972).
282. U.S. Pat. 3,547,584 (1970) (to Celanese Corp.).
283. U.S. Pat. 3,576,769 (1971) (to Monsanto Co.).
284. (a) Japan. Pat. 72,054,33 (1972) (to Monsanto Co.); (b) U.S. Pat. 3,549,307 (1970) (to Monsanto).
285. Japan. Pat. 72,054,36 (1972) (to Monsanto Co.).
286. (a) Brit. Pat. 1,214,554 (1970) (to Monsanto Co.); (b) U.S. Pat. 3,607,798 (1971) (to Monsanto Co.).

7

Relationship Between Chemical Structure and Flammability Resistance of Polyurethanes

Kurt C. Frisch and Sidney L. Reegen

1. Introduction

Many techniques have been utilized to impart flame retardancy into polyurethane foams. These include the use of polyols and polyisocyanates that contain halogen and phosphorus atoms, as well as the use of additives containing various elements, including halogens and phosphorus. For a number of reasons, including cost, compatability, and ease of handling or processing, early work with polyester-based rigid urethane foams was based on additives, using inert organic compounds containing these elements, as well as inorganic compounds such as diammonium phosphate. Large quantities of such flame retardants are in use today and halogenated phosphate esters, in particular, have been shown to be very effective in reducing the flammability of urethane foams. However, the low degree of permanency and significant reduction in the physical properties of these foams made it imperative to develop other and more effective techniques.

Kurt C. Frisch and Sidney L. Reegen · Polymer Institute, University of Detroit, Detroit, Michigan.

Major research emphasis has shifted to flame-resistant compositions which are integral parts of the polyurethane chemical structure. These, of course, can be utilized by either the one-shot or prepolymer methods in the preparation of foams or other polyurethanes. Most of these urethane reactants, polyols or polyisocyanates, also contain halogens and/or phosphorus. Recent trends in the development of low-flammability polyurethane foams have been based on copolymers containing heat-resistant groups, such as isocyanurates, and cyclic imides or other nitrogen-containing heterocycles.

Rigid urethane foams prepared with reactive-type flame retardants comprise roughly 75% of total flame-retardant rigid foam sales. Sales of flame-retardant rigid foams for the construction industry have been lower than earlier predictions. However, it is still thought that sales will increase sharply when the performance of flame-retardant rigid foams meets the requirements of that industry. A high percentage of research and development activity is directed toward the construction market, but the transportation industry still remains one of the biggest outlets for urethane foams. Flame retardants for flexible foams also appear to be an area of increasing importance; an increasing amount of recent legislative activity, as well as certain safety codes (e.g., MVSS 302), provide great incentives for this increased activity.

A thorough study of the mechanisms of flammability would be of considerable aid in understanding the details of the problems and in the research designed to develop nonflammable foams. This has been very largely neglected, however, and except for some early investigations, flame retardants are still, in the main, empirically tested. The flame retardation of plastics by halogens has usually been interpreted primarily as the inhibition of a gas-phase free-radical chain-propagation mechanism. It is believed that high-energy hydroxyl and hydrogen free radicals induce competitive low-energy chain processes by interaction with hydrogen halides. The latter are generated by the decomposition of organic halogen compounds. The mechanism of the retardation of oxidation by organophosphorus compounds is described as inhibition of carbon oxidation (in the solid state) by a low-energy process. Phosphorus acids are reduced by carbon to phosphorus-containing free radicals. They can then be reoxidized back to phosphorus acids. A char barrier is also produced from the substrate when combustible gases are starved of oxygen. The mechanisms by which low concentrations of halogens and phosphorus compounds induce flame retardation have not been fully developed.

The effectiveness of phosphorus compounds as fire retardants has also been described by Lyons.[1,2] Phosphorus compounds decompose to acid fragments under the influence of heat. These acids form polymeric species which act as effective dehydrating and esterifying agents. Some of the equations proposed by Lyons demonstrate these reaction sequences:

1. Formation of phosphorus acid by heat:

$$R\text{—}O\text{—}\overset{\displaystyle O}{\underset{\displaystyle |}{\overset{\displaystyle \|}{P}}}\text{—} \xrightarrow{\Delta} \text{Alkene} + HO\text{—}\overset{\displaystyle O}{\underset{\displaystyle |}{\overset{\displaystyle \|}{P}}}\text{—}$$

2. Polymerization to form polyacids:

$$HO\text{—}\overset{\displaystyle O}{\underset{\displaystyle |}{\overset{\displaystyle \|}{P}}}\text{—} \xrightarrow{\Delta} HO\text{—}\overset{\displaystyle O}{\underset{\displaystyle |}{\overset{\displaystyle \|}{P}}}\text{—}O\text{—}\overset{\displaystyle O}{\underset{\displaystyle |}{\overset{\displaystyle \|}{P}}}\text{\Large\char`\~}O\text{—}\overset{\displaystyle O}{\underset{\displaystyle |}{\overset{\displaystyle \|}{P}}}\text{—}OH + H_2O$$

3. Esterification and dehydration:

$$R\text{—}CH_2\text{—}CH_2\text{—}OH + \text{\Large\char`\~}\overset{\displaystyle O}{\underset{\displaystyle OH}{\overset{\displaystyle \|}{P}}}\text{—}O\text{—}\overset{\displaystyle O}{\underset{\displaystyle OH}{\overset{\displaystyle \|}{P}}}\text{—}OH \longrightarrow$$

$$R\text{—}CH_2CH_2\text{—}O\text{—}PO_3H_2 + \text{—}\overset{\displaystyle O}{\underset{\displaystyle OH}{\overset{\displaystyle \|}{P}}}\text{—}OH \longrightarrow R\text{—}CH{=}CH_2 + H_3PO_4$$

Lyons also proposed a carbonium ion mechanism as follows:

$$R\text{—}CH_2\text{—}CH_2OH \xrightarrow[\Delta]{H^+} (R-CH_2\text{—}CH_2OH_2{}^+) \longrightarrow R\text{—}CH{=}CH_2 + H_2O + H^+$$

The resulting compounds may then undergo cross-linking or other reactions to form char. The formation of char provides insulation for the remaining foam beneath it and retards further degradation.

Many investigators have analyzed the structural features of polyurethanes which might influence their flammability. Energy factors, such as cohesive energy, hydrogen bonding, heat of combustion and dissociation energies, effects of elemental constituents, including synergism, and of molecular structure, have been reviewed by Dickert and Toon,[3] Frisch,[4] and others.[1,5,6] Flammability, during the initial reactions, is thought to be dependent on the formation of flammable gaseous decomposition products. Flame propagation is dependent on the rate of these reactions. It was thought that fire-resistant urethane foams could be prepared by incorporating thermally stable structural units or units which decompose to products that are not combustible.

The problem of smoke and the toxicity of gases formed during the combustion of urethanes has come under public scrutiny as a result of various fires and has been the subject of many investigations. Reviews have been written that give some understanding of the nature and scope of the problems. Study of the effect of the use of flame retardants for urethane foams on the amount of smoke developed indicates that the amount of smoke increases with the concentration of flame retardants containing halogens and phosphorus.

Relatively few efficient commercial flame retardants have been designed for flexible foams. Their open-cell structure, low degree of cross-linking, and basic chemical structure make the flameproofing of flexible urethane foams more difficult than the highly cross-linked rigid foams. It is also more difficult to retain the physical properties of the flexible foams. Not only will technical problems have to be overcome, but they will have to be economical.

A review of the basic physical, processing, and composition parameters that might influence flammability properties and requirements of both rigid and flexible urethane foams is required. Flexible foams are usually produced by the one-shot process by the metal–base catalysis of a polyether polyol with the 80/20 TDI in the presence of water and a surfactant. Carbon dioxide is produced by reacting the water and isocyanate groups, aiding in the expansion of the polymerizing resin and forming primary aromatic amines which react with other isocyanate groups to produce N,N'-diphenylurea linkages. Besides the predominant concentration of urethane groups, minimal levels of biuret and allophanate groups are formed. These flexible foams consist primarily of polyether backbone structures which are connected by urethane and urea groups. Glass-transition temperatures for these polymers range from -30 to $-50°C$. Some of the polymer will melt, making it difficult even to design meaningful standard laboratory tests. The melted polymer might cause rapid flame spread or might drip from the burning area and cause the flame to be extinguished. These problems are similar to those encountered with thermoplastic polystyrene foams. The open-cell structure of flexible urethane foams allows a steady supply of air to diffuse through the burning structure.

The introduction of high-resilience (HR) flexible foams has brought about some changes in the basic formulations of these materials, particularly for use in automotive interiors. The use of "crude" polyisocyanates, either alone or in combination with TDI, the incorporation of a liquid form of 3,3'-dichloro-4,4'-diaminodiphenylmethane (LD-813, Du Pont), or the addition of a special flame-retardant additive, and the use of graft copolymers of polyether polyols with acrylonitrile and styrene has led to the development of flexible foams which pass the MVSS 302 test. However, recent Office of Safety and Health Administration regulations have led to the elimination of the chlorinated diamine (LD-813) in automotive and other foam applications.

Rigid urethane foams utilize polyisocyanates having a functionality of

2.3–3.0; the polyols used generally have an average functionality of at least 4. The resultant higher cross-link density results in little melting when burning and formation of volatile gases, a carbonaceous residue, and a resinous distillate. The higher degree of aromaticity in rigid foams also aids in retarding burning. Higher concentrations of fluorocarbon blowing agent in closed cells further retard burning. Parrish and Pruit[7] have noted that foams that contain larger amounts of phosphorus-containing additives were good in resistance to flammability; however, they also became less thermally stable, as detected by TGA measurements.

2. Effect of Structure on Flammability of Urethane Foams

Both equivalent weight (or hydroxyl number) and functionality of polyols affect the flame properties of urethane foams. The effect of the hydroxyl number of polyether on the burning rate of one-shot rigid urethane foams is shown in Fig. 1.[3] It is apparent that the burning rate as well as the weight loss (amount burned) decreases as the hydroxyl number increases (or equivalent weight decreases).

Anderson[6] also found that a decrease in equivalent weight improved the flame resistance in a homologous series of foams based on sucrose polyethers. Likewise, the flame resistance was generally found to increase as the polyol functionality increased. However, in addition to functionality and hydroxyl number, other structural features must be taken into account. These factors include cyclic *vs.* open chain structures and neutral *vs.* basic polyols. Dickert and Toon,[3] Anderson,[6] and Frisch *et al.*[4,8] reported a higher degree of flame resistance afforded by cyclic polyethers (e.g., based on α-methylglucoside or sucrose) than by polyethers based on open chain polyols (e.g., pentaerythritol or sorbitol) having the same hydroxyl number and functionality (e.g., α-methylglucoside vs. pentaerythritol).

Table 1[8] shows a comparison of flammability between α-methyl-glucoside and pentaerythritol-based one-shot rigid polyether foams at two different levels of isocyanate index (105 and 115). Table 2[8] lists the Butler chimney data for the same types of foams shown in Table 1 but containing a flame-retardant additive (Vircol VC-611, Mobile Chemical Co.) at three different levels of phosphorus. It is apparent that the cyclic structure of the α-methylglucoside imparts considerably greater flame resistance than the pentaerythritol polyether-based foam, even at a low phosphorus level.

Aliphatic or aromatic amine-based polyether polyols impart a greater degree of flame resistance than neutral polyol-based polyethers, particularly in combination with phosphorus- and/or halogen-containing polyols,

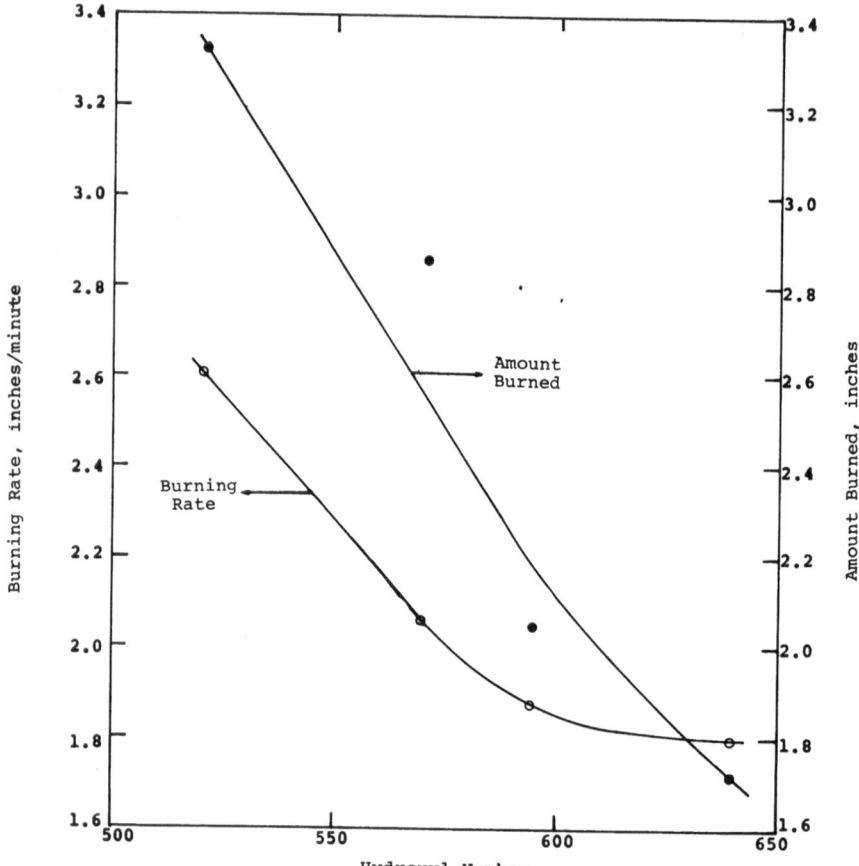

FIGURE 1. Effect of hydroxyl number of polyether on burning rate of one-shot rigid urethane foams.

presumably because of the synergistic effect between nitrogen and phosphorus and nitrogen, phosphorus, and halogens.

Polyols containing halogens (bromine or chlorine) also offer a suitable technique for flameproofing polyurethane foams. The bromine analogs appear to provide a greater flame retardancy than do the chlorine-containing polyols. Aromatic chlorides and bromides are usually too heat stable, decomposing far above the ignition temperature of the foams. Some aliphatic bromides may decompose too readily, below the ignition temperature of the foams, sometimes during the foaming reaction itself, resulting in scorch. Organic halides offer stability during foam processing and the polyester polyols modified with chlorendic acid have been among the most successful.[9-12] Their flame retardancy is increased by phosphorus compounds.[12,14]

TABLE 1
Flammability of Foams (No Flame Retardant Added)

Type of polyol	Name	Index NCO/OH	Flammability ASTM 1692, cm/min	Butler chimney, % weight retained
α-Methylglucoside	Cord 370	105	25.1	27.3
		115	24.3	27.2
	Cord 450	105	26.2	41.9
		115	24.8	41.9
Pentaerythritol	Pluracol PeP 650	105	35.5	23.4
		115	31.1	29.1
	Pluracol PeP 550	105	32.4	33.5
		115	27.8	39.1

Polyester resins based on chlorendic anhydride, a Diels-Alder adduct of hexachlorocyclopentadiene and maleic anhydride, and containing about 35% chlorine, have been utilized in the preparation of rigid urethane foams (Hetrofoam, Hooker Chem. Corp.).[9-11] In addition to excellent flame resistance, these foams also exhibited high thermal stability and were resistant to water exposure.[11] It was also found that the addition of antimony oxide (1-5%) greatly improved the flame resistance of these foams.[10]

Similarly, halogen-containing polyesters, particularly those derived from chlorendic acid, glycerol, and adipic acid, were reacted with a polyisocyanate in the presence of tertiary alcohols (e.g., *tert*-butyl alcohol and *tert*-amyl alcohol) as foaming agents to yield flame-resistant urethane foams.[12] Small amounts of a concentrated acid such as sulfuric or phosphoric acid increased the effectiveness of the tertiary alcohols as foaming agents.

Hindersinn and Worsley[13] described the use of polyols based on

TABLE 2
Butler Chimney—Flame Retardant (VC-611) Added

Type of polyol	Name	Index NCO/OH	% weight retained		
			0.5% P	1.0% P	1.5% P
α-Methyl glucoside	Cord 370	105	46.8	47.7	53.3
	Cord 450	105	52.5	58.9	55.4
Pentaerythritol	Pluracol PeP 650	105	28.0	38.4	52.5
	Pluracol PeP 550	105	35.2	48.0	50.2

hexahalocyclopentadiene in the preparation of both flexible and rigid foams by reaction with a polyisocyanate and subsequently with a polyester or polyether polyol.

In a somewhat similar procedure, Lyon and Applewhite[15] prepared Diels-Alder adducts from hexachloro- or hexabromocyclopentadiene and ricinoleic acid esters (e.g., castor oil). Flame-resistant rigid foams were prepared from these adducts by reaction with polymethylene–polyphenyl isocyanate in the presence of triisopropanolamine and trichlorofluoro-methane blowing agent.

The use of hydroxyl-containing terpolymers made by a one-step copolymerization of vinyl chloride, vinyl oxyethanol, and a monomeric fumarate ester (e.g., dibutyl fumarate) was described by Slocombe.[16] The resulting terpolymer was then reacted with TDI in the presence of water, tertiary amine catalysts, and a silicone surfactant yielding urethane foam which exhibited flame resistance as well as good solvent resistance.

Kuryla and Knopf[17] reported the preparation of quasi-prepolymers by the reaction of polyisocyanates (e.g., PAPI or Mondur MR) with monofunc-tional halogen-containing alcohols (e.g., 2,3-dibromo-1-propanol) or halo-gen-containing amines or thiols. The quasi-prepolymers were then further reacted with a sucrose-based polyether or an aniline–formaldehyde-based polyether resulting in flame-resistant rigid foams.

Brominated allyl glucoside polyethers were reported recently by Othey, Westhoff, and Mehltretter[18] for use in flame-retardant rigid urethane foams. The structure of the brominated polyether can be represented as follows:

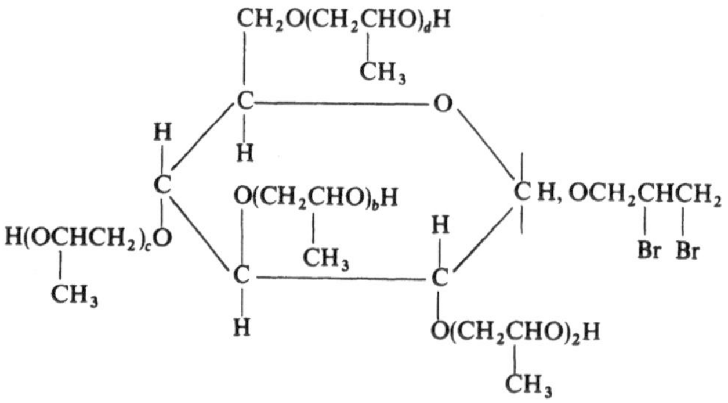

Good flame resistance and good humidity aging was noted for the foams, particularly when the polyethers were formulated with four to five moles of propylene oxide per allyl glucoside.

Jolles[19] has compared the effect of halogens, both with and without Sb (Tables 3 and 4).

These tables indicate not only that bromine is more effective than

TABLE 3

Halogen Required to Produce 2–3 sec Burning Time in a Rigid Polyurethane Foam[19]

Compound	Halogen, %
Hexachlorobenzene	15.8
Hexachlorocyclohexanol	5.2
Hexabromobenzene	6.2
Pentabromoethylbenzene	6.0
1,2,3,4-Tetrabromobutane	3.1

chlorine, but aliphatic halogens (which are more readily decomposed) are more effective than aromatic halogens.

Lyn and Applewhite[20] also gave comparative data on chlorine- and bromine-containing foams; the effect of antimony was also dramatic.

A diisocyanate derived from chlorendic acid dichloride via the Curtius reaction was described by Hoch.[21]

Holmquist[22] utilized undistilled, ring-halogenated tolylene diisocyanate for the manufacture of flame-resistant flexible foams. Preferred materials were diisocyanates containing a minimum of 32% halogen with at least 50 mole % of the halogen in the form of bromine, the remainder being chlorine. In order to ensure that the diisocyanate component remained liquid during the preparation of the foams, at least 10 parts tolylene diisocyanate were added to the halogenated, undistilled tolylene diisocyanates.

Trichlorobromoalkyl isocyanates were reported by Farrissey, Recchia, and Sayigh.[23] These isocyanates were prepared by reacting an unsaturated isocyanate such as allyl isocyanate with a haloalkane, e.g., bromotrichloromethane. In a typical example, a 4,4,4-trichloro-2-bromobutyl isocyanate (1 part) was blended with PAPI (3 parts), and this mixture was then used in the preparation of rigid urethane foams.

TABLE 4

Comparisons of Fire-Retardant Effectiveness of Chlorine (as Trichlorphenol) and Br (as Tribromophenol) in a Rigid Urethane Foam (by ASTM 1692)

	Fire retardant, %	Halogen content in foams, %	Antimony oxide, %	Burning time, sec	Extent of burning, in.	Burning rate, in./min	ASTM Class
Trichloro- phenol	11.7	6.3 Cl	5.0	66	1.5	1.4	SE
	10.8	5.8 Cl	2.8	86	2.5	1.6	SE
	8.0	5.8 Br	2.5	13	0.3	1.1	SE
Tribromo- phenol	7.8	5.7 Br	5.1	0	0	—	NB
	5.6	4.0 Br	5.0	1	1.4	1.4	SE

Monobrominated tolylene diisocyanate has been employed in combination with antimony oxide in the preparation of flame-resistant rigid urethane foams.[24] It was also found that the foams prepared from the brominated isocyanate were not significantly different from the TDI-based foams with respect to reactivity, density, cell structure, or resistance to aging, but exhibited somewhat lower compressive strengths.

Very recently Argabright and Phillips[25] reported the preparation of brominated polyisocyanates containing isocyanurate rings. These polyisocyanates are characterized by having the isocyanate groups not linked directly to aromatic rings but rather via alkylene groups such as methylene groups. Excellent flame-resistant rigid foams have been produced from these brominated polyisocyanates.

Trichloroalkylene oxides have been employed in the preparation of polyols as reactive intermediates for flame-retardant urethane foams.[26-29] These polyether polyols were prepared by the acid-catalyzed addition of trichloroalkylene oxides to a variety of base polyols. Vogt and Davis[29] described the addition of 3,3,3-trichloropropylene oxide (TCPO) to diol or polyol initiators employing BF_3 as a catalyst, e.g.:

$$\underset{\underset{O}{\diagdown\diagup}}{HC\!-\!CH_2} + HO\!-\!\underset{CCl_3}{\overset{|}{CH}}\!-\!CH_2OH \xrightarrow[BF_3]{HX} HO\!-\!CH_2\underset{CCl_3}{\overset{|}{CH}}\!-\!O\!-\!(CH_2\!-\!\underset{CCl_3}{\overset{|}{CH}}\!-\!O)_nH$$

The TCPO-based diols and polyols can be used either alone or in combination with phosphorus-containing compounds for the preparation of flame-retardant polyurethanes.

Recently Pitts, Fuzesi, and Andrews[28] described the use of 4,4,4-trichloro-1,2-butylene oxide (TCBO) in the preparation of polyethers based on polyols such as glycerol, pentaerythritol, and oxyethylated sucrose using BF_3 etherate as catalyst. The representative properties of TCBO-containing polyols are shown in Table 5. Flame-retardant rigid foams were prepared from these polyols using the formulations given in Table 6. The relative flame retardancy of these foams is illustrated in Table 7. The corresponding tests with some commercially available rigid urethane foams are included for comparison (Table 8). These tests show the TCBO-containing foams to be at least comparable to commercial foams as indicated by several tests. Only Kode 25 exhibited better values in the Bureau of Mines flame-penetration test.

Phosphorus is one of the elements most commonly utilized, usually as part of the polyol structure, bonded to carbon and oxygen. These polyols have been prepared in many combinations of these elements, and are blended with other polyols during the foam production. Anderson[6] has found, as

TABLE 5

Typical Physical Properties for Various TCBO-Containing Polyols

Polyol	Initiators	Moles TCBO	OH No.	% H₂O	Acid No.	Apparent pH	% Cl	Viscosity at 25°C
RF-4	Pentaerythritol–glycerol (1:1 Molar ratio)	7.5	253	0.1	0.9	4.5	51.8	1,010,000
RF-13	Oxyethylated sucrose (2.5:1)	6.5	277	0.5	1.5	3.9	43.0	1,250,000
RF-14	Oxyethylated sucrose (2.5:1)	2.4:7.25 PO	376	0.13	1.5	3.4	19.9	82,200
RF-19	Oxyethylated sucrose (2.5:1)	2.2(modified): 6.65 PO	350	0.06	2.3	4.4	18.8	26,000
RF-20	Oxyethylated sucrose	4.6(modified)	343	0.07	2.9	4.5	38.8	736,000

TABLE 6

Machine Formulations for Flame-Retardant Rigid Foam Containing TCBO

Ingredients	RF-13 A	RF-13 B	RF-14 A	RF-14 B	RF-19 A	RF-19 B	RF-20 A	RF-20 B
Oxytrichlorbutylated polyol	100	100	100	100	100	100	100	100
Dow Corning 113 silicone surfactant			1.5	1.5	1.5	2.0	2.0	1.5
DC-193 Surfactant	1.5							
DC-201 Surfactant		2.0						
General Electric SF 1066 Surfactant	1.5	2.3						
Union Carbide TMBDA Amine Catalyst			1.26	1.6				
DMEA Amine Catalyst								
DMEA/Houdry DABCO 33-LV (3.3:1)					2.8	2.0	2.0	2.8
Refrigerant-11 Blowing Agent[a]	26	29	31	31	29.5	30	30	28.5
Upjohn PAPI[b]	75		94.1	94.1	87.1		87.1	
Mobay Mondur MR[b]		95				88.2		88.2

[a] Adjusted if necessary to give uniform 2 lb/ft³ density foam.
[b] 105 Index used.

TABLE 7
Relative Flame Retardancy of TCBO-Containing Rigid Urethane Foams

Rigid foam	% Cl in Foam[a]	ASTM D-1692-59T	Bureau of Mines flame penetration test, min[b]	ASTM E-84-68 25-ft flame spread rating	Tunnel test smoke development
RF-4/TDI	35.0	SE 1.6 in.			
RF 4/PAPI	24.8	NB			
RF 13/PAPI	21.1	NB	10–11		
RF 14/PAPI	8.7	NB	21	35	211
RF 19/PAPI	8.5	NB	15–18	50	144
RF 19/MR	8.5	NB	14–15	50	217
RF 19/V-88 (50%) MR	4.2(6.5 Br, 0.7 P)	NB	24–27	25	99
RF 20/PAPI	18.0	NB	15–16	40	435
RF 20/MR	18.0	NB	19–25	40	370
RF 20/V-88 (30%) MR	12.2 (3.9 Br, 0.4 P)	NB	18–19	20	77

[a] From polyol contribution only.
[b] 1900 F—automated (see Ref. 28a).

have other investigators[62] that structurally bound phosphorus polyols are more effective than the use of nonreactive additives. Foams formed from such polyols were found to be more effective in retaining their flame retardancy after aging.

Phosphite polyols have been formed by the transesterification reaction of the corresponding esters and polyhydric alcohols.[63-65] Triaryl phos-

TABLE 8
Relative Flame Retardancy of Various Commercial Rigid Urethane Foams

Rigid foam	ASTM D-1692-59T	Bureau of Mines flame penetration test, min[a]	ASTM E-84-68 25-ft flame spread rating	Tunnel test smoke development
Kode 25[b]	NB	38 (45)[c]	25	119
Hetrofoam[d]	NB	1	40 (35)[e]	960 (419)[e]
Barfire[f]	NB	1	35	765
			30	330
Pluracol 463[g]/NCO 20[h]	NB	2(> 15)[i]	30(25)[i] (25)[i]	336(150)[i](450)

[a] Automated—1900°F (see Ref. 23).
[b] Upjohn Company, CPR Div.
[c] Data taken from Ref. 20.
[d] Hooker Chemical Corp., Durez Div.
[e] Test performed by Southwest Research Institute, El Paso, Tex.
[f] Allied Chemical Co., Barrett Div.
[g] Wyandotte Chemicals Corp.
[h] Kaiser Chemicals Div.
[i] Data taken from Ref. 28b.

phites, as well as chlorinated aliphatic phosphites such as tris(2-chloroethyl)-phosphite have been used successfully.[66]

The hydrolysis of phosphorus polyols may be caused by the presence of even small amounts of water in the polyether polyols. Care must be taken with the use of prepolymers of these phosphorus polyols to avoid this problem; many phosphorus compounds are catalysts for the formation of carbodiimides.[30] Products of the esterification of phosphoric or polyphosphoric acids with a polyhydric alcohol can be reacted with alkylene oxides. Products for flexible or rigid foam applications can be obtained in this manner. Phosphoric acids containing as much as 84% phosphorus pentoxide can be used. The phosphorus anhydrides can be cleaved by alcohols to the esters and acids,[67] or solid phosphorous pentoxide can be used as the phosphorous acid.[68] Polyols can be prepared by reaction of the polyols with phosphoric acid, but complex phosphoric acid esters may be formed. The conversion of all phosphorous acid to the orthophosphate form can be accomplished by reaction with alcohols in such a manner that the required functionality is obtained. The desired hydroxyl number is obtained by using a reaction temperature of 50–150°C. The products actually formed include:

$$\underset{\text{(I)}}{\text{HO}-\overset{\displaystyle O}{\overset{\|}{P}}-\text{ORO}-\overset{\displaystyle O}{\overset{\|}{P}}-\text{OROH}}$$
$$\text{(OH)}\quad\text{(OH)}$$

$$\underset{\text{(II)}}{\text{HO}-\overset{\displaystyle O}{\overset{\|}{P}}-\text{O}-\overset{\displaystyle O}{\overset{\|}{P}}-\text{O}-\overset{\displaystyle O}{\overset{\|}{P}}-\text{OH}}$$

$$\underset{\text{(III)}}{\text{HO}-\overset{\displaystyle O}{\overset{\|}{P}}-\text{ORO}-\overset{\displaystyle O}{\overset{\|}{P}}-\text{OH}}$$

Although other alkylene oxides can be utilized, propylene oxide is usually preferred, forming

$$\left[\text{H}-(\text{OCHCH}_2)_n\text{O}\atop\text{CH}_3\right]_2 \ \overset{\displaystyle O\ \ \ O}{\overset{\|\quad\|}{\text{POROP}}}-\text{OROH}$$
$$\text{O}(\text{CH}_2\text{CHO})_n\text{H}$$
$$\text{CH}_3$$

Direct alkoxylation of mono- and dialkyl esters of pyrophosphoric acid produced from monofunctional alcohols and a polyphosphoric acid also can produce important phosphate polyols.[69,70]

Modifications of this reaction include the use of a carboxylic acid or carboxylic acid anhydride, together with the phosphorous acid and epoxide, to produce the ester of the carboxylic acid as well as the phosphate ester.[71]

Phosphonate polyols have been formed by the Mannich reaction between dialkyl phosphites, formaldehyde, and an alkanolamine. Other preparative methods for phosphonate polyols include the Arbuzov reaction and the addition of dialkyl phosphites to olefins.

Among the phosphonates that have been used as flame-retardants, the aminomethyl phosphonate:

$$
\begin{array}{c}
O \\
\parallel \\
(C_2H_5O)\,P\!-\!CH_2N(C_2H_4\!-\!OH)_2
\end{array}
$$

has proved to be very effective.[31] Because it is a reactive diol, the phosphorus becomes part of the polyurethane, but not part of the polymer backbone. Being in a pendant position, any thermal degradation that forms an acidic phosphorus fragment does not cause scission of the backbone. A typical formulation (tested by ASTM D-1692) gave a burning length of 1/4 in. or less and was self-extinguishing in 15 sec. Approximately 15% of the fire retardant (2% phosphorus) was required.

As in the case for most phosphorus-containing urethanes, the thermal decomposition temperature was substantially reduced. The acidic fragments from the phosphorus act as catalysts for dehydration to char. It has been assumed, therefore, that the lower decomposition temperature is a necessity for the flame-retardant activity of phosphorus. Similarly, an increase in smoke density usually occurs.

Other phosphonates that contain phosphorus connected to hydroxyl groups put the phosphorus in the polymer backbone, such as [32]

$$
\begin{array}{c}
O \\
\parallel \\
(HORO)_2P\!-\!CH_2NR_2
\end{array}
$$

Such polyols were reported to be somewhat less effective. Other polyols have been made by ethoxylation of nitrilotris(methylenephosphonic acid), $N(CH_2PO_3H_2)_3$. The resultant hexols have been claimed to produce urethanes with good fire retardancy.[33]

Usually, about 1.5% phosphorus is required to produce fire-retardant polyurethanes, when no halogens are present. This can be reduced to 1%, but it requires 10–15% Cl to do so; addition of 4–7% Br may lower the need for P to about 0.5%. The halogenated phosphonate esters must be stored with isocyanates in two-part premix systems, to avoid inactivation of the amine catalyst that is usually packaged with the polyol; this is indicative of the instability problems caused by the presence of halogens.

The bromine–phosphorus synergistic systems are more efficient than chlorine–phosphorus systems, but more expensive.

THPC[(HOCH$_2$)$_4$P − Cl]] can be used as an additive or prereacted with a polyol.[34] Phosphorus–sulfur additives have also been suggested for urethanes; they include P$_4$S$_{10}$, P$_4$S$_7$, and P$_4$S$_3$.[35] Approximately 20% Cl or 12% Br is required in urethanes to produce nonburning products, but when antimony is also used, these levels are reduced considerably.[36,37] Dezzinger, Dickert, and Wiles[38] suggested that 4.4% Sb$_4$O$_6$ plus 3.8% Cl is adequate; others reported that 6.3% Sb$_4$O$_6$ plus 7% Cl[39] or 2.5% Sb$_4$O$_6$ plus 2.4% Br[40] produce nonflammable urethanes. Pape *et al.*[71] reported a series of results with a polyol from a tetrabromophthalic anhydride polyester. Unfortunately, Sb$_4$O$_6$ is insoluble, and it is difficult to use in prepackaged foam systems because of its tendency to settle.

It has been reported that it hardly mattered what form the phosphorus was in, when used as a fire retardant for urethanes, so long as the compound can decompose to acids when heated.[41] However, the amount of fire retardants necessary for nonburning properties varies with the type of element used,[3] as seen in Table 9.

Polyurethanes containing phosphorus have been prepared from phosphorus-containing diisocyanates,[42] triisocyanates,[43,44] or triols[45] such as RP(O)(NCO)$_2$, P(O)(NCO)$_3$, P(O)(OC$_6$H$_4$NCO)$_3$, or P(O)(CH$_2$OH)$_3$. In addition, hydroxy-terminated phosphonopolyester-based prepolymers were also used.[46] These urethane polymers containing P–O–C or P–N–C bonds are susceptible to hydrolytic aging.[47]

Fielding[48] described the preparation of phosphoryl triisocyanate, PO(NCO)$_3$, by first reacting phosphorus trichloride with sodium or potassium cyanate in liquid sulfur dioxide medium, yielding a solution of phosphorus triisocyanate, P(NCO)$_3$. This solution, when treated with

TABLE 9
Amount of Flame-Retardant Elements
for Nonburning Foams
(ASTM D-1692)^3

| Foam | Amount of Elements, % | | | |
	Cl	Br	Sb	P
A	13			
B		10		
C			6	
D				2
E	6.5			
F	6.5		3	
G		5		1
H		5	3	

sulfur trioxide, leads to the formation of phosphoryl triisocyanate:

$$PCl_3 + 3NaOCN \longrightarrow P(NCO)_3 + 3NaCl$$

$$P(NCO)_3 + SO_3 \longrightarrow PO(NCO)_3 + SO_2$$

Phosphoryl triisocyanates can also be obtained by oxidation of the corresponding phosphorus triisocyanates of the general formula $R_nP(NCO)_{3-n}$ with nitrogen dioxide:[49]

$$R_nP(NCO)_{3-n} + NO_2 \longrightarrow R_nPO(NCO)_{3-n} + NO$$

Alternately, the oxidation of phosphorus triisocyanates may be carried out by means of a stoichiometric amount of oxygen in the presence of a minor proportion of nitrogen dioxide:[49]

$$2P(NCO)_3 + O_2 \longrightarrow 2PO(NCO)_3$$

In addition to phosphoryl triisocyanate, substituted phosphoryl triisocyanates, prepared by oxidation of the corresponding phosphorus triisocyanates, include diethyl phosphorus monoisocyanate, ethyl phosphorus diisocyanate, ethoxy phosphorus diisocyanate, phenyl phosphorus diisocyanate, and phenoxy phosphorus diisocyanate.

The phosphoryl isocyanates react with glycols and diamines to form the corresponding phosphoryl urethanes and phosphoryl ureas. It is reported that these polymers can be molded and shaped as well as used in the preparation of laminates.[49] However, the urethanes derived from phosphoryl or phosphorus isocyanates generally are susceptible to hydrolysis.

Dyer and Dunbar[50] prepared polyurethanes with P–C bonds in the backbone with the objective of imparting both hydrolytic stability as well as flame resistance. These polymers were prepared from phosphorus-containing diamines [alkyl bis(*m*-aminophenyl)phosphine oxides] and bischloroformates (1,4-butanebischloroformate or *p*-xylyene α,α'-bischloroformate) by means of interfacial polycondensation:

$$H_2N-C_6H_4-\underset{\underset{O}{\|}}{P}(R)-C_6H_4NH_2 + Cl\underset{\underset{O}{\|}}{C}OCH_2-X-CH_2O\underset{\underset{O}{\|}}{C}Cl \longrightarrow$$

$$\left[NH-C_6H_4-\underset{\underset{O}{\|}}{P}(R)-C_6H_4NH\underset{\underset{O}{\|}}{C}OCH_2-X-CH_2O\underset{\underset{O}{\|}}{C}\right]_n$$

These polymers were found to have comparable stability towards alkaline hydrolysis (9% NaOH solution at 50°C for extended periods) as compared to the corresponding non-phosphorus-containing polymers ($-CH_2-$ group instead of $-\underset{\underset{O}{\|}}{P}-CH_3$ group) and a film from one of the polymers ($R = CH_3$, $X = 1$, $4\text{-}C_6H_4$) exhibited self-extinguishing properties.

Bakhitov, Kuznetsov, and Obryadina[51] described the preparation of phosphorus-containing polyurethanes from equimolar amounts of tris-(hydroxymethyl) phosphine (THMP), tris(hydroxymethyl)phosphine oxide (THMPO), tetrakis(hydroxymethyl)phosphonium chloride, hydroxymethyl-phosphonic acid, and diisocyanates. The formation of polyurethanes by the reaction of THMPO with diisocyanates can be represented as follows:

$$
n\text{HOCH}_2\!-\!\overset{\displaystyle\overset{O}{\|}}{\underset{\displaystyle\underset{|}{CH_2OH}}{P}}\!-\!\text{CH}_2\text{OH} + n\text{OCN}\!-\!\text{R}\!-\!\text{NCO} \longrightarrow
$$

$$
\text{\small\wedge\wedge\wedge\wedge}\; \text{OCH}_2\!-\!\overset{\displaystyle\overset{O}{\|}}{\underset{\displaystyle\underset{|}{CH_2OH}}{P}}\!-\!\text{CH}_2\text{OCONH}\!-\!\text{R}\!-\!\text{NH}\!-\!\text{CO}\;\text{\small\wedge\wedge\wedge\wedge}
$$

It was also shown that THMPO reacts with isocyanates more slowly than THMP. Tetrakis(hydroxymethyl)phosphonium chloride reacts readily with diisocyanates to form initially soluble products which become insoluble upon further polymerization. In the copolymerization press, the tetrakis-(hydroxymethyl)phosphonium chloride is partly converted to THMPO by splitting off formaldehyde and HCl, as shown by the low chlorine values in the resulting polyurethanes.

Hydroxymethylphosphonic acid also reacts readily with diisocyanates with evolution of carbon dioxide (2 mol CO_2 for 1 mol diisocyanate and 1 mol acid).

The polyurethanes resulting from the above-described reaction can be used to impart fire resistance to wood, paper, or fabrics by impregnating them with solutions of the initial copolymerization stage, followed by completion of polymerization on the specimen.

Self-extinguishing, linear, phosphorus-containing polyurethanes were prepared by the reaction of butyl bis(hydroxymethyl)phosphines and their oxides with hexamethylene diisocyanate in dimethylformamide or in the absence of a solvent.[52]

The copolycondensation of ethyl bis(hydroxymethyl)phosphine and ethylene glycol with HMDI was carried out without a solvent over a range of mole ratios of ethyl bis(hydroxymethyl) phosphine and the glycol from 1:5 to 5:1 employing an overall NCO/OH ratio of 1:1.[52] The melting points of the resulting copolymers increased regularly with increase in the amount of ethyl bis(hydroxymethyl) phosphine in the starting hydroxy-containing monomer mixture.

Studies have been carried out on the effect of phosphonitriles, $(PNX_2)_{m'}$ especially derivatives where X may be partially replaced by a number of groups;[53] if one of the groups is a polyol, the retardant becomes a reactive component.

Papa and Proops[54] studied the structural effects of halogen- and phosphorus-containing polyol mixtures on flame retardancy of flexible urethane foams.

Two bromine-containing polyols, an aliphatic bromide, dibromoneopentyl glycol [2,2-bis(bromomethyl)-1,3-propanediol, DBNG], and an aromatic bromide, prepared by reacting tetrabromophthalic anhydride with a flexible polyether polyol [poly(oxypropylene) adduct of glycerine, mol wt 3000, Niax Polyol LC-60 (Union Carbide Corp.)], were used in this investigation.

Three different phosphorus-containing polyols were used: a phosphate, a phosphonate, and a phosphite. The hydroxyl number and phosphorus content of these polyols are shown in Table 10. Standard formulations (in parts by weight) consisted of polyol(s), 100; water, 4; silicone surfactant, 0.5; Niax Catalyst A-1, 0.1; TDI, varied; and stannous octoate, varied. The isocyanate index was 105. Oxygen index (OI) values as a function of bromine content and the amount of char formed at 300°C (for 45 min) and at 500°C (for 30 min) are compared in Fig. 2 for foams containing dibromoneopentyl glycol. From these data it is obvious that the degree of flame resistance is not a linear function of the bromine content. The plots pass through an initial maximum at 2% bromine, then fall off and exhibit finally an increase in oxygen index at 5% bromine. From Fig. 2, it can be seen that the amount of char formation at 300°C followed about the same pattern as the flammability curve. Of the foams examined, only those containing DBNG at 3% and 5% bromine were rated self-extinguishing by ASTM D-1692-67T. A substantial portion of the bromine from both aliphatic and aromatic bromide flame retardants was accounted for in the char. In the presence of phoshorus, a constant level of about 1% ionic bromine was found in the chars, regardless of the initial concentration of bromine in the foams or the type of phosphorus

TABLE 10
Properties of Phosphorus Polyols

Chemical identification	OH No., mg KOH/g	P, %
Phosphate-containing polyol	300	11.1
Diethyl-*N*,*N*-bis (2-hydroxyethyl)-aminomethyl phosphonate	440	12.2
Tris(dipropylene glycol)phosphite	395	7.1–7.3

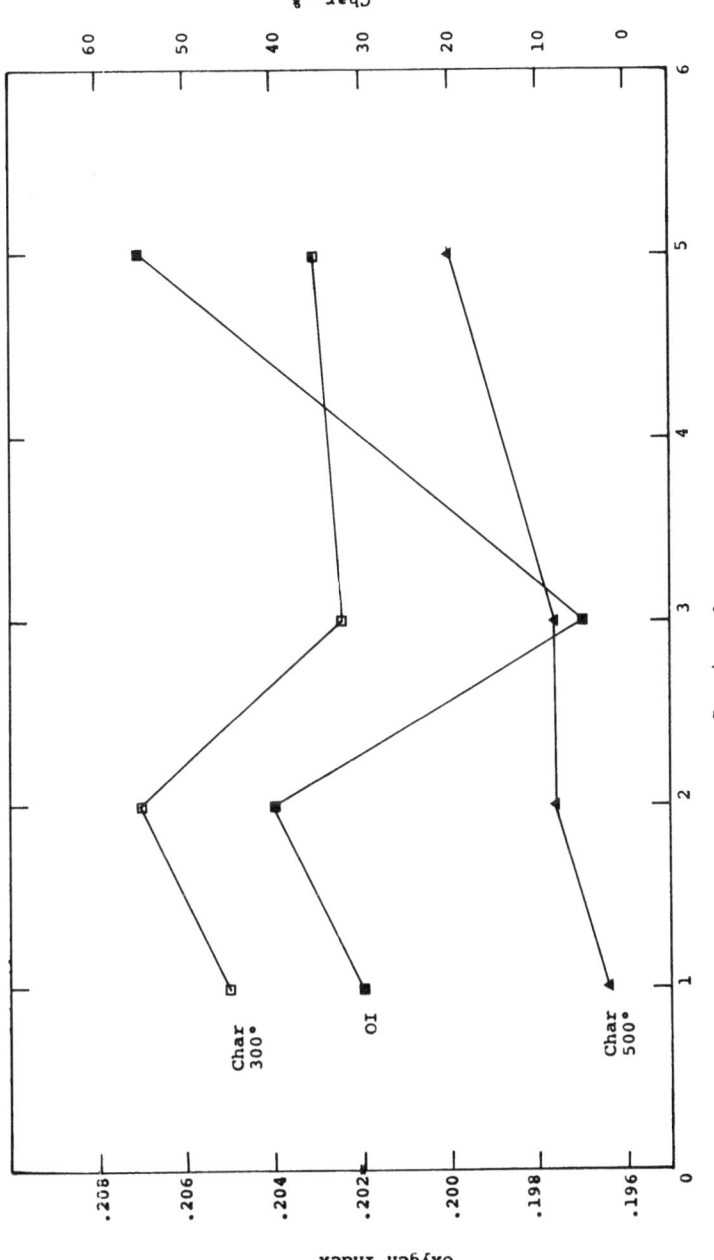

FIGURE 2. Degree of char and flammability resistance of flexible foams containing dibromoneopentyl glycol.

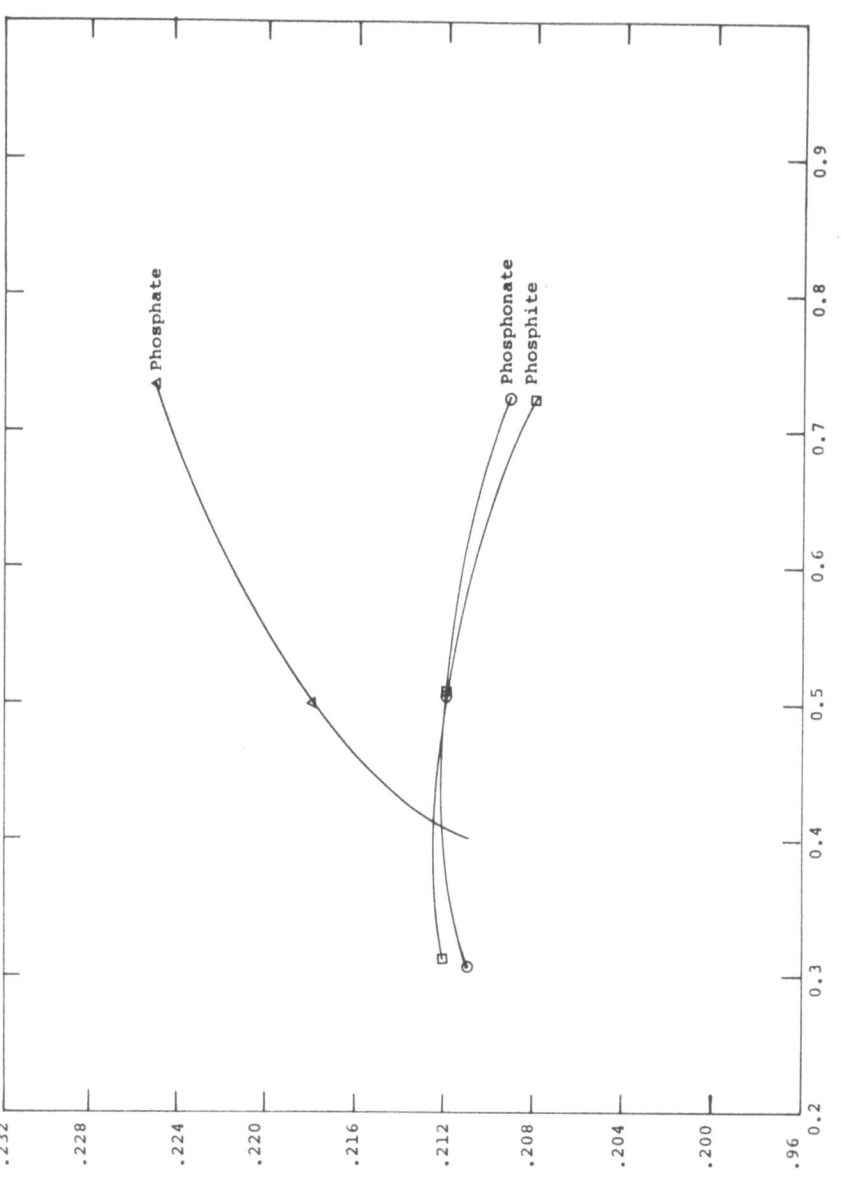

FIGURE 3. Relationships of oxygen index with phosphorus content of foams.

compound used. A uniform phosphorus-to-ionic-bromine ratio of 1 was observed for all cases.

Flexible foams were also prepared with phosphorus contents of 0.3–1.0% employing phosphate, phosphite, and phosphonate polyols, as described in Table 10.[54] The oxygen index values as a function of phosphorus content of the foams are shown in Fig. 3. At a lower level of phosphorus (0.3%), the phosphate polyol-based foam exhibited a significantly lower OI than the phosphonate or phosphite polyol-based foams. However, at a level of more than 0.4% phosphorus, a considerable increase in OI was observed for the phosphate, whereas the flame resistance of the phosphonate and phosphite remained essentially unchanged. Good correlation between flame resistance, as determined by OI and ASTM D-1692-67T, was formed for all phosphorus compounds investigated. A study of the char from these foams indicated that most of the phosphorus originally present in the foam was found in the char. The TGA curves for the foams containing the various phosphorus polyols are given in Fig. 4,[54] showing marked differences in their initial decomposition as well as rate of weight loss.

In order to determine phosphorus–bromine synergism in these foams, foams were prepared containing 0.5% phosphorus from both the phosphate and phosphonate polyol and varying the bromine (DBNG) level. In another set of experiments foams were made containing 0.3% phosphorus from the phosphite polyol and varying the bromine level. The OI values of these foams as a function of the bromine level are shown in Fig. 5[54]

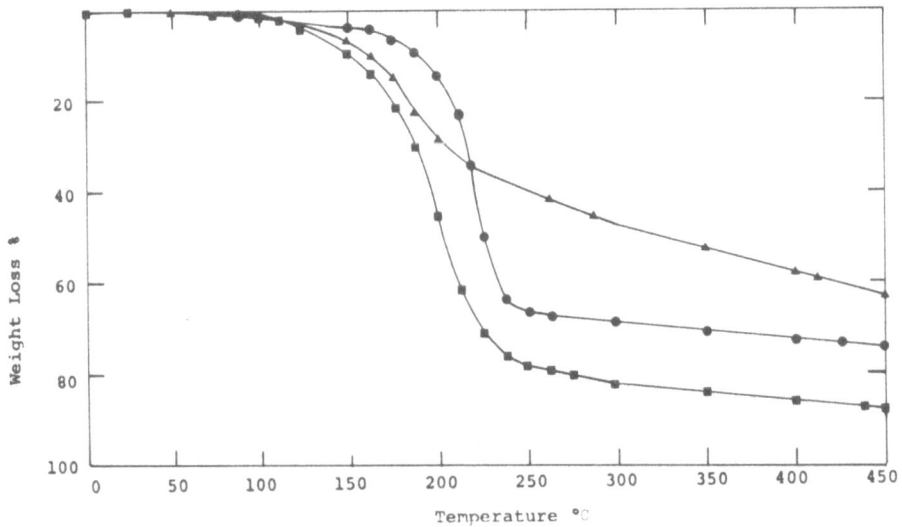

FIGURE 4. Comparative TGA for various phosphorus alcohols.

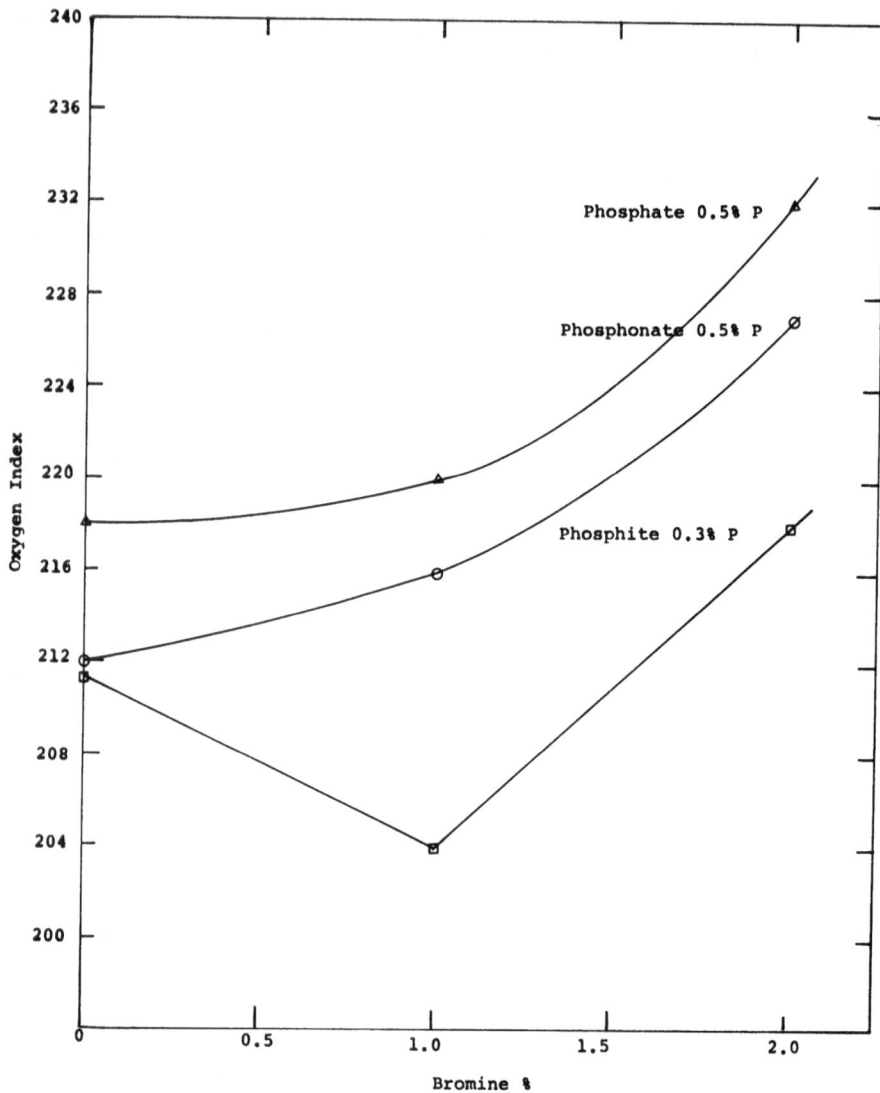

FIGURE 5. Flammability and bromine content of foams at two levels of phosphorus.

These curves indicate that bromine is more effective in combination with phosphates and phosphonates than with phosphites.

Correlations of flame resistance with char yields are depicted in Fig. 6[54] Increasing amounts of bromine have the greatest influence on char yields in combination with phosphates or phosphonates. Maximum char yields were reached at 0.5% phosphorus from both the phosphate and phosphonate

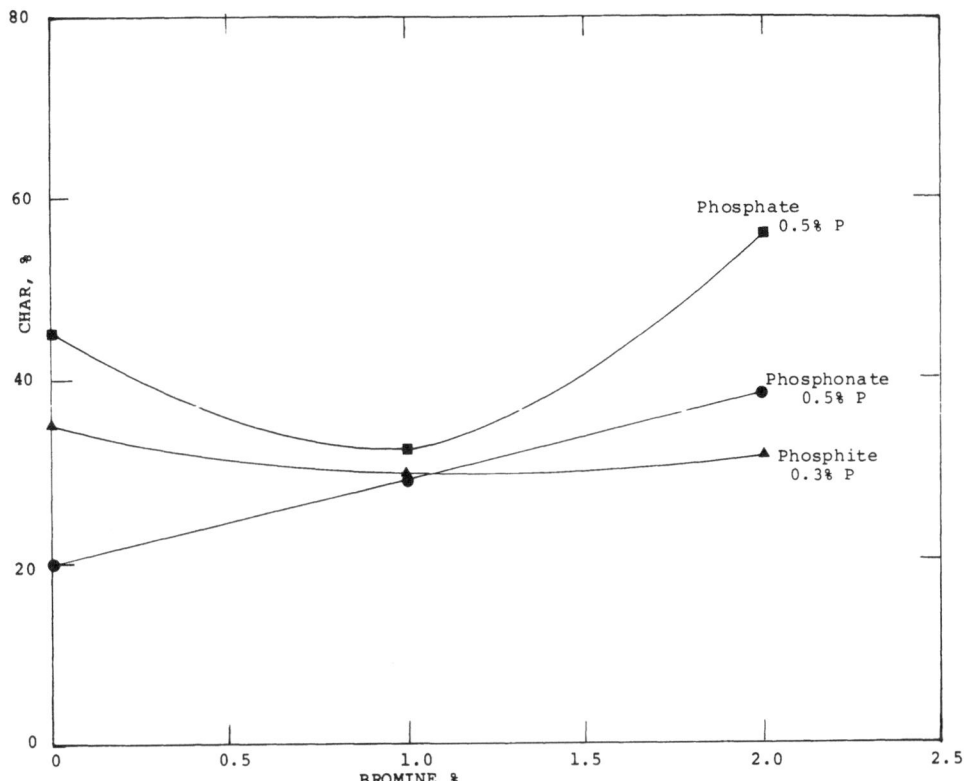

FIGURE 6. Effect of bromine on char yields at two levels of phosphorus at 300 C.

with 2% bromine, where maximum flame resistance was also observed. Foams containing phosphite produced about 30% char and were found to be insensitive to the addition of bromine. At 2% bromine level, these foams exhibited flame retardancy just sufficient to pass the ASTM D-1692-67T flame test.

Many phosphorus–halogen compositions are additives rather than reactive flame retardants. Tris(chloroethyl) phosphate and especially tris-(dibromopropyl) phosphate are among the most widely used. Addition of lithium salts[54] was found to improve heat and humidity resistance.

Phosphorus-containing, nonreactive additives were found to be less effective in their permanence, as found with tris(2-chloroethyl) phosphate and tris(2,3-dichloropropyl) phosphate. It was also found[54] that flame resistance does not increase proportionally with increased phosphorus content, and a maximum fire retardancy may be reached at 1–2% P. Not all the phosphorus will escape as volatile decomposition products, but its continued presence in the char suggested a mechanism involving solid meta-phosphoric acid formation. Anderson[6] suggested that foam char reduced flame heat transfer to the unburned polymer beneath the char. However, phosphates as plasticizers can lead to poor heat-distortion properties and may interfere with foam production when blended with polyether polyols.

Baumann and Szabat[55] recently reported the use of halogen-containing phosphate ester (Phosgard 2XC20, Monsanto Chemical Co.) in high resilience flexible urethane foams. The structure of this phosphate is shown below:

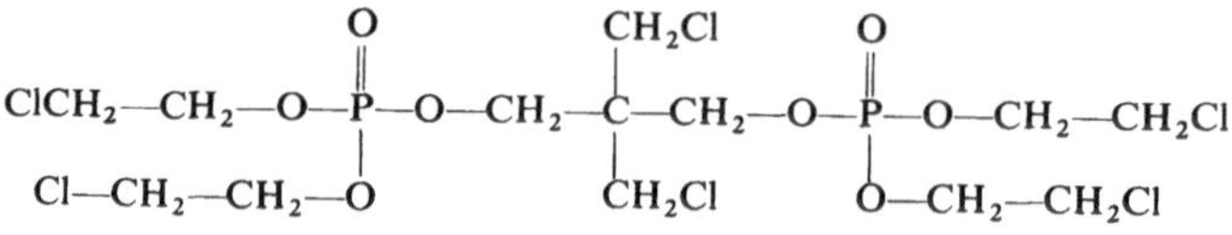

$$\underset{Cl-CH_2-CH_2-O}{ClCH_2-CH_2-O-\overset{\overset{\displaystyle O}{\|}}{P}-O-CH_2-}\underset{CH_2Cl}{\overset{\overset{\displaystyle CH_2Cl}{|}}{C}-CH_2-O-}\underset{O-CH_2-CH_2Cl}{\overset{\overset{\displaystyle O}{\|}}{P}-O-CH_2-CH_2Cl}$$

This additive was used in ethylene oxide-capped polyether triol-based foams in concentrations ranging from 9 to 11 parts/100 parts polyether in order to pass the MVSS 302 flame test for automotive interior cushioning applications.

The use of diallyl phenyl phosphonate,

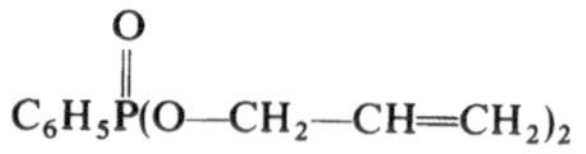

$$C_6H_5\overset{\overset{\displaystyle O}{\|}}{P}(O-CH_2-CH=CH_2)_2$$

bis(methallyl)benzene phosphonate,

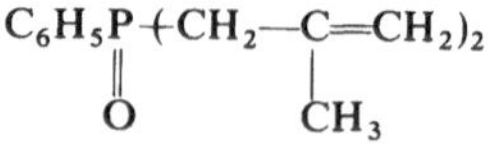

$$C_6H_5\underset{\overset{\displaystyle O}{\|}}{P}(CH_2-\underset{\overset{\displaystyle CH_3}{|}}{C}=CH_2)_2$$

and diallyl toluene phosphonate,

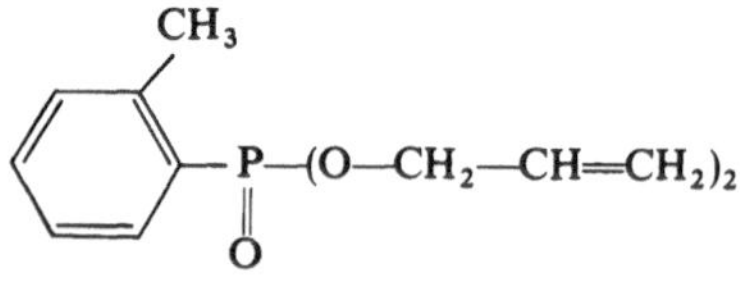

in the preparation of flame-retardant urethane foams was described by Simon and Thomas.[56] These foams were based on alkyd resins and TDI. Small amounts of a peroxide, such as benzoyl peroxide, were added to the mixture to permit polymerization of the flame-retardant additives during the foaming reaction.

Other phosphorus-containing, flame-retardant additives were employed by Simon and Thomas.[56] These included trichloroalkyl phosphates, such as tris(β-chloroethyl) phosphate, trichloropropyl phosphate, and unsaturated alkyl esters of alkenyl phosphoric acids. The latter included diallyl isobutene phosphonate, diallyl styrene phosphonate,

$$\langle C_6H_5 \rangle - CH = CH$$
$$O = P(OCH_2 - CH = CH_2)_2$$

and bis(methallyl)styrene phosphonate.

A comparison of some nonreactive inorganic phosphorus-containing fire retardants with some organophosphorus fire retardants in sucrose polyether polyol-based rigid foams has been reported by Miles and Lyons.[57] Vertical-bar flammability tests with these foams, all containing 1.8% phosphorus with the exception of the control foam, are shown in Table 11. These data indicate that the foams containing the liquid organic fire retardants lose about 15% of their weight on burning compared to about 10% for the solid inorganic fire retardants. Of special interest are the flammability data of the same foams after immersion in water for 7 days (see Table 11). It can be observed that the highly insoluble ammonium polyphosphate (Phos-Chek P/30, Monsanto Chemical Co.) exhibited little change in flame spread while diammonium and triammonium phosphate appear to have been extracted from the foam, giving flame-spread values approaching that of the control.

TABLE 11
Immersion Tests on Flame-Retardant Foams[a]

	Flame Spread	
	Before immersion	After 7 days immersion in water
1. Control (no retardant)	100	100
2. Ammonium polyphosphate	33	36
3. Commercial "reactive" organophosphorus compound	39	42
4. Diammonium phosphate	36	90
5. "Nonreactive" halogenated organophosphorus compound	33	30
6. "Nonreactive" nonhalogenated organophosphorus compound	39	83
7. Triammonium pyrophosphate	37	83
8. Phosphorylated polyol	43	37

[a] Water immersion leached out the diammonium phosphate and the triammonium pyrophosphate as well as the nonhalogenated additive.

TABLE 12
Flame-Retardant Flexible Foam

Materials	Foam A	Foam B
Poly(oxypropylene)triol, 3000 mol wt	100	100
Silicone surfactant	1.0	1.0
tert-Amine catalyst	0.15	0.15
Stannous octoate	0.40	0.40
Water	4.0	4.0
TDI (80/20)	52	52
Polyvinyl chloride resin	22	22
Antimony trioxide	7	14
Zinc oxide, French process	3	0

Low-density, flexible and semiflexible polyether urethane foams were made self-extinguishing by adding into the foam mixture prior to expansion a mixture of a finely divided halogen-containing polymer such as polyvinyl chloride, zinc oxide, and antimony trioxide.[58] The synergistic action of this mixture is demonstrated by foams of the composition shown in Table 12. The results of the ASTM D-1692 tests are seen in Table 13.

Parker and Riccitiello[59] used a somewhat similar approach in producing flame-resistant rigid and semirigid urethane foams. They employed polyether polyols (e.g., poly(oxypropylene) adducts of α-methyl glucoside or sucrose) with a polymeric isocyanate (Mondur MR) in combination with a polyvinyl chloride resin or a PVC copolymer. The preferred vinyl polymer was a terpolymer consisting of vinyl chloride, vinyl acetate, and maleic anhydride (VMCH, Union Carbide). Polyvinylidene chloride or neoprene could also be used. In addition to halogenated polymers, an inorganic salt, preferably potassium borofluoride, was employed.

It is believed that the mechanism of the charring process is as follows:[59] The urethane foam is highly cross-linked and highly branched. When heated, volatile material is expelled, and there is further cross-linking and condensation of rings. This cross-linking and ring condensation is catalyzed by the HCl which is split off from the PVC polymer. The PVC polymer also gives

TABLE 13
ASTM D-1692-59T Results

Foam	Self-extinguishing time, sec	Distance burned, in.
A	25.0	1.60
B	70.0	2.40

rise to a conjugated polyene, which enters into condensation reactions and contributes to the yield and stability of the char. The KBF_4 decomposes further into elemental boron and fluorine. The boron enters into the char structure and stabilizes it. The fluorine has the desirable effect of reacting with free radicals, such as hydroxyl radicals, which are chain carriers that act to sustain and propagate combustion. The reaction of fluorine with these chain-carrier free radicals suppresses combustion.

Doerge and Wismer[60,61] recently described a new approach to reduce the flame spread and smoke development in rigid urethane foams. It involves the incorporation of solid dicarboxylic acids into the urethane foams. The acids selected in this study were fumaric, maleic, succinic, isophthalic, and HET acid (chlorendic acid). The foams were prepared by mixing the solid acid in the polyisocyanate prior to adding it to the premix consisting of a sucrose-based polyether polyol (Selectrofoam RS-6406, PPG Industries), dibutyltin diacetate catalyst, silicone surfactant, and Fluorocarbon 11 blowing agent. It can be seen from Table 14 that the foams containing acids were self-extinguishing according to the ASTM D-1692 test, as compared to the unmodified foam which burned. The foam without acid had a maximum specific optical density D_m of 121. The addition of fumaric acid (15% by weight of the total formula) led to a 29% reduction to a value of 86. The corresponding values for maleic acid was $D_m = 105$ or a 13% reduction. Succinic acid actually produced a higher value, $D_m = 146$ or an increase

TABLE 14
Effect of Acids in Non-Fire-Retardant Formulation

Acid used	ASTM D-1692 test	NBS smoke chamber test	
		D_m	% change
(none)	Burned	121	
Fumaric	SE	86	−29
Maleic	SE	105	−13
Succinic	SE	146	+21
Isophthalic	SE	106	−12
Chlorendic	SE	119	−2
Foam Formulations			
		I	II
Polyisocyanate Side			
Crude MDI		90	90
Acid			32
Premix side			
Selectrofoam RS-6406		63.6	63.6
Dibutyltin diacetate		0.3	0.4
Si-emulsifier		1.0	1.0
Fluorocarbon 11		29.1	25.0

of 21%. The aromatic dicarboxylic acids, isophthalic acid, and the highly chlorinated chlorendic acid had D_m values of 106 and 119, respectively, corresponding to a reduction of 12% and 2% respectively.

The effect of these acids in a fire-retardant foam formulation (Fyrol 6, diethyl-N,N-bis(2-hydroxyethyl)aminomethyl phosphonate added) is shown in Table 15.[60] The foam containing fumaric acid had a flame-spread rating of 92, representing a reduction of 35%. The smoke value was reduced by 39%. The maleic acid-containing foam had a 11% reduction in flame spread, but exhibited a 42% reduction in smoke density. The aromatic acids showed a considerable reduction in flame spread (35%) but exhibited only a moderate reduction in smoke density.

Doerge and Wismer[60] also reported on ASTM E-84 tunnel tests with foams based on a fire-retardant spray formulation with and without fumaric acid. The percentages of phosphorus and chlorine in these foams, together with the results of the flammability tests, are shown in Table 16.

It was also established by these investigators that fumaric acid did not appear to react with polyisocyanate in the presence of 0.3% dibutyltin diacetate catalyst until the mixture was heated to 150°C for about 1 min.

3. Modified Urethane Foams

Conventional methods for improving flame resistance of urethane foams include the addition of phosphorus- and/or halogen-containing compounds either in the form of an additive, a reactive component, or to a lesser extent, in the form of a protective coating. While these methods are satisfactory for many industrial applications, they have some obvious shortcomings. These are relatively high smoke evolution, limited heat resistance, and in many cases, reduced hydrolytic stability. Hence, numerous efforts have been made to modify urethane foams by the incorporation of more heat-stable groups.

Significant improvements in the flame resistance and heat stability of urethane foams were achieved by incorporation of isocyanurate rings. The trimerization reaction of isocyanates to yield isocyanurates has been extensively investigated using many different catalyst systems.[72-77]

The preparation of rigid foams by trimerization of an isocyanate-terminated prepolymer was first described by Burkus[74] in a U.S. patent in 1961.

Nicholas and Gmitter[76] described the preparation of heat-resistant rigid isocyanurate foams by trimerization and simultaneous blowing of isocyanate-terminated prepolymers, made from 6.3 equivalents of TDI and 1 equivalent of a hydroxyl-terminated triol adipate. A preferred trimerization catalyst used was N,N',N''-tris(dimethylaminomethyl) phenol (DMP-30, Rohm & Haas). The isocyanurate-modified urethane foams were found to retain dimensional stability and considerable strength up to about 232°C (450°F).

One-shot, flame-resistant isocyanurate foams were described by Bernard, Backus, and Darr[78] They were prepared from crude MDI, a trimerization catalyst (e.g., DMP-30), a silicone surfactant, and a trichloro-fluoromethane blowing agent. These foams exhibited excellent flame resistance. (One foam had a flame-spread rating of 25 according to ASTM E-84, a fuel contribution of 15, and a smoke density rating of 25). However, these

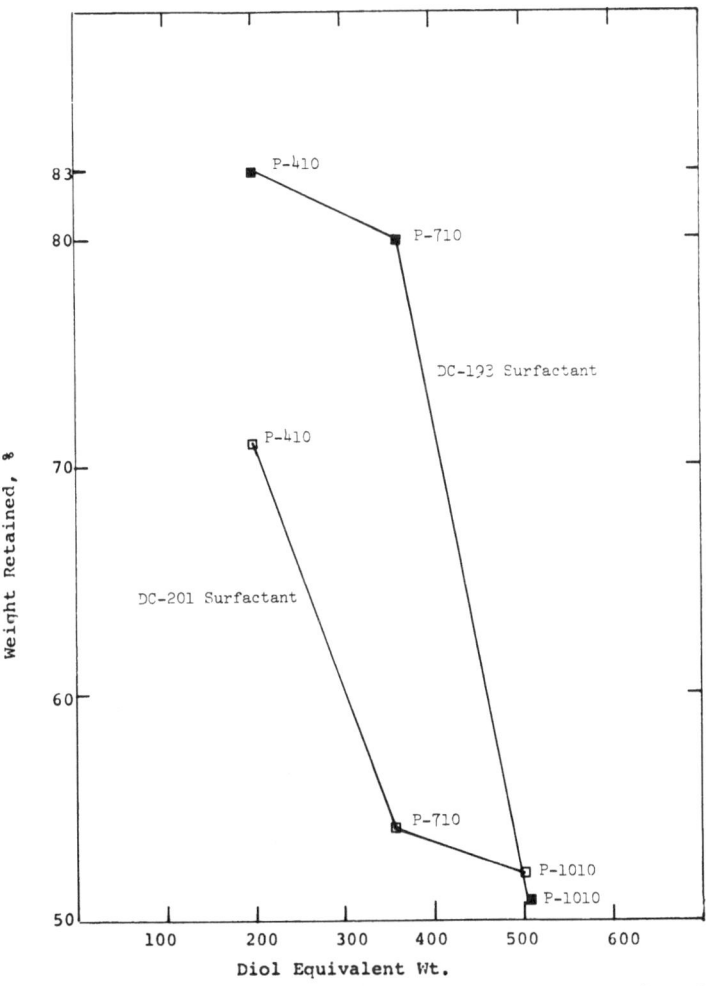

FIGURE 7. Butler chimney test. Effect of diol equivalent weight and surfactant. DMP-30 catalyst.

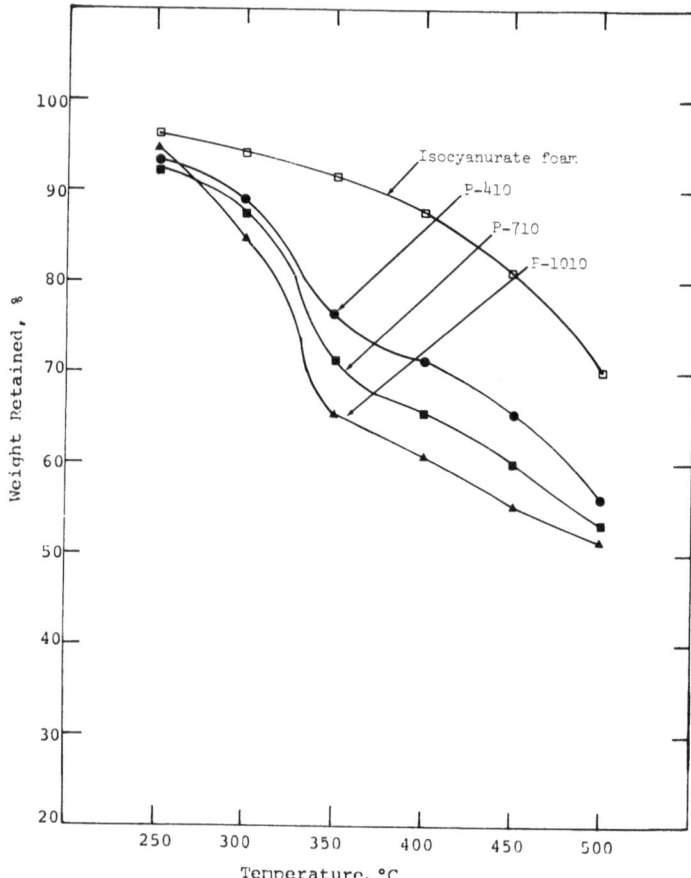

FIGURE 8.   TGA curves for isocyanurate–urethane foams. Effect of diol equivalent weight. Heating rate = 20°C/min; air flow = 1.0 SCFH.

foams displayed considerable friability, and hence, certain amounts of polyols were used to introduce urethane groups, thereby improving the friability. Isocyanurate–urethane foams were reported by Bernard, Backus, and Doheny[79] Ashida and Yagi,[77] and others. Ball et al.[80] described the preparation of urethane-modified isocyanurate foams from crude MDI, stressing the excellent fire resistance of these foams, as demonstrated by the Bureau of Mines flame penetration test.

Highly flame-resistant isocyanurate foams, modified by a polyether polyol of more than 300 mol wt, were described by Ashida and Yagi[77] employing either one-shot or prepolymer methods.

Frisch, Patel, and Marsh[81] employed polyether diols and triols to reduce the inherent brittleness of isocyanurate foams. While the flame resistance, as measured by the Butler chimney test, was somewhat reduced by the incorporation of the polyether polyols, the smoke density decreased significantly. Increasing the polyether chain length led to a reduction in flame resistance and a decrease in friability. The resistance to degradation, as measured by TGA, decreased with increasing equivalent weight and concentration of polyether. Good correlation was obtained between the TGA measurements and the Butler chimney test data. These results are shown in Figs. 7 and 8.

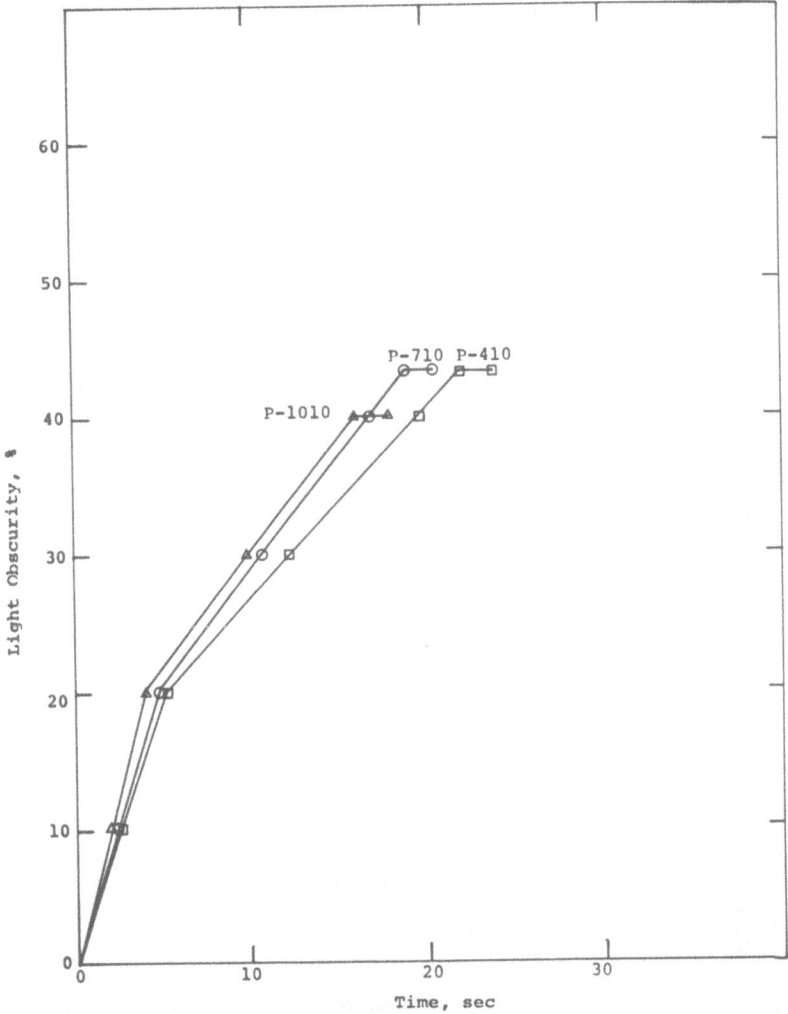

FIGURE 9. Smoke density test. Effect of diol equivalent weight. DMP-30 catalyst; DC-201 surfactant.

It is also interesting to note that the nature of the surfactant had a pronounced effect on the flammability characteristics of the foams. This was apparent in both the Butler chimney weight loss measurements and the smoke density data (Figs. 9 and 10).[81] The concentration of the surfactant also had a pronounced effect on the flame spread of these foams as shown in Fig. 11.

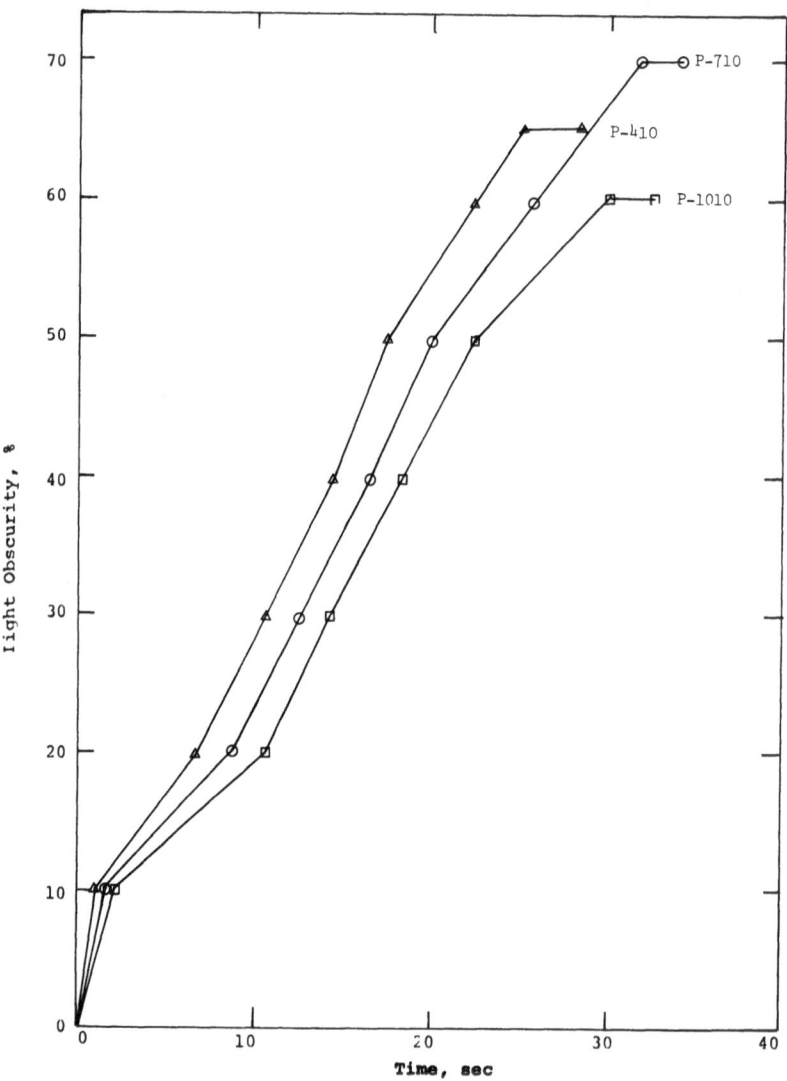

FIGURE 10. Smoke density test. Effect of diol equivalent weight. DMP-30 catalyst; DC-193 surfactant.

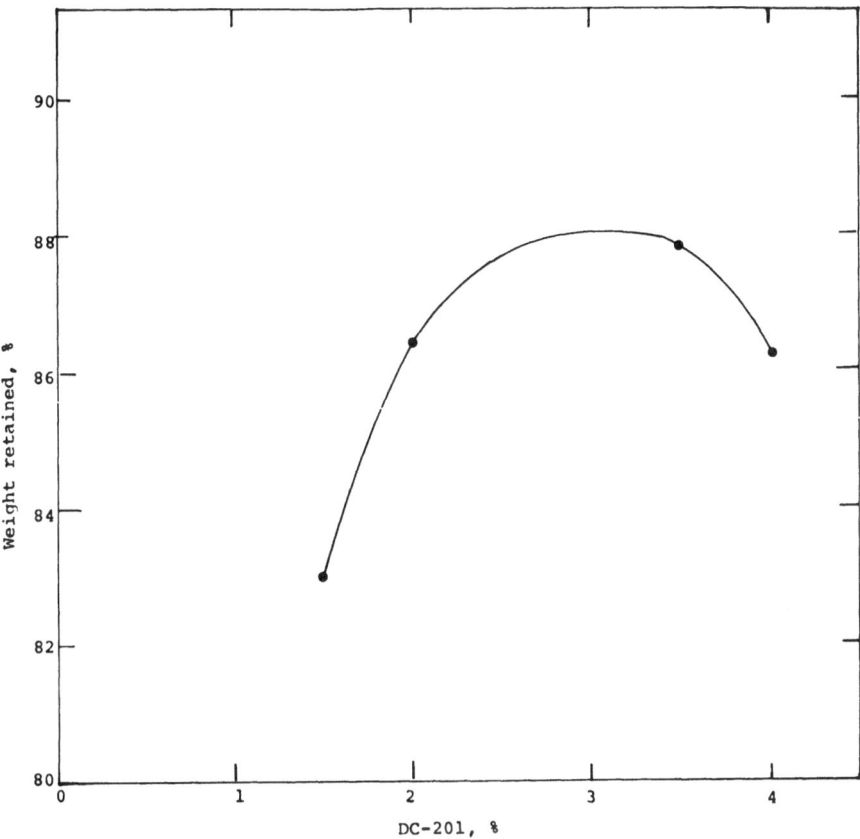

FIGURE 11. Butler chimney test. Effect of surfactant DC-201 concentration.

Isocyanurate foams also were modified by the incorporation of urethane groups from "flame-retardant" polyols containing phosphorus or a combination of phosphorus with halogens (chlorine and bromine).[82] The flame resistance of these foams increased with increasing NCO/OH ratio as did the thermal resistance, as indicated by TGA. The smoke density of these foams decreased with increasing NCO/OH ratio, i.e., with decreasing amounts of phosphorus and halogens in the foams.

Recently, Riccitiello *et al.*[83] have reported the preparation of rigid urethane-modified isocyanurate foams utilizing an acrylonitrile-grafted polyether polyol based on glycerol. The acrylonitrile groups begin to cyclize at temperatures around 200°C, ultimately yielding heterocyclic structures (quinizarine rings) as follows:

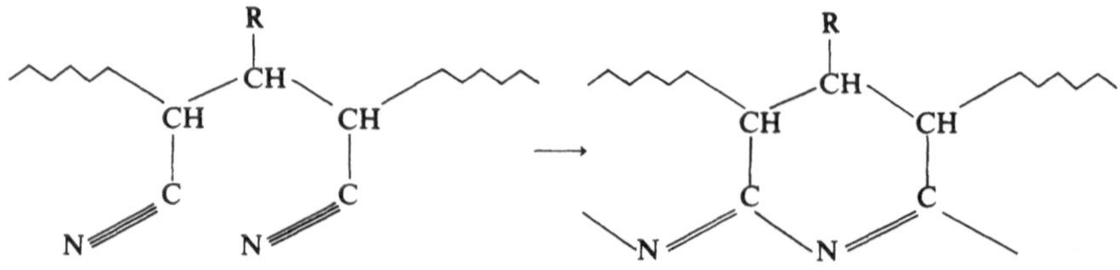

Polyamide-modified isocyanurate foams, prepared by the one-shot method, were reported by Ashida and Yagi[77] and Grieve.[84]

Carbodiimide-containing isocyanurate foams have been disclosed by Bernard and Doheny[85] and by Kan.[86] The latter used 2,4,6-tris(dialkanol-amino)-s-triazine as selective catalyst for carbodiimide formation.

Epoxy resins as cocatalysts for the trimerization of isocyanates have been reported in literature.[76,77,87]

Epoxy resins as a major component of modified isocyanurate foams were disclosed by Hayash, Reymore, and Sayigh.[88] These foams were prepared by reacting a polyisocyanate and a monomeric polyepoxide in approximately stoichiometric proportions, in the presence or absence of a polyol, and employing a tertiary amine catalyst. The resulting high-tempera-ture-resistant foams were flame retardant, and the authors claimed as the principal reaction the formation of polyoxazolidones as follows:

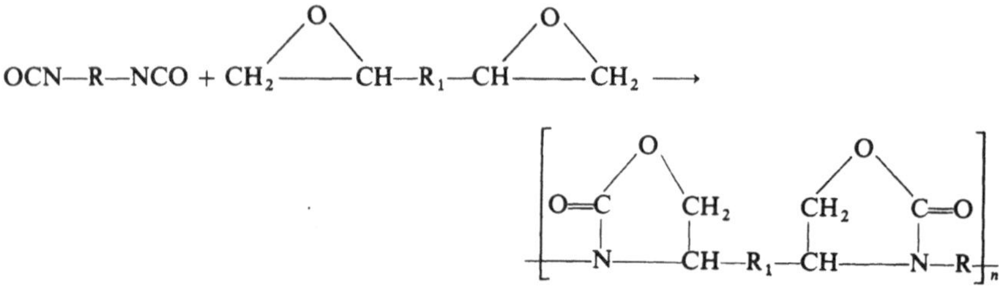

However, other reactions may also take place, such as trimerization of the polyisocyanate and the polymerization of the polyepoxide to form ether linkages.

Ashida and Frisch[89,90] reported the preparation of epoxy-modified, isocyanurate foams, either by a one-shot or prepolymer method. Typical formulations for one-shot epoxy–isocyanurate and epoxy–urethane-modi-fied isocyanurate foams are given in Tables 17 and 18, respectively. DMP-30 was used as the sole trimerization catalyst in these foams. In order to deter-mine the optimum composition of the three principal components in these foams, i.e., polyether polyol, epoxy, and polyisocyanate, with regard to flammability characteristics (flame spread as measured by the Butler chimney test, flame penetration as measured by the Bureau of Mines pene-tration test, and smoke density as determined in a Rohm & Haas XP-2

TABLE 15
Effect of Acids in Fire-Retardant Formulations

Acid used	Modified Monsanto test		NBS smoke chamber test	
	Flame spread	% change	D_m	% change
(none)	142		212	
Fumaric	92	−35	129	−39
Maleic	127	−11	122	−42
Succinic	113	−20	178	−16
Isophthalic	92	−35	177	−16
Chlorendic	92	−35	197	−7

Foam Formulations	I	II
Polyisocyanate Side		
Crude MDI	90	90
Acid		32
Premix Side		
RS-6406	57.2	57.2
Fyrol 6	6.4	6.4
Dibutyltin diacetate	0.3	0.4
Si-emulsifier	1.0	1.0
Fluorocarbon 11	29.1	35.0

Smoke Chamber) and foam friability (ASTM C-421), triangular plots were made delineating the preferred range of foam composition to impart specific optimum properties. A composite of these optimum regions is shown in Fig. 12, as indicated by the shaded area which should meet the four requirements: relatively low friability, high flame resistivity (burn-through time), low surface flammability, and low smoke generation.

Since the one-shot procedure for epoxy-modified isocyanurate foams yields only a limited concentration of oxazolidone groups, owing to the

TABLE 16
Fire-Retardant Urethane Spray Foam Data

Composition and Properties	Foam A FR spray foam	Foam B FR spray foam + 15% fumaric acid
% P in formula	0.75	0.65
% Cl in formula	1.7	1.4
Modified Monsanto flame spread rating	100	86
ASTM E-84 flame spread rating	65	20
ASTM E-84 smoke rating	450	287

TABLE 17
Effects of EPON 828/Isonate 135 Ratio and DMP-30
Concentration

Composition and Properties	Parts by Weight				
	1	2	3	4	5
Isonate 135	135	120	105	90	75
Epon 828	15	30	45	60	75
R-11	22	–	–	–	–
L-5340	1.5	–	–	–	–
DMP-30	13.5	12	9.0	7.5	5.0
Cream time, sec	10	8	8	8	8
Rise time, sec	130	133	175	185	455
Tack-free time, sec	130	130	175	204	455
Density, lb/ft^3	2.9	1.9	1.8	2.4	2.2

reactivity of isocyanate groups to form other groups, primarily isocyanurate, ether, and urethane groups, a two-step procedure was developed by Ashida and Frisch.[90] This involved first the preparation of isocyanate-terminated polyoxazolidones as shown in the equation

The resulting prepolymers were then trimerized to yield oxazolidone-modified isocyanurate foams.

The effect of the chemical structure of epoxy resins on the physical properties of oxazolidone–isocyanurate foams is seen in Table 19.[89] The Bureau of Mines flame penetration tests gave the best results with foams containing bisphenol A-type epoxy (Epon 828) followed by the novolak-type epoxy (Epon 152), with the cycloaliphatic epoxy resin-containing foams (Bakelite ERL-4221 and ERL-4206) exhibiting the shortest burn-through time.

Isocyanurate foams were also modified by the incorporation of urethane groups from "flame-retardant" polyols containing phosphorus or chlorine and combinations of phosphorus with chlorine.[91] The synergistic effect of N, P, and Cl on the flammability of polyisocyanurate–urethane (PIU) foams was investigated and compared to that of the corresponding poly-urethane (PU) foams. The raw materials used in these foams are listed in Tables 20 and 21.

TABLE 18
General Formula for Epoxy–Urethane-Modified
Isocyanurate Foams

Composition	Parts by weight
1. Isonate 135	Varied
2. Epon 828	Varied
3. Pluracol 463	Varied
4. R-11	15 parts/100 parts of a mixture of a + b + c
5. DMP-30	2.25 parts/100 parts of a mixture of a + b + c
6. L-5340	1.0 parts/100 parts of a mixture of a + b + c

Basic foam formulations are listed in Table 22. The effect of the chlorine-containing polyols on the flammability of foams as measured by the oxygen index test (ASTM D-2863) is presented in Fig. 8. It can be seen that the oxygen index of the PIU and PU foams increase linearly with the chlorine content in the foams. No significant difference was observed between the two polyols used in this study. Polyol RF-23 is structurally different from the HMF-246. In polyol RF-23 chlorine is present in the form of pendant —CH_2—CCl_3 groups attached to the polyether chain. In the case of the HMF-246 polyol, chlorine is bonded through chlorendic acid units.

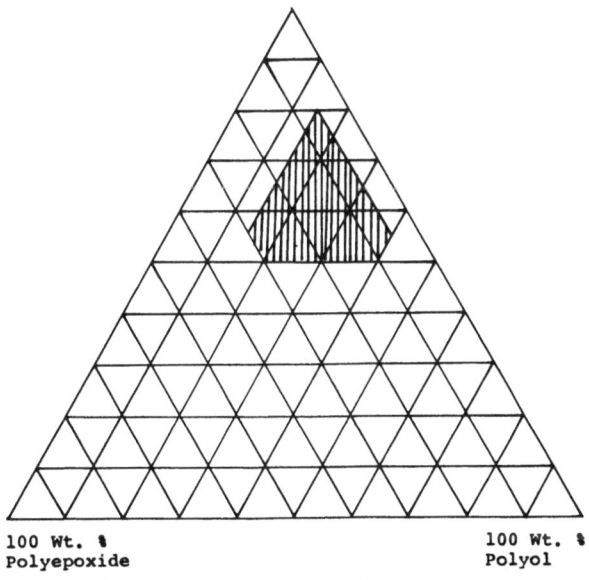

100 Wt. % Polyisocyanate

100 Wt. %
Polyepoxide

100 Wt. %
Polyol

FIGURE 12. Optimum composition of foams in ternary system.

TABLE 19
Isocyanurate Foams Modified with Different Epoxy Resins

Composition and Properties	Parts by weight			
	1	2	3	4
Isonate 901	91	91	91	91
Epon 828	9	0	0	0
Epon 152	0	9	0	0
Bakelite ERL-4221	0	0	9	0
Bakelite ERL-4206	0	0	0	9
Ucon 11B	15	15	20	15
L-5340	1.0	1.0	1.0	1.0
DMP-30	7.0	10.0	10.0	5.0
Niax 3CF	0	0	17.0	10.0
Cream time, sec	2	2	2	5
Rise time, sec	40	20	45	50
Density, lb/ft^3	2.5	2.9	2.5	3.0
Friability, % wt loss	18.6	10.0	14.7	22.0
Burn-through time, min	120	48	27	32
Butler chimney, % wt retained	76.3	79.4	78.2	81.1
Smoke density,				
50% obs	8.4	12	12	14
90% obs	18.0	20	25	30
100% obs	24.6	26	37	41
Closed cell, %	96.5			
K factor at 75°F	0.132			

TABLE 20
Flame-Retardant Polyols

Polyol	Chemical identification	Supplier	OH No., mg KOH/g	P	Cl
Isonol FRP-8	Phosphate-containing polyol	Upjohn	395	7.3	
VC-611	Phosphate-containing polyol	Mobil	145	16.0	
Niax RO 350	Phosphate-containing polyol	Union Carbide	300	11	
Fyrol 6	Diethyl-*N*,*N*-bis(2-hydroxyethyl)-aminomethyl phosphonate	Stauffer	450	12.2	
Weston DPG phosphonate	Bis(dipropylene glycol)dipropylene glycol phosphonate	Weston	395	7.2	
Olin RF-23	Poly(trichloroxybutylene) adduct of sucrose	Olin	372		47
Hetrofoam HMF-246	Chlorine-containing polyol (derived from chlorendic anhydride)	Hooker	380		28.5

TABLE 21
Other Foam Components

Composition	Chemical identification	Supplier	OH No., mg KOH/g	N	Isocyanurate eq wt
Cord CS 124	Poly(oxypropylene) adduct of α-methylglucoside modified with an aliphatic amine based polyol	CPC International	541	2.34	133.5
PAPI	Polymethylene polyphenylisocyanate	Upjohn			
DC-193	Silicone copolymer	Dow Corning			
F-11 B	Trichlorofluoromethane	du Pont			
Dabco R 80-20	20 1,4 Diazabicyclo[2,2,2]octane/80 dimethylethanolamine	Houdry			
DMP-30	2,3,6-Tris(dimethylaminomethyl)phenol	Rohm and Haas			

TABLE 22

Formulations for Polyurethane (PU) and Polyisocyanurate–Urethane (PIU) Foams

	PU	PIU
Polyol Cord CS 124 + FR Polyols	100	100
Surfactant DC-193	1.5	1.5
Catalyst	Variable	Variable
	⎰20 1,4-Diazabicyclo[2,2,2]- octane	2,4,6,-Tris(dimethyl- aminomethyl)-
	⎱80 Dimethylethanol amine	phenol
Trichlorofluoromethane	Adjusted for a foam density of 1.8 ± 1	
Polymeric isocyanate (PAPI)	Isocyanate index 105	Isocyanate index 200

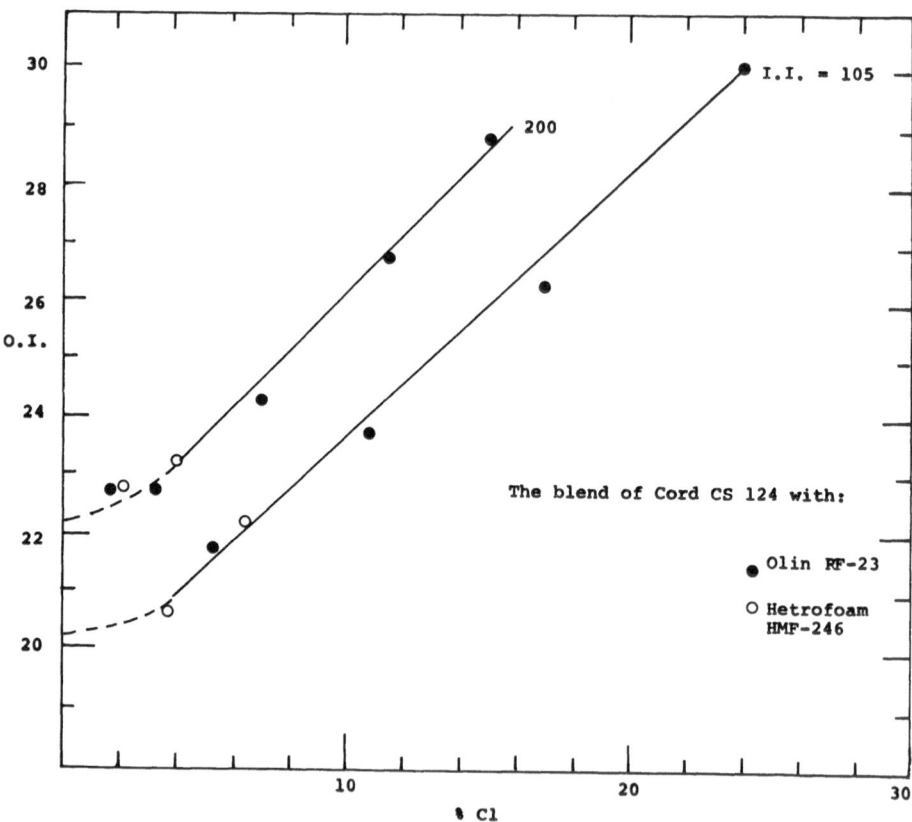

FIGURE 13. Effect of chlorine content on oxygen index.

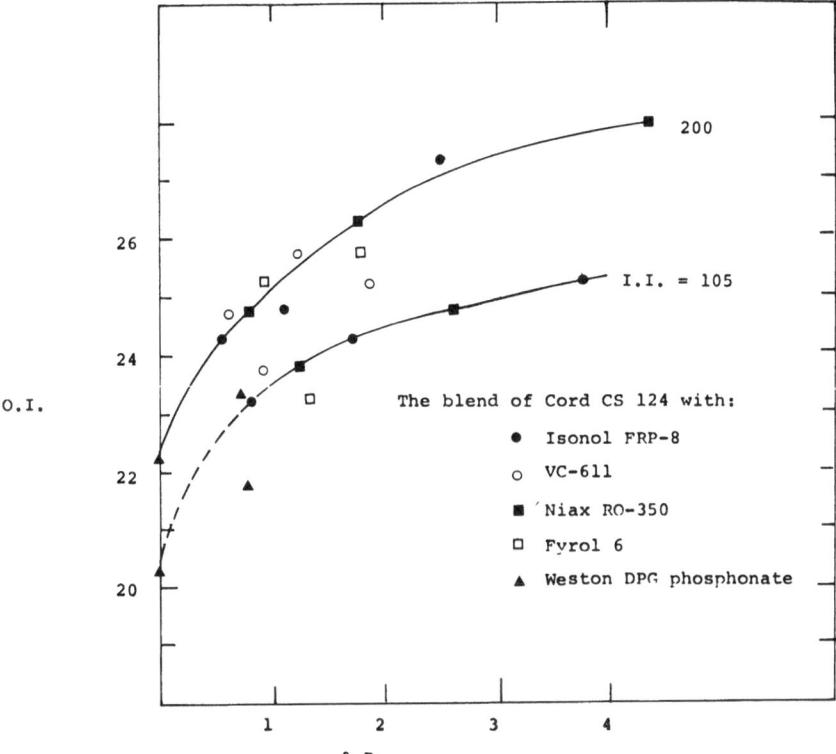

FIGURE 14. Effect of phosphorus content on oxygen index.

The influence of isocyanurate groups in these foams is apparent by the shift of the oxygen index (OI) to higher values at the same chlorine level (see Fig. 13). The effect of phosphorus on the flammability of the PUI and PU foams is presented in Fig. 14. As can be seen from Fig. 14, the dependence of the OI on the phosphorus content in foams is different from that shown in Fig. 13. The exponential shape of the OI *vs.* P curve shows, that in this case the surface reactions played an important role in the flame-retardancy mechanism. It was found that the effect of phosphorus on the OI can generally be correlated by the following equation[91]:

$$[OI]_2 = [OI]_1 + k \ln \frac{[P_2]}{[P_1]}$$

where $[P_1]$ and $[P_2]$ = weight concentration of phosphorus in foams at two concentration levels, and k = slope. In contrast to the halogen-containing polyols, where the structure of the Cl-containing groups did not

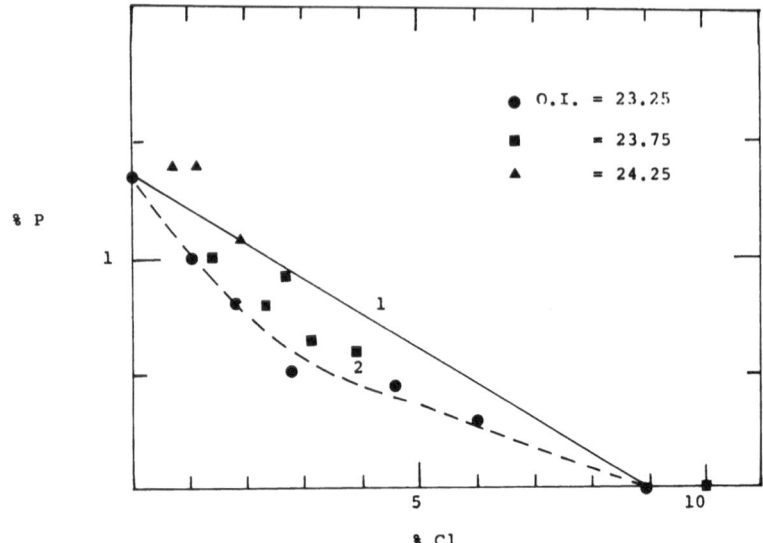

FIGURE 15. Phosphorus–chlorine "synergism."

play an important role, the effectiveness of the phosphorus-containing polyols depends on the type of bonding of phosphorus. Polyols based on phosphates (Isonol FRP-8, VC-611, and Niax RO-350) are more effective as flame retardants in PIU and PU foams than polyols based on phosphonates (Fyrol 6 and Weston DPG). Similar results for polyurethane foams were found by Papa and Proops,[54] who investigated the effect of various phosphorus-containing polyols in flexible urethane foams. At higher temperatures (above 250°C) phosphate-based polyols decompose faster than phosphonates into the catalytically active species, which cause char formation.

The influence of the isocyanurate structure on flame resistance, as compared to the corresponding urethanes, in the presence of chlorine- or phosphorus-containing polyols is also apparent from Figs. 13 and 14, which show consistently higher oxygen index values for the isocyanurate-modified urethane foams. The phosphorus–chlorine synergism was studied with the poly(trichlorooxybutylene) adduct of sucrose and diethyl-N,N-bis(2-hydroxyethyl)aminomethyl phosphonate in PU foams. The flame-retardant efficiency based on the concentration of P and Cl in PU foams, as measured by the oxygen index, is presented in Fig. 15. If the effects of the individual P- and Cl-containing flame-retardant polyols are cumulative in nature, the dependence of the oxygen index values at constant level of 23.25 on the P and Cl concentration in foams is represented by a straight line, 1 (Fig. 15). As can be seen from Fig. 10, the "Isooxygen Index Values" follow curve 2.

The deviation of curve 2 from curve 1 would indicate the effect of synergism between Cl- and P-containing polyols.[91] However, the effect of the concentration of phosphorus on the OI values is not a linear one but is represented by a logarithmic function. Hence, plots of the isooxygen values in linear chlorine–phosphorus coordinates create "pseudo-synergistic" effects between phosphorus and chlorine. As can be seen from Fig. 16, the isooxygen index values, correlated in a semilog plot, follow a straight line in the one order range.

Another type of high-temperature-resistant and flame-retardant foam is prepared by the reaction of di- and polyfunctional isocyanates with dianhydrides. Farrissey and coworkers[92] have reported the preparation of polymide foams using dimethylsulfoxide as reaction medium and employing a catalytic amount of an active hydrogen-containing compound:

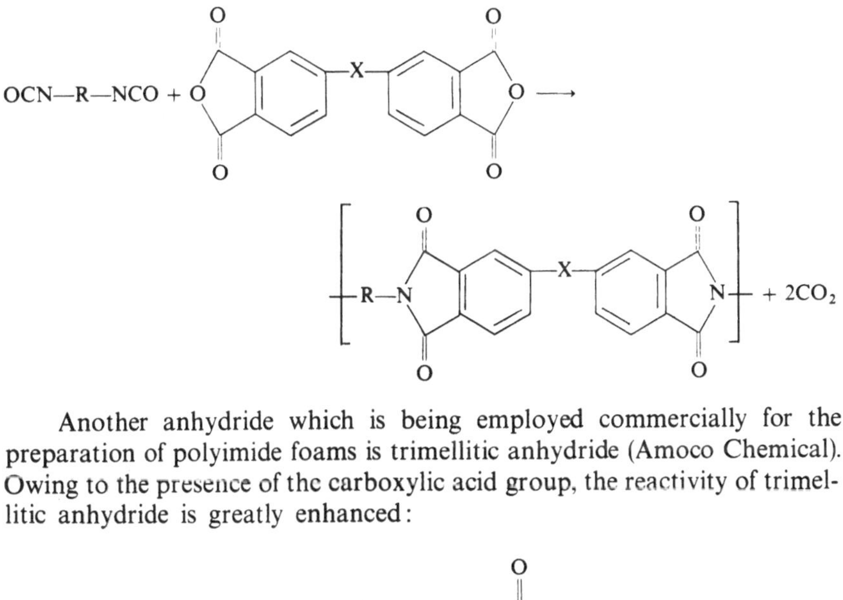

Another anhydride which is being employed commercially for the preparation of polyimide foams is trimellitic anhydride (Amoco Chemical). Owing to the presence of the carboxylic acid group, the reactivity of trimellitic anhydride is greatly enhanced:

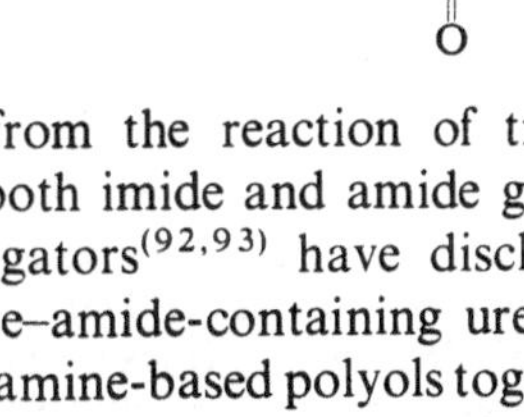

Polymers resulting from the reaction of trimellitic anhydride with polyisocyanates contain both imide and amide groups.[93]

A number of investigators[92,93] have disclosed the preparation of one-shot imide- and imide–amide-containing urethane foams by incorporating neutral polyols and amine-based polyols together with acid anhydrides and polyisocyanates into the foam mixture. These foams, which can be prepared at room temperature, retain their strength and structural integrity in excess of 250°C for extended periods of time, and exhibit outstanding flame resistance. However, care must be taken to remove any traces of aprotic solvent, e.g., dimethylsulfoxide, used in the preparation of these foams. However, foams employing trimellitic anhydride have been prepared without the use of any solvent.

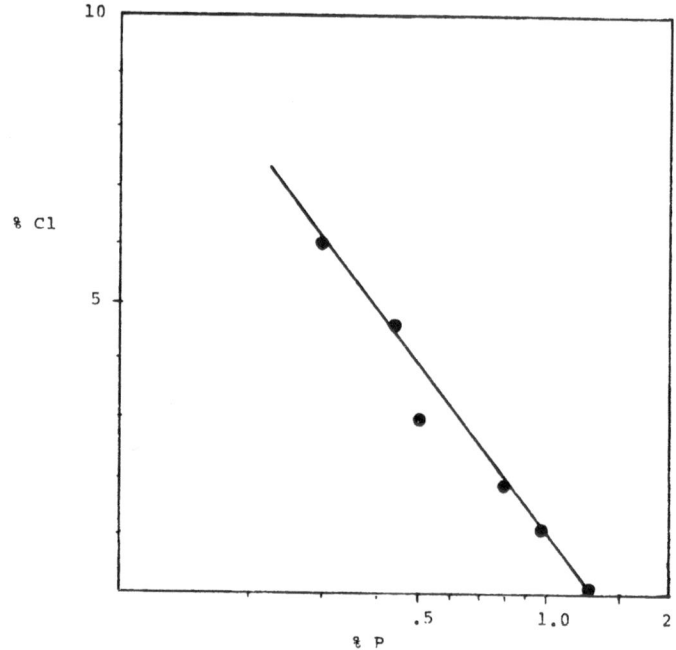

FIGURE 16. Synergism of phosphorus and chlorine in foams.

4. *References*

1. J. W. Lyons, *The Chemistry and Uses of Fire Retardants*, Wiley, New York (1970).
2. J. W. Lyons, *J. Cell. Plast.* **6**(6), 302 (1970).
3. E. A. Dickert and G. C. Toone, *Mod. Plast.* **42**(5), 197 (1967).
4. K. C. Frisch, *J. Cell. Plast.* **1**(2), 3 (1965).
5. K. C. Frisch, S. L. Reegen, and E. J. Robertson, in *Cellular Plastics, Proc.* Conference at Natick, Mass., April, 1966, National Academy of Sciences–National Research Council (1967).
6. J. J. Anderson, *Ind. Eng. Chem., Prod. Res. Dev.* **2**(4), 261 (1963).
7. D. B. Parrish and R. M. Pruitt, *J. Cell. Plast.* **5**(6), 348 (1969).
8. J. Kresta and K. C. Frisch, unpublished results (1972).
9. P. Robitschek, U.S. Pat. 3,058,925 (1962) (to Hooker Chemical Corp.).
10. P. Robitschek and J. L. Olmstead, U.S. Pat. 2,090,501 (1937) (to Hooker Chemical Corp.).

11. P. Robitschek, *J. Cell. Plast.* **3**, 395 (1967).
12. R. R. Hindersinn and S. M. Creighton, U.S. Pat. 2,865,869 (1958) (to Hooker Chemical Corp.).
13. R. R. Hindersinn and M. Worsley, U.S. Pat. 3,055,849 (1962) (to Hooker Chemical Corp.).
14. E. V. Gouinlock, Jr., F. W. Long, and S. M. Creighton, *Plast. Technol.* **8**(12), 40 (1962).
15. C. K. Lyon and T. H. Applewhite, U.S. Pat. 3,466,253 (1962) (to U.S. Secretary of Agriculture).
16. R. J. Slocombe, U.S. Pat. 3,451,925 (1969) (to Monsanto Co.).
17. Kuryla and Knopf, U.S. Pat. 3,467,607 (1969) (to Union Carbide Corp.).
18. F. H. Othey, R. P. Westhoff, and C. L. Mehltretter, *J. Cell. Plast.* **8**(3), 156 (1972).
19. Z. E. Jolles, *Plast. Inst. Trans. J. Conf. Suppl., No. 2, 3* (1967).
20. C. K. Lyon and T. H Applewhite, *J. Cell. Plast.* **3**, 91 (1967).
21. P. E. Hoch, U.S. Pat. 3,151,143 (1964) (to Hooker Chemical Corp.).
22. H. E. Holmquist, U.S. Pat. 3,479,304 (1969) (to E. I. du Pont de Nemours and Co.).
23. W. J. Farrissey, Jr., F. P. Recchia, and A. A. R. Sayigh, U.S. Pat. 3,437,680 (1969) (to the Upjohn Co.)
24. R. C. Nametz, R. D. Deanin, and P. M. Lambert, *Mod. Plast.* **41**(1), 166 (1963).
25. P. A. Argabright and B. L. Phillips, U.S. Pat. 3,627,689 (1971) (to Marathon Oil Co.).
26. D. R. Jackson, U.S. Pat. 3,419,532 (1964) (to Wyandotte Chemicals Corp.).
27. H. A. Bruson and J. S. Rose, U.S. Pat. 3,244,754 and U.S. Pat. 3,269,961 (1966) (to Olin Mathheson Chemical Corp.).
28. J. J. Pitts, S. Fuzesi, and W. R. Andrews, *J. Cell. Plast.* **8**(5), 274 (1972).
28a. Wyandotte Chemicals Corp., Pluracol 463 for 25 Flame-Spread Rigid Urethane Foams, in *Tech. Data Bull. U-91*, Mar. 11, 1969.
28b. W. R. Andrews, A. D. Cianciolo, E. G. Miller and W. L. Thompson, *J. Cell. Plast.* **4**(3), 102 (1968).
29. H. C. Vogt and P. Davis, *Polym. Eng. Sci.* **11**, 312 (1971).
30. D. L. Bernard, J. K. Backus, and A. J. Doheny, Belg. Pat. 712,731 (1968) (to Mobay Chemical Co.).
31. T. M. Beck and E. N. Walsh, U.S. Pat. 3,235,517 (1966) (to Stauffer Chemical Co.).
32. L. Friedman, U.S. Pat. 3,309,342 (1967) (to Union Carbide Corp.).
33. Neth. Appl. 6,613,257 (April 24, 1967) (to Stauffer Chemical Co.).
34. P. A. T. Hoge and H. Coates, Brit. Pat. 974,033 (1964) (to Albright and Wilson, Ltd.).
35. N. J. Clark and F. McCollough, Jr., U.S. Pat. 3,294,712 (1966) (to Stauffer Chemical Co.).
36. W. M. Lanham, U.S. Pat. 3,075,927 (1963) (to Union Carbide Corp.).
37. Z. E. Jolles, Brit. Pat. 994,087 (1965) (to F. W. Berke & Co., Ltd.).
38. E. R. Dezzinger, E. A. Dickert, and R. A. Wiles, Belg. Pat. 657,423 (1965) (to Allied Chemical Corp.).
39. Neth. Appl. 6,414,132 (June 8, 1965) (to Diamond Alkali Co.).
40. Z. E. Jolles and B. J. Riley, Brit. Pat. 1,079,984 (1967) (to F. W. Berk & Co., Ltd.).
41. H. Piechota, *J. Cell. Plast.* **1**, 186 (1965).
42. A. C. Haven, Jr., *J. Am. Chem. Soc.* **78**, 842 (1956).
43. H. C. Fielding, Brit. Pat. 892,931 (1965).
44. H. Holtschmidt and G. Oertel, *Angew. Chem., Internatl. Ed.* **1**, 617 (1962).
45. L. M. Kendley, H. E. Podall, and H. Filipescu, *SPE Trans.*, **1962**, 122.
46. R. L. McConnell and H. W. Coover, Jr., U.S. Pat. 2,926,145 (1960).
47. A. F. Childs and H. Coates, *Kunstoffe* **54**, 501 (1964).
48. H. C. Fielding, U.S. Pat. 3,145,077 (1964) to Imperial Chemical Ind.).
49. H. C. Fielding, U.S. Pat. 3,144,302 (1964) (to Imperial Chemical Ind.).
50. E. Dyer and R. A. Dunbar, *J. Polym. Sci., Part A-1* **8**, 629 (1970).
51. M. I. Bakhitov, E. V. Kuznetsov, and M. Ya. Obryadina, in *Soviet-Urethane Technology* (A. M. Schiller, ed.), Vol. 1, Chap. 12, p. 79, Technomic Publishing Co., Westport, Conn. (1973).

52. R. K. Valetdinov, E. V. Kuznetsov, M. Kh. Khasanov and T. Kh. Valeeva, in Ref. 51, p. 94.
53. R. W. Ashworth and A. V. Mercer, Brit. Pat. 1,020,254 (1966) (to Shell International).
54. A. J. Papa and W. R. Proops, *J. Appl. Polym. Sci.* **16**, 2361 (1972).
55. G. F. Baumann and J. F. Szabat, Fire Retardant Flexible Urethane Foam in *Advances of Fire Retardants*, Part 1, (V. M. Bhatnagar, ed.) Technomic Publishing Co., Westport, Conn. (1972).
56. E. Simon and F. W. Thomas, U.S. Pat. 2,577,281 (1951) (to Lockheed Aircraft Corp.).
57. C. E. Miles and J. W. Lyons, *J. Cell. Plast.* **3**, 539 (1967).
58. J. T. Harrington, U.S. Pat. 3,574,149 (1971) (to General Tire & Rubber Co.).
59. J. A. Parker and S. R. Riccitiello, U.S. Pat. 3,549,564 (1970) (to NASA).
60. H. P. Doerge and M. Wismer, *J. Cell Plast.* **8**(6), 311 (1972).
61. H. P. Doerge and M. Wismer, U.S. Pat. 3,039,307 (1972) (to PPG Industries).
62. L. E. Miles and W. Lyons, *J. Cell. Plast.* **3**, 539 (1967).
63. L. Friedman, U.S. Pat. 3,084,939 (March 12, 1963) (To Weston Chemical Corp.).
64. L. Friedman, U.S. Pat. 3,003,939 (Nov. 21, 1961) (to Weston Chemical Corp.).
65. L. Friedman, U.S. Pat. 3,359,348 (Dec. 19, 1967) (to Weston Chemical Corp.).
66. V. Baker, G. Braun, and G. Nischk, Ger. Pat. 1,170,636 (May 21, 1964) (to Farbenfabriken Bayer A.G.).
67. F. D. Popp and W. E. McEwen, *Chem. Rev.* **58**, 321 (1958).
68. D. Shaw and B. W. Greenwald, U.S. Pat. 3,291,867 (Dec. 12, 1966) (to Merck & Co.).
69. Brit. Pat. 951,792 (Apr. 8, 1964) (to Virginia-Carolina Chem. Corp.).
70. J. J. Anderson, Brit. Pat. 1,034,489 (Dec 13, 1967) (to Mobil Oil Corp.).
71. P. G. Pape, J. E. Sanger, and R. C. Nametz, *J. Cell Plast.* **4**, 438 (1968).
72. B. D. Beitchman, *Ind. Eng. Chem., Prod. Res. Dev.* **5**, 35 (1966).
73. S. R. Sandler, *J. Appl. Polym. Sci.* **11**, 811 (1967).
74. J. Burkus, U.S. Pat. 2,979,485 (1961) (to U.S. Rubber Co.).
75. I. I. Jones and N. G. Savill, *J. Chem. Soc.* **1957**, 4392.
76. L. Nicholas and G. T. Gmitter, *J. Cell. Plast.* **1**(1), 85 (1965).
77. K. Ashida and T. Yagi, Brit. Pat. 1,155,768 (1969) (to Nisshin Spinning Co.).
78. D. L. Bernard, J. K. Backus, and W. C. Darr, U.S. Pat. 3,644,232 (1972) (to Mobay Chem. Co.).
79. D. L. Bernard, J. K. Backus, and A. J. Doheny, Belg. Pat. 712,731 (1968) (to Mobay Chem. Co.).
80. G. W. Ball, G. A. Haggis, R. Hurd, and J. F. Wood, *J. Cell. Plast.* **4**, 248 (1968).
81. K. C. Frisch, K. J. Patel, and R. D. Marsh, *J. Cell. Plast.* **6**(5), 203 (1970).
82. K. C. Frisch and S. K. Mukherjee, First Conference of the Mexican Urethane Industry, Mexico City (Aug. 1970), Reprint Book, pp. 77–100.
83. S. R. Riccitiello, R. M. Fish, J. A. Parker, and E. J. Gustafson, *J. Cell. Plast.* **7**, 91 (1971).
84. R. L. Grieve, Belg. Pat. 1,080,487 (1969) (to the Upjohn Co.).
85. D. L. Bernard and A. J. Doheny, Belg. Pat. 723,153 (1967) (to Mobay Chem. Co.).
86. P. T. Kan, U.S. Pat. 3,645,923 (1972) (to Wyandotte Chemicals Corp.).
87. L. Nicholas, Jap. Pat. Publication No. Sho44-1666 (1964) (to General Tire & Rubber Co.).
88. E. F. Hayash, Jr., H. E. Reymore, Jr., and A. A. R. Sayigh, U.S. Pat. 3,673,128 (1972) (to the Upjohn Co.).
89. K. Ashida and K. C. Frisch, *J. Cell. Plast.* **8**(3), 160 (1972).
90. K. Ashida and K. C. Frisch, *J. Cell. Plast.* **8**(4), 194 (1972).
91. J. E. Kresta and K. C. Frisch, *J. Cell. Plast.* **11**(2), 1 (1975).
92. W. J. Farrissey, Jr., J. S. Rose, and P. S. Carleton, *J. Appl. Polym. Sci.* **14**, 1093 (1970).
93. H. E. Reymore, Jr., and A. A. R. Sayigh, U.S. Pat. 3,431,223 (1969) (to the Upjohn Co.).

8

Retardation of Combustion of Phenolic, Urea–Formaldehyde, Epoxy, and Related Resin Systems

Robert T. Conley and Daniel F. Quinn

1. Introduction

The flame resistance of phenolic, urea–formaldehyde, epoxy, and related resin systems appears to represent examples of polymers which are intermediate in their combustion behavior between most of the vinyl systems and the so-called nonflammible, heat-resistant polymers. In fact, the phenolic resin system in fiber form has been reported to exhibit a high degree of flame resistance.[1,2] In a chemical sense, these materials can be considered to exhibit properties which can be termed self-retarding with regard to their combustion behavior. The combustion process, in general terms, can be described in terms of the degradation of these materials. As you have already seen in previous chapters in this volume, the combustion of polymeric materials can be considered to take place in a number of stages. Such generalized stages are closely intertwined owing to variations in the type of external heat applied to the material, the geometry of the combusting polymer, the atmosphere in which the combustion is occurring, and the morphological and chemical characteristics of the specific material.

Robert T. Conley and Daniel F. Quinn · Department of Chemistry, Wright State University, Dayton, Ohio.

As the heat from the external source is applied, the temperature of the combustible polymer increases. At the outset, in air, the primary processes occurring at the surface of the material are oxidative reactions (resulting in deterioration) in the temperature range from approximately 120 to 450°C. The rate of temperature increase to the bulk of the material is dependent, of course, on the basic characteristics of the polymer which control heat transfer through the polymer system. In any case, the heat energy to the bulk phase is increased by the chemical reactions which produce thermal energy by exothermic decomposition of oxygenated surface structures. We must recognize that, while rapid exothermic oxidation reactions may be taking place at the surface, the thermal decomposition processes which require no oxygen are taking place at the elevated temperature regions immediately below the surface of the polymer mass. In such a case, polymers such as the phenolic resins undergo simultaneous thermal and thermal-oxidative degradation. As thermal degradation proceeds, the bulk phase produces combustible volatile products and forms a residual carbon char which is more resistant to further decomposition, either thermal or oxidative, than the nondegraded resin. In this sense, the thermal cross-linking and char-forming processes produce materials which are combustion resistant.

Heat transfer through the char layer may be quite different from that of the resin itself, and the reactions causing further erosion of the materials through volatile product formation become extremely complex. For example, the char itself is reactive, forming more carbon-rich materials which are usually radical in nature, and the polymer itself is producing new char and additional volatile species which must percolate through the char layer into the combustion zone. Radical trapping reactions can reduce the amount of combustible material reaching the combustion zone. In terms of these general aspects of the combustion processes, we should consider the degradation processes characteristic of these polymers and then, where possible, how chemical modifications of these materials affect the retardation of combustion reactions.

In these resin systems it would appear that there are two predominant ways to retard the combustion processes. The first is to control or eliminate the production of combustile volatiles. The second is to chemically modify the system in such a fashion that the diffusion of volatiles to the combustion zone is retarded. In a general sense, it should be pointed out that the majority of the chemistry of combustion is the chemistry of the bulk phase of the polymer under 500°C. That is to say, the processes occurring below 500°C are the processes which provide the chemical changes which result in more (or less) char and less (or more) combustible volatile products. It is through an understanding of this chemistry that the eventual control of flammability will be accomplished.

The reader should understand that much of the presentation here is speculative. Confirmation of the reactions proposed still remain to be provided by future observations. Some may even be erroneous, and newer more reasonable routes proposed. We hope, however, the material presented here is ample food for thought for additional research in these particular classes of polymers.

2. Phenolic Resins

The degradation, both thermal and thermal-oxidative, has been examined in the temperature range from curing at just above room temperature through 1000°C. Conley and Bieron[3,4] reported a detailed study of the oxidation of cured films of phenol–formaldehyde resins. Later, Conley[5] extended these initial studies to include additional products of the early, low-temperature oxidation processes. Infrared spectral analysis was used extensively in the investigations of phenolic-resin decompositions, as well as thermogravimetric analyses,[6] elemental composition studies,[6] gas chromatographic studies,[7,8] model-compound and degraded-polymer syntheses,[3,4] and various curing procedures, to examine the effect of the degree of cure on the degradation of the phenolic-resin system.[5-8] Shulman and Lochte[9,10] provided additional information concerning the thermal and oxidative behavior of the phenolic resins using mass spectral thermal analysis. Although several different polymers were examined in these studies,[10] the same base-catalyzed phenol–formaldehyde resin was used as a standard for comparison. Therefore, at least, a single phenolic resin has been subjected to close scrutiny during degradation using modern instrumental analysis techniques.

The first step in the oxidation of phenolic resins can be visualized as shown:

These reactions agree with previously reported findings[11] regarding the spontaneous oxidation of these materials. Although the peroxide intermediates were detected by spectral analysis, chemical detection provided evidence for their formation at temperatures between 120 and 220°C. Further oxidation resulted in evidence that was consistent with the following scheme which results in the formation of chain scission products:

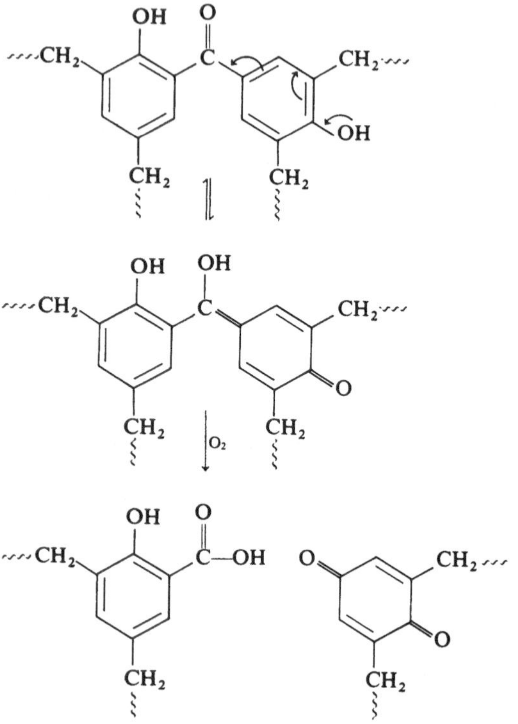

Previous evidence favors an interpretation of this type. Kharasch and Joshi[12] reported that sterically hindered phenols are oxidized to quinone-type structures. Cook, Kuhn, and Fianu[13] have observed that quinone-type structures are not formed if hydrogens are located alpha to the aromatic ring. Once the carbonyl structure is formed by oxidation of the methylene group, the conditions required for quinoid intermediates are fulfilled, thus lending support to the intervention of quinoid species in the secondary oxidation processes. Spectral evidence for reactions taking place beyond the acid-formation stages was not found possible to obtain. Thus spectral observations lending support for further reactions are not found to be feasible.

It is known that both the methylol species and the dibenzyl ether linkage also play a role in phenolic-resin chemistry. The methylol group is a primary reaction product formed during base-catalyzed resin formation, and the ether group has been shown to be formed either by heating methylol group-containing resins to over 120°C or by treatment under acid-catalyzed condensation conditions. The ether linkage appears to play only a minor role, if any at all, in the oxidative degradation of phenolic resins. Methylol groups, on the other hand, have been shown to react in the temperature range between those used for curing and 450°C, where they can no longer be detected as discrete chemical moieties in the resin system. The three possible reactions of methylol groups have been observed: the reversal of the condensation process forms formaldehyde; the heating of the methylol group results in the formation of additional methylene linkages (curing); and under oxidative conditions the methylol group is oxidized to acid groups.

Figure 1 summarizes the initial degradation reactions and competitive methylol reactions which appear, from reported data, to best account for

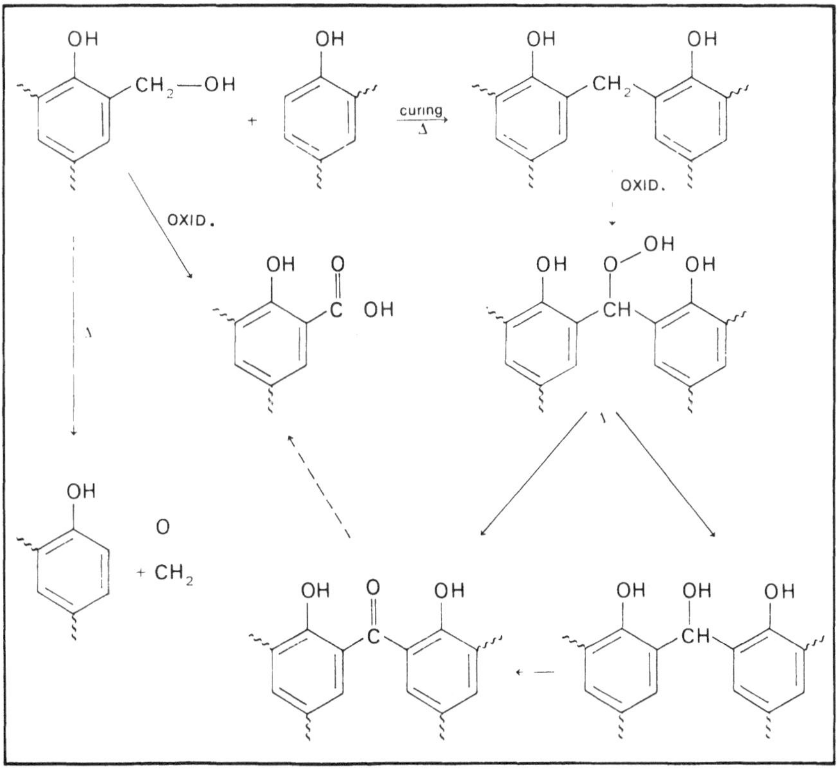

FIGURE 1. Summary of the initial degradation reactions and competitive curing reactions for phenolic resins.

the chemical changes taking place on heating phenolic resins in air at temperatures between 120 and 250°C.

Degradation above 250°C and as high as 1000°C has been examined.[6-10] Gas chromatographic monitoring of the volatile products indicated that the formation of carbon char during resin decomposition parallels the evolution of carbon monoxide. Examination of the volatile products as a function of temperature and degree of pyrolysis showed that the chief products were carbon dioxide, carbon monoxide, and water, as expected for the extended oxidation of the resin. Paraformaldehyde, methane, and aromatic products were also formed. In addition, higher-molecular-weight fragments were noted on walls and exit lines of the pyrolysis chamber. Water and paraformaldehyde were among the major products formed on heating the resin to 400°C. The formation of these products is consistent with our previous discussions concerning the reactions of methylol groups.

If one considers the possible routes for the formation of volatile products, two processes are in competition: thermal bond rupture and oxidation. From the available data, it seems clear that oxidation is the primary mode of phenolic-resin decomposition through approximately 700°C, regardless of whether the resin was decomposed in air, argon, or nitrogen atmospheres. At temperatures between 700 and 1000°C, the products of thermal pyrolysis are observed. The reactions which have been proposed to account for the observed product types are shown in Fig. 2. These proposals are generally consistent with the mass spectral thermal analysis data obtained by the high-temperature pyrolysis of phenolic resins in vacuo (Fig. 3).[7]

Parker and coworkers[14] have proposed an alternative nonoxidative pyrolysis mechanism for novolac-type phenolic resins. The mechanism involves the pyrolytic cleavage of the diphenylmethane linkage to produce o- and p-hydroxy-substituted phenylmethyl radicals in the 300°C temperature range. Further, at 500°C diphenylether linkages are formed. Neither proposal is in agreement with the previous discussion based upon most of the available data and would be difficult to verify experimentally, particularly since the available data in the 500°C range seems to indicate a loss of oxygenated species from the cured resin in that temperature range. However, the underlying reasoning presented by the Parker and coworkers[14] is consistent with the prior discussion.

From the data reported concerning phenolic resin degradation, it is possible to suggest that regardless of the particular resin species present before curing, the degradation processes occurring at elevated temperatures are dependent upon the thermal and oxidative stability of the dihydroxydiphenylmethane unit, and, therefore, these polymers all behave similarly from a chemical viewpoint once the curing processes are complete.

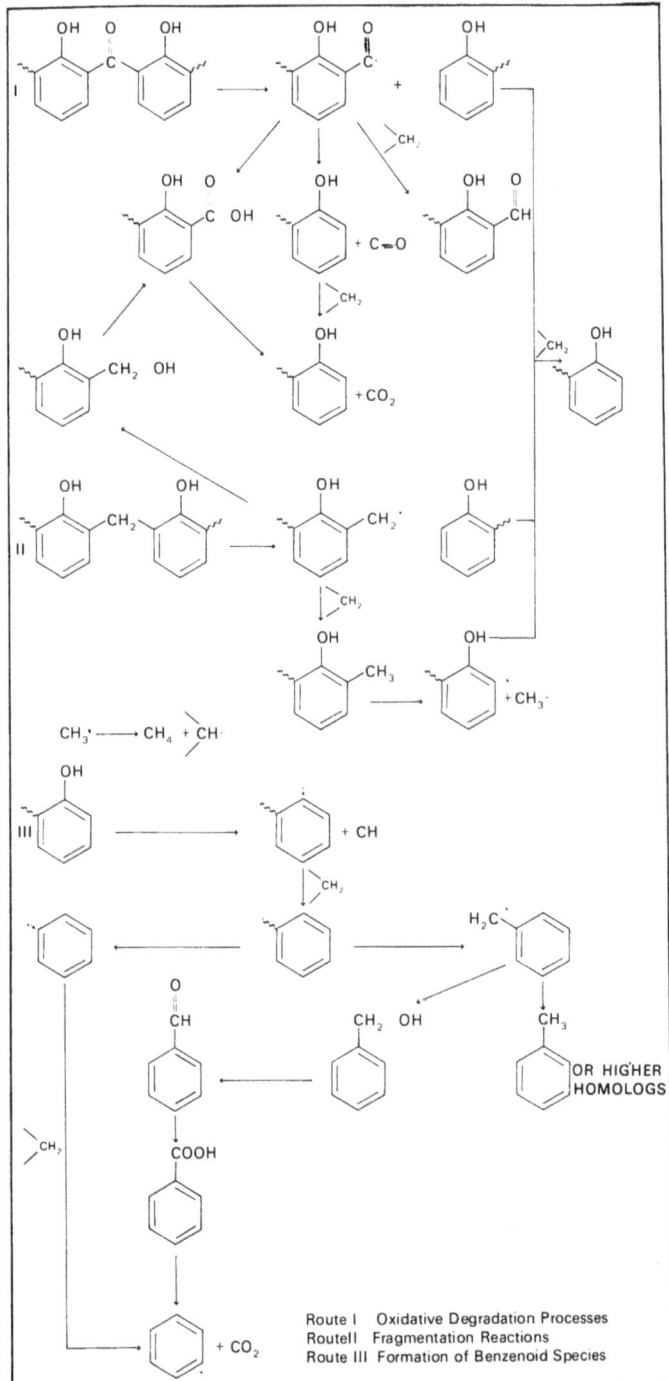

FIGURE 2. Summary of the proposed degradation processes, both thermal and thermo-oxidative types, for the deterioration of phenolic resins at high temperatures.

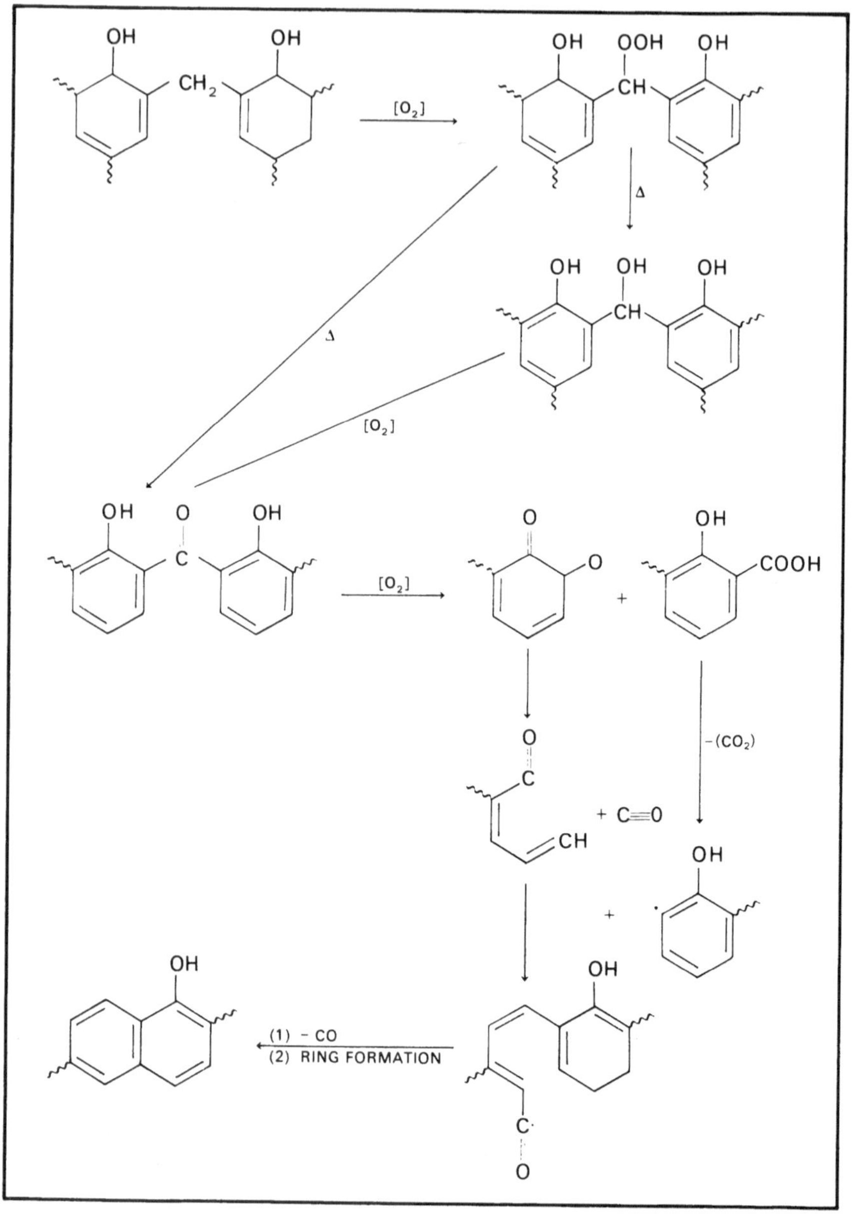

FIGURE 3. Proposed char-forming reactions in phenolic resin systems.

2.1. Flame Retardation in Phenolic Resins

Examination of the available literature reveals a number of interesting points concerning phenolic resins. For example, the phenolic resins used have been urea–formaldehyde,[15] melamine,[16] amino aldehyde,[17] and cellulose.[18] In certain forms, such as foams[19] and fibers,[1,2] phenolic resins themselves seem to exhibit good to excellent flame resistance. These latter observations seem to suggest that the rapid loss of volatiles below the ignition temperature leaves a carbonaceous residue which is quite stable (i.e., forms a char structure which does not undergo further combustion in normal environments). There is reason to suspect that this is indeed the case since phenolic chars, exposed to high temperature in air using an external heat source, absorb oxygen on removal of the heat source and cooling. Within a short time the chars undergo a highly exothermic reaction which raises the internal temperature of the material dramatically. The carbon foam or fiber actually glows (700–1000°C). Such reaction characteristics strongly suggest the formation of peroxide species in high concentration by a carbon-radical savenging process during the cooling stage. Decomposition of the peroxidic species, in turn, is so highly exothermic that the char becomes incandescent.

However, most of the technology concerning flame retardation of phenolic resins involves either the incorporation of substances known to exhibit flame-retarding properties into the backbone structure of the organic polymers or the addition of various compounds or combinations of compounds into the resin system to impart flame-resistant characteristics. Both chlorine and bromine have been reported to provide enhanced flame resistance to phenolic resin systems. In the polymer, halogen is expected to add structural rigidity and promote hydrogen bonding. When incorporated into the polymer structure, halogen atoms take part in dehalogenation reactions at the elevated temperatures required for decomposition providing cross-linking sites for increased char formation. Upon pyrolysis, the gaseous halogen products are decomposed in the ignition gases and tend to smother flame. As for relative general behavior characteristics, about one-half the amount of bromine is required to achieve the same flame-retardation level as a given amount of chlorine.

Halogen has been incorporated into the phenolic resin system partially by the synthesis of halogenated resins and by the addition of halogen-containing materials to the preformed phenolic resin. For example a mixture of 50% phenol and 50% of a mixture of predominantly 2,3-dichlorophenol, 2,4-dichlorophenol, and 2,6-dichlorophenol with formaldehyde formed a chlorinated phenolic resin which when used as a paper-impregnating material was found to be self-extinguishing and nonpunking when ignited.[20,21] In this case the comonomers were phenol and halogenated phenols. In a somewhat different fashion, it has been reported that direct halogenation of preformed phenolic resins results in halogen-containing polymers with improved flame-retardance properties.[22] This kind of flame-retarding behavior provided by direct halogenation is not simply restricted to phenolic resins. A variety of fibrous organic materials containing phenolic substances such as lignin and tannins exhibit improved flame resistance when halogenated. Interestingly in this regard, there is no doubt that wood halogenation produces a potentially highly cross-linked char not attainable from the wood alone when ignited. At 5% chlorine levels a 55% residual weight char was formed 3–5 min after ignition in contrast to a 1% char for untreated samples.[23]

Modification of the aldehydic portion with halogen-containing moieties also produced polymers with self-extinguishing characteristics.[24] For example, the condensation of phenol with p-bromobenzaldehyde at elevated temperatures produced such a resin.

Chlorinated polyphenyls[25] and other chlorinated phenyl-ring compounds[26] when mixed with phenolic resin also produce self-extinguishing resins and foams.[27] As we shall see, halogen is often accompanied by added phosphorus-containing materials to further enhance the flame retardance of phenolic-resin systems.

A number of patents have been issued which describe the formation of noncombustible phenolic resins when modified with boric acid, boric oxide, or borate salts (sodium, zinc, etc.).[16,28-30] Resins modified in this manner have been reported to fail to ignite in a flame and to produce excellent flame resistance when used as an impregnating material for paper. It is thought that the boron oxides retard flammability through both a surface-coating effect and as substances which smother flame. The major effect of these compounds in improving the flame resistance of phenolic resins is most probably their action in the condensed phase. These materials, by forming a surface coating, promote resin char reactions and thereby reduce the amount and rate of diffusion of combustible volatiles to the ignition zone. One might postulate that boric acid, when added to the liquid resin phase or when impregnated into partially cured phenolic resin systems, reacts with the phenolic hydroxyl groups to form borate esters with the elimination of water. The esters so produced would be expected to enhance the char-formation decomposition reactions, as well as the retention of the boric oxide surface coating during high-temperature exposure of the phenolic mass. Although we have little evidence concerning the role of the boric oxide in phenolic resin decomposition processes, it might be expected that this area would most fruitfully be studied with trimeric model compounds and low-molecular-weight novolacs where the decomposing species can be readily examined by available analytical techniques, particularly spectrophotometric methods.

As in the case of many other polymers, compounds of nitrogen, phosphorus, antimony, arsenic, and bismuth all enhance the flame-resistance properties of phenolic resins. Phosphoric acid or ammonium phosphate, when added to phenolic resin lacquer, have been reported to produce a fire-resistant impregnating resin for paper. Similarly when phosphoric acid or its ammonium salt in combination with phenolic resin is impregnated into wood, a notable increase in flame resistance was observed.[17,31] A wide variety of phosphorus-containing materials have been utilized to improve the fire-retardation properties of phenolic resins. For example, when glycerol phosphate acidic ester, formed on mixing one equivalent of phosphoric acid with two equivalents of glycerol, was mixed with 20 wt % phenolic resins, a flame-resistant film or solid resinous piece was prepared.[32] The literature is replete with the number and types of organophosphorus materials which have been used for this purpose.[32-44] In virtually all cases esters of the phosphorus compound and the phenolic resin system were undoubtedly produced. Similar behavior was found for the antimony, arsenic, and bismuth.[33,34] All of these materials have been reported to produce flame-retardant resin products. As in the case of phosphoric acid, phosphorus-containing esters of the phenolic hydroxyl group are undoubtedly the products which result from heating the resinous product at temperatures much in excess of 100°C regardless of the specific additive used. That is to say, the mixed esters of the groups already attached to the phosphorus-containing additive and the phenolic hydroxyl group would be anticipated as the initial reaction products when the resin-additive mixture is heated. Since many additives contain halogen or nitrogen (amines or amides), the chemistry involved in secondary thermal processes must be expected to be very complex. It is doubtful that the additive used in resin systems similar to the phenolics and the related polymers under discussion in this chapter are unreactive.

If we consider flammability of a phenolic polymer to be a series of complex reactions similar to those described in the previous section (i.e., thermal and oxidative degradation), then it seems quite reasonable to consider the stabilization by phosphorus to occur in two very general ways, both of which involve the formation of a comparatively high-cross-link-density structure as degradation proceeds. Since the majority of phosphorus-containing additives decompose at temperatures between 200 and 500°C, we would expect to form phosphoric acid and a cross-linked carbon skeleton. With diffusion of phosphoric acid and organophosphorus fragments to the surface, it is reasonable to propose a phosphorus oxide coating very similar to that proposed for boric oxides on the polymer surface. This would, of course, have a profound effect on the rate of diffusion of volatiles to the surface and their subsequent evaporation into the pyrolysis zone. Since the diffusion rate would be slowed down, radical trapping processes to form carbon char (from cross-linking) should be more favorable. Therefore we might project that the stabilization by phosphorus takes place primarily, but perhaps not exclusively, first, by the formation of organophosphorus–phenolic ester moieties which promote thermal polymer cross-linking thereby retarding volatile formation and, second, by the surface-coating behavior of phosphorus oxides, which would be expected to render the polymer self-extinguishing.

It has been noted that aluminum chloride[45] and zinc salts[46] confer increased flame and heat resistance and self-extinguishing characteristics to acid-cured phenolic-resin systems. There are a number of interesting speculative processes which might be considered as possibilities to account for this behavior. These substances are known to function as complex-forming materials. Although not extensively used as stabilizers, they may impart stabilization to the phenolic system through salt formation and actually act as catalytic agents in promoting cross-linking reactions, thereby increasing char formation. Indeed, complex formation may be a very important process for the phosphorus-containing additives as well. These particular observations suggest that one of the key areas of future research to gain information concerning the chemistry involved in flame retardation may be detailed study of the thermal behavior of complex-forming substances in phenolic systems.

Finally, a brief consideration of the role of nitrogenous materials, both as additives and reactants, in phenolic-resin systems seems in order. It is well known that the amino resins are less flammable than phenolic resins. Therefore, we might expect that a variety of nitrogen-containing materials would provide improved flammability characteristics when incorporated into phenolic resins. The incorporation of amines such as aniline into phenolic resins produces generally higher char yields, and the chars are found to be nitrogen rich. Examination of the thermal-oxidative behavior of aniline-modified phenolic resins at temperatures up to 200°C has indicated a lesser amount of oxidation at methylene linkages (together with reduced oxidation rates) and higher retention of the nitrogen portion of the system.[47] The incorporation of formaldehyde–dicyandiamide condensate and hydrogen chloride into a phenolic-resin foam composition produces a fire-resistant resin foam.[48] Similarly, the urea–dicyandiamide–phenol–formaldehyde resin mixture forms a coating which is nonburning.[55] In each of these cases the nitrogen additive or reactant appears to function as a cross-linking agent through radical trapping reactions. The overall behavior of nitrogen in these systems is believed to be similar to that discussed for phosphorus and related substances.

3. Furan Resins

The self-resinification of furfuryl alcohol is the primary commercial process forming "furan resin," substances which have excellent resistance to corrosive chemicals, acids, alkalies, and solvents. The furfuryl alcohol is first partially polymerized to form a liquid resin, using an acid catalyst. In the presence of concentrated mineral acids or strong organic acids, the reaction is strongly exothermic, and often the polymerization occurs with explosive violence unless carefully controlled by regulation of the temperature and acid concentration. Hachihama and Shono[49] concluded that the viscous liquid resin consists of 2-oxymethyl-5-furfuryl furan, 2-oxymethyl-5-(5′-furfuryl)furfuryl furan, difuryl ether, di-2-furylmethane, formaldehyde, and levulinic acid. The resin structure can best be summarized as shown below:

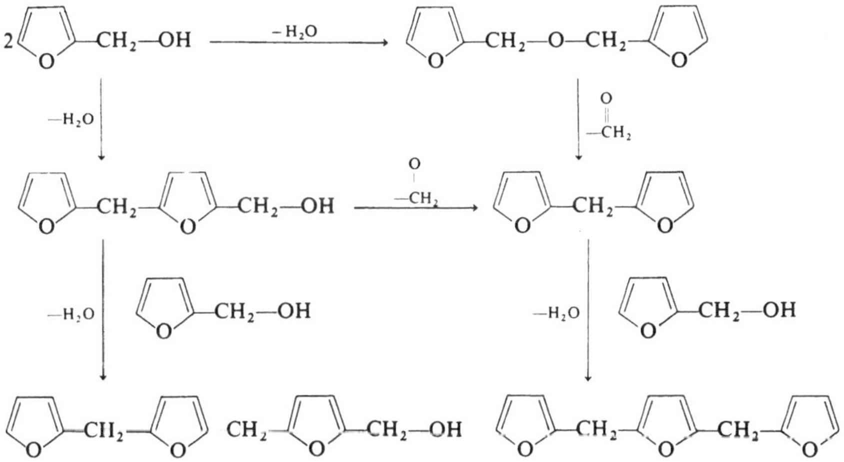

The amount of terminal methylol groups depends upon the temperature of the polymerization and the reaction time. On curing, the following polymer structure has been proposed for the thermoset product:[50]

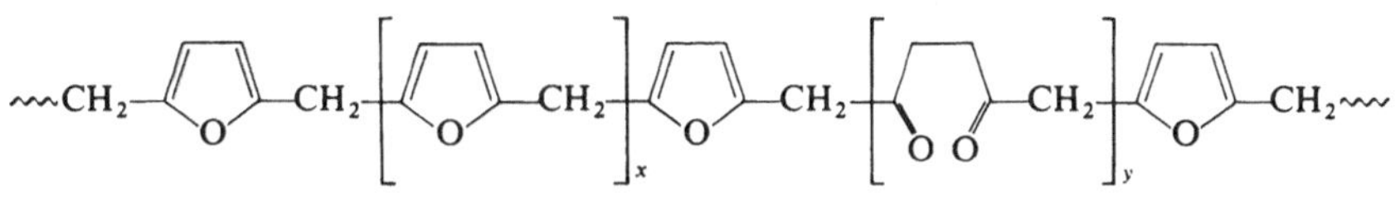

Thermal decomposition of the resin at temperatures up to 300°C gives a residue which is quite stable, infusible, and a good insulator. Under more severe conditions, products such as furoic acid, dehydromucic acid, acetic acid, and low-polymer fragments are formed. Like other thermosetting resins of the methylene bridge type, the thermal reactions which take place in the 100–500°C temperature range are complicated by competing oxidation reactions which result in the eventual formation of carbon monoxide and carbon dioxide, together with volatile low-polymer and monomeric fragments very similar in structure to those resulting from thermal processes.[51]

Oxidation of the methylene groups would be expected to form first ketonic species, then in further oxidative steps acid species with chain rupture. The initial oxidation steps forming ketonic groups can be summarized as shown below:

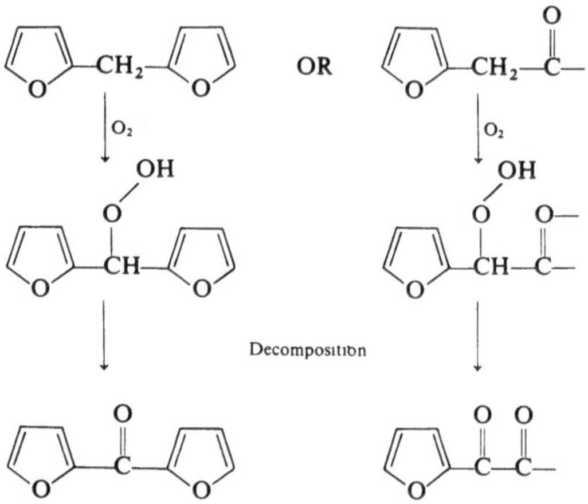

The formation of acidic species can arise from either of two reactions: (1) Methylol groups, not yet condensed with a neighboring furan nucleus, might be expected to oxidize to carboxyl groups (forming furoic acids). (2) The bifuryl ketone linkages would be expected to undergo oxidative chain scission to form substituted furoic acids. These processes are shown below:

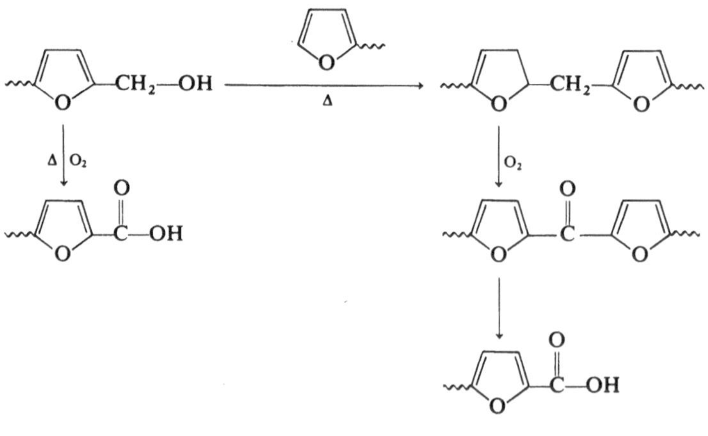

Subsequent steps stemming from these reaction products would be expected to form combustible volatile fragments. At elevated temperatures (above 600°C), thermal rupture processes similar to those of other aromatic ring-containing polymers occur to produce small carbon fragments such as acetylene. As the polymer body is heated to high temperatures, these latter reactions provide the volatile organic fuels to sustain flame.

3.1. Flame Retardation in Furan Resins

Very little is known about the chemistry of retardation processes in the furan-resin system. It has been reported that the condensation of furfuryl alcohol and trimethylolphosphinic acid results in a self-extinguishing resin.[52] In addition, the polymerization of 5-chloromethylfurfural forms a fire-retardant polymer.[53] Furfuryl alcohol with urea, formaldehyde, and phosphoric acid forms an adhesive for glass mat laminates which is reported to be fire retardant.[54] These three examples of retardation effects in the furan system suggest that the behavior is quite similar to the phenomena described in the previous section for phenolic resins.

4. Urea–Formaldehyde Resins

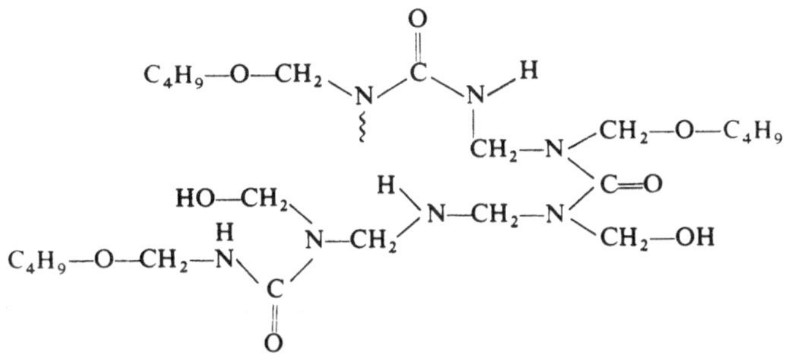

In contrast to the information available and pertinent to the phenolic-resin systems, the literature describing the thermal and thermal-oxidation chemistry of urea–formaldehyde resins is quite limited. Lady, Adams, and Kesse[55] and Conley[56] have examined the degradation of film samples of butylated urea–formaldehyde resins using infrared spectrophotometric techniques. Volatile-product formation from thermal-oxidative deterioration has been examined by vapor-phase chromatography.[56] The resin structure, as exemplified by the gross features shown below, has been postulated to undergo deterioration through reactions taking place at the N—CH$_2$O methylene group based upon the above-mentioned studies.

The most apparent initial change observed at temperatures between 150 and 200°C was the loss of butanol as a volatile fragment and the formation of carbon dioxide. The data suggest oxidation at the N—CH$_2$—O methylene group. This interpretation was reinforced by concurrent changes in the infrared spectrum as deterioration progressed. At temperatures above 250°C, the polymer decomposed more extensively to products which would be expected from oxidation and thermal rupture of the N—CH$_2$—O bond system, namely, butene, butyric acid, paraformaldehyde, and water, together with carbon dioxide and butanol. From these data the degradation route shown in Fig. 4 was postulated. The key feature in the proposed scheme is the initial attack by oxygen at the N—CH$_2$—O methylene group to form the predominant products, carbon dioxide and butanol.

Although little information can be derived from the above chemistry

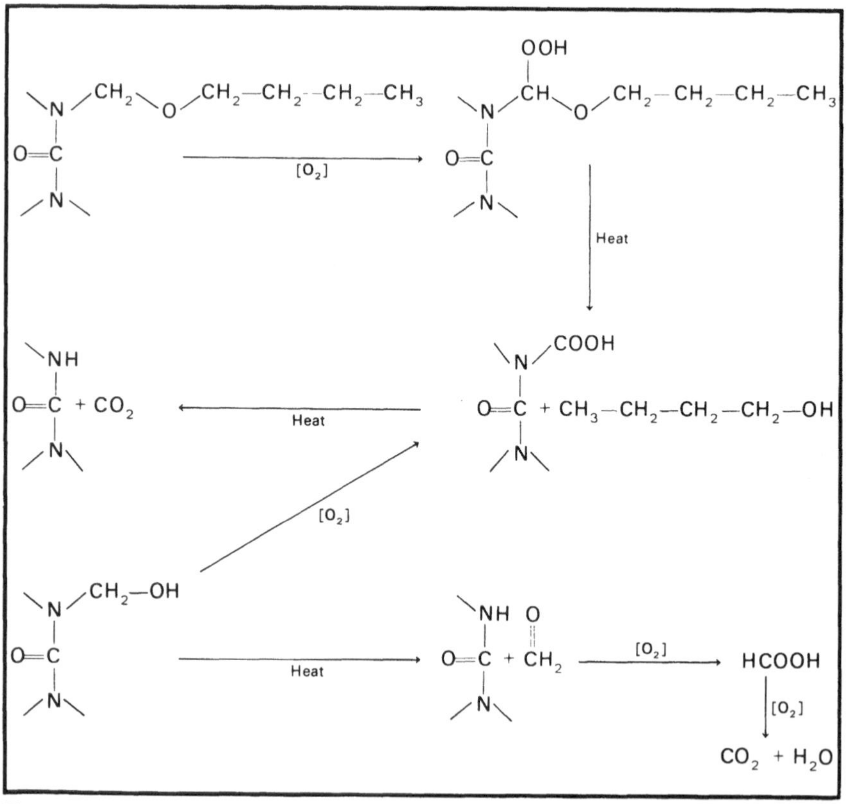

FIGURE 4. Summary of the initial degradation reactions for urea–formaldehyde resins.

(or for that matter from the available literature regarding the application of these materials) concerning the nature of the presumably complex secondary degradation steps, the chemistry of flame retardation appears to be quite similar to that described for phenolic resins.

4.1. Flame-Retardation Applications of Urea–Formaldehyde Resins

Urea–formaldehyde resins are considered to be difficult to ignite. For this reason they have found application in areas where it was deemed appropriate to impart flame-retardant properties to other materials.[57] For example, a urea–formaldehyde and sodium bisulfite mixture has been found to be an effective flame-resistant material for textile fabrics, particularly nylon netting.[58] Urea and thiourea mixtures have been found to exhibit wide application in flame-resistant coatings for fabrics in general[58-65] but specifically for use with nylons[61] and cottons.[62] The use of organophosphorus compounds[63] and phosphoric acid or phosphate salts[62,63] has been reported to improve the flame-retardation properties of the urea–formaldehyde coating systems.

It seems reasonable to project that the chemical processes responsible for flame retardation are associated with the ability of the urea–formaldehyde system to promote cross-linking reactions in the basic fabric system to which it is applied. The higher the nitrogen content in the coating system, apparently the greater the flame resistance imparted. One would also project that, in combination with phosphates, the reactive urea–formaldehyde resin would cure at an enhanced rate and, as well, provide reaction sites (via esterification) which undoubtedly would result in a highly cross-linked residue from this system. There seems to be no direct evidence available to substantiate the chemical validity of such a postulation. However, it would be expected that phosphate esters of the pendant methylol groups would be very reactive species at elevated temperatures.

As an impregnant for paper and wood, urea–formaldehyde resins have received considerable attention and use. Although it has been reported[31] that wood first treated with ammonium phosphate and then phenolic resin is somewhat superior to the product similarly prepared but using urea–formaldehyde resin as the impregnating component, numerous fire-resistant wood and paper compositions have been reported using the latter materials.[66-74] A typical example of this type of application is the acid-catalyzed polymerization of urea and formaldehyde after impregnation of the wood product.[71] Timber, veneer, particle board, fiber board, and other wood products are first treated with the reaction product mixture obtained from equimolar amounts of urea and phosphoric acid. The impregnated product is then sprayed with formaldehyde solution and dried.

The treated product is fire resistant and exhibits no afterglow during a 7-hr burning test. In applications such as that described above, the urea–formaldehyde resin undoubtedly reacts as previously described to provide a more highly cross-linked material. On heating this exhibits increased stability toward chemical reactions which result in the formation of flammable volatiles.

5. Melamine Resins

There are remarkable similarities between the melamine system and the urea–formaldehyde resin system. The initial decomposition chemistry

in air appears to be virtually the same; the flame resistance is quite similar; and the materials used to improve the flame resistance of melamines more often than not are compounds found to be suitable for urea–formaldehyde resin systems. Indeed one could include the phenolic resin system as well, since the three, although they vary somewhat in inherent flame resistance, seem to respond very similarly to flame retardants.

The oxidative decomposition of butylated melamines has been examined.[55,56] From the data collected in the 150–300°C temperature region, it was possible to ascertain that the thermal-oxidative deterioration proceeds at the initial stages as shown in Fig. 5.

This route is identical to that proposed for the urea–formaldehyde system in that the $N-CH_2-O$ methylene group is the oxidatively sensitive unit within the structure framework of the melamine polymer system. The cross-linked residue produced on heating is clearly a triazine ring-containing residue. This residue is quite resistant to further oxidation at lower temperatures (350°C and below) producing, therefore, a stable nitrogen-rich, charlike residue.

5.1. Flame-Retardation Applications of Melamine Resins

As a dry powder, a mixture of 50–98% free melamine and magnesium stearate has been found to be an effective fire-extinguishing powder for all

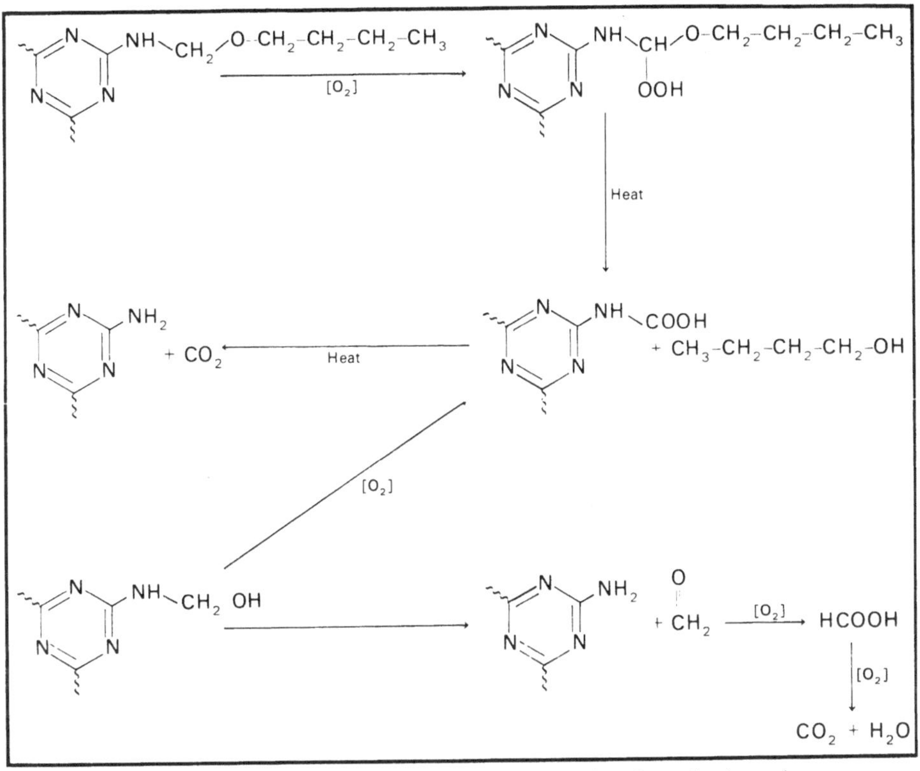

FIGURE 5. Summary of the initial degradation reactions for melamine resins.

types of fires. It has been reported[75] that it is particularly effective for fires involving such metals as magnesium. It would be reasonable to project that during decomposition in a flame the mixture forms a layer or coating which is resistant to combustion.

Melamine resin has been used in combination with both phenolics[67,76,77] and urea–formaldehyde resins[78] to form fire-resistant coatings and adhesives. In addition to the melamine additive, phosphoric acid, phosphates, and borates have added to the fire-resistant composite system. For example a mixture of phenolic resin containing 33% melamine and 5% sodium borate imparted high flame resistance when used as an impregnant for kraft paper. In addition other polymers such as nylon 6 or 6-6,[79] polychloroprene,[80] and rubber,[80] as well as polyurethane prepolymers,[81] have been rendered fire resistant or self-extinguishing by admixture with melamine resin together with other additives such as phosphoric acid or organophosphorus compounds. When 5–8% melamine was blended with nylon 6 or 6-6, a self-extinguishing molding compound has been reported.[79]

Further, a 1:1 by weight mixture of melamine and the condensation product of 3-(dimethylphosphono)propionamide and formaldehyde produces a fire-retardant coating for a variety of fibers including nylon.[82] In these cases a nitrogen- and phosphorus-rich layer, formed by initial decomposition of the paper, fiber, and molding system, is proposed to be the mode of fire retardation by the melamine–phosphorus-containing additives.

Numerous reports of the use of melamine–phosphorus-containing additive systems as fire-resistant mixtures for textile fabrics and fibers can be found in the literature.[62-64,83-87] In virtually all cases exceptional flameproofing was reported. Similar reports of melamine compositions permeate the wood fireproofing literature.[88-90] Construction sheets, decorative laminates, shingles, and adhesives for plywood have been fireproofed using melamines in admixture with phosphoric acid, phosphates, organophosphorus compounds, and borates, to illustrate a few of these applications.

Interestingly, melamines have been used for such diverse flame-retardant applications as the incorporation of melamine pyrophosphate in the preparation of fire-extinguishing, swelling coatings in conjunction with a vinyl acetate latex[91] and in combination with magnesium carbonate and phenolics for improved fire resistance of electrical insulations.[92]

As might be anticipated from the above discussion, there is very little that can be said concerning the chemical processes taking place which would aid in our understanding of the fire retardance by melamine systems. However, it is possible to speculate that the melamine-additive systems react at elevated temperatures to provide a highly cross-linked residue at the surface of the coated or impregnated body, which acts both as a scavenger for thermally degraded fragments within the body itself and as a surface coating which, similar to the phenolic and urea–formaldehyde systems, exhibits increased stability toward chemical reactions which result in the formation of combustible volatiles.

6. Epoxy Resins

In contrast to the resins treated in the foregoing discussion, normal epoxy resins are flammable materials. It is not unexpected, then, to find extensive literature concerned with retarding combustion of epoxy resin systems both by chemical modification of the resin structure itself and by the incorporation of a wide variety of additives to the resin, usually prior to curing. Structurally the resins in the uncured state can be visualized as polymeric materials containing an oxirane ring. For example, the products of the condensation of epichlorohydrin and bisphenol A and phenolic resin (a novolac) typify the basic resin structure prior to curing. In practice, these materials are cured with difunctional materials possessing available active

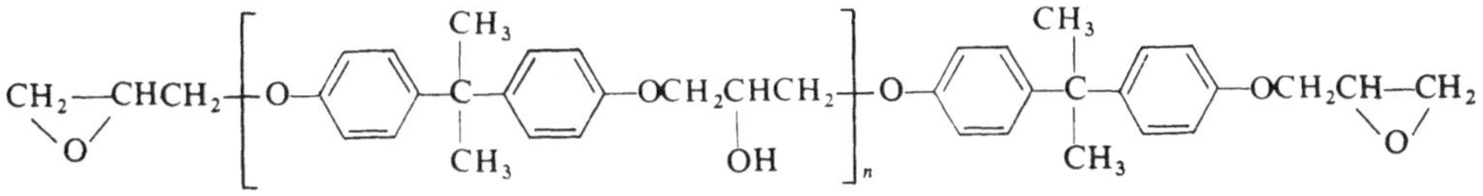

$$CH_2—CHCH_2 \left[O—\underset{CH_3}{\overset{CH_3}{C}}—OCH_2CHCH_2 \right]_n O—\underset{CH_3}{\overset{CH_3}{C}}—OCH_2CH—CH_2$$

hydrogen atoms (such as amines, alcohols, carboxylic acids) or anhydrides of acids.[93,94]

On thermal-oxidation of epoxy film samples, it has been observed that the initial oxidation processes are surface reactions.[95] Spectral examination of the films indicated a loss of hydroxyl groups, and the development of carbonyl functional species resulted from oxidation at temperatures below approximately 300°C. It has further been reported[96,97] that the thermal oxidation was an autocatalytic process, radical in type, and presumably dependent upon the formation of hydroperoxide groups in the cured resin. In addition, electron spin resonance studies[98] have indicated that the semiquinone or aryloxy radicals (shown below) are

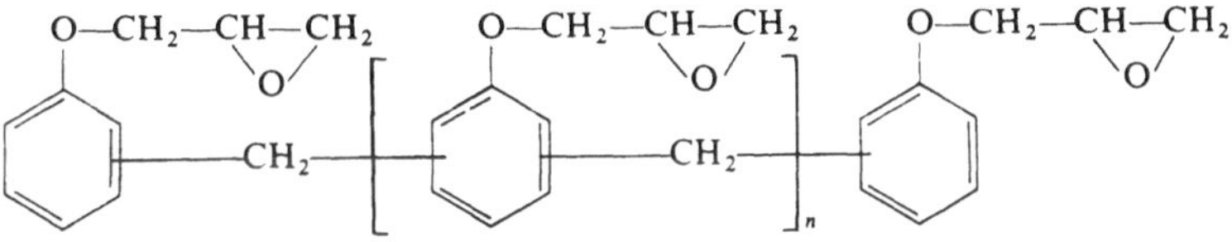

$$O—CH_2—CH—CH_2 \qquad O—CH_2—CH—CH_2 \qquad O—CH_2—CH—CH_2$$

possible radical species, resulting from thermal oxidation, rather than aliphatic, peroxy, or alkoxy types. Infrared studies[99] of the thermal-oxidative deterioration of bisphenol A-based epoxy resin films from samples with molecular weights (uncured) between 340 and 10,130 and cured with both amine and anhydride reagents (i.e., m-phenylenediamine and phthatic anhydride) in the temperature region between 100 and 250°C indicated the following pertinent points: (1) The hydrogen bonding within the resin changed from long range to short range. (2) The carbonyl species formed on oxidation decreased in quantity as the molecular weight of the base resin (uncured resin) increased. (3) With an aliphatic curing agent (diethylenetriamine), the cured resin from a noncured sample (mol wt 340) oxidatively degraded more rapidly than the same resin cured with m-phenylenediamine. These data clearly support an oxidative breakdown in the polymer at the amino-alcohol portion of the cured resin backbone.

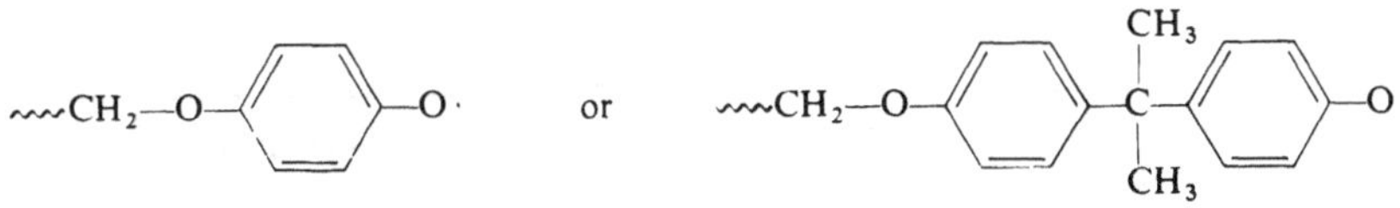

$$\sim\!\!\sim\!CH_2—O—\!\!\!\bigcirc\!\!\!—O· \qquad or \qquad \sim\!\!\sim\!CH_2—O—\!\!\!\bigcirc\!\!\!—\underset{CH_3}{\overset{CH_3}{C}}—\!\!\!\bigcirc\!\!\!—O·$$

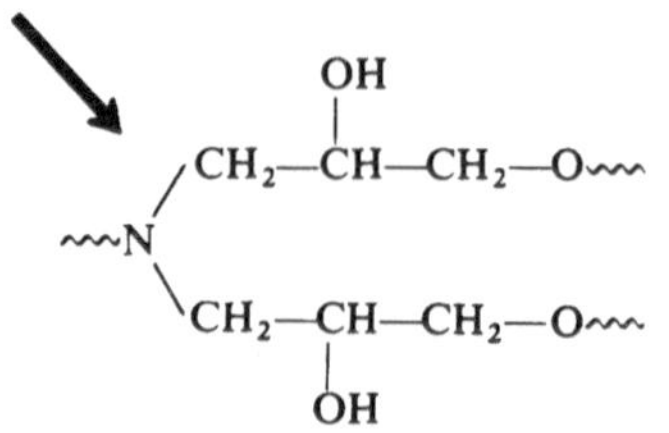

$$
\begin{array}{c}
\text{OH} \\
| \\
\diagup \text{CH}_2{-}\text{CH}{-}\text{CH}_2{-}\text{O}\text{\small\leadsto} \\
\text{\small\leadsto}\text{N} \\
\diagdown \text{CH}_2{-}\text{CH}{-}\text{CH}_2{-}\text{O}\text{\small\leadsto} \\
| \\
\text{OH}
\end{array}
$$

These data have been interpreted as shown in Fig. 6. It must be emphasized that at the present time this interpretation is speculative. However, undoubtedly the breakdown is dependent on the nature of the curing agent used, since the anhydride-cured resins were markedly more resistent to oxidation then those cured with amines.[99]

Insofar as thermal stability of epoxy resins is concerned, the most definitive and comprehensive report has resulted from studies conducted by Lee.[100] Other studies[101-110] should be consulted for related information. For example, the thermal-stability relationships of various epoxy resins has been reported,[101] as has the effect of varying curing agents[102] and the chlorine content of the resin.[103] Using thermogravimetric analysis,

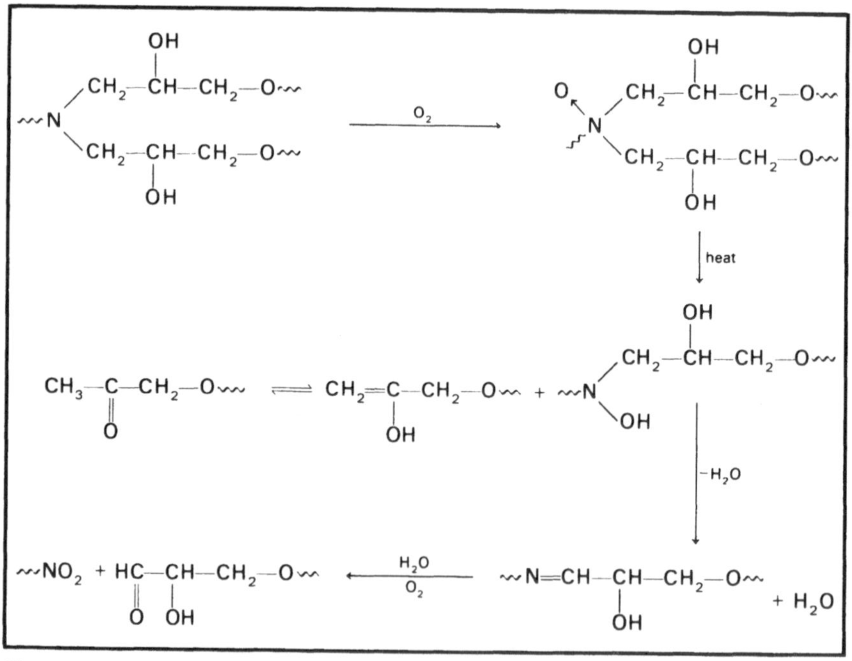

FIGURE 6. Proposed thermooxidative degradation processes taking place in the initial deterioration of epoxy resins.

differential thermal analysis, vapor-phase chromatography, and mass spectrometry, Lee[100] examined the thermal behavior of both cured and uncured resins. In the temperature range from 300 to 380°C, the differential analysis data obtained from examination of cured resins, uncured resins, and uncured resins containing curing agents indicated that the exothermic reactions occurring were reactions of the epoxy group. The types of processes occurring were proposed to be as shown in Fig. 7.

This route seems reasonable, since the volatile products from pyrolysis contained a significant amount of ethane and carbon monoxide. In addition allyl chloride and a number of aldehydic and ketonic components were detected. The formation of such products can be visualized as resulting from the cross-linking of the epoxide group with the hydroxyl group of an adjacent molecule to form a 1,2,3-triether. Thermal decomposition of

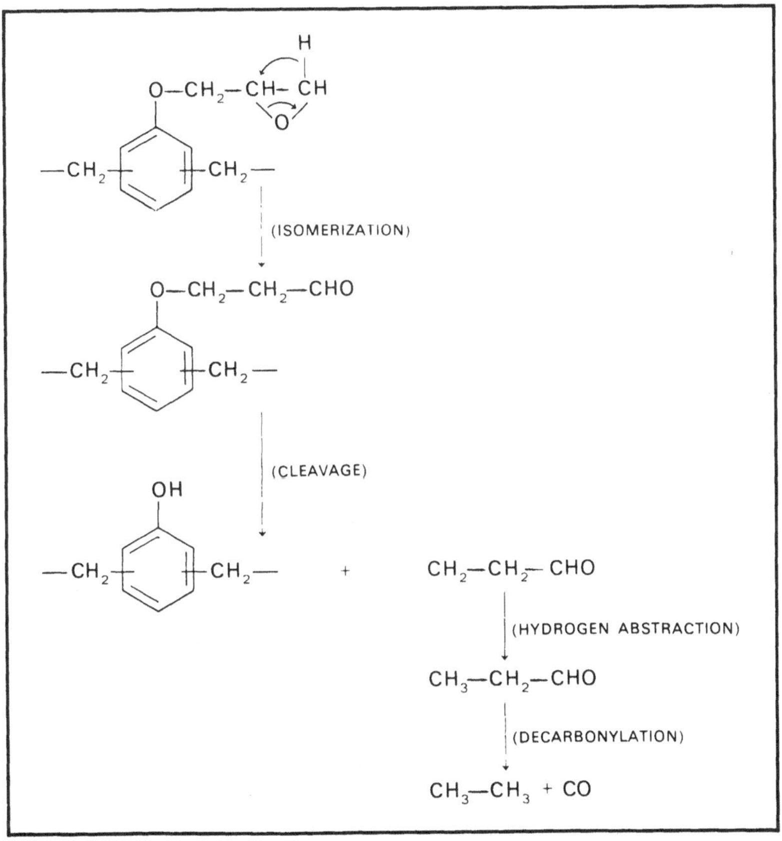

FIGURE 7. Proposed exothermic degradation reactions for the initial thermal deterioration of epoxy resins.

such linkage as illustrated in Fig. 8 would result in the formation of acrolein, acetaldehyde, formaldehyde, acetone, and/or secondary products from the propylene, ethylene, methane, and hydrogen moieties.

Spectral data indicated that amine-cured epoxy resins had a significant number of non-hydrogen-bonded hydroxyl groups. Figure 9 indicates a

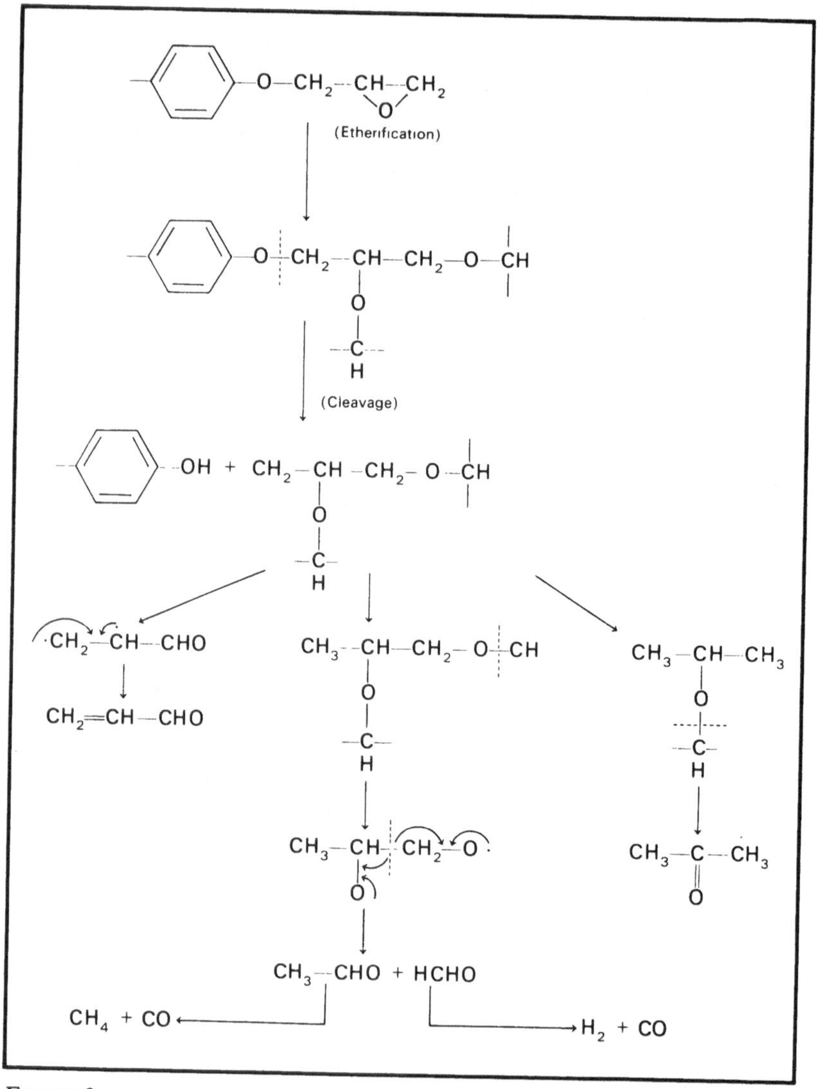

FIGURE 8. Proposed routes to the volatile products formed during the thermal degradation of epoxy resins.

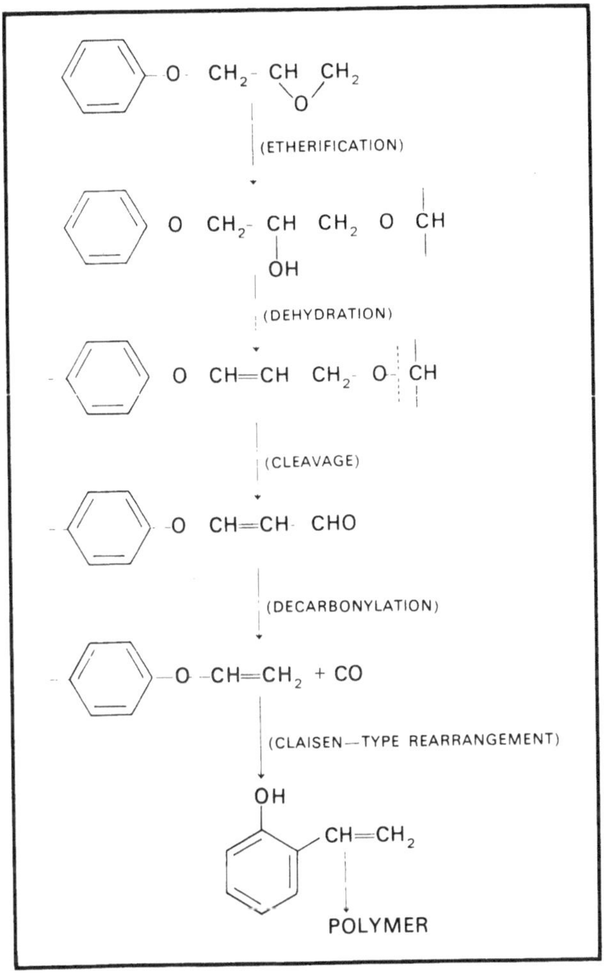

O   CH$_2$ CH CH$_2$
                \O/

(ETHERIFICATION)

O   CH$_2$- CH CH$_2$ O CH
                |
                OH

(DEHYDRATION)

O   CH=CH CH$_2$ O CH

(CLEAVAGE)

O CH=CH CHO

(DECARBONYLATION)

O CH=CH$_2$ + CO

(CLAISEN—TYPE REARRANGEMENT)

OH

CH=CH$_2$

POLYMER

FIGURE 9. Proposed alternative route to the formation of carbon monoxide and the formation of non-hydrogen-bonded hydroxyl groups.

reaction sequence which seems quite plausible as a possible reaction sequence to account for this observation and also as an alternative source of carbon monoxide (see Figs. 7 and 8).

Although these suggestions for both thermal oxidation and thermal degradation of epoxy resin systems are largely speculative, they should be considered as possible reactions which occur during the initial steps of deterioration and which result in the eventual combustion of epoxy resin systems. Both structural modifications and flame-retarding additives must, then, influence these types of chemical processes to be effective.

6.1. Flame Retardation in Epoxy Resins

There is extensive literature describing the effect of a wide variety of phosphorus compounds in flame retardation of epoxy resins.[111-131] The reports indicate that the epoxy resin system is not significantly different from other polymers described in this chapter. Ammonium phosphate, phosphorus–halogen mixtures, organophosphorus compounds, and compositions of organophosphorus compounds containing chlorine and bromine all impart flame resistance to epoxy resins. Approximately 5–6% phosphorus is required to attain significant flame retardation. A variety of phosphorus–halogen compositions containing about 2% phosphorus and 6% chlorine produce self-extinguishing epoxy resins. Halogen can be attached to either the aromatic or aliphatic portions of the resin. However, the effect of halogen on the aliphatic part of the resin backbone seems more pronounced than when the halogen is attached to an aromatic ring.

Antimony oxide, like phosphorus compounds, reduces the halogen-content requirements when added to the resin composition, suggesting that the effect of such additives is more extensive than simply providing gas-phase species which inhibit flame propagation.[131-137]

In the absence of phosphorus or antimony the levels of halogen required to effect flame retardation are dramatically increased.[131,138-158] Fluorine is relatively ineffective.[158] Chlorine is less effective than bromine; 30% chlorine is required to produce the same relative level of retardation as 15% bromine.

As noted in previous sections, boron as boric oxide, boric acid, sodium and calcium borate, or organic esters of boric acid produces self-extinguishing epoxy resin systems.[159-161] Again these data suggest an overall similarity in general stabilization chemistry for the resins described in this chapter and in many other sections of this volume. However, the specific chemical processes describing how these additives function in epoxy resins remains speculative.[162-164]

7. Summary

Retardation of combustion, flame resistance, fire retardance, and fireproof remain relative terms. Indeed, at this time, we have few, if any, operational definitions which adequately address the chemistry of combustion and the retardation of the chemical processes which generate flame. The uncertainty of literature interpretation is further complicated by the widely varying tests and measures for both the relative flammability and retardance of flammability for the materials treated in this chapter. It must be realized that all of these polymers are capable of producing combustible volatiles upon pyrolysis and will burn if the conditions are stringent enough. It seems, therefore, that the retardation of combustion centers on the quenching of combustion on a relatively small scale and/or removal from the environment of any combustible volatiles produced either during initial thermal-oxidative degradation or during the thermal deterioration which occurs competitively as the temperature of the system increases.

There appear to be two ways to retard combustion and, hence, flame in these resins: the first is to modify the system so that little or no combustible volatiles are produced (i.e., promote cross-linking to a maximum extent); the second is to furnish components which will retard diffusion of volatiles to the hot oxygen environment or vice versa.

In these resin systems the technology has been focused on two main approaches to flame retardation. The first and less frequently encountered is the direct incorporation of flame-retarding substances into the polymer backbone during synthesis. This method should be quite permanent, since there is complete dispersal of the retarding species throughout the polymer matrix whether it is in fiber, film, or molded shape. However, this approach suffers from the obvious problem, all too often encountered in practice, that a decrease in other polymer properties mitigates against the desired structural modification. For example, the incorporation of halogen atoms into the aromatic rings of epoxy resins significantly decreases the thermal stability of the resin. However, when such property changes are not of critical concern, effective retardation can be accomplished. Most retardation by synthetic means has involved the incorporation of halogen atoms into one of the species involved in the polycondensation step. In epoxy resins, for instance, the use of tetrabromobisphenol A; in phenolic resins, the use of chlorinated phenols; or in urea–formaldehyde and melamine resins, the use of halogenated butyl groups illustrates this approach. Apparently the halogenation of an alkyl portion of the resin is preferred over aryl-substituted halogen systems. We might wish to include in this category the inherent flame resistance of urea–formaldehyde and melamine resins resulting from the high nitrogen content of these systems and the retardation effects observed in high-nitrogen-content phenolic resins (from aniline modification as an example).

The second and most frequently employed method of retardation in these resins involves the use of various additives or combinations of additives. The explanations of the chemical phenomena responsible for retardation are largely speculative. The systems used are often complex, and little is known about the additive-polymer reactions which occur at elevated temperatures. However, such reactions undoubtedly do occur, and the elucidation of the nature of these reactions is the key to our future understanding of retardation phenomena. When we refer to additives, we must broaden our definitions somewhat in relation to the compounds most frequently cited and include polymers impregnated in or added to other polymers as well. For example, the use of urea–formaldehydes as part of flame-retardant systems for textile fibers, both natural and synthetic, and wood illustrates this kind of an additive. The retardation effect can be presumed to be due to increased nitrogen content of the system and the capability of nitrogen species to promote char formation through their greatly increased radical savenging capability.

The available literature emphasizes the following additive systems for use in phenolic, furan, urea–formaldehyde, melamine, and epoxy resin systems: Phosphorus, undoubtedly the oldest of the flame-retarding agents, as phosphoric acid undoubtedly is an ester- and anhydride-forming reactant in these systems which promotes cross-linking through thermal bond ruptures (the phosphate ester is a good "leaving group"). In addition, in the presence of heat and oxygen, a polyphosphate surface layer crusts over the polymer mass retarding combustion. Antimony, arsenic, and bismuth, other members of group Va of the periodic table, are also believed to exhibit similar retardation chemistry, although this is largely speculation at the present time. The presence of these elements combined with oxygen in the polymer mass provide additional sites for hydrogen bonding; in the case of multiply-bonded units (i.e., $P{=}O$), an increase in thermal stability should be attained since dissociation of these groups is a highly endothermic process; Phosphorus–carbon bonds, such as found in organophosphorus compounds, also add stability to the system and are reported to be very effective flame-retardation agents. With halogens and phosphorus or antimony a synergistic effect has been noted. The separate actions of these agents do not account for the retardation effect realized by them in combination. One explanation of this effect, a rather straightforward one but speculative, is that the antimony and phosphorus not only stabilize in the condensed phase but also, in the presence of halogen, form heavy gaseous products which aid in preventing gas-phase combustion and on decomposition form effective radical scavengers.

Boron most likely reacts much like phosphorus, since it also is an effective ester forming material. At elevated temperatures it is expected to function as a good cross-linking agent, thus promoting char formation. In addition, as an oxide, boron would exhibit a surface crust over the polymer mass, retarding combustion.

Halogens, namely chlorine and bromine, when incorporated into the polymer mass probably enter into dehydrohalogenation processes at elevated temperatures, thus promoting cross-linking reactions. On pyrolysis the halogen gases formed are further decomposed in the oxidation area to form radical scavengers or, in the presence of such materials as phosphorus and antimony, heavy gases which act as flame-smothering agents as noted above.

At this time most of the chemical bases for flame retardation are speculative. This area should be a fruitful future research endeavor. Solution of the basic problem seems to require a more detailed definition of the mechanism of both oxidative degradation and thermal degradation, together with the respective temperature profiles for these polymers. In combination with information concerning the reactions of additives with the polymer mass over the temperature region between room temperature and 700–800 C, the postulation of a decomposition schema to describe the chemical details of retardation should be possible. The problem is a formidable one. However, it appears to be within the capability of available scientific methodology.

8. References

1   J. Economy and L. C. Wohrer. *Spinner. Weber. Textilveredl.* **88**, 1223 (1970).
2. J. Economy. F. J. Frechette, and L. C. Wohrer. Ger. Pat. 1.948,412 (April 23, 1970).
3. R. T. Conley and J. F. Bieron, *J. Appl. Polym. Sci.* **7**, 103 (1963).
4. R. T. Conley and J. F. Bieron, *J. Appl. Polym. Sci.* **7**, 171 (1963).
5. R. T. Conley, *J. Appl. Polym. Sci.* **9**, 1117 (1965).
6   H. W Lochte. E. L. Strass. and R. T. Conley. *J. Appl. Polym. Sci.* **9**, 2799 (1965).
7. W. M. Jackson and R. T. Conley. *J. Appl. Polym. Sci.* **8**, 2163 (1964).
8. R. T. Conley, *Development of Applied Spectroscopy*. Vol. 4. pp. 377–402. Plenum Press. New York (1965).
9. G. P. Shulman and H. W. Lochte. *Polym. Prepr.* **6**, 36 (1965).
10. G. P. Shulman and H. W. Lochte. *J. Appl. Polym. Sci.* **10**, 619 (1966).
11. R. Q. Hall. *Chem. Ind. (London)*. **1952**, 693.
12. M. S. Kharasch and B. S. Joshi. *J. Org. Chem.* **22**, 1439 (1957).
13. C. D. Cook. D. A. Kuhn. and P. Fianu. *J. Am. Chem. Soc.* **78**, 2002 (1956).
14. J. A. Parker. E. L. Winkler, B. H. Miles, and L. F. Sonnabend. *Rep. Org. Coatings Plast Chem.* **27**, 132 (1967).
15. B. P. Barth. U. S. Pat. 3,076,780 (Feb. 5, 1963).
16. D. C. Lowe and A. S. Carello. Brit. Pat. 901,663 (July 25, 1962).
17. Societe Francaise Albert, Fr. Addn. 88,149 (to Fr. Pat. 1,389,825) (Dec. 16, 1966).
18. A. G. Levitas. P. V. Evseev, P. A. Nuss. A. N. Yampolskii, and Yu. I. Koltunov. U.S.S.R. Pat. 157,604 (Oct. 5, 1963).
19. R. G. B. Mitchell, *Plastics (London)* **24**, 44, 85 (1959).
20. G. M. Wagner, U.S. Pat. 2,814,607 (Nov. 26, 1957).
21. E. C. Soule, L. S. Burnett, and G. M. Wagner. U.S. Pat. 3,038,822 (June 12, 1962).
22. G. Gavlin and W. M. Boyer, U.S. Pat. 3,038,822 (June 12, 1962).
23. T. G. Brandts. U.S. Pat. 3,092,537 (June 4, 1963).
24. H. A. Smith, U.S. Pat. 3,320,213 (May 16, 1967).
25. Brit. Pat. 969,095 (Sept. 9, 1964) (to Monsanto Company).
26. M. E. Hull. III. and E. W. Simpson. Fr. Pat. 1,364,821 (June 26, 1964).
27. P. Dorier and B. Robinet. Fr. Pat. 1,482,967 (June 2, 1967).
28. S. Shimizu, Japan. Pat. 13,073 (June 24, 1965).
29 R. W. Quailes and J. A Baumann. U.S. Pat. 3,298,973 (Jan. 17, 1967).
30. C. Elmer and J. J. Mestdagh. U.S. Pat. 3,352,744 (Nov. 14, 1967).
31. A. A. Konstantnaya. *Isv. Vyssh. Ucheb. Zaved.. Les. Zh.* **12**, 165 (1969).

32. L. E. Hoxie, U.S. Pat. 3,377,317 (April 9, 1968).
33. A. F. Shepard and B. F. Dannels, U.S. Pat. 3,409,571 (Nov. 5, 1968).
34. B. F. Dannels and A. F. Shepard, *Rep. Org. Coatings Plast. Chem.* **27**, 125 (1967).
35. B. D. Davis, E. Elvis, and R. E. Morgan, Jr., Ger. Pat. 1,960,201 (June 18, 1970).
36. D. A. Brown, H. Coates, J. Girard, A. Thiot, and G. Quesnel, Ger. Pat. 1,300,294 (July 31, 1969).
37. A. Taniuchi and M. Koide, Japan. Pat. 70–11,432 (April 24, 1970).
38. V. K. Ninin, S. V. Shner, N. V. Shorygina, I. K. Rubtsova, and L. P. Bocharova, U.S.S.R. Pat. 200,149 (July 13, 1967).
39. N. V. Shorygina, V. K. Ninin, L. Z. Soborovskii, A. B. Bruker, and Kh. R. Raver, U.S.S.R. Pat. 179,920 (Feb. 28, 1966).
40. J. Girard, A. Thiot, and G. Quesnel, U.S. Pat. 3,121,697 (Feb. 18, 1964).
41. Brit. Pat. 931,146 (July 10, 1963 (to Albright and Wilson, Ltd.).
42. J. M. Pollock and N. B. Ray, Brit. Pat. 920,722 (March 13, 1963).
43. H. Grahmann and N. Pusch, Ger. Pat. 1,065,116 (Sept. 10, 1959).
44. G. L. Quesnel, J. Girard, and A. Thiot, Fr. Pat. 1,213,894 (April 5, 1960).
45. P. N. Erickson and A. N. Erickson, U.S. Pat. 3,256,216 (June 14, 1966).
46. A. Heslinga and P. J. Napjus, Ger. Pat. 1,197,614 (July 29, 1965).
47. R. T. Conley and L. L. Chow, *Rep. Org. Coatings Plast. Chem.* **27**, 133 (1967).
48. D. Hanton, Fr. Pat. 1,588,803 (May 9, 1968).
49. Y. Hachihama and T. Shono, Studies on Furfuryl Alcohol Resin, I, Tech. Report, Osaka Univ., Japan (1954).
50. R. T. Conley and I. Metil, *J. Appl. Polym. Sci.* **7**, 37 (1963).
51. R. T. Conley and I. Metil, *J. Appl. Polym. Sci.* **7**, 1083 (1963).
52. L. M. Prutkov, F. M. Gurdzhi, and O. A. Pochernikova, U.S.S.R. Pat. 260,175 (Dec. 22, 1969).
53. F. O. Mamedov, I. V. Kamenskii, and V. I. Itinskii, U.S.S.R. Pat. 149,434 (Feb. 26, 1965).
54. L. H. Brown and D. D. Watson, Ger. Pat. 1,927,776 (Dec. 4, 1969).
55. J. H. Lady, R. E. Adams, and I. Kesse, *J. Appl. Polym. Sci.* **3**, 65 (1960).
56. R. T. Conley, Proc. Battelle Mem. Inst. Symp., Thermal Stability of Polymers, Columbus, Ohio (Dec. 1963), pp. E1–E63.
57. C. J. Hilado, *Flammability Handbook for Plastics,* p. 31, Technomic Publishing Co., Stamford, Conn. (1969).
58. R. Polansky and W. F. Herbes, U.S. Pat. 2,854,437 (Sept. 30, 1958).
59. R. Polansky and W. F. Herbes, U.S. Pat. 2,859,206 (Nov. 4, 1958).
60. W. F. Herbes, U.S. Pat. 2,923,644 (Feb. 2, 1960).
61. J. J. Nemes, R. Polansky and W. F. Herbes, U.S. Pat. 3,308,098 (March 7, 1967).
62. Z. Pawlowska and J. Linkowicz, *Pr. Cent. Inst. Ochrony Pr.* **15**, 147 (1965).
63. R. A. Corai, Fr. Pat. 1,399,173 (May 14, 1965).
64. L. Krema, *Textil* **21**, 230 (1966).
65. V. Duroux, Fr. Pat. 993,047 (Oct. 25, 1951).
66. F. A. M. Cotelle, Fr. Pat. 1,228,230 (Aug. 29, 1960).
67. J. Urbanek, Brit. Pat. 864,099 (March 29, 1961).
68. Belg. Pat. 665,047 (Dec. 8, 1965) (to Chemische Werke Albert).
69. K. Beltzig and A. Becker, Ger (East) Pat. 32,580 (Nov. 15, 1964).
70. A. G. Zabrodkin, L. I. Khitrova and N. S. Solomatina, U.S.S.R. Pat. 270,980 (May 12, 1970).
71. S. Iwasa, Y. Wada and T. Hayakawa, Japan. Pat. 70-05,674 (Feb. 25, 1970).
72. G. M. Wagner, Belg. Pat. 629,820 (Oct. 21, 1963).
73. S. K. Kasymbekov, U.S.S.R. Pat. 285,214 (Oct. 29, 1970).
74. W. Schulenburg, U.S. Pat. 2,881,088 (April 7, 1959).

75. H. Frieser, Ger. Pat. 1,138,323 (Oct. 18, 1962).
76. D. C. Lowe and A. S. Carello, Brit. Pat. 901,663 (July 25, 1962).
77. Fr. Pat. 1,532,540 (July 12, 1968) (to Compagnie de Saint-Gobain).
78. S. Koshima, N. Nakao and H. Horimuki. Japan. Pat. 70-17,078 (June 12, 1970).
79. Fr. Pat. 2,004,073 (Nov. 21, 1969) (to Farbenfabriken Bayer, A. G.).
80. Ger. Pat. 1,120,127 (Dec. 21, 1971) (to Firestone Tire and Rubber Co.).
81. S. Okada, Y. Hori and K. Yamagami. Japan. Pat. 70-09,957 (April 10, 1970).
82. Y. Yamata, A. Iida, O. Kobayashi, M. Nomura, and M. Yoshizawa. Japan. Pat. 71-04,358 (Feb. 3, 1971).
83. H Malz and F. Kassack, Ger. Pat. 1,102,095 (March 16, 1961).
84. D. M. Soignet, R. R. Benerito, and R. H. Berni. U.S. Pat. 3,494,719 (Feb. 10, 1970).
85 D. J. Dargle and D. J. Donaldson, *Text. Chem. Color.* 1, 534 (1969).
86. I. S. Goldstein, U.S. Pat. 3,479,211 (Nov. 18, 1969).
87. G. C. Tesoro, Ger. Pat. 1,913,137 (Oct. 2, 1969).
88. W. Scholles and H. Koddebusch, Ger. Pat. 1,453,377 (April 9, 1970).
89 C. A. Holmes, U.S. Forest Serv. Res. Pap., FPL 158 (1971), pp. 1 27.
90. Societe Française Albert, Fr. Addn. 87,009 (to Fr. Pat. 1,389,825) (May 27, 1966).
91. R. G Fessler and B. C. Tredinnick, Ger. Pat. 2,018,565 (Nov. 5, 1970).
92. J J. Millane, Brit. Pat. 1,213,625 (Nov. 25, 1970).
93. H. Lee and K. Neville, *Epoxy Resins, Their Applications and Technology*. Chap. 2, McGraw-Hill, New York (1957).
94. Bordon Award Symposium on Epoxy Resins, *Rep. Org. Coatings Plast. Chem.* 29, 1 (1969).
95. W. R. R. Park and J. Blount, *Rep. Div. Paint, Plast. Print. Ink Chem.* 16(2), 56 (1956); W. R. R. Park and J. Blount, *Ind. Eng. Chem.* 49, 1897 (1957).
96. M. B. Neiman, B. M. Kovarskaya, M. P. Yazvikova, A. I. Sidnev, and M. S. Akutin. *Vysokomolekul. Soedin.* 3, 602 (1961).
97. M. B. Neiman, B. M. Kovarskaya, L. I. Golubenkova, A. S. Strizhkova, I. I. Levantovskaya, and M. S. Akutin, *J. Polym. Sci.* 56, 383 (1962).
98. D. W. Ovenall, *Polym. Lett.* 1, 37 (1963).
99. R. T. Conley and M. F. Dante, Stability of Plastics, Tech. Conf. Soc. Plastics Eng., Washington, D.C. (June, 1964).
100. L. H. Lee. Proc. Battelle Memorial Inst. Symp. Thermal Stability of Polymers. Columbus, Ohio (Dec. 1963), pp. F1–F23; L. H. Lee, *J. Polym. Sci.* 3, 859 (1965).
101. J. Wynstra, A. G. Farnham, N. H. Reinking, and J. S. Fry, *Rep. Div. Paint, Plast. Print. Ink Chem* 19(2), 84 (1959).
102. J. Delmonte, *Chem. Eng. Prog.* 58, 51 (1962).
103. W. J. Belanger and S. A. Schulte, *Mod. Plast.* 37, 154 (1959).
104. M. B. Neiman, L. I. Golubenkova, B. M. Kovarskaya, A. S. Strizhkova, I. I. Levantovskaya, M. S. Akutin, and V. D. Moiseev, *Vysokomolekul. Soedin.* 1, 1531 (1959).
105. M. M. Neiman, B. M. Kovarskaya, I. I. Levantovskaya, A. S. Strizhkova, and M. S. Akutin, *Sov. Plast.* 7, 18 (1960).
106. M. B. Neiman, B. M. Kovarskaya, A. S. Strizhkova, I. I. Levantovskaya, and M. S. Akutin, *Dokl. Akad. Nauk. U.S.S.R.* 135, 1147 (1960).
107. H C. Anderson, *Rep. Div. Paint, Plast. Print. Ink Chem.* 19(2), 104 (1959).
108. H. C. Anderson, *Anal. Chem.* 32, 1592 (1960).
109. H. C. Anderson, *Polymer* 2, 452 (1961); H. C. Anderson, *Kolloid 2* 184, 26 (1962).
110. S. L. Madorsky and S. Straus, *Mod. Plast.* 36, 134 (1961).
111. S. P. Edwards and P. H. Franke, Jr., U.S. Pat. 3,597,476 (Aug. 3, 1971).
112. Brit. Pat. 1,231,814 (May 12, 1971) (to Chemische Werke Albert).
113. T. J. Dijkstra and E. J. W. Vogelzang, Brit. Pat. 1,112,139 (May 1, 1968).
114. B. J. Bremmer, U.S. Pat. 3,399,174 (Aug. 27, 1968).

115. J. J. Rizzo, U.S. Pat. 3,312,636 (April 4, 1967).
116. W. L. Baak, U.S. Pat. 3,352,947 (Nov. 14, 1967).
117. B. B. Levin, E. I. Goldshtein, N. I. Telegina, G. V. Markova, M. I. Kiseleva, and R. V. Shaton, U.S.S.R. Pat. 191,790 (Jan. 26, 1967).
118. S. Horun, *Mater. Plast.* **5**(4), (1968).
119. Y. Tanaka, M. Seki, and M. Murata, *Shikizai Kyokaishi* **40**(1), 19 (1967).
120. Neth. Appl. 6,602,522 (Aug. 29, 1966) (to Dynamit-Noble A.-G.).
121. Brit. Pat. 1,061,616 (March 15, 1967) (to Canadian Industries Ltd.).
122. A. F. Nikolaev, T. A. Zyryanova, G. A. Balaev, N. A. Voronova, and G. M. Grigoreva, U.S.S.R. Pat. 184,443 (July 21, 1966).
123. V. I. Kirilovich, S. M. Shner, I.-K. Robtsova, A. E. Rabkina, and M. A. Tikhonova, U.S.S.R. Pat. 183,379 (June 17, 1966).
124. Neth. Appl. 6,515,176 (May 25, 1966) (to Dynamit-Noble A.-G.).
125. P. L. Smith and C. W. McGary, Jr., U.S. Pat. 3,236,863 (Feb. 22, 1966).
126. V. D. Valgin, E. A. Vasil'eva, V. A. Sergeeva, E. L. Gefter, and A. Yuldashev, U.S.S.R. Pat. 168,881 (Feb. 26, 1965).
127. A. L. Bullock, W. A. Reeves, and J. D. Guthrie, U.S. Pat. 2,916,473 (Dec. 8, 1959).
128. A. F. Nikolaev, K. A. Makarov, L. N. Mashlyakovskii, M. S. Trizno, and S. I. Kulicheva, U.S.S.R. Pat. 255,553 (Oct. 28, 1969).
129. S. R. Hargis, Jr., U.S. Pat. 3,524,903 (Aug. 18, 1970).
130. O. Mauz, F. Rochlitz, and D. Schleede, Ger. Pat. 1,113,826 (Feb. 13, 1958).
131. J. Riera Tuebols, *Color. Pinturas* **18**(124), 5 (1969).
132. S. Porejko, Z. Brzozowski, and J. Kusmierek, Fr. Pat. 2,042,118 (Feb. 5, 1971).
133. A. M. Partansky, Br. Pat. 1,125,632 (Aug. 28, 1968).
134. M. H. Nickerson, H. S. Schnitaer, J. E. Curtis, and G. D. Patterson, U.S. Pat. 3,223,654 (Dec. 14, 1965).
135. V. D. Elarde, U.S. Pat. 2,885,380 (May 5, 1959).
136. Br. Pat. 967,259 (Aug. 19, 1964) (to DeBell and Richardson, Inc.).
137. R. Sidlow and L. Williams, Brit. Pat. 953,206 (May 25, 1964).
138. C. M. Hayes, U.S. Pat. 3,391,112 (July 2, 1968).
139. R. M. Lusskin, F. Backer, and J. R. Larson, U.S. Pat. 3,396,147 (Aug. 1968).
140. E. E. Gilbert, O. A. Barton, R. M. Hetterly, E. R. Degginer, and C. R. McArthur, U.S. Pat. 3,361,717 (Jan. 2, 1968).
141. Yu A. Muravev, A. D. Valgin, D. F. Kutepov, and V. V. Korshak, U.S.S.R. Pat. 204,572 (Oct. 20, 1967).
142. F. Andreas and St. Skora, Ger. (East) Pat. 51,710 (Dec. 5, 1966).
143. B. J. Bremmer, U.S. Pat. 3,294,742 (Dec. 27, 1966).
144. F. Andreas and St. Skora, *Plast. Kaut.* **13**(8), 451 (1966).
145. R. A. Skiff, U.S. Pat. 3,350,334 (Oct. 31, 1967).
146. B. J. Bremmer, *Ind. Eng. Chem., Prod. Res. Dev.* **5**(4), 340, (1966).
147. A. M. Partansky, U.S. Pat. 3,252,850 (May 24, 1966).
148. Brit. Pat. 962,020 (June 24, 1964) (to Ciba, Ltd.).
149. G. L. Popova, G. L. Khromov, I. P. Khoroshilova, and O. A. Kochurenkova, *Plast. Massy* **11**, 11 (1965).
150. R. C. Doss, U.S. Pat. 3,477,966 (Nov. 11, 1969).
151. M. S. Ogii, A. S. Dzhoi, A. P. Kuzyakov, and N. K. Moshchinskaya, U.S.S.R. Pat. 230,415 (Oct. 30, 1968).
152. H. A. Newey, U.S. Pat. 3,449,375 (June 10, 1969).
153. M. Wismer, U.S. Pat. 3,043,881 (July 10, 1962).
154. J. Vuillemenot, Fr. Pat. 1,323,734 (April 12, 1963).
155. M. E. Chiddix and R. W. Wynn, U.S. Pat. 2,951,829 (Sept. 6, 1960).
156. W. E. Prescott and W. W. Bressler, U.S. Pat. 2,989,502 (June 20, 1961).

157. P. Robitschek and S. J. Nelson, *Ind. Eng. Chem.* **48**, 1951 (1956).
158. S. J. Nelson, J. S. Sconce, and P. Robitschek, U.S. Pat. 2,833,681 (May 6, 1958).
159. H. H. Chen and A. C. Nixon, *Soc. Plast. Eng. Trans.* **5**(2), 90 (1965).
160. R. A. Skiff, Ger. Pat. 1,089,167 (Sept. 15, 1960).
161. R. C. Namez, U.S. Pat. 3,058,946 (Oct. 16, 1962).
162. P. C. Warren, Soc. Plast. Eng. Tech. Paper, p. 362 (1970).
163. C. J. Hilado, *Flammability Handbook for Plastics,* pp. 31, 40, 81ff, Technomic Publishing Co., Stamford, Conn. (1969).
164. J. W. Lyons, *The Chemistry and Uses of Fire Retardants,* pp. 401–412, Wiley-Interscience, New York (1970).

9

Candle-Type Test for Flammability of Polymers

C. P. Fenimore

1. Flammability Measurements

The flammability of a gaseous fuel can be investigated by finding the limiting mixture with air which just propagates a flame. For example,[1] mixtures of hydrogen and air burn only if the molecular ratio, $O_2/2H_2$, lies in the range 0.15–10. Mixtures of n-hexane and air burn only over a narrower range of compositions containing 0.3–1.8 times the stoichiometric amount of oxygen. The hydrocarbon may therefore be considered less flammable than hydrogen.

If the fuel cannot be premixed with air, one needs a different measure of flammability. For diffusion flames, in which the fuel and oxidant mix as they burn, the limiting oxygen index (LOI) has been found useful. This concept is based on the fact that many substances burn in pure oxygen, but nothing burns in inert nitrogen. A mixture of oxygen and nitrogen can usually be found in which a given fuel is just able to burn. The mole fraction of oxygen in this critical atmosphere is called the critical oxygen index (COI) of the material and can be taken as a measure of flammability.

$$\text{Oxygen index} = n = \frac{O_2}{(O_2 + N_2)}$$

The larger n, the less flammable the material.

C. P. Fenimore · General Electric Research and Development Center, Schenectady, New York.

Simmons and Wolfhard[2] published such *n* values for diffusion flames of some light fuels. They supplied gases or liquids countercurrent to a gently rising stream of oxidant in the apparatus sketched in Fig. 1. After igniting the fuel issuing from the porous hemisphere, they adjusted the oxygen content in the stream of oxidant until the LOI was found. Some of their results are listed in Table 1. One notes that *n* is smaller for hydrogen than for *n*-hexane, so the latter is less flammable by this criterion also.

The critical atmosphere contains $(1 - n)/n$ times as much nitrogen as oxygen, and the heat of combustion must warm this inert nitrogen along with the reaction products to whatever flame temperature is attained. The larger *n*, the less the nitrogen diluent and the hotter the flame. A relatively nonflammable material is therefore one which needs a high flame temperature in order to burn.

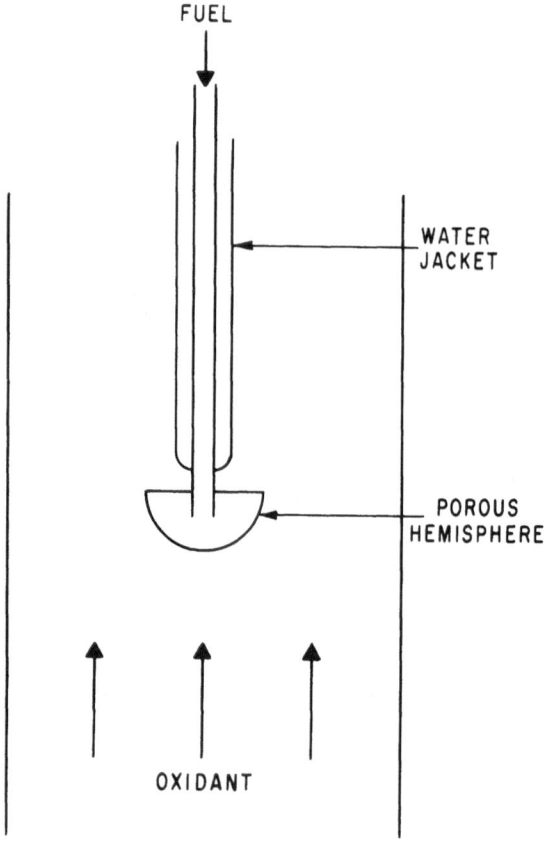

FIGURE 1. Schematic arrangement for measuring oxygen indexes of light fuels.

TABLE 1
Oxygen Indexes of Light Fuels
Burning as Diffusion Flames

Fuel	Index No.
Hydrogen	0.054
Formaldehyde[a]	0.071
Acetylene	0.085
Ethylene	0.105
Methyl alcohol	0.111
Ethyl alcohol	0.126
n-paraffins	
(propane to decane)	0.127–0.135
Cyclohexane	0.134
Benzene	0.133
Methane	0.139

[a] Formaldehyde vapor preheated to 420 K [3]

There are reasons to believe[2] that the maximum temperatures in diffusion flames burning at their COIs are about the same as the flame temperatures of the same fuels when premixed with air and burned at their fuel-lean composition limits. That is, in either limiting condition where burning is just possible, a hydrogen flame needs a temperature of around $1080°K$; a flame of less flammable *n*-hexane needs a higher temperature of around $1770°K$.

The effect of inert diluents other than nitrogen is often compatible with the view that some approximately constant minimum flame temperature is required for a given fuel. In fact, if the quenching action of an additive is no more than expected in consequence of its heat capacity (and possibly its thermal conductivity) under the assumption that the minimum flame temperature remains constant, one usually considers it proven that the additive is chemically inert. A flame can be diluted more with argon than with nitrogen because argon has the smaller heat capacity.[4]

Some additives are considered not to be inert because flames containing them require a considerably higher temperature to burn than would suffice in the absence of the additive. Such materials are called flame inhibitors. For example, Lask and Wagner[5] showed that a mixture of about 1/2 mol Br_2 and 1 mol of *n*-hexane just failed to form a flammable premixture with air. The addition of bromine raised the minimum flame temperature required from around 1770 to about $2100°K$ or more. Another way of expressing the effect of bromine is to put it in terms of the COI. If no premixture of (*n*-hexane + 1/2 Br_2) with air can burn, neither can a diffusion flame of (*n*-hexane + 1/2 Br_2) burn in air, of oxygen index 0.21. Therefore, the addition of bromine increases the oxygen index of *n*-hexane by about 60%, from the $n = 0.13$ given in Table 1 to $n = 0.21$. In a subsequent section we will compare the inhibition of hydrocarbon flames by bromine with the inhibition of polymer burning by bromine incorporated in the polymer.

2. Critical Oxygen Indexes of Polymers

Oxygen indexes can be measured for polymers in the apparatus sketched in Fig. 2.[6] Atmospheres of known composition are prepared by metering nitrogen and oxygen through critical flow orifices, mixing the gases, and feeding them to the bottom of a glass tube about 60 cm long by 8 cm diameter. The gas passes through a bed of glass beads to smooth its flow, then up the tube at a velocity of 3–12 cm/sec, and out of the open top. A sample of polymer, often a rod or strip 0.3–1 cm thick and about 8 cm long, is clamped in the center of the tube. It is ignited at the upper end with a hydrogen flame; the igniter flame is withdrawn; and the atmosphere found in which the sample just burns for its entire length.

The apparatus is easy to construct if one has a good gas-metering system. Critical flow orifices such as are described by Anderson and

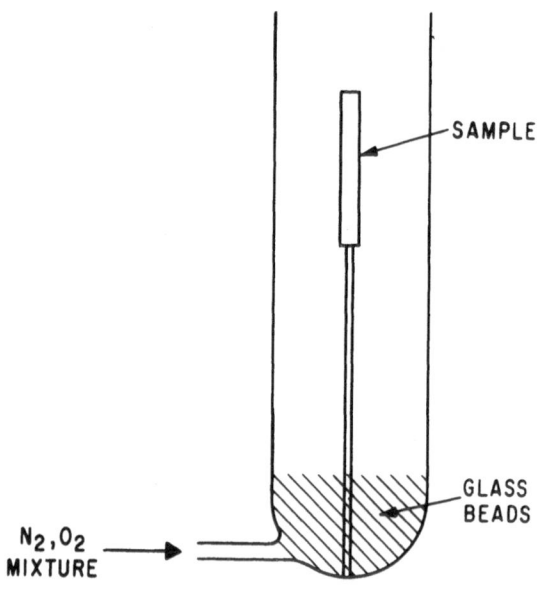

FIGURE 2. Schematic arrangement for measuring oxygen indexes of plastics.

Friedman[7] work very well. The flow through each orifice is controlled and measured by controlling the upstream pressure with precision regulators and bourdon gages of 0–100 psi range. The oxygen and nitrogen can be drawn from gas cylinders, and the calibration curves of flow as a function of upstream pressure can be made with a wet test meter for flows up to about 200 cc/sec. For larger flows it is better to use another method of calibration, such as the pressure rise in a known volume during a timed flow. The pressure ratio across the orifice hole should be two or more, for then the mass flow rate through the orifice is directly proportional to the absolute upstream pressure and independent of the downstream pressure.*

The main difference between the apparatus sketched in Fig. 2 and the arrangement used by Simmons and Wolfhard for light fuels is that the polymers are burned from the top downward to minimize convective heating of unburned polymer by the hot products. Simmons and Wolfhard supplied their light fuels countercurrent to the ascending flow of oxidant and burned from the bottom up. Different burning directions give different indexes. Polymethylmethacrylate has an index of $n = 0.174 \pm 0.001$ in the standard configuration, but if it is supported from the top and ignited at the bottom so that convective heating is maximized, the index falls to 0.151 ± 0.004. The reason for using the configuration shown in Fig. 2 is that many polymers drip so badly when ignited at the bottom that a reproducible result is impossible.†

Figure 3 is a photograph of 0.8-cm-thick polyethylene burning near its extinction limit. A clear molten dome of polymer at the top of the specimen has lost a few drops which run down the side and freeze. The dome is enclosed by a thin blue flame and topped by a yellow cap owing to thermal emission from carbon particles. As the oxygen content of the atmosphere is decreased, the flame grows smaller, loses its yellow cap, and eventually extinguishes at the critical index. Similar extinction is observed for other polymers which burn without residue, such as polymethylmethacrylate and polytetrafluoroethylene.

A filled material, such as silica-filled silicone, has little visible flame; a zone of glowing silica works down the specimen or ceases to propagate and cools at extinction. Other materials may go out less smoothly; polycarbonate and polyvinyl chloride char as they burn and seem to smother their flames in char at extinction.

Some materials melt too easily to be measured even in the standard

---

* A flammability tester, including the calibrated gas-metering system, is sold by General Electric Co., Instrument Dept., West Lynn, Mass.
† Samuel Johnson, to the annoyance of his hostess, demonstrated a similar drip from candles. "His uncouth habits, such as turning the candles with their heads downwards when they did not burn bright enough, and letting the wax drop upon the carpet, could not but be disagreeable to a lady." Boswell's *Life of Johnson*.

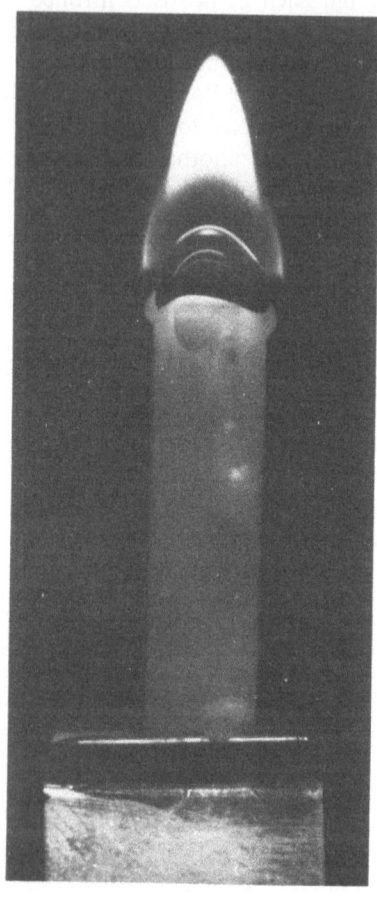

FIGURE 3. Polyethylene burning in an oxygen–nitrogen mixture near the critical oxygen index. Width of specimen is 0.8 cm.

configuration. For example, a proper end-burning specimen of poly-α-methylstyrene was never found. Low-molecular-weight polypropylene, polyethylene, etc., tend to give larger n values than those listed in Table 2, presumably because they melt too easily. Much more molten polymer is lost from the burning region of easily melting substances than is shown by Fig. 3, and this abstracts more heat from the flame than is required merely to gasify the fraction of the material which burns. Hindersinn[8] and others[9] showed that a more fluid melt can make thermoplastic polymers appear more flame resistant, and this is discussed in a subsequent section.

A list of oxygen indexes is given in Table 2. The values reported to three figures are usually reproducible to $\pm 1\%$. On comparing the table with the "slow burning" or "self-extinguishing" ratings which are often applied to some of the materials, one would judge that "slow burning"

TABLE 2

Oxygen Indexes of Plastics and Fabrics at Atmospheric Pressure

	Index no.
Polyoxymethylene (Delrin or Celcon)	0.157
Kitchen candle	0.16
Polymethylmethacrylate (Rohm and Haas Plexiglas)	0.174
Polypropylene (Hercules Profax 6505)	0.174
Asbestos-filled "slow-burning" polypropylene (Union Carbide JMDC-4400)	0.205
"Self-extinguishing" polypropylene (Union Carbide JMDA-9490)	0.282
Polyethylene	
Allied Chem 1220 or Phillips Marlex 5002	0.174
Marlex filled with 50 wt $\%$ Al_2O_3	0.196
Marlex filled with 60 wt $\%$ $Al_2O_3 \cdot 3H_2O$	0.302
Polystyrene	0.181
Epoxy[a]	
Unfilled	0.198
Filled with 50 $\%$ Al_2O_3 (Norton 38900)	0.250
Filled with 60 $\%$ $Al_2O_3 \cdot 3H_2O$ (Alcoa C-333)	0.408
Polyvinyl alcohol (Du Pont Elvanol 70-05)	0.225
Polyvinyl fluoride (Du Pont Tedlar)	0.226
Chlorinated polyether (Hercules Penton) $[-CH_2C(CH_2Cl)_2CH_2O-]$	0.232
Polycarbonate (General Electric various clear Lexans)	0.26–0.28
Polyphenylene oxide (General Electric) $[-C_6H_2(CH_3)_2O-]$	0.28–0.29
Silicone rubber (Silica-filled General Electric SE9029)	0.30
Polyvinylidene fluoride (Kynar)	0.437
Polyvinyl chloride (Geon-101)	0.47
Polyvinylidene chloride (Dow, Saran)	0.60
Polytetrafluoroethylene (Du Pont, Teflon)	0.95
Loosely woven fabrics	
Cotton	$0.185 \pm .005$
Wool	0.238
Synthetic 60/40 = vinyl chloride/acrylonitrile (Dynel)	0.298

[a] Epoxy was 100 parts Epon-826, 75 parts hexahydrophthalic anhydride, 1 part DMP30, cured 16 hr at 400 K.

means $n = 0.20$–0.26 and "self-extinguishing" means $n \simeq 0.27$ or higher on the oxygen index scale.

2.1. Effect of Some Variables on the Oxygen Index

The indexes of polymers are independent of the gas velocity supplied to the burning tube in the range 3–12 cm/sec, possibly because convection currents generated by the flame in the neighborhood of the specimens always exceed 3–12 cm/sec anyway.

For polyethylene, polymethylmethacrylate, polystyrene, and possibly for most other unfilled thermoplastic polymers, the indexes are independent of sample thickness in the range 0.2–1.0 cm and decrease only slightly for thicknesses less than 0.2 cm. Polycarbonate sheet less than 0.2 cm thick has a pronounced dependence of the index on about the 0.13 power of thickness (see Fig. 4), and polytetrafluoroethylene rods have a similar

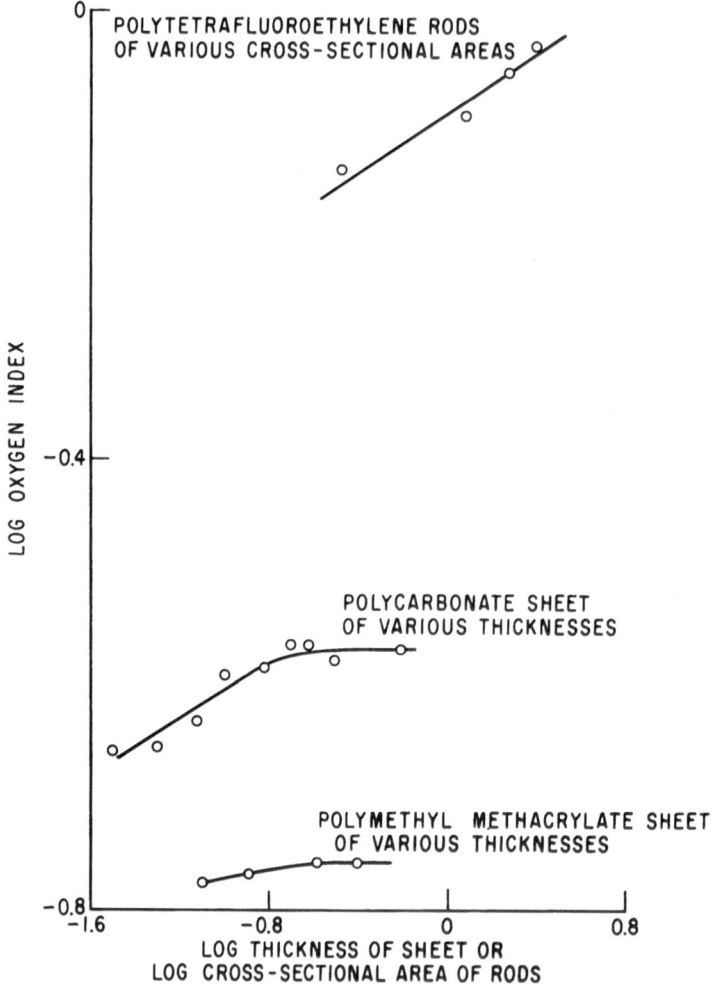

FIGURE 4. Oxygen indexes for polytetrafluoroethylene rods,[10] for polycarbonate sheets (Lexan; data from K. B. Goldblum, personal communication), or for polymethyl methacrylate sheets.[11] Sheets were 2.5 × 10-cm strips of varying thickness. Specimens burned in O_2–N_2 initially at room temperature except for polytetrafluoroethylene which was preheated to 400°K and burned in O_2–Ar initially at 400°K.

dependence of the index on the cross-sectional area even for large specimens. It is possible that a more marked dependence of index on sample size is associated with a heterogeneous mode of burning; it will be shown in the next section that polytetrafluoroethylene burning probably involves a heterogeneous attack of gas-phase species on the surface of the condensed phase, but polyethylene or polymethylmethacrylate burning probably does not.

Whatever the size effect, it does not prevent measuring reproducible values for quite thin specimens. Martin[11] obtained reproducible results on fabrics by clamping strips of cloth, 15 × 2.5 cm, in a light metal U frame which held the long edges of the specimen. After igniting the fabric at the open top of the U frame, he found the atmosphere in which the cloth just burned except along the edges where the metal frame quenched it. A few of his results are given at the end of Table 2.

By jacketing the apparatus of Fig. 2 and precooling or preheating the gas and polymer sample, one can determine the effect of initial temperature on the index. Some results in Fig. 5 show a strong dependence for polytetrafluoroethylene, a moderate dependence for the filled resin, and a weaker dependence for other materials.

By working with a closed apparatus, one can vary the pressure. Figure 6 shows that the index for polytetrafluoroethylene depends inversely on about the 1/2 power of pressure. Polyoxymethylene, polyethylene, and polymethylmethacrylate have a much weaker dependence, the variation for the last of these being shown at the bottom of the fifure.

One can use other inert gases than nitrogen to dilute the oxygen, and the index then depends on the inert gas used. Martin[12] investigated this dependence, with results given in Table 3. The order of inhibition of the different diluents is the order usually observed for flammability limits in gas mixtures where it is interpreted on thermal grounds, and Martin discussed his data in similar terms as indicating that some minimum critical temperature must be attained by the gas flame for the polymer to burn.

The atmosphere in which the polymers burn might also be changed

TABLE 3

Oxygen Indexes, $O_2/(O_2 + \text{Inert Gas})$, for Various Inert Gases

	Ar	N_2	CO_2	He
Polyoxymethylene	0.121	0.157	0.217	
Polymethylmethacrylate	0.135	0.173	0.253	0.191
Polyethylene	0.133	0.175	0.253	0.186
Chlorinated polyethylene				
$C_2H_{3\,8}Cl_{0.2}$	0.169	0.211	0.276	0.231
Polycarbonate (Lexan)	0.272	0.280	0.348	0.286

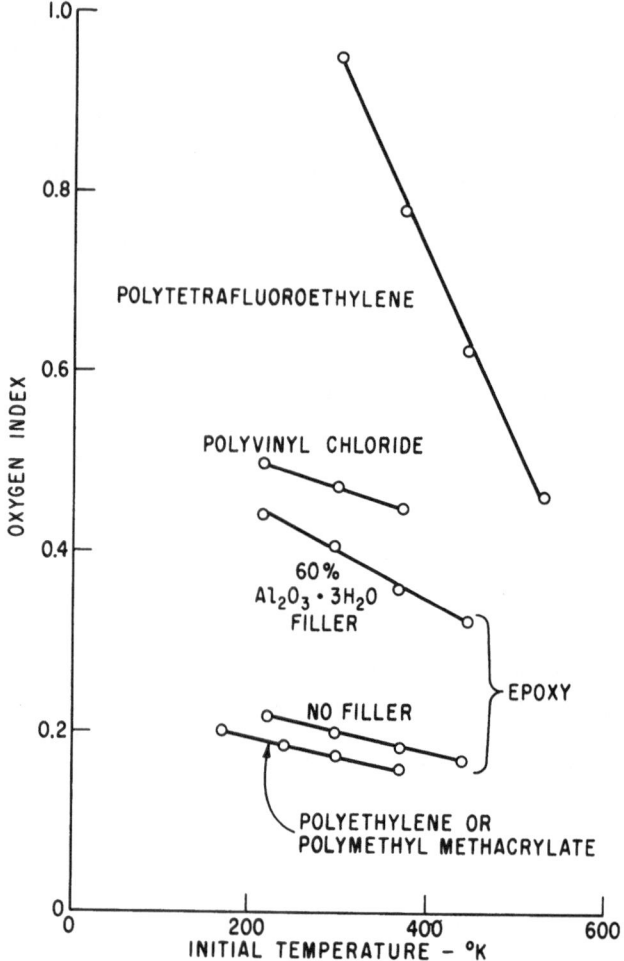

FIGURE 5. Effect of initial temperature on the oxygen index of various materials.

by substituting another oxidant for oxygen. This has been done but a discussion of the change is deferred.

3. A Candle Model of a Burning Polymer

When a candle burns, the fuel vaporizes from the wick in the heat of the surrounding diffusion flame and continually feeds the flame with fresh hydrocarbon. About 2–3% of the total heat of combustion is required to

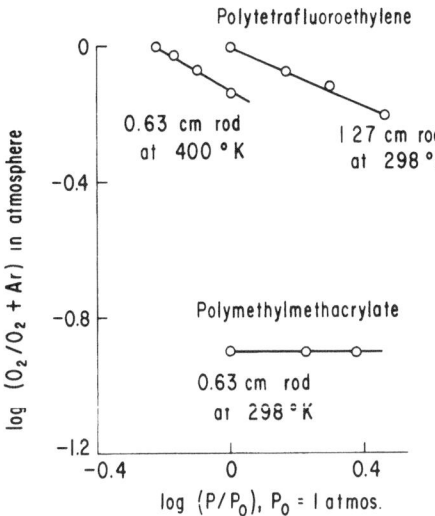

FIGURE 6. Effect of pressure on the oxygen index in O_2–Ar atmospheres.

vaporize the fuel. Except that it has no wick, the same model might serve for polymers, although the gasification of polymers requires a larger fraction of the total heat of combustion.

One would like to test the candle model with polymers for which the gasification process could be taken as known. To select such materials, one turns to Madorsky's book,[13] which reviews much of the experimental data on the pyrolysis of polymers, although most of the work was at temperatures lower than the surface temperatures of burning polymers. On the basis of his book the following hypothesis is assumed: if polymethylmethacrylate, polyoxymethylene, or polytetrafluoroethylene gasify merely by heating when they burn, the primary gaseous product will be the monomer in each case. On this hypothesis one can test the validity of the candle-burning model for the three polymers by asking whether large concentrations of monomer are evolved from the decomposing surface and whether a heat flow of the correct order is observed from the flame to the condensed phase. The heat flow is expected to be

$$\int_{T_i}^{T_s} C \, dt + L_{vap}$$

where C is the heat capacity of the polymer, T_i and T_s are the initial and surface temperatures, and L_{vap} is the heat of depolymerization to gaseous monomer at T_s.

The heat capacities of polymers are usually not known at temperatures in the vicinity of the burning surface temperature. It is easier and, thanks to Hess' law, equivalent to compute the expected heat flow as the sum of the heat of gasification at T_i and the estimated heat required to warm the gaseous monomer from T_i to T_s.

Experiments to test the candle model have been carried out.[10,14] Vertical rods were burned from the top downwards, and the gases above the burning tip sampled through a microprobe. The rods were mounted on a shaft which could be racked up, as the specimens burned, to maintain any desired distance between the probe and the burning tip. Fine thermocouples were usually used to measure the gas temperature above the tip, and the measurements were corrected for radiation losses. Sometimes thermocouples were embedded in the condensed phase instead, and the polymer allowed to burn back over them. In either case the object was to measure the surface temperature and the thermal gradient in the gas just above the surface.

Figure 7 shows some results for polymethylmethacrylate. There is little doubt that this material gasified mainly by depolymerization; the observed mole fraction of 0.11 monomer at a distance 0.1 cm above the burning tip accounts for 70% of the carbon in the gas at this point, and even more monomer is present closer to the surface.

The heat flow to the condensed phase was estimated as follows. The temperature gradient just above the tip was $\simeq 2500°$K. If the thermal conductivity of the gas was $\sim 2 \times 10^{-4}$ cal/cm sec °K, the local heat flux due to thermal conduction was $\sim 2500 \times 2 \times 10^{-4} = 0.5$ cal/cm^2 sec. If a flux of the same order occurred over the entire burning surface (approximately a hemisphere of 0.3 cm radius), the heat conducted to the condensed phase, of known burning rate, was of order 30 kcal/monomer unit burned.

The expected heat required is 21.4 kcal/monomer unit to gasify the condensed phase,[15] plus ~ 17 kcal to warm the gas to the observed surface temperature of 850°K. Thus, ~ 38 kcal is expected as compared to ~ 30 kcal found; the discrepancy is within the errors of the comparison. The fraction of the total heat of combustion of the polymer which is conducted back into the condensed phase is ~ 0.05.

Table 4 summarizes the results just given and includes similar data for other polymers. In all cases, the observed heat conducted to the polymer was based on the assumption that the local heat flux was 2×10^{-4} of the observed temperature gradient in the gas just above the condensed phase. This constant factor was considered a reasonable value for the thermal conductivity of the gas at the temperatures encountered. The burning rates were corrected for the fraction of the polymer which dripped — except in the case of polytetrafluoroethylene which did not drip.

Polyoxymethylene also exhibited agreement between the heat flows observed and calculated. The fraction of carbon recovered as monomeric formaldehyde from the gas was less than the fraction of monomer recovered above polymethylmethacrylate, but formaldehyde itself doubtlessly reacts

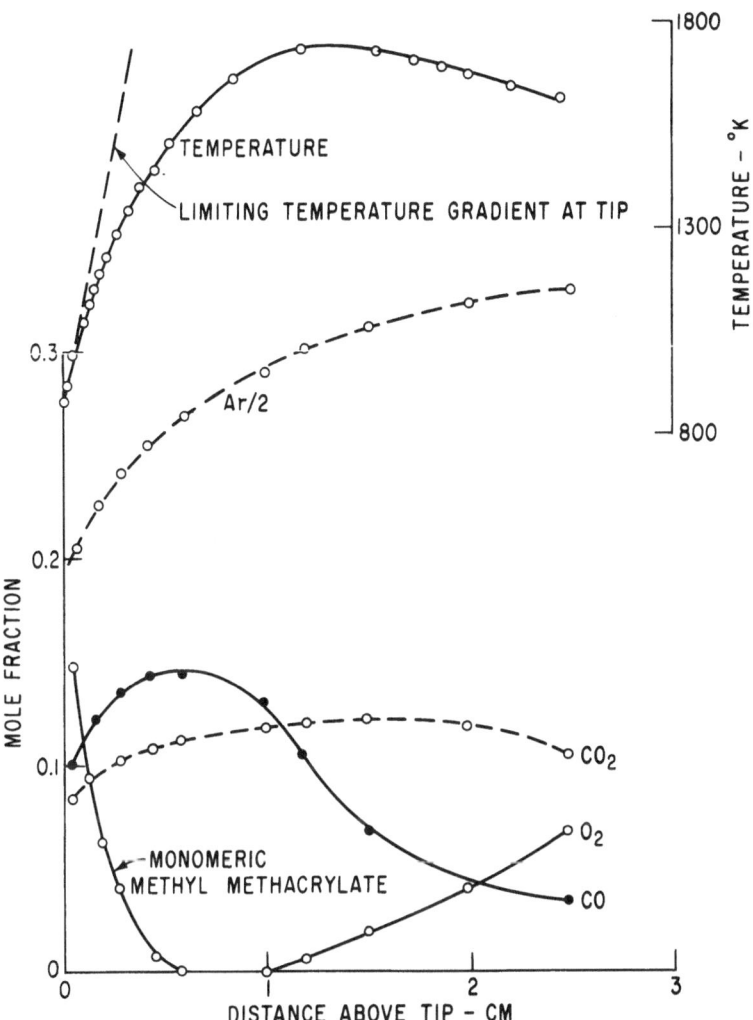

FIGURE 7. Gas composition and temperature on the central axis above burning polymethylmethacrylate. Polymer was 0.36 cm² cross section and burned at 5 cm Hg pressure in $O_2/(O_2 + Ar) = 0.20$. Burning rate (corrected for fraction which dripped without burning) 2.6×10^{-3} cm/sec. Limiting temperature gradient $\simeq 2500$ K/cm. Gas constituents not shown include H_2 (max. mol fraction $= 0.06$) and C_2 hydrocarbons (max. mol fraction $= 0.03$). CH_4 and C_3 hydrocarbons less than C_2.

TABLE 4

Heat Flow from Flame to the Condensed Phase for Polymers Burning in Ar–O$_2$ Atmospheres[a]

Polymer (cross section of rod)	Surface temp., K	Fraction of carbon as monomer, 0.1 cm above tip	Burning rate, cm/sec × 10³	Heat conducted to polymer, kcal/monomer unit		Observed heat conducted as fraction of total heat of combustion
				Obs.	Calc[b]	
Methylmethacrylate (0.36 cm²)	850	70	2.6	30	38	0.05
Formaldehyde (0.36 cm²)	750	33	2.3	15	17	0.14
Tetrafluoroethylene						
11.3 cm²	920	21	5		50	
4.9 cm²	920	13	7		50	
2.8 cm²	920	5	7	13	50	0.12
1.2 cm²	920	~0	8		50	
Ethylene (0.36 cm²)	900	~30	1.8	20		0.07

[a] Burned at 0.07 atm P in 20–22% O$_2$, except for polytetrafluoroethylene which was burned at 1 atm in 88–100% O$_2$ after being preheated to 400 K.

[b] Heat of gasification to monomer (21.4 kcal for methylmethacrylate,[15] 13 for formaldehyde,[16] 41.1 for tetrafluoroethylene[17]), plus estimated heat to warm to surface temperature gives this column.

and diffuses more readily than gaseous methylmethacrylate, and it is reasonable to conclude that polyoxymethylene fits the proposed model.

Polytetrafluoroethylene does not fit the candle-burning model because the observed conduction of heat from the gas into the condensed phase was only about 1/4 of the 50 kcal/monomer unit required to warm and depolymerize the polymer. Figure 8 shows the appearance of a steadily burning rod of 2.8 cm² cross section as measured from a photograph. The inner cone at 600 K is the known transition from crystalline polymer to a clear rigid liquid, and the measurement of the correct temperature proves good thermal contact between the polymer and the thermocouples embedded in it. The indicated temperatures are only a few point values from the continuous trace of a recording potentiometer as the polymer burned back over the thermocouple. The temperature gradient in the gas normal to the decomposing surface was not more than 3500 K/cm either at the tip or at a point 0.65 cm off the axis of the rod, and the heat flow by conduction worked out to be not more than ~13 kcal/monomer unit burned. The heat flow which could be reasonably attributed to radiation was far less than that resulting from conduction.

Figure 9 shows the distribution of carbon in gas samples collected, through corroding probes of quartz or alumina, above burning poly-

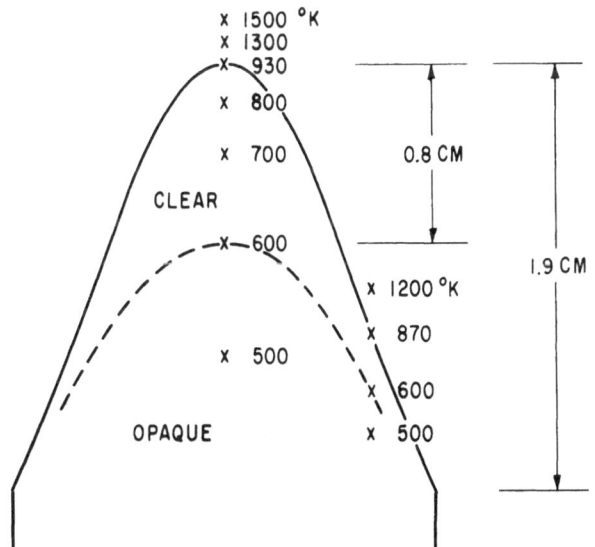

FIGURE 8. Appearance of 1.9-cm-diam polytetrafluoroethylene rod burning in 93% O₂, 7% Ar at 1 atm pressure. Polymer and oxidant preheated to 400 K. Temperatures are local values obtained as the polymer burned back over thermocouples embedded in the rod.

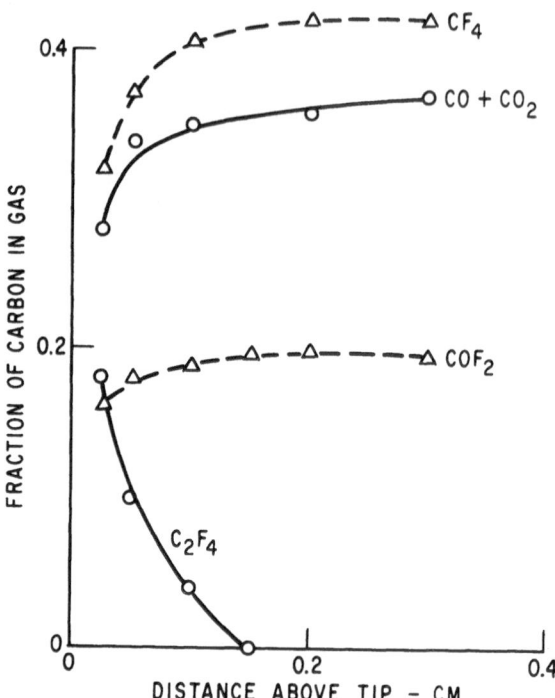

FIGURE 9. Fraction of carbon recovered in various forms above the tip of the rod described in Fig. 8. A little C_2F_6 and C_3F_6 were also found but are not indicated.

tetrafluoroethylene. The distribution depended on the size of the rod; molecular oxygen penetrated all the way to the burning tip of small rods and little, if any, monomer was then observed. Oxygen did not penetrate to the tip of larger rods, and the fraction of monomer, as indicated in Table 4, was larger, the larger the rod.

In view of the small heat flow, the decomposition of burning polytetrafluoroethylene probably involves heterogeneous reactions. However, it is not obvious what gas-phase species reacts at the surface; O_2 does not penetrate to the tip of the larger rods, so in some cases at least the heterogeneous reaction need not involve O_2. It seems likely that flame-generated species other than O_2 or O atoms attack the surface and aid its decomposition. Fluorine atoms are likely candidates for this active species; calculations of the equilibrium adiabatic products of polytetrafluoroethylene burning in O_2 indicate that gas having the distribution of carbon shown by Fig. 9 among carbon oxides, COF_2, and CF_4 (and containing neither O_2 nor O) would contain 6–10% of free F atoms. The presence of at least some F atoms in the hot gas was proven spectroscopically, and it was suggested that a diffusive flux of reactive F atoms from the flame to the decomposing surface might be more important than the thermal flux of heat in decomposing the condensed phase.

In summary, the candle-burning model seems applicable to two of the three polymers investigated. The right-hand column of Table 4 shows that the heat flow by conduction from the flame back into the condensed phase is ~ 0.05–0.14 of the heat of combustion of the polymers. When a fraction of this order is sufficient to gasify the polymer completely, the candle-burning model has some validity. When 0.10 ± 0.05 of the heat of combustion of the polymer is far too little to gasify the condensed phase, as with polytetrafluoroethylene, the model fails.

4. Polyethylene Burning and a Possible Limitation of the Candle Model

Polyethylene, the last entry in Table 4, cannot be checked against the model in the same way as the other entries because the primary pyrolysis products are uncertain. Molecular oxygen was absent from gas sampled just above the burning tip, however, and if no flame-generated species attacked the surface, the polymer must have gasified by pure pyrolysis.

Under somewhat different experimental conditions, Burge and Tipper[18] reached an opposite conclusion. They placed a loosely fitting glass mantle, with a 1.5-cm-diameter hole in its end, over the top of a vertical 2.5-cm-diameter polyethylene rod, and ignited the polymer through the hole. The upper 1–2 cm of polymer melted inside the mantle, and part of the melt was forced through the hole and burned in the air without overflowing. Their surface temperature was $\simeq 670°K$ (as compared to $\simeq 900°K$ in Table 4). They found 1.1% O_2 in gas samples collected as close to the surface as possible (as compared to $\simeq$ zero O_2 in the work summarized in Table 4). They concluded that heterogeneous, oxidative degradation occurred in their surface layers, that is, that the candle-burning model did not apply.

Different experimental conditions might possibly account for the apparent contradiction. By capping their rod, Burge and Tipper must have quenched the reactions to a greater extent than was the case at the tip of a rod burning freely without a glass mantle. An oxidative degradation could be important at surface temperatures so low that a purely pyrolytic decomposition was unimportant. However, if additional quenching can alter the mode of decomposition of the condensed phase, one wonders what kind of burning occurs on even a freely burning rod in regions where the leading edge of the hot gases encounters previously unexposed polymer. In such regions, the quenching may also be greater than at the burning tip.

It is possible that the decomposition of the condensed phase may differ on different parts of the surface, and that the candle model of burning is an oversimplification. Nevertheless, we use the model to discuss inhibition of flammability in the next few sections.

5. Modes of Inhibiting the Flammability of Candle-Burning Polymers

The gasification of a candle-burning polymer requires a sufficient flux of heat from the flame to the condensed phase. If the condensed phase could be modified so that a greater heat flux was required, the polymer would appear less flammable. The polymer would also be less flammable if the flame reactions could be poisoned and the heat flux to the condensed phase decreased. In either case a higher value of the oxygen index would be required for the inhibited than for the uninhibited polymer to burn.

The two modes of inhibition ought to possess different characteristics. An inhibitor which modifies the condensed phase might work well in only certain classes of substances, its action depending critically on the nature of the polymer. Phosphorus compounds exhibit such a selective inhibition; Table 5 shows only a mild inhibition of polyethylene but a marked inhibition of an epoxy resin.

TABLE 5

Inhibition of Polyethylene and Epoxy by Phosphorus
Compounds, Polymers Burned in N_2,O_2 Atmospheres[a]

Polymer	Additive	Atom ratio in plastic, ρ/C	η
Polyethylene	———	0	0.175
	Tricresyl phosphate	0.0013	0.182
		0.005	0.192
		0.010	0.200
	Triphenyl phosphine	0.010	0.193
	Phosphonitritic chloride	0.010	0.200
Epoxy	———	0	0.196
	Triphenyl phosphine	0.0014	0.241
		0.0041	0.305
		0.0065	0.330

[a] Polyethylene data from Ref. 6, epoxy from Ref. 19.

On the other hand, an inhibitor which poisons the flame ought to work about equally well in polymers of different types because the gas flames are often similar. In fact, a flame poison should work well even if the gaseous fuel is supplied from some other than a pyrolyzing polymer. Bromine is an effective flame poison, and it appears to work equally well in *n*-hexane, oxygen flames, in polyethylene, or in epoxy, as will now be shown.

In the first section of this chapter it was shown that the addition of bromine to *n*-hexane to give an atom ratio of $Br/C = 0.17$ increases the index by about 60%, from $n \simeq 0.13$ to $n \simeq 0.21$. In Fig. 10, this result is compared with the percentage increase in the index of polyethylene when it is inhibited by added tetrabromobenzene or tetrabromobisphenol A.[6] A point is also plotted for an epoxy polymer, originally of index $n = 0.196$, which was altered by making it partly from a tetrabromobisphenol A resin.[19] The effect in the different systems is comparable, presumably because bromine works mainly by poisoning the flames and the flames themselves are quite similar.

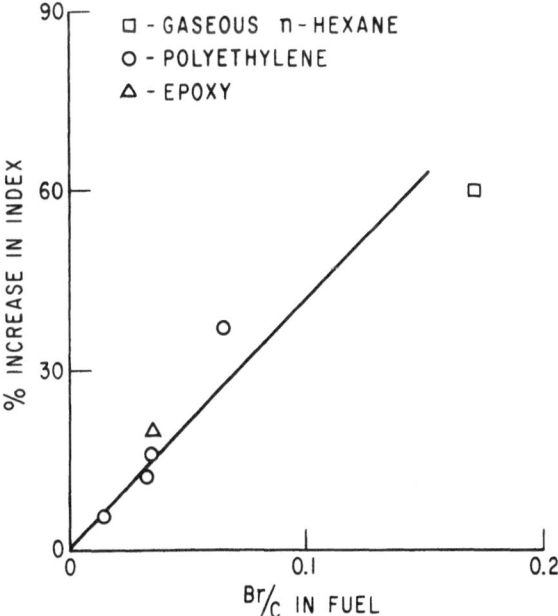

FIGURE 10. Percentage increase in oxygen index as a function of atom ratio in fuel for a gaseous hydrocarbon or two polymers.

6. Inhibition by Antimony

Figure 11 shows that antimony trioxide milled into polyethylene does not affect the critical oxygen index substantially, but that the index of partly chlorinated polyethylene or of polyethylene containing bromine is markedly increased by antimony.[6] Similar effects are observed when antimony is added as triphenylstibine rather than as the oxide. The inhibition only occurs when the polymers are burned in O_2–N_2 atmospheres; in N_2O–N_2 atmospheres, the index of chlorinated polyethylene of composition $C_2H_{3.94}Cl_{0.06}$ is $N_2O/(N_2 + N_2O) = 0.40$, whether Sb_2O_3 is added or not.[14] Presumably, therefore, antimony works in O_2–N_2 atmospheres because it poisons the flame with oxygen.

The role of halogen is indicated by Table 6, which lists the results of analyzing some residues from burning the compositions shown in Fig. 11. A comparison of the last two columns of the table shows that antimony is only effective when it vaporizes, and even then only if the burning is in oxygen atmospheres. Evidently halogen aids the vaporization.

It should be added that the index of polyethylene containing both halogen and phosphorus, added as polymeric $PNCl_2$ or as trisdichloropropyl phosphate, is not increased by added Sb_2O_3. Nor is the antimony vaporized

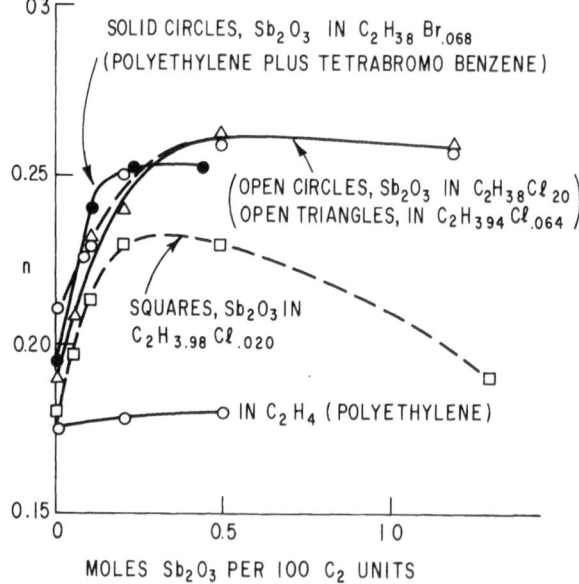

FIGURE 11. Oxygen indexes as a function of added Sb_2O_3 in polyethylene containing chlorine or bromine.

TABLE 6
Fraction of Antimony Which Vaporized from Burning
Polymers Containing Sb_2O_3

Base polymer	Sb C ratio. from added Sb_2O_3	n^a	Fraction of Sb which vaporized in burning
C_2H_4	0	0.175	
C_2H_4	0.005	0.180	~ 5
$C_2H_{3.98}Cl_{0.02}$	0.013	0.190	~ 10
$C_2H_{3.98}Cl_{0.02}$	0.005	0.229	31
$C_2H_{3.94}Cl_{0.06}$	0.005	0.261	75
$C_2H_{3.94}Cl_{0.06}$	0	0.40	
$C_2H_{3.94}Cl_{0.06}$	0.005	0.40	75

^{a} First 5 indexes are ratios of O_2 $(N_2 + O_2)$. the last two $(n = 0.40)$ are ratios of N_2O $(N_2 + N_2O)$ at which polymers extinguish in N_2O N_2 atmospheres

when the compositions burn. Phosphorus kills the antimony–halogen synergism in polyethylene by preventing the vaporization of antimony.

Martin and Price[19] compared the inhibition of epoxy by phosphorus and by antimony, both added as the triphenyl compound. Their results (see Fig. 12), indicate that small additions of antimony are more effective than phosphorus when the epoxy burns in O_2 but considerably less so when the oxidant is N_2O. If phosphorus inhibits by working in the condensed phase, as suggested in the last section, it might inhibit equally well when the polymer burns in any oxidant. But it seems likely that antimony mainly poisons the flame with oxygen. The main difference between the action of antimony in epoxy and polyethylene is that halogen is unnecessary in epoxy for the inhibition to occur. An analysis for adventitious halogen gave only 0.15% chlorine by weight in the epoxy, and this is too little to matter.

The implication is that antimony added as triphenylstibine vaporizes from burning epoxy without the aid of halogen but not from burning polyethylene. Martin and Price speculated how this might be possible, but have not tested their suggestion.

7. Inhibition by Chlorine

Figure 13 shows the increase of oxygen index, or of the analogous nitrous oxide index, when chlorine is substituted in polyethylene. The inhibition is effective in both oxidants and, therefore, probably does not involve poisoning the flame with oxygen. That flame poisoning is not the

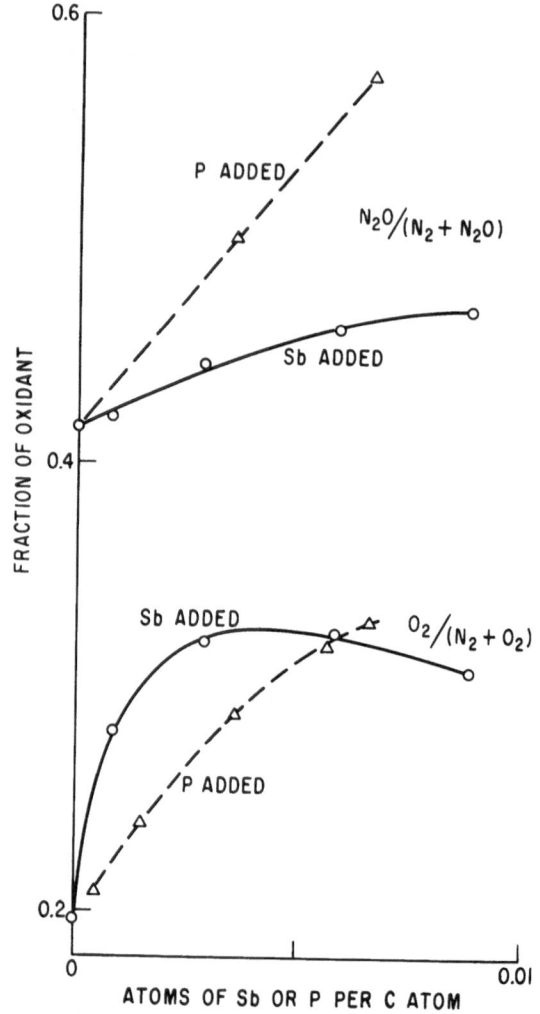

FIGURE 12. Effect of added $Sb(C_6H_5)_3$ or of $P(C_6H_5)_3$ on the flammability of an epoxy resin burning in N_2–O_2 or in N_2–N_2O atmospheres.

major effect is also indicated by the slight increase in the oxygen index of polyethylene when the N_2–O_2 atmosphere is loaded with Cl_2 or HCl. Compare the bottom two curves in Fig. 13.

As might be expected for an inhibitor which works in the condensed phase, the effect of chlorine differs in different polymers. Polyethylene chlorinated to a ratio of Cl/C = 0.4 has an index of $n \sim 0.39$ according to Fig. 13, but Table 2 shows that the same Cl/C ratio in the chlorinated polyether, penton, gives an n of only 0.232.

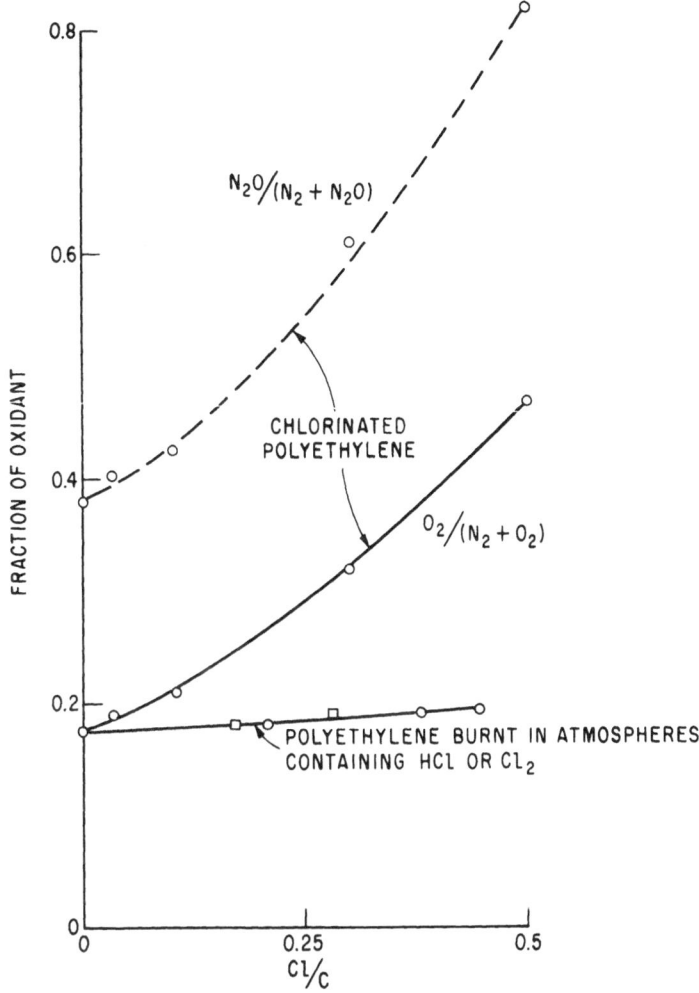

FIGURE 13. Effect of chlorine. Top two curves give fraction of oxidant required to burn chlorinated polyethylene or polyvinyl chloride in N_2O–N_2 and O_2–N_2 atmospheres. Bottom curve gives ratio of $O_2/(O_2 + N_2)$ required to burn unchlorinated polyethylene in O_2, HCl. N_2, or O_2, Cl_2, N_2 atmospheres.

The vacuum pyrolysis of polyvinyl chloride has been studied at lower temperatures. Stromberg and coworkers[20] believe that HCl, the main product at temperatures below about 570°K, is evolved in a radical chain process which produces polyenes and leads to rapid cross-linking of the polymer system. A little benzene is consistently evolved, and this is considered to be a secondary product of cyclization at the ends of the conjugated chains, produced in the dehydrochlorination. At higher temperatures, vari-

TABLE 7
Fraction of Carbon Found as Various Species in Gas Samples Collected 0.1 cm
above Polymers Burning at 15 cm Hg Pressure[a]

	Carbon oxides	$CH_4 +$ C_2H_6	Unsaturates				
			C_2H_2	C_2H_4	C_3	C_4	Benzene
Polyethylene	68	3	6	14	5	3	1
Chlorinated to							
$C_2H_{3.8}Cl_{0.2}$	69	2	9	8	4	4	4
Chlorinated to							
$C_2H_{3.4}Cl_{0.6}$	70	2	10	3	2	3	10
Penton $C_5H_8Cl_2O$	77	2	3	2	3	13	None

[a] First two runs burned in $Ar + 0.25O_2$, last two in $Ar + 2O_2$. Last two formed considerable char and were difficult to sample. Penton, but not polyethylenes, gave chlorinated hydrocarbons; most of the C-4 carbon species found above Penton were C_4H_7Cl.

ous hydrocarbons, including aromatics, are produced and a residue of char remains.

One expects that the same processes occur in the condensed phase when chlorinated polyethylenes burn. Samples of the flame gases collected just above the decomposing surface[10] contain no molecular oxygen, no chlorinated hydrocarbon, but more benzene and acetylene the greater the degree of chlorination of the polyethylene (see Table 7). If considerable cross-linking occurs in the chlorinated polyethylenes, the gasification ought to become more difficult and the flammability less, the greater the degree of chlorination.

The chlorinated polymers form much char when they burn and one might think that this would insulate the undecomposed polymers from the flame. The importance of such a barrier in increasing the oxygen index is uncertain. It is not necessarily very important because penton and polyvinyl alcohol, which also form considerable char, are much more flammable than chlorinated polyethylene.

8. Inhibition by Dripping

A polymer which melts and runs easily abstracts more heat from the flame than is required to gasify the fraction of the condensed phase which burns. The additional heat abstraction can make the polymer appear less flammable than if it did not melt so easily. For example, a polypropylene which did not exhibit excessive melting (Hercules Profax 6505) had an oxygen index of 0.174, but a low-molecular-weight polypropylene (Epolene

N15) had $n \sim 0.23$. The latter burned with a light blue flame which rode down the specimen on a wave of molten polymer, leaving the center of the specimen unconsumed. Its high index reflected the extra heat loss resulting from easy melting and dripping because the same material gave only $n = 0.177 \pm 0.005$ when it was cast about a ceramic wick and burned off the wick as a candle.[6]

Another example of inhibition by dripping was discussed by Hindersinn.[8] The inhibition of polystyrene by added bromine compounds, which must be due in part at least to flame poisoning by bromine, is further accentuated by including dicumyl peroxide or other free-radical initiators in the plastic.[21] Hindersinn explained this by showing that the peroxide degraded the hot polymer, gave a more fluid melt, and that more heat was then required to ignite the polymer or to keep it alight because more molten material dripped from the neighborhood of the flame. Dicumyl peroxide was ineffective when dripping was repressed.

His view has been supported by other experiments[9] which indicated that an added bromine compound inhibited polystyrene whether the polymer dripped or not, but that the additional effect conferred by dicumyl peroxide only appeared when dripping was permitted.

Excessive dripping makes a polymer look less flammable in the standard configuration used to determine the oxygen index. However, it is possible that in other configurations and with a different criterion for flammability, the same quality could make a material look more flammable. In some flammability tests, one ignites a polymer at the bottom and worries about the possibility of starting secondary fires by flaming drip falling onto cotton batting. Increased dripping might facilitate secondary fires and lead to a more flammable rating in such a test.

9. Inhibition by Fillers

The addition of a little filler, insufficient to leave a standing ash when the substance burns, usually imposes a cyclic character on the burning because the solid residue repeatedly accumulates and falls away from the burning specimen. With larger proportions of filler, a standing ash is possible and steady burning processes are observed. The candle-burning model may be irrelevant to materials containing much filler, however. There may be no important gas-phase reactions, all of the important reactions being condensed-phase or heterogeneous processes.

Figure 14 shows some results of Martin and Price[19] on the effect of fillers in epoxy. The observation that increasing amounts of anhydrous Al_2O_3, in the range 40–60% filler by weight, do not increase the oxygen

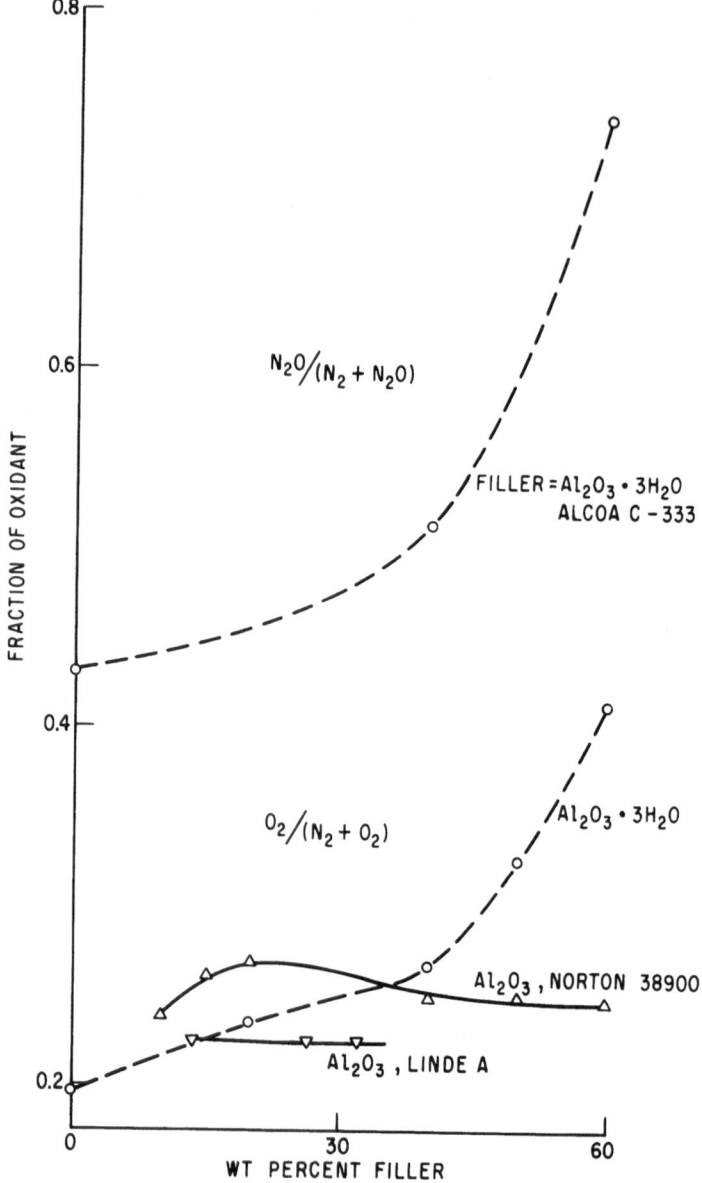

FIGURE 14. Fraction of oxidant required to burn filled epoxy resins as a function of content of filler. Composition of epoxy given at end of Table 2.

index certainly suggests that a thermal model is irrelevant. On any thermal model, an increased content of filler ought to inhibit the burning by soaking up heat and lowering the temperatures attained in the system.

A sufficiently large proportion of filler, and particularly of a filler which decomposes endothermally, does inhibit the burning. The entries in Table 2 for polyethylene and epoxy, and Fig. 14, show that 60% of $Al_2O_3 \cdot 3H_2O$ forces an increase in the index which is not observed with 50% of anhydrous Al_2O_3, although both compositions contain the same proportion of Al_2O_3 to plastic. The increase is presumably due mostly to the heat absorbed in the partial dehydration of $Al_2O_3 \cdot 3H_2O$. The similar effects in N_2O-N_2 as in O_2-N_2 atmospheres suggest that the additional inhibition need not involve gas-phase oxidation reactions with O_2, but there may be no important purely gas-phase reactions anyway.

ACKNOWLEDGMENT

The writer is grateful to F. J. Martin for many helpful discussions of the subject and for the photograph, Fig. 3.

10. References

1. H. F. Coward and G. W. Jones, Limits of Flammability of Gases and Vapors, Bulletin 503, U.S. Bureau of Mines (1952).
2. R. F. Simmons and H. G. Wolfhard, *Combust. Flame* **1**, 155 (1957).
3. A. R. Hall, J. C. McCoubrey, and H. G. Wolfhard, *Combust. Flame* **1**, 53 (1957).
4. C. E. Mellish and J. W. Linnett, Fourth Symposium on Combustion, William and Wilkins, (1953), p. 413.
5. G. Lask and H. G. Wagner, Eighth Symposium on Combustion, William and Wilkins, (1962), p. 432.
6. C. P. Fenimore and F. J. Martin, *Combust. Flame* **10**, 135 (1966).
7. J. W. Anderson and R. Friedman, *Rev. Sci. Instr.* **20**, 61 (1949).
8. R. R. Hindersinn, Polymer Conference Series, Wayne State University, Detroit (June 1966).
9. C. P. Fenimore, *Combust. Flame* **12**, 155 (1968).
10. C. P. Fenimore and G. W. Jones, *J. Appl. Polym. Sci.* **13**, 285 (1969).
11. F. J. Martin, General Electric Co. Report 68-C-422 (Nov. 1968).
12. F. J. Martin, *Combust. Flame* **12**, 125 (1968).
13. S. L. Madorsky, *Thermal Degradation of Organic Polymers*, Wiley-Interscience, New York (1964).
14. C. P. Fenimore and G. W. Jones, *Combust. Flame* **10**, 295 (1966).
15. K. J. Ivin, *Trans. For. Soc.* **51**, 1273 (1955).
16. J. F. Walker, *Formaldehyde*, Reinhold, New York (1951).
17. J. C. Siegle, L. T. Muus, Tung-Po Lin, and H. A. Larsen, *J. Polym. Sci.* **A2**, 391 (1964).
18. S. J. Burge and C. F. H. Tipper, *Chem. Ind* **1967**, 362.
19. F. J. Martin and K. R. Price, *J. Appl. Polym. Sci.* **12**, 143 (1968).
20. R. R. Stromberg, S. Straus, and B. G. Achhamer, *J. Polym. Sci.* **35**, 355 (1959).
21. J. Eichhorn, *J. Appl. Polym. Sci.* **8**, 2497 (1964).

10

Flame-Retardant Organic Coatings

Wen-Hsuan Chang, Roger L. Scriven,
and Ronald Blair Ross

1. Introduction

Many ways to decrease the flammability of polymers and coatings have
been discovered as a result of the research done on various mechanisms
of flame retardation. Currently, the most effective flame-retardant materials
contain halogens, phosphorus, various transition metal oxides, strong acids,
or some amine and metal salts. These materials can be used to produce
acceptable flame retardance in coatings when the substances are either
chemically combined with the film-forming polymer or used as additives.

In order to produce a coating with suitable performance properties,
many additives are included in a coating formulation to achieve specific
property modifications or improvements; however, those which are added
to produce suitable flame-retardance and performance properties may cause
some sacrifice in other coating properties. The factors that determine which
materials are included in a formulation depend upon the particular end
use and its requirements.

Some of the important physical properties and characteristics which

Wen-Hsuan Chang · Coatings and Resins Division, PPG Industries, Incorporated, *Roger L.*
Scriven · Coatings and Resins Division, PPG Industries, Incorporated, and *Ronald Blair Ross* ·
Coatings and Resins Division, PPG Industries, Incorporated, Allison Park, Pennsylvania.

must be considered by the formulator of flame retardant coatings include:

Flexibility	Stain resistance
Adhesion	Ease of application
Thermal stability	Tolerable solvents
Exterior durability	Hardness
Cleanability	Gloss
Scrub resistance	Color purity

and what is often the deciding factor—cost. Therefore, a coating must have acceptable performance properties at a low cost.

The most successful formulations are proprietary information and rarely published. Patent disclosures are the main source of new flame-retardant coating formulations, and this information is not only very limited, but also contains little or no data on the degree of flame retardance achieved. Some trade literature from manufacturers of flame-retardant chemicals or polymers provides sources of information or formulations, but most of such data is not commercially acceptable owing to various unspecified shortcomings.

This chapter includes general information on flame-retardant organic coatings and important current developments. Because of space limitations, no special effort has been made to review all recent publications on these coatings. This chapter is divided on the basis of primary types of flame retardant coatings: architectural, fabric, and heat-resistant coatings.

2. Architectural Coatings

Architectural coatings designate those used primarily for painting interiors of public buildings, homes, and marine vessels. They may also be used to a lesser degree in mobile-home and airplane interiors. Spurred by recognized social need, the use of flame-retardant coatings for such painting has and will be greatly expanded. For example, in 1974 the city of St. Louis passed legislation requiring flame-retardant coatings for interior painting in schools and public buildings. Other cities can be expected to do the same.

The demand for flame-retardant coatings has come as a result of building-material manufacturers becoming more and more aware that the flammability characteristics of their products may govern their acceptance under building codes. This has prompted coatings manufacturers to experiment with new approaches and applications of coatings in order for the coating-substrate composite to meet specific requirements.

What protection will flame-retardant architectural coatings provide to occupants of a building in the event of a fire? The degree of protection is largely dependent on the means used to achieve flame retardance. The most effective coatings are those that intumesce (i.e., bubble and foam) to form an insulating multicellular layer when subjected to high temperatures. This char protects the substrate from heat damage. Other types of coatings are nonintumescent and provide flame retardance by inhibiting the rate of flame spread. These are admittedly much less effective than intumescent coatings and provide little, if any, protection to the substrate. They are in demand primarily because they are less expensive, easier to apply, and usually have better resistance to soiling and wear than intumescent coatings.

Flame retardance must be achieved within the framework of other architectural considerations. These include adhesion, hardness, gloss, color purity, scrub and stain resistance, ease of application, and, of course, cost. A coating can not be said to be marketable unless it has acceptable performance at a favorable cost.

In order to achieve the desired degree of performance, many additives are included in a particular coating formulation, but these compounds sometimes interact with flame-retardant additives and pose additional problems to the coatings formulator because the results of such interaction can often not be predicted. For example, many chemical additives which reduce flammability cause larger quantities of smoke to be developed.

Some architectural coatings which are not expressly formulated for flame retardance may slightly reduce the flammability of the surfaces to which they are applied. For example, a highly reflective heat-resistant coating might delay flame spread as long as the temperature of the substrate does not reach ignition point. Other coatings may be termed flame retardant if they are antagonistic to flame spread, that is, if they do not enhance burning.

Research work on flame-retardant systems specifically designed for architectural applications must necessarily fall into two categories—first, that of understanding the chemical processes involved in combustion and secondly, the more difficult task of developing flame-retardant compositions.

The difficulty of discerning what goes on in combustion is apparent when one considers that 150 organic compounds[1] have been identified in chemical compounds of smoke from cellulose. Added to this is the fact that the actual mechanisms of burning are not fully understood. The manufacturer of organic coatings for architectural applications must be aware of the nature of the burning phenomenon, particularly of the composite coating and substrate, at the same time pressing for the realization that while flame-retardant coatings can contribute much to fire safety, they can play only a limited part in materials design.

Flame-Retardance Testing. Various laboratory testing procedures have been developed to evaluate the effectiveness of a coating in providing flame retardance.[1a] Unfortunately, there are no standard fire conditions, and any test can at best only approximate what a coating might do under a very narrow range of fire conditions. The test considered most acceptable to date is the ASTM E-84-70 Test for Surface Burning Characteristics of Building Materials.[2] This test measures the flammability of various materials under controlled laboratory conditions and is used for the evaluation of flame retardance of coatings and coating-substrate composites. However, ASTM has stated that this and other ASTM standards and methods should not be considered to describe the hazard of materials under actual fire conditions.[2a]

The test consists of a 25-ft-long horizontal flue with the sample to be tested forming the top. The sample size is 25 ft × 20 in. The chamber is sealed with an air-tight covering. In the test, a controlled gas-fed flame impinges on one end of the sample and ignites any combustibles present. This flame is applied for the entire 10-min test period, and the advance of the flame front is recorded during the test. In order to provide reproducible results, it is also necessary to carefully control the draft conditions.

ASTM E-84-70 measures the "flame spread" of various surfaces exposed to fire. The term "flame spread" is an arbitrary rating based on a control test sample of asbestos–cement board (0 rating) and a conditioned piece of red oak flooring (100 rating). This rating is often used in building codes to set specifications for flammability requirements. A flame-spread rating of 25 or less is frequently required for critical areas such as emergency exits in buildings used by large groups of people. A flame-spread rating of 75 or less is usually required for materials used in stairways, corridors, etc. Of course the flame-spread requirements may become more severe if public concern over fire safety of dwellings continues to increase.

The ASTM E-84-70 test also sets standards for smoke-developed and fuel-contributed factors. The smoke-developed factor is based on measurement of the smoke at the end of the tunnel and is monitored throughout the test with a special photoelectric cell. The smoke-developed factor is also based on red oak flooring as the standard and is assigned arbitrarily a 100 rating. The fuel-contributed factor is a measure of the amount of combustible materials the tested specimen provides, and it, too, uses red oak flooring as the standard and is given an arbitrary rating of 100.

A flame-retardant coating should not greatly contribute fuel to a fire as compared with the uncoated sample, nor should it contribute to the hazards of large amounts of smoke which may be generated during a fire. Increased concern over these hazards has resulted in more stringent requirements on building materials in these critical areas.

The ASTM E-84-70, often called the "tunnel test," is expensive (total costs may run as high as $1000 for one sample) and time consuming. As a result, many tests similar to the tunnel test have been devised to predict performance in the tunnel test. These consist of short tunnels, 2,[3] 4,[4] and 8-ft[5] in length, which are less expensive and more convenient for rapid screening of flame-retardant coating formulations. However, the results obtained may not always correlate directly with the result of the 25-ft tunnel test.

Another test that is useful in evaluating flame-retardant coatings is the oxygen index test.[6-8] The results of this test are expressed in terms of the limiting oxygen index (LOI). In this test, a sample of a film of a coating is suspended in a vertical position in the test chamber. The percent of oxygen in the test chamber is adjusted to permit ignition and combustion of the sample After ignition. the percent oxygen is reduced at a prespecified rate until the flame extinguishes. The minimum level of oxygen required for combustion is the LOI value expressed as the fraction of oxygen in an atmosphere of nitrogen.

Other tests have also been used to evaluate flame retardance under various conditions. The cabinet test (ASTM D-1360),[9] stick and wick test (ASTM D-1361),[10] and radiant panel test (ASTM E-162)[11] have been used extensively for coating evaluation instead of the ASTM E-84-70 test. Unfortunately there is usually no exact correlation of results of the different methods used to evaluate the flame retardance of a coating.

2.1. Nonintumescent Flame-Retardant Coatings

Nonintumescent coatings. usually applied 1–2 mil thick, are designed to function in the same capacity as other non-flame-retardant coatings. As such. these coatings must show comparable performance properties such as stain resistance, washability, and ease of application at a comparable cost.

Nonintumescent coatings prove particularly useful in marine applications. This constitutes one of the largest uses of such coatings because ships receive repeated paintings in an effort to provide corrosion protection. As the layers of paint build, they pose a fire hazard even though the substrate is steel. In the event of fire, the paint may catch fire, melt, drip. and cause severe injuries and increased damage to the vessel. To prevent this. coatings were formulated which would not sustain combustion, even when several layers were exposed to fire.

Nonintumescent coatings do not provide the same degree of flame protection to the substrate as intumescent coatings: nevertheless, they should not enhance the spread of flame by rapid combustion nor contribute a significant amount of fuel to the fire.

It should be pointed out that non-flame-retardant coatings usually give low flame-spread ratings (except for highly flammable coatings such as nitrocellulose) over asbestos–cement board, steel, or cement block. When the coatings are tested over wood and other flammable materials, flame-spread ratings are similar to those of the substrate. As shown in Table 1, the three conventional non-flame-retardant coatings tested had similar flame-spread ratings when applied on Douglas fir planks and on asbestos boards. All three coatings showed an increase in fuel contribution factors over Douglas fir, but no increase over asbestos–cement board. There is also no increase of flame spread when the coatings are applied on the two substrates. The lowered smoke-contribution factors in Table 1 are probably a result of noncombustible inorganic fillers and pigments restricting the flow of gases from the burning substrate.

The nonintumescent, flame-retardant coatings showing the largest sales are based on chlorinated alkyds. As with most alkyd coatings, the vehicle commonly used is based on an unsaturated fatty acid or oil which is reacted with a polyol (compound containing two or more OH groups) and a polyacid (compound containing two or more COOH groups) or anhydride. When the coating is applied, the unsaturated sites react with the oxygen in the air and "dry" or cure to form a hard, cross-linked coating.

Catalysts (driers) which accelerate the reaction with oxygen are added to hasten drying of the coating. Various fillers and pigments are added to lower cost and improve appearance of the coating. Introduction of chlorine into an alkyd resin results in a flame-retardant coating.

Very little recent work has been done toward preparing new types of

TABLE 1

Regular (Non-Flame-Retardant) Coatings in the E-84 Tunnel Test[12]

Test specimen	Flame-spread rate	Fuel contributed factor	Smoke density factor
Without coating			
Asbestos–cement board	0	0	0
Red oak flooring	100	100	100
Douglas Fir	75	45	75
Coating on Douglas Fir			
Semigloss enamel (26-2)	70	60	30
Semigloss latex enamel (26-3)	75	60	65
Flat Latex (26-4)	70	60	30
Coatings on asbestos–cement board			
Semigloss enamel (26-2)	0	0	0
Semigloss latex enamel (26-3)	0	0	0
Flat latex (26-4)	0	0	0

flame-retardant alkyds. Cost is a predominant factor in these relatively inexpensive coatings, and any new coating would have to provide similar performance properties at a lower price.

The most common method of preparing flame-retardant alkyd coatings is to use chlorinated diacids or anhydrides such as chlorendic anhydride,[13] tetrachlorophthalic anhydride, or Chloran.[14] These materials can be used in standard alkyd preparation methods, but temperatures in excess of 200°C should be avoided during preparation or dark-colored products may result. A typical flame-retardant alkyd resin[15] may be prepared from chlorendic anhydride, 0.740 mol; glycerine, 0.835 mol; and soybean fatty acid, 0.917 mol.

By using the proper chlorinated acids, coatings can be made which have properties comparable to those of conventional coatings in addition to being flame retardant. It is this high performance at a relatively low cost which has made chlorinated alkyd coatings so successful.

The use of chlorinated anhydrides as the only source of halogen limits the amount that can be added. There are some halogen- and phosphorus-containing polyols available, such as 2,2-bis(bromoethyl)-1,3-propanediol, dibromoneopentyl glycol, polyhalogenated bisphenol polyols, and alkylene oxide adducts of phosphoric acid, which can also be used to improve flame retardance. However, these compounds are either too expensive or they detract from physical properties, thus preventing any large usage. As a result, the addition of fire-retardant additives to alkyd resins is the preferred means of increasing flame retardance of alkyd coatings.

The use of fire-retardant additives in coatings has been recently reviewed by Lyons,[16] Thiery,[17] and others.[18] Not all the flame-retardant additives are suitable for inclusion in alkyd formulations because many of them are not compatible with the vehicle or substantially reduce physical properties of the coatings. In general, phosphorus- and halogen-containing flame-retardant additives enhance conventional coating properties only when they act as plasticizers.

Halogenated materials such as chlorinated paraffins have been used extensively because of their low cost and minimal effect on regular properties of the coatings. These chlorinated paraffins by themselves do not greatly improve flame retardance—they require metal oxides to facilitate decomposition. The difference[19] in performance can be illustrated by considering that a chlorine content of over 40% is needed to make an alkyd film self-extinguishing (LOI = 21), while only 20% chlorine is needed if 7% antimony oxide is present.[20] This is the result of synergism, a cooperative action between two compounds such that together the total effect is greater than either used singly. Antimony oxide without halogenated compound also does not improve flame retardance greatly.

Other metal salts are also capable of synergism with halogen. Lower-cost materials such as stannic oxide and zinc borate have recently been used[21] to partially replace antimony oxide with no loss in flame retardance; however, total replacement will probably detract from flame retardance.

The use of mineral fillers in a coating formulation also reduces the flammability of the coating. The organic material is responsible for burning, and mineral fillers, in many cases, simply perform a dilution effect on the combustible organic material. Therefore, relatively high filler loadings must be used before any significant improvement is seen. This approach is limited to essentially opaque, low-gloss coatings.

Some fillers do have more than a dilution effect in reducing flammability. Aluminum hydrate ($Al_2O_3 \cdot 3H_2O$) is dehydrated in a fire through a very endothermic reaction that lowers the oxidation rate. In addition, water vapor is given off which causes the flammable gases to be diluted. When the temperature is lowered sufficiently below the ignition temperature, the combustion is retarded. Calcium carbonate is also flame retardant because it produces a noncombustible gas (CO_2) under high temperatures. However, when the flame is extinguished, the resulting calcium oxide is capable, in some cases, of undergoing an exothermic reaction with carbon dioxide which may lead to afterglow.[15]

The main disadvantage of using high filler loadings is the detrimental effect on scrub and humidity resistance of the coating because of the high porosity of the paint film.

2.1.2. Polyvinyl Chloride Coatings

Water-based, nonintumescent, flame-retardant coatings (i.e., coatings in which the majority of diluent is water) have not been developed to any significant degree. The only coatings showing any potential development in this area are based on emulsions of copolymers of vinyl chloride and other monomers. Coatings based on copolymer emulsions have been known and used for many years, but not as flame-retardant coatings. Vinyl chloride copolymer emulsions have been used extensively to prepare interior coatings because of their low cost and good performance. These materials are thermoplastic, and the emulsion particles flow together or coalesce to form a continuous film as the water and other diluents evaporate.

Polyvinyl chloride, containing 57% by weight chlorine, is self-extinguishing.[22] It is, however, not a good vehicle for architectural flame-retardant coatings because of its high melting point. The melting point can be drastically lowered by copolymerization with other vinyl monomers such as vinyl acetate. In order to make the copolymers useful, addition of plasticizers and coalescing solvents is often necessary in order to give suitable application and performance properties. These additions dilute the overall concentration of chlorine and thereby reduce flame retardance.

One polyvinyl chloride copolymer emulsion with flame-retardant properties is made with a terpolymer of vinyl chloride, vinyl acetate, and ethylene with an undisclosed quantitative composition. It produces a coating with good flexibility and scrub resistance. The terpolymer has a 24 LOI rating, compared to 18–19 for vinyl acetate and acrylic emulsions, and is considered self-extinguishing.[23] When the terpolymer emulsion is formulated in coatings containing antimony oxide, the LOI value is 50% higher than a coating not based on vinyl chloride-containing polymers.

2.1.3. Polyurethane Coatings

Polyurethane coatings are coatings based on polymers which contain two or more urethane groups. The incorporation of urethane groups results in increased wear and abrasion resistance.

There are four types of polyurethane coatings commonly used in architectural coatings: urethane alkyds, urethane lacquers, moisture-cure coatings, and two-package systems. The mechanisms of cure of these coatings have recently been reviewed.[24]

Urethane alkyds generally undergo air oxidation of double bonds to form cross-linked films. They are prepared by reacting an alcoholized drying oil (i.e., the reaction product of polyol and unsaturated fatty acid) and a polyisocyanate to form urethane groups. The alcoholized oil may also be first esterified with a diacid or anhydride, followed by reaction with a polyisocyanate.

The urethane lacquers are prepared from saturated polyols, dry by solvent evaporation, and do not cross-link. As a result, the coatings often have poor solvent resistance. They are widely used as clear coatings for wood panels and furniture.

Moisture-cure coatings are based on the reaction product of an excess of polyisocyanates with polyols. The product contains unreacted NCO groups which, after application of the coating, react with atmospheric moisture to produce the corresponding amine (RNH_2). The amine then rapidly reacts with another NCO group to form a urea cross-link.

$$\text{—RNCO} + H_2O \longrightarrow -[\text{RNH}\overset{\overset{\displaystyle O}{\|}}{\text{C}}\text{OH}]- \xrightarrow{-CO_2} \text{—RNH}_2$$

$$\text{—RNH}_2 + \text{—RNCO} \longrightarrow \text{—RNH}\overset{\overset{\displaystyle O}{\|}}{\text{C}}\text{NHR—}$$

Moisture-cure coatings are used most often on seamless floors and on wood products demanding coatings with good chemical and abrasion resistance.

Two-package polyurethane coatings are those which have the two reactive components packaged separately for stability. One package contains an isocyanate prepolymer which is the reaction product of one OH equivalent of a polyol with an excess of polyisocyanate. The other package is a polyol or polyamine. When the two packages are mixed together, they rapidly react to form a tough cross-linked film. The excellent physical properties of the two-package systems frequently overcome the mixing inconvenience.

The most common way to achieve flame-retardant polyurethane coatings is by addition of halogen- or phosphorus-containing polyols. Such flame-retardant polyurethane coatings have good abrasion and wear resistance. They are usually relatively expensive, especially the nonyellowing types based on aliphatic polyisocyanates instead of the inexpensive aromatic polyisocyanate. The aromatic polyisocyanate-based material causes film to yellow upon aging.

A flame-retardant polyurethane coating suitable for seamless flooring and wood coating has been obtained from a mixture of epoxy reaction products of halogenated aromatic hydroxy compounds and phosphoric acids.[26] Another uses 1,2,5,6-tetrabromo-3,4-dihydroxyhexane as the polyol.[25]

Flame retardance of polyurethane is often enhanced by incorporating aromatic and cyclic groups into the polymer chain. This can be accomplished by trimerization of a prepolymer of methylene diphenyldiisocyanate to form isocyanurate rings in the presence of a catalyst such as triethyl phosphine. Much work related to this subject has been reported.[27-34,34a]

Isocyanurate ring

The isocyanurate rings provide flame retardance by increasing the thermal stability of the coating. Better flame retardance is obtained from coatings containing large amounts of isocyanurate rings. However, coatings containing such large amounts (ca. 20–30%) of isocyanurate groups are usually hard and inflexible due to the highly cross-linked and crystalline nature of the isocyanurate group. They often do not have the excellent abrasion resistance of polyurethane coatings, and they are expensive. Such a coating is not yet commercially acceptable.

Little information is available concerning the effectiveness of flame-retardant polyurethane coatings for architectural applications. In one case the extent of flame retardation is very dependent upon film thickness.[35] A chlorinated paraffin (43%) was used in a two-package polyurethane coating. At 4 mils on wood, the coating is flame retardant and formed a thin, dense char. When the coating is thinner than 4 mils, it did give enough flame retardation in a 2-ft tunnel test to be classed as a nonintumescent flame-retardant coating. When the coating is more than 6 mils in thickness, it intumesces and should be classified as an intumescent coating.

2.1.4. Epoxy Coatings

Architectural epoxy coatings are based on thermosetting polymeric systems that can cross-link at low temperatures. The cross-linking reaction consists of a reactive group reacting with the epoxy ring. A primary amine can react with two epoxy rings to provide cross-linking and a polyamine can react with more than two epoxy rings to provide cross-linking as shown below:

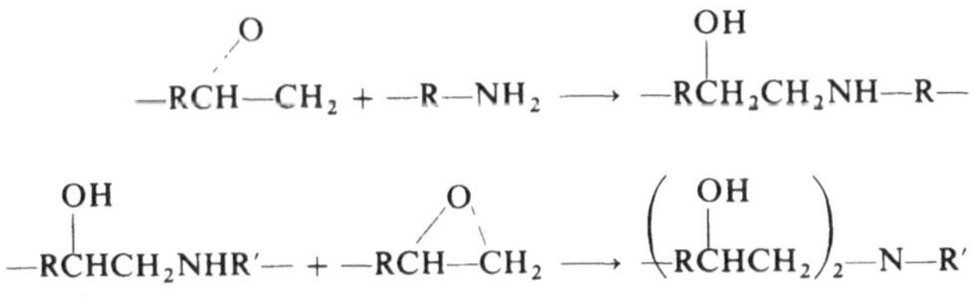

$$-R\overset{O}{\overset{/\backslash}{CH}}-CH_2 + -R-NH_2 \longrightarrow -R\overset{OH}{\overset{|}{CH}}_2CH_2NH-R-$$

$$-R\overset{OH}{\overset{|}{CH}}CH_2NHR'- + -R\overset{O}{\overset{/\backslash}{CH}}-CH_2 \longrightarrow \left(-R\overset{OH}{\overset{|}{CH}}CH_2\right)_2-N-R'$$

Epoxy coatings for low-temperature curing can also be affected by polymers containing multiple acid groups derived from strong acids. One acid group usually reacts with one epoxy group as shown below:

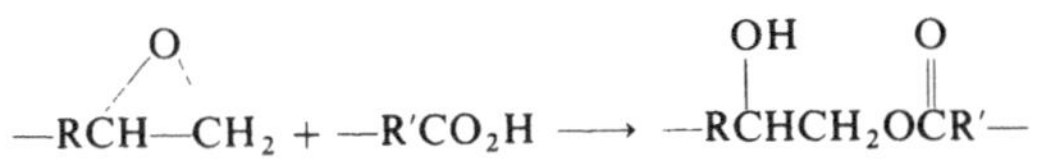

$$-R\overset{O}{\overset{/\backslash}{CH}}-CH_2 + -R'CO_2H \longrightarrow -R\overset{OH}{\overset{|}{CH}}CH_2O\overset{O}{\overset{||}{C}}R'-$$

In this application, many halogenated acids (chlorendic acid, tetrabromophthalic acid, tetrachlorophthalic acid, etc.) and their monoesters (the reaction product of a polyol and an excess amount of acid or anhydride) have suitable reactivity.

Although it is relatively simple to formulate flame-retardant nonintumescent epoxy coatings, high cost prevents the epoxy coatings from being used commercially. This is because nonintumescent flame-retardant coatings are low performance in flame retardancy and must be very inexpensive. However, the synergism of phosphorus–halogen in epoxy systems for flame retardancy has been reported.[36] The technology of flame reretardation of epoxy resins has been recently reviewed.[37]

2.1.5. Miscellaneous Types of Coatings

As mentioned earlier, most architectural coatings are applied in the field and thus must cure under ambient conditions. However, there is an increasing demand for manufactured wood paneling and simulated wood materials with flame retardancy better than unpainted wood. Since these materials are usually coated in factories, elevated cure temperatures can be used. Flame-retardant coatings of this type usually also have improved stain and solvent resistance.

Most common heat-cured coatings utilize urea–formaldehyde resins as cross-linking agents. These resins cure polymeric polyols by reacting the OH groups with methylol or alkoxymethylol groups. This reaction

$$
\begin{array}{c}
\qquad\qquad \overset{\displaystyle O}{\overset{\displaystyle \|}{\text{HOCH}_2\text{NHCNH}-}} \\[2mm]
\{\text{—OH} + \quad \text{or} \qquad\qquad\qquad \overset{\Delta}{\longrightarrow} \quad \{\text{—OCH}_2\text{NHC}\overset{O}{\overset{\|}{}}\text{NH—} + \text{H}_2\text{O} \quad \text{or} \quad \text{ROH} \\[2mm]
\qquad\qquad\qquad \overset{\displaystyle O}{\overset{\displaystyle \|}{\text{ROCH}_2\text{NHCNH}-}}
\end{array}
$$

may be catalyzed by acids (e.g., *p*-toluenesulfonic acid). The polymeric polyols may be any polymeric material containing a significant amount of OH groups (at least 2–3/molecule). The most common flame-retardant polyols used are based on halogen-containing alkyds and vinyl copolymers.

Coatings containing large amounts of aromatic and cyclic groups such as isocyanurates, melamine–formaldehyde resins, and phenol–formaldehyde resins all usually have a certain degree of flame resistance. Indeed many synthetic polymers provide better flame retardance as wood coatings than wood coated with regular alkyd coatings.

It is likely that all these coatings will see increasing use, as flame retardance assumes greater priority in coating specifications.

2.2. Intumescent Coatings

The growing enthusiasm for intumescent coatings in architectural applications is typified by a report [37a] from a tavern owner who credited the use of flame-retardant intumescent coatings with helping to save his life when fire broke out in the building which housed his tavern and apartment.

"If it hadn't been for·the fire retardant paint on the tavern ceiling, we wouldn't be alive now," he reported. Fire officials who inspected the building after the blaze concurred. The ceiling paint had a significant effect on containment of the flames.

The fire-retardant paint was an intumescent coating which puffs when it comes in contact with fire. This puff or char insulates the structural material under it against combustion for a variable period of time, depending on the heat of the flames and the type of material.

2.2.1. Mechanism of Coating Intumescence

The mechanism of intumescence acts in the following way:

1. When the coating comes in contact with intense heat, it starts to decompose thermally, generating strongly acidic materials such as partially dehydrated phosphoric acid or its salts. At the same time, the coating is softened, either by melting or chemically decomposing the film-forming material, to form a highly viscous molten mass.
2. The acid catalyst enhances the decomposition rate of the mass, thereby generating gases. The gas generated causes the molten mass to rise as a foam very much like the preparation of a urethane or polystyrene foam.
3. The viscosity of the molten mass increases as the foam rises and cross-linking takes place during this period. By the time the mass has expanded to a carbonaceous foam 50–100 times greater than its original height or to about 0.5–1.0 in. in thickness, it has been transformed by the intense heat into a rigid intumesced coating.
4. The coating is now flame resistant and much like graphite. Only continued intense heat in the presence of oxygen can cause the surface to be slowly oxidized, while the bulk of intumescent coating will continue to withhold the destructive intense heat from the substrate for a long period of time.

The following types of ingredients are necessary to achieve proper intumescence:

1. A catalyst that causes the decomposition of carbon-containing compounds and aids in the gas-formation reaction

2. A carbonaceous material which will produce enough carbonlike mass to permit an adequate density and thickness of carbonaceous foam
3. A spumific or blowing agent which decomposes to produce abundant quantities of noncombustible gases
4. An additive which aids the flame retardation

A coating must have a binder or vehicle to incorporate all the coating ingredients to form strong films. The binder for intumescent coatings must not be highly cross-linked with heat-stable links since this will prevent the coating from being transformed into a viscous molten mass. On the other hand, should the binder be too fluid at high temperatures, the coating will flow away and/or gas bubbles will escape before proper foam formation. It is most important that the liquefaction of the binder, the gas generation, and the cross-linking reactions to form the foam char occur almost simultaneously, but in proper sequence. Should the cross-linked char occur much before gas formation, the gas will not be able to cause a rigid char to rise into a foam structure. Should the cross-linked char occur much after gas is released, the gas will escape and the expanded foam structure will collapse.

The surface tension of the molten mass at high temperatures is important. If the mass has a high surface tension, it encourages the combination of small gas bubbles to large ones. The gases have a greater tendency to escape, and this in turn produces an unacceptable low-rise, high-density foam structure with little insulating effect.

Proper foam formation in intumescent coatings involves a complicated mechanism which is not fully understood. The authors have demonstrated in their laboratory that the performance of such a coating is largely dependent on intensity of the flame and heat which contacts the coating. Their experiments show that during one set of heating conditions an intumescent coating will produce a thick foam char of very low density. Under certain different fire conditions, the same coating will intumesce to a much thinner carbonaceous foam at a much higher density.

Since there are no standard conditions for actual fires, the determination of a good or poor intumescent coating must be determined by a controlled test procedure. As with nonintumescent coatings, the ASTM E-84-70 25-ft tunnel test is widely accepted and most often used for final performance testing. It is understood, however, that a coating performing well in this test may not do the same under certain real fire conditions.

Intumescent coatings, applied in the recommended manner, are expected to keep the substrate relatively cool and protected from extensive fire damage for extended periods of time during a fire. The value of such protection to the substrate can be seen in the case of structural steel. Steel has a significant decrease in strength above 600°C, and this decrease in strength often causes a building to collapse during a fire. When an intumescent coating is used to protect structural steel beams, it slows the heat transfer to the beam and may prevent a total building collapse during a fire.

In a similar manner intumescent coatings are also used to protect the outside of pressurized gas tanks, military bombs, and railroad tank cars by extending the "cook-off" time of the coating and thus delaying or preventing the explosion of the container's contents.

To achieve good protection with intumescent coatings, a thickness about four times that of conventional paints is necessary. This thickness can only be applied in two or three coats or the coating will craze owing to shrinkage from drying. Although spray equipment may be used, rolling and brushing are preferred, since the large crystals formed from the intumescent chemicals often interfere with spray application. These thick coatings will often have a rough or textured appearance owing to the presence of these large particles.

As a result of the required film thickness, the coverage rate to achieve a 6-mil dry film is approximately 150 ft^2/gal at 50% volume solids. Nonintumescent coatings which require $1\frac{1}{2}$–3 mils films thickness have a coverage rate of approximately 500 ft^2/gal. Therefore, the amount of coating to achieve the extraordinary thickness required (6–15 mils) for intumescent coatings is 3–4 times that of conventional paints.

There are a number of inherent problems in intumescent coatings in addition to low coverage, need for multicoats, and high cost. These include poor hiding power, poor can stability, and uncertain adhesion. Then, too, some are water sensitive, permitting active ingredients to be removed by leaching during washing or weathering. Unless they are overcoated with water-resistant coatings, most intumescent coatings are not suitable for outdoor use. Besides, the overcoating often degrades the flame retardancy of the intumescent coating.

Many improvements have been made over the first intumescent coatings developed around 1938,[39] which were two-package aqueous systems containing monoammonium phosphate. These coatings are now obsolete. The current one-package aqueous coatings have better scrub resistance and film-forming properties.

2.2.2. Current Status

In spite of many advances in coating properties, the relatively high cost of intumescent coatings is probably the main drawback to their increased use. In 1972, about $10 million worth of these coatings were sold. Of that amount 75% was for water-based systems, while the solvent-based systems accounted for only 25%. Most sales were made to contractors or government agencies with little material going to consumers.

Vandersall has written a detailed account of the development and chemistry of both water-based and solvent-based intumescent coating systems.[40] The vehicle materials for water-based intumescent coatings are usually amino formaldehyde resins, vinyl acetate, vinyl chloride co-polymer, or acrylic latexes. The solvent-based coatings may be epoxies, polyurethanes, phenolics, alkyds, vinyl toluene–acrylate copolymers, and chlorinated rubbers. The advantages of water-based coatings over solvent-based systems are lower cost, absence of large amounts of offensive organic solvents, and the fact that the diluent (water) is not a flame hazard, but they have many disadvantages compared with solvent-based coatings.

The solvent-based coatings are generally superior to the water-based coatings. Normally, they provide good adhesion to the substrate, high water resistance, and better can stability. Of course, they have the problem of employing combustible and toxic solvents. Although much better than the water-based intumescent paints, they still do not have the brushing and flow properties expected of a good paint.

The most serious deterrent to the use of solvent-based intumescent coatings is cost—solvent-based intumescents may sell for $15–20/gal, and good epoxy coatings may sell as high as $50/gal. Although their use has increased rapidly in the past several years, it is estimated that the total sales of intumescent coatings are still only about 0.1% of the entire coatings market.

2.2.3. Typical Methods of Formulation

Formulations of conventional coatings comprise the vehicle, pigments or colorants, and a solvent or carrier. Various additives are also used for specific purposes such as fillers, antioxidants, defoamers, antisagging agents, thickeners, coalescing agents, and many more whose uses are detailed elsewhere.[43] Formulation of intumescent paint may be considered as adding intumescent materials into regular paint. However, since the primary concern of such a coating system is flame retardancy, it is generally necessary that the intumescent chemicals and colorants constitute 80% of the dry film.[40] The coating's composition and basic properties, therefore, may be quite different from those of decorative coatings. There are several articles reviewing the general formulation and composition of intumescent coatings.[40-42] A particular coating formulation may be quite involved and often guarded as a trade secret; only principal ingredients are usually published. For the present study, the following is an example of an aqueous, intumescent coating formulation:[44,45]

Compound	Quantity, %
Ammonium phosphate	22.9
Dipentaerythritol	3.8
Melamine	7.7
Melamine–formaldehyde resin	1.9
Chloroparaffin (70% Cl)	4.6
Coalescing agent	2.0
Titanium dioxide	7.7
Polyvinyl acetate resin	9.5
Water, thickener, etc.	39.9

and a solvent-based formulation:[46]

Compound	Quantity, %
Melamine	5.00
Zinc borate	11.93
Magnesium silicate	0.60
Lead sulfate	6.80
Lead carbonate	6.80
Zinc oxide	3.82
Tris(2,3-dibromopropyl)phosphate	10.00
Titanium dioxide	4.05
Pentaerythritol polyurethane	15.00
Chlorendic anhydride tung oil alkyd	25.00
Chlorinated paraffin	5.54
Polyamide resin	4.44
Driers, etc.	1.02
Solvents	—

When the above formulations are examined, it is easy to visualize that many materials in the formulations are used for more than one function such as acid catalyst, carbonaceous material, blowing agent, flame-retardant additive, and binder as described below:

1. *Inorganic catalysts*: Ammonium phosphate, zinc borate, tris(2,3-dibromopropyl)phosphate are used in the above two formulations. These ingredients cause the decomposition of carbon-containing compounds and aid in the gas formation reaction. Strong inorganic acids without chemical modifications are not commonly used. The sources of inorganic acids are those acid derivatives which yield inorganic acids at temperatures between 100 and 250°C. For this purpose the inorganic acids customarily are used in their neutralized forms. Neutralization is accomplished by the formation of amides from amines, by the formation of esters from alcohols and/or

polyols, or by the formation of salts of metals and basic nitrogen-containing compounds.

The basic nitrogen-containing compounds commonly used are ammonia, urea, guanyl urea, melamine, analine, and nitroaniline. The alcohols most often chosen are substituted phenols, simple alcohols and halogenated alcohols such as bromohydrin. When polyols are used, it is possible to include film-forming materials in the intumescent ingredients. The inorganic acids suitable for inclusion are phosphoric acids, polyphosphoric acids, sulfuric acids, boric acids, and hydrogen halides. The metal is usually zinc or antimony in the form of zinc borate or antimony oxides and other salts of antimony.

The neutralized acids most frequently used are ammonium phosphate, trisbromopropyl phosphate, melamine phosphate, potassium fluoroborate, and zinc borate.

2. *Carbonaceous materials*: In the above formulations, dipentaerythritol, pentaerythritol polyurethane, chlorendic anhydride tung oil alkyd, polyamide resin, and polyvinyl acetate resin are used for this purpose. These types of materials produce a large amount of carbonaceous residue, especially by dehydration in the presence of inorganic acids. Compounds of this type are starches, dextrins, sugars, pentaerythritol, and/or its self-condensed products, such as dipentaerythritol, phenols, phenol–formaldehyde resins, polyurethane, and various other polyols.

3. *Blowing agents*: In the above formulations, melamine, pentaerythritol polyurethane, polyamide resin, ammonium phosphate, and melamine–formaldehyde resin are used as the blowing agents.

These are almost always basic nitrogen-containing materials. Compounds in this group are urea, methylol urea, butyl urea, dicyandiamide, benzene sulfonylhydrazide, melamine, glycine, casein, polyamide resin, and wool flock. Many of these compounds are used in neutralized or prereacted forms. Amine–formaldehyde resins are most effective when used with hydroxyalkyl-substituted phosphorus compounds. They can function both as film formers and/or blowing agents.

4. *Flame-retardant additives*: In the above formulations, chloroparaffin and tris(2,3-dibromopropyl)phosphate are used as the additives. These additives usually are in the form of halogenated compounds. The function of these compounds depends on the production of HCl—a noncombustible gas as well as a flame-propagation arrestor. HCl is known to inhibit flame propagation by a free radical mechanism.[41] Compounds and polymers of this type are chlorinated paraffin, chlorinated rubber, halogenated phenol, pentachlorophenoxyglycerol, and halogenated polyacids.

5. *Vehicles*: The vehicle used in the above formulations consists of melamine–formaldehyde resin, polyvinyl acetate resin, pentaerythritol

polyurethane, chlorendic anhydride tung oil alkyd, and polyamide resin.

The vehicle is the principal ingredient of a coating formulation, and so most of the coating properties are dependent upon the type of vehicle used. Of the two general types, thermosetting and thermoplastic, the latter is more useful for intumescent coatings as aqueous emulsions. These emulsions use polymers with relatively high molecular weights which provides good film formation. Once the dry coating is formed, the coalesced polymer is no longer dispersible in water. Solvent solutions of thermoplastic materials have very high viscosities, and application of the coatings is very difficult. For this reason, the aqueous emulsions are preferred for thermoplastic intumescent coating vehicles.

Thermosetting film formers are made of low-molecular-weight polymers, which, when deposited as a film, react further to form high-molecular-weight cross-linked polymers either by air oxidation, by reaction with moisture in the air, or by coupling of internally reactive groups. Here too, the cross-linked coating is insoluble in the solvent from which it was applied.

This type of coating is packaged in either a one- or two-package system. In the two-package system, both ingredients are mixed before application. The one-package system requires no premixing; however, both thermosetting coating systems have the advantage of being easy to apply at relatively high solids, thereby forming thick coatings. They are often also superior to thermoplastic polymers in scrub resistance.

Good intumescent coatings should not be highly cross-linked because such a system will not form a molten mass under intense heat thereby preventing the formation of an acceptable intumesced foam.[47] With this in mind, one can conclude that in the two previous formulations the film formation for water-based coating is the result of coalescence of polymer particles after water evaporation. The organic solvent-based coating is mainly a thermosetting air-drying chlorinated alkyd system. The cross-linking of the film results from air oxidation of the alkyd.

It should be noted, however, that even if a coating contains all of the above five types of ingredients and is not highly cross-linked, it does not necessarily produce an intumescent coating. For intumescence to occur, several distinct steps must occur in proper sequence as has been discussed in the mechanism of carbonaceous foam formation. The importance of correct sequence of the steps of intumescence can be illustrated by the following experiment.[40] A mixture of monoammonium phosphate and starch, both having relatively low decomposition temperatures, will intumesce under intense heat. Ammonium polyphosphate and dipentaerythritol, with much higher decomposition temperatures will also intumesce upon intense heating. However, when a mixture of ammonium polyphosphate and starch is heated, there is *no* intumescence—the sequence of decomposition is not effective.

It is not effective because the starch, in the second case, decomposes and produces a rather large quantity of flammable tar which becomes hard before the ammonium polyphosphate decomposes. When the ammonium polyphosphate decomposes to release gases, there is no molten mass present for intumescent foam formation. The acid catalyst produced is also essentially useless since the starch has already decomposed to form flammable tar; the acid cannot assist in producing large amounts of carbonaceous residue by dehydration.

2.2.4. Recent Advancements in Technology of Intumescence

Prior to 1960, most intumescent coatings consisted of aqueous two-package systems of which one package was a urea–formaldehyde resin, and the other package a curing agent and coating additives. These had many shortcomings. They had to be mixed before application, films were water sensitive, had poor brushability, and clear coatings were not possible.

In the last decade major changes in intumescent-coating formulations helped overcome some of those shortcomings. With the advent of one-package emulsion systems, the coatings became much more acceptable. The discovery of flame-retardant additives which had little or no solubility in water resulted in coatings with better scrubbability. The development of solvent-based systems made available intumescent coatings with improved application and weathering properties.

In the past, intumescent coatings always suffered from the use of a very high concentration of additives. The use of vehicles which also act as intumescent agents has often made it possible to use less intumescent additives. This has resulted in coatings with better application and film properties. Particular success has been obtained with this method in two-package solvent-based coatings such as epoxy or polyurethane coatings. Although these materials are more expensive, they have much better overall coating properties and can be used to make clear intumescent coatings by using less solid additives and eliminating materials of very high refractive index.

Two interesting new methods of producing intumescent coatings have been reported recently. In one case, a self-intumescent monomer acts as the sole intumescent additive; the other uses a mixture of powdered glass and a blowing agent to form an intumesced glass foam instead of the typical carbonaceous intumescing agents. These two discoveries may direct intumescent research into quite new approaches and will be discussed in more detail later in the chapter.

While these innovations point to solutions of major problems, they may also create other less serious ones. The challenge for the intumescent coating chemist will continue to be the search for coatings which incorporate the best intumescent properties and at the same time lower the cost of such coatings.

2.2.4.1. One-Package Aqueous Systems. As a result of the discovery that polyvinyl acetate emulsion[48] can tolerate flame-retardant additives without being precipitated, many other stable latex systems were developed. Currently almost all water-based intumescent coatings are based on latexes.

2.2.4.2. Additives with Improved Water Resistance.

1. *Ammonium polyphosphate.* Before 1965 only very water-soluble ammonium phosphates were available as flame-retardant additives for intumescent coatings. Currently, most intumescent coatings use Phos-Chek P/30[44,47,49,50] (Monsanto Chemical Co.), an ammonium polyphosphate additive which was introduced commercially in 1965. This material exhibits excellent hydrolytic stability[51] and a fairly neutral pH in aqueous media. A coating containing this material can also be formulated with other suitable intumescing agents to obtain good properties such as package stability and scrub resistance.[52] The compound is represented by the following formula (mol wt 1000–3000):

$$\left(\begin{array}{c} O \\ \| \\ -P-O- \\ | \\ O^-NH_4^+ \end{array}\right)_n$$

2. *Melamine phosphate.* The use of melamine phosphate, although first reported in 1953,[53] has gained prominence only recently. This additive, consisting of 1 mol melamine and 2 mol phosphoric acid, has the advantage of having an even lower water solubility than ammonium polyphosphate. A large proportion of solvent-based intumescent coatings which have unusually high moisture and scrub resistance with good adhesion to substrates and excellent flame retardance and intumescent properties uses this material.[40]

A series of four papers[46,54-56] discusses the properties of coatings using melamine phosphate and halogenated phosphate esters as additives. However, this material cannot be used in aqueous systems, because, although it is insoluble in water, it is slowly swollen by water to form a gel-like structure.[40]

3. *Phosphoryl trianilide.* Phosphoryl-trianilide,[57] an excellent intumescent agent, was reported in 1969. Unlike most flame-retardant additives, it is not a salt. This material has a high aromatic ring content, is water insoluble, and is not swollen by water. With proper formulation, 6–10 mils

phosphoryl-trianilide-containing coating can be made to intumesce forming a stable foam 1 in. thick. Although this material is not yet reported in commercial coatings, it is expected that it will be used in water-based coatings to give even better water scrubbability than coatings containing ammonium polyphosphate.

2.2.4.3. Clear Coatings and Vehicles as Intumescent Agents. Clear intumescent coatings are rather difficult to obtain: most intumescent additives give opaque coating because of their insolubility and their very different refractive indexes from the vehicle. The earliest reported clear coating[58] makes use of urea–phenol–formaldehyde resin as the vehicle as well as the intumescent ingredient. The flame-retardant additives are mainly ammonium mono- and dibutyl phosphate. The same report notes another clear coating, based on a proprietary phosphorus-containing amide.[58]

The use of vehicles as intumescent agents is also desirable because it results in the reduction of the normally high additive loadings which are often detrimental to needed coating properties. The various types of vehicles that intumesce are reported in many patent disclosures as outlined below.

Using various epoxy resins, Blair, Witschard, and Hindersinn[59] obtained clear as well as opaque coatings. The resins were diluted with a variety of phosphorus compounds and cured with amines to obtain a coating 10–15 mils thick which intumesced to form a foam 1 in. thick. As expected, the coatings were not water sensitive, and the flame retardant properties remained unchanged after immersion in water for 24 hr at 50°C.

Another coating in which the vehicle performs as an intumescent component consists of an epoxy resin which is reacted with diaminodiphenyl sulfone in the presence of melamine hydrobromide.[60] In like manner, a nitrosubstituted epoxy polymer is used in the intumescent coating.[61] In another case, an intumescent coating is made from the reaction of compounds such as

$$Br-CH_2-CH_2-O-\underset{\underset{O}{\parallel}}{P}-Cl_2$$

with epoxy resins.[62]

Various types of phosphorous polyols can also be cured with amino resins to form solvent-resistant intumescent coatings. The cross-linking reaction is shown below:

$$-ROH + HOCH_2NH\overset{O}{\overset{\parallel}{C}}- \xrightarrow{-H_2O} -ROCH_2NH\overset{O}{\overset{\parallel}{C}}-$$

or

$$-ROH + R'OCH_2NH\overset{O}{\overset{\parallel}{C}}- \xrightarrow{-R'OH} -ROCH_2NH\overset{O}{\overset{\parallel}{C}}-$$

Clear coatings[63] using a hydroxyalkylated pyrophosphate (shown below), methylol melamine compound, an alkylated amine, and a halogenated hydrocarbon[63] have been used to formulate intumescent coatings. Other new, clear intumescent coatings feature the incorporation of an amide binder,[65] and in another application an epoxy resin[64] was used.

Other advances in the technology of intumescent coatings use flexible coatings[57] which are formed by the reaction of toluene diisocyanate with the condensate of pentachlorophenyl glycidyl ether and phosphorus-containing polyols. Also using a vehicle as intumescent agent is a patent[66] which claims the addition of a halogenated polyphosphate (below) to an unsaturated polyurethane resin to give flexible coatings.

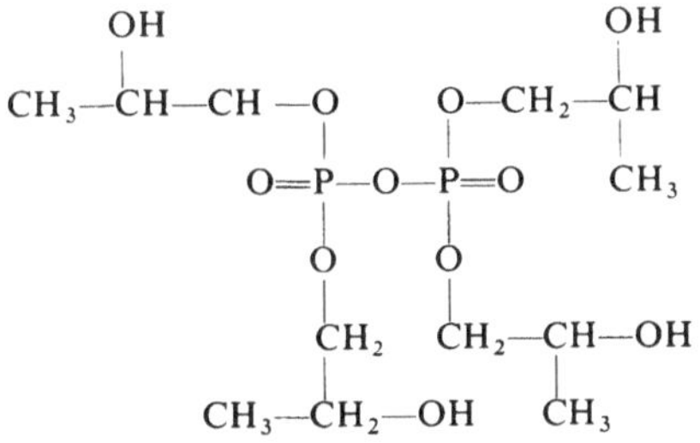

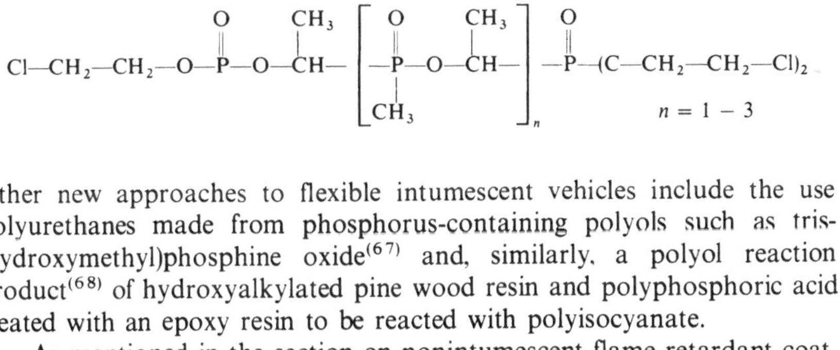

$$OH \qquad\qquad OH$$
$$CH_3-CH-CH-O \qquad O-CH_2-CH$$
$$O=P-O-P=O \qquad CH_3$$
$$O \qquad O$$
$$CH_2 \quad CH_2-CH-OH$$
$$CH_3-CH_2-OH \qquad CH_3$$

$$Cl-CH_2-CH_2-O-\overset{O}{\overset{\|}{P}}-O-\overset{CH_3}{\overset{|}{CH}}-\left[-\overset{O}{\overset{\|}{P}}-O-\overset{CH_3}{\overset{|}{CH}}-\right]_n-\overset{O}{\overset{\|}{P}}-(C-CH_2-CH_2-Cl)_2$$
$$CH_3 \qquad\qquad n = 1 - 3$$

Other new approaches to flexible intumescent vehicles include the use polyurethanes made from phosphorus-containing polyols such as tris-(hydroxymethyl)phosphine oxide[67] and, similarly, a polyol reaction product[68] of hydroxyalkylated pine wood resin and polyphosphoric acid treated with an epoxy resin to be reacted with polyisocyanate.

As mentioned in the section on nonintumescent flame-retardant coatings, it is possible to make coatings in which the vehicle is cured with atmospheric moisture. These can be used for intumescent coatings also. One such coating[69] is made from the reaction product of a haloalkyl phosphate such as 2-chloroethyl phosphate, an alkylene diaminetetraacetic acid, a polyhydric alcohol, and an aromatic diisocyanate. The product is then moisture cured to give intumescent properties. Another is formulated from the reaction product of an aromatic diisocyanate, pentachlorophenoxyglyceryl ether diol, and triethylene glycol which is moisture cured to give intumescent coatings. The intumescent character of the coating is improved by the addition of haloalkyl phosphates.

2.2.2.4. Self-Intumescent Monomers. The U.S. National Aeronautics and Space Administration disclosed in 1968 a new type of intumescent agent which is called a self-intumescent monomer. The compound often acts as the sole additive in making intumescent coatings, instead of using many additives as previously discussed.

The first self-intumescent monomers[71-73] were the bisulfate salts of nitrosubstituted aromatic amines. They were dispersed in a vehicle which intumesces when heated to 350–500°F. Although the bisulfates are water soluble, they can be formulated to produce acceptable coatings.[74] The same source reported using the ammonium salt of 4-nitroaniline-2-sulfonic acid with a mercaptan-terminated polymer and other ingredients to form an intumescent coating with high water resistivity.[75] Coatings of this type show promise in military applications because they can, in some instances, delay detonation of bombs or missiles. One such coating[76] applied to the exterior of loaded weapons increases the time of detonation in the presence of fire from 3 min to 10–13 min. This coating uses 4-nitro-aniline-2-sulfonic acid and a copolymer of a polysulfide epoxy resin as the vehicle, and low-density phenolic and silica balloons and silica fiber as filler additives.

The above concepts were actively pursued by workers at the Cities Service Company and a series of patents were issued in 1972 and 1973. The following compounds were claimed as self-intumescent monomers for coating systems:[77-81] aminobenzoic acid sulfate,[77] 5-amino-2-nitro-benzoic acid,[78] *p*-aminobenzene sulfonamide,[79] *p,p'*-oxy-bis(aromatic sulfonamide),[80,81] and analogs.

2.2.4.5. Miscellaneous Intumescent Coatings. There are several published innovations in intumescent coatings that do not belong to any of the above classifications, the most interesting of these is the use of blown-glass foam to act as a protective barrier[82] instead of foam char. The coating contains powdered glass, nucleating agents, and blowing agents which intumesce at 1020°C to a glass foam.

Although this coating would be limited to use with materials that could withstand the high temperatures necessary for the intumescence, the foams would be completely oxidation resistant and would withstand flame much better than carbonaceous foams. Another innovation[83] is the use of predispersed glass fibers to improve the integrity of the coating, making it easier to apply, increasing hiding power, and enhancing the mechanical stability and fracture resistance of the char foam. Another patent claimed[84] that the use of glass fibers will cause a fourfold increase in the coating thickness on heating at 400°C.

One of the problems with intumescent coatings has been the corrosion of the substrate by intumescent additives. To avoid substrate corrosion problems, microspheres have been used[85] to encapsulate corrosive ingredients, thus rendering the coating noncorrosive. Another patent[86] describes a method for overcoming the problem of water sensitivity of many aqueous intumescent coatings by applying a solvent-based overcoating in order to protect its water sensitivity.

In order to increase fire protection for construction materials such as plywood, polyester foam, and fabric sheeting, a patent[87] suggests the use of aluminum foil as the undercoating which is then covered with an intumescent coating.

3. Fabric Coatings

The development of flame-retardant coatings for fabrics requires considerable knowledge in all areas of technology of flame retardancy. It also requires an understanding of the design and application of fabrics and fabric processing and of performance requirements of the finished fabrics.

At present the main uses for fire-retardant fabric coatings are in transportation vehicles. An estimated 86 million pounds of such coatings were used in 1972. Second in volume are various types of wallpaper; 60 million pounds were used in 1972. Some relatively small amounts are also used for clothing, tents, furniture covers, handbags, shoes, and artificial leather goods, as well as very special uses such as space suits and conveyer belts.[88]

Flame-retardant fabric coatings are currently applied by a number of techniques. One of these involves applying the coating from a roller, first applying an adhesive to the fabric, if necessary. Then a pattern or textured surface may be applied to the coating by passing the material through a lithographic press. A clear top coat may be applied if good wear resistance high luster, or stain and weather resistance are desired. These coatings are usually applied as a solution or emulsion, after which the coated fabric is passed through drying or curing equipment such as heated rolls, vertical or horizontal ovens.

Water-thinned latex coatings are preferred because they offer less fire hazard; however, where very thin coating films are applied by spray, solvent-based coatings are used. Very thick fabric coatings, usually from organosols or plastisols, may be applied to the front and/or to the back of a fabric.[90] Sometimes a foam is sandwiched[91] between the coating and the fabric.

The degree of flame retardancy needed for materials used in transportation vehicles is more or less based on the premise that "although the qualities of the interior material of a vehicle cannot by themselves make occupants safe from the hazards of fuel-fed fires, it is important when fire occurs in the interior of the vehicle from such sources as matches, cigarettes, or short circuits in the interior wiring that there be sufficient time for the driver to stop the vehicle and, if necessary, for occupants to leave before injury occurs."[92]

The above premise suggests only relatively mild flammability test conditions, and these are usually conducted on coated fabrics rather than on free films of the coatings. Although different tests are recommended for different usages,[93] all tests on the coated fabric are ordinarily conducted without a continuous external supply of heat and fuel. Such tests are much milder than tests performed on architectural coatings for construction materials used in buildings, where an intensive external heat and a large amount of fuel and air is continuously supplied to the tested materials during the entire test period.

Since most flame-retardant coatings are applied to upholstery fabrics of transportation vehicles, it is appropriate to use the testing method for this usage to illustrate the typical relatively mild testing procedure. The test[94] is as follows:

A coated fabric is cut into a 4 × 14-in. strip. The strip is supported horizontally, and a bunsen burner flame fueled with natural gas is applied for 15 sec to one 4-in. end of the strip. The rate of burning is measured from the time when the flame reaches a point 1.5 in. from the start. The material should not burn or transmit a flame front across the surface at a rate of more than 4 in./min.

If a material stops burning before it has burned for 60 sec from the start of timing, and has not burned more than 2 in. from the point where timing was started, it is considered to have met the burn-rate requirements.

Various automobile manufacturers impose different but often somewhat more rigorous requirements than that above. For example, in 1971, American Motors Corporation required a burning rate of 3 in./min; Fisher Body of General Motors required the fabric to be zero-burning.[95] The amount of smoke generated by a coated fabric during a fire has not yet been regulated by the U.S. Department of Transportation. No tests have been required and no quantitative limits have been set for smoke generation, but in general, the lower the amount of fire-retardant additives, the lower the amount of smoke produced during burning.

It would be well to define some of the terms used in the test just described. The material is termed combustible if the test specimen burns completely; self-extinguishing if it burns but self-extinguishes before the flame travels a predetermined distance; and nonburning if it does not burn after the flame is removed. However, it must be noted that the terms "self-extinguishing" and "nonburning" may be misleading since they do not necessarily reflect performance under actual fire conditions, and various results reported in the literature are based on different testing conditions which may not be completely identical with the testing method just described. Thus, the classification of the flammability of a material is defined according to the original testing conditions.

All fabric coatings must be flexible; the chief consideration in formulating coatings for textiles is their end usage. Most elastomeric polymers and some plasticized nonelastomeric polymers can be formulated to give coated fabrics with acceptable flame retardancy. The predominant material found to give a commercially acceptable coating at low cost is plasticized polyvinyl chloride, with polyurethane next in importance. Other coating materials have minor commercial importance and very little information is available.

3.1. Polyvinyl Chloride Fabric Coatings

Polyvinyl chloride (PVC) is made by the polymerization of vinyl chloride. The raw material currently sells for about $.07/lb and after polymerization, costs $.13–25/lb depending upon the purity and form. Very often PVC is made with comonomers such as vinylidene chloride, vinyl acetate, and other vinyl monomers to increase processability, flame retardance, and other uses by means of bulk, solution, emulsion, or suspension polymerization.[96] At the present time, suspension polymerization is the most important commercial process, but the material produced by emulsion polymerization is often preferred for fabric coatings.

Pure PVC contains 57% chlorine, is combustible only in intense heat, and burns slowly with a smoky flame.[97] The heat of combustion (4300 cal/g) is comparable to that of cotton (4000 cal/g), and is much smaller than that of polyisobutylene rubber (11,180 cal/g). The flash-ignition and self-ignition temperatures of PVC are 391°C and 354°C, respectively,[98] with an oxygen index of 45.0–49.0.

Polyvinyl chloride is not flammable when it is in contact with a small flame (hence the high oxygen index) and is classified as "nonburning." The toxic effects from combustion products of PVC, when studied for the relationship of death incidence of animals exposed to the combustion products and the carboxyhemoglobin level in the animals' blood,[99] revealed that the major source of danger resulted from the relatively high concentration of carbon monoxide in the combustion gas.

When used in fabric coatings, polyvinyl chloride is formulated with various plasticizers and other additives such as heat stabilizers, extenders, pigments, etc., to improve overall properties; but in the presence of such additives and without extra flame-retardant chemicals being added, PVC-coated fabrics usually are not "self-extinguishing." The proper combination of PVC, fabric substrate, plasticizers, pigments, flame-retardant additives, and extenders must be chosen to formulate acceptable flame-retardant PVC-coated fabrics.

3.1.1. Types of Polyvinyl Chloride

The polymerization of vinyl chloride requires special pressurized equipment because vinyl chloride is a gas at room temperature. While large coating companies produce most of the polymers used in their coatings, PVC users normally do not produce this compound but obtain it from commercial sources. This trend will continue, especially since vinyl chloride monomer has been found to be a human carcinogen.[100] PVC is usually supplied as a free flowing powder or a latex, in many varieties all with different properties. The variety of types of PVC may be noted in Penn's[101] tabulation of more than 50 suppliers in the Western Hemisphere, each of whom may have more than 10 different grades of PVC designed for different applications. The main considerations[102] in choosing PVC for formulations can be summarised as follows:

1. The method of polymerization determines the particle structure and the amount of impurities present in the polymer. PVC from emulsion polymerization has a small particle size, is easy to process, and gives water emulsions; suspension polymerization gives dry powders, and solution polymerization gives PVC solutions. Coatings made from PVC emulsions have little static accumulation and relatively high electrical conductivity. This is desirable when generation of sparks from static electricity would be dangerous. It is not desirable, for example, as wire-coating insulators, where high resistivity is needed. PVC produced by emulsion polymerization also has a tendency to yellow more than material based on other methods of polymerization.

2. The molecular weight of the PVC has a great influence on the properties of flame-retardant fabric coatings. High-molecular-weight PVC, while giving better film properties than low-molecular-weight materials, is more difficult to process. Molecular weight can be expressed in terms of intrinsic viscosity, but it is more accurately expressed as a relative viscosity of the polymer in a specified solvent at a concentration of 0.4–0.5 g/100 ml solvent.

3. The type and amount of comonomers used also affects the PVC copolymer in many ways. In order to modify PVC during polymerization, various amounts of vinyl acetate, vinylidine chloride, and other monomers[103] are introduced. The incorporation of a second monomer lowers the processing temperature, increases the solubility of PVC, and changes the percentage of chlorine and, hence, the flame retardancy of the product. The formulator of flame-retardant fabric coatings from PVC has a wide range of such copolymers to choose from. To illustrate, vinyl acetate, a widely used comonomer in polymers specially designed for coating use, is available from one company[104] in nine or more different copolymer compositions with various viscosities, specific gravities, solubility characteristics, and potential uses.

4. Particle shape and size distribution of the PVC also affect processing conditions. The polymerization technique makes it possible to obtain particles with many shapes other than the usual spheres. Particle size distribution can be uniform or drastically nonuniform. For example, Imperial Chemical Industries, Ltd., offers a series of PVC polymers having the shape of hollow-centered spheres (ICI Corvic R65/33) as well as polymers with a wide particle size distribution (ICI Corvic R48/82). These types of resins are especially desirable for plasticized powders because their structures facilitate plasticizer swelling of the polymer during formation of the coating.

3.1.2. The Choice of Fabric

The flame retardancy of a coated fabric is controlled to a large extent by the flammability of the fabric. Thus, the type of fabric used, such as cotton, nylon, polyamide, polyester, rayon, or glass, must be taken into consideration in formulating the coating.[105,106] The weight of the fabric as well as the weight of the coating are also important factors in making the coating–fabric composite flame retardant. If the substrate is easily combustible, more flame-retardant chemicals must be added to the coating formulation. This is because the burning of a polymer composite depends greatly upon the weight/surface ratio. The amount of surface area available, and possibly the wicking effect, is important in determining the rate of combustion, and therefore, the intensity of the flame.[107]

3.1.3. Formulating Fabric Coatings

All polyvinyl chloride materials used in fabric coatings require plasticizers, heat stabilizers, extenders, and other additives to make commercially acceptable products. For fabric coatings, plasticizers are used in the range of 30–100 parts plasticizer per 100 parts PVC. The plasticizers are the most important ingredients among the additives. The regular non-flame-retardant plasticizers are available in about 500 different generic types.[95] They are usually high-boiling polyesters of dibasic acids, but sometimes epoxy resins and other resinous materials are used.[108] Since all plasticizers plasticize PVC to a degree, and more than one plasticizer may be used in any formulation, the choice is often determined by cost.

In formulating these coatings, flame-retardant additives are also important ingredients. Among the hundreds of available commercial flame-retardant chemicals tabulated by Kuryla,[22] there are only a few which are used extensively for PVC coatings. These flame-retardant additives are discussed below:

3.1.3.1. Antimony Oxide. This is the most important and most efficient flame retardant used with PVC. As with other coatings, its synergistic effect with chlorine greatly enhances the flame retardance of the coatings. This insoluble pigment powder is available in an ultrafine high-purity form or as a coating on a carrier such as silica.[110] The latter grade is slightly more efficient in flame retardation than pure antimony oxide.

The exact mechanism of the synergism of antimony oxide and halogen compounds is not known. The various proposed flame retardation mechanisms have been reviewed by Pitts[111] and can be summarized as follows:

1. In the presence of chlorine-containing materials, antimony oxide creates $SbCl_3$ which is an effective free-radical trap or chain breaker very much like HCl or HBr in the gas phase of the burning process.
2. Trivalent antimony facilitates the generation of halogen radicals which then interfere with the normal free-radical mechanism of the flame propagation.
3. The formation of SbOCl and/or $SbCl_3$ simply delays the rate of escape of the halogen from the flame and thus increases the chance of reaction with the propagating species.
4. The formation of volatile $SbCl_3$ inhibits flame spread by blanketing the combustible material and cutting off the air supply.
5. Condensed liquid or solid $SbCl_3$ particles reduce the energy of the flame by the "wall" or surface effect.

Although a relatively small amount of antimony oxide, 2–4 phr, is usually needed to impart flame resistance, an amount of 15 parts per hundred rubber (phr) may be tolerated since the physical properties of the coating change very little with or without the addition of antimony oxide. Another advantage in the use of this odorless compound as a flame retardant is that it will not be leached out by washing or other types of cleaning since it is not soluble in water or organic solvents. Then too, it has no vapor pressure, and it will not be lost through sublimation upon aging. Furthermore, it will not fog surfaces located near the coating as many plasticizers do.

The uncertain supply of antimony oxide, especially in large quantities, coupled with price fluctuations in the last few years of \$.70–2.40/lb, is a disadvantage in its use as a flame-retardant additive. Another drawback is the fact that coatings raw materials are usually bought by weight, but the products are sold by volume; thus the high density (5.3–5.6) of antimony oxide increases the cost of the product.

The opacity of antimony oxide, caused by its insolubility in PVC and the large differences in refractive indexes, make it unsuitable for use in deep-colored or transparent coatings. Such opacity tends to make grayish colors and causes problems in color matching when the amount of antimony oxide is varied to modify flame retardancy.

3.1.3.2. Phosphate Polyesters. Another important method of increasing flame retardation of PVC fabric coatings is by using phosphate esters (e.g., trioctyl phosphate). The following mechanisms have been proposed as explanations of flame retardation by phosphates.

1. A lower fuel contribution results since phosphorus cannot be further oxidized.
2. It forms a very viscous coating upon heating which protects the substrate from penetration of air and heat.[36]
3. The acidity at high temperature induces char formation[114] which protects the substrate.
4. The main combustion products are carbonaceous char and CO instead of CO_2.[113] The heat of formation of CO_2 is 94.0 kcal/mol but the heat of formation of CO is only 26.4 kcal/mol. This represents a great reduction of heat of combustion per C atom during the burning stage. The heat of formation of char is unknown, but the heat of reaction to form char must be much less than that of the formation of CO per carbon atom.

The main advantage in using phosphate esters over other types of flame-retardant additives, such as antimony oxide, is their lower cost. They have the ability to produce clear and deep colored coatings and are low in toxicity and volatility. In addition, the plasticizing effect of the phosphate esters improves the processability of the formulated coatings.

Some factors[95] which limit the use of phosphate as plasticizers are their higher cost and lower plasticizer efficiency when compared to non-flame-retardant plasticizers. Then too, relatively high levels of phosphate esters are necessary for the same flame retardance possible with antimony oxide. Furthermore, compounded coatings may have decreased heat stability during processing, but this often can be corrected by the addition of an appropriate stabilizer such as an epoxy resin.

There are over 200 phosphorus-containing flame retardants available,[112] but only a very few are of great commercial importance as plasticizers for PVC. The most widely used are trioctyl phosphate, isodecyl diphenyl phosphate, tolyl phenyl phosphate (mixture), tritolyl phosphate, and trixylyl phosphate.[115,116] More than one plasticizer is often used to achieve greater flexibility or better overall properties. For example, bis(2-ethylhexyl)phenyl phosphate[117] is used with a mixture of dioctyl phthalate, dioctyl sebacate, and antimony oxide or ammonium molybdate. This mixture is blended with PVC to produce a coating for tents that is flexible at $-40°C$.

Halogenated phosphates such as tris(chloroethyl) phosphate induce a greater degree of flame retardancy than regular phosphates because of the presence of halogens. However, halogenated phosphates are more expensive, less compatible with PVC, and decompose at lower temperatures compared to regular phosphate plasticizers.

3.1.3.3. Chlorinated Paraffins and Aromatics. These materials represent another type of plasticizer used for PVC whose flame-retardation is enhanced by the high chlorine content (40–70%). However, they are generally classified as secondary plasticizers since they spew and exude to the surface when used at concentrations above 30% of the total plasticizer. These types of plasticizers generally have low viscosities and vapor pressures which make them easy to process. Poor plasticizer efficiency—almost double amounts are needed to plasticize PVC—and decreased heat and light stability of plasticized formulations are offset by their low cost.

3.1.3.4. Other Flame Retardants. Many of the same materials used to improve the flame retardance of architectural coatings are used with fabric coatings. Zinc borate, for example, shows a similar synergistic effect with chlorine, although it is not as strong as antimony oxide. Aluminum hydrate is also used to improve the flame retardance of architectural coatings. It loses water by an endothermic reaction and absorbs heat from the flame. In addition, the water vapor dilutes the flame and helps to develop char,[118,119] but processing temperatures must be kept below 350–400°F to avoid dehydration of the additives. Flame retardance is improved by pigments and colorants[120] used to modify color and by inorganic filler materials used to lower cost, reduce shrinkage, and increase hardness. These additives "dilute" the combustible material present in the coating.

TABLE 2
Flame-Retardant Coatings

Ingredients	Parts by weight
PVC clear coating[121]	
Opalon 650 PVC Resin (Monsanto)	100.00
Dioctyl phthalate	26.00
Di-n-octyl, n-decyl/adipate	7.00
Santicizer 141, Octyldiphenyl phosphate (Monsanto)	12.00
Santicizer 409, phosphate plasticizer (Monsanto)	10.00
Calcium carbonate filler	20.00
Epoxy stabilizing plasticizer	3.00
PVC plastisol opaque coating[122]	
PVC 71 (Diamond Shamrock)	100.00
Dioctyl phthalate	57.00
Chlorowax 500 (Diamond Shamrock)	31.00
Paraplex G-62 (Rohm and Haas)	5.00
Dymstab SL-262 (Nopco)	3.00
Antimony oxide	30.00
Surfex MM (Diamond Shamrock)	30.00

3.1.3.5. PVC Formulation Recipes. A clear coating and an opaque coating formulation composition are shown in Table 2.

3.2. Polyurethane Fabric Coatings

The elastomeric polyurethane fabric coatings market has shown a substantial increase in the last decade as a result of the increased popularity of polyurethane-coated fabrics. Such fabrics have high tensile and tear strength, good chemical and abrasion resistance, and they are lightweight. Coatings can be made with a "wet" look and/or an excellent hand (i.e., how the material feels when handled) similar to leather. Since much of the coated fabric is used for clothing and upholstery in motor vehicles,[123] the demand for flame-retardant polyurethane fabric coatings has also greatly increased.

Many of the techniques[24] used to produce architectural polyurethane coatings can also be used for fabric coatings. Polyurethane lacquers are available in both water-base systems or solvent systems. The latter require only solvent evaporation to produce a dry, nontacky coating. Thermoset coating systems which are available include one-package coatings such as polyurethane polyols cured with urea–formaldehyde resins, isocyanate-containing polymers cured with atmospheric moisture, and polyester or polyether polyols cured with blocked isocyanate prepolymers and two-package systems such as polyol plus isocyanate prepolymer.

Most polyurethane coatings are not highly flammable as a result of the relatively high nitrogen content. As with architectural polyurethane coatings, an increase in flame retardance usually can be obtained by in-corporating in the coating flame-retardant additives such as polyhalogenated compounds, antimony oxide, and/or phosphorus compounds. A wide variety of additives, nonreactive or reactive, may be used, but the reactive ones require relatively low temperatures for processing and curing the coatings.

The recent literature regarding the flame retardance of polyurethanes has been reviewed by Frisch[27] and Lyons.[124] While most of the information deals with polyurethane foams, much of the technology is also applicable to fabric coatings. The main differences are that the coatings are relatively thin and the fabric support may have a substantial influence on burning characteristics as with PVC coatings.

3.2.1. Nonreactive Flame-Retardant Additives

Nonreactive additives are often used in polyurethane lacquers because the lacquers are based on high-molecular-weight polymers that require pure difunctional raw materials.[125] Most reactive flame-retardant polyols and polyisocyanates do not meet these requirements, hence additives are usually added after polymer preparation. This allows incorporation of a greater variety of additives which may be used to maximize performance and lower cost. For these same reasons, many nonreactive additives are also used in thermoset coatings.

As in the case of PVC and other coatings, only a relatively small number of the hundreds of flame-retardant additives available[109] have shown commercial importance. The most commonly used additives are listed.

1. Metal oxides, such as antimony oxide, zinc oxide, and potassium fluoroborate, are particularly used with halogenated additives.
2. Halogenated compounds of PVC in both powder form[126] and solution[127] may be used with polyurethane coatings, and chloroparaffins are sometimes used as a chlorine source.
3. Phosphorus compounds have been extensively used in polyurethane foams as flame-retardant additives and to a lesser degree in polyurethane coatings.

The halogenated phosphorus compounds are highly efficient flame retarders, but they are generally sensitive to processing temperatures. However, since many polyurethane fabric coatings can be processed at low temperatures, heat stability is often not a serious problem. The addition of lithium salts improves the heat and humidity resistance of polyurethanes containing phosphorus compounds.[128] In experiments with foams a maximum flame-retarding effect is reached at phosphorus levels of $1-2\%$. There is not a proportional increase in flame retardance with increase in phosphorus content, and similar behavior would be expected in coatings. Tris(chloroethyl) phosphate and tris(dibromopropyl) phosphate are most often used because of their relatively low cost and good performance properties. Another similar compound[129] is Phosguard 2XC20 (Monsanto Chemical Co.) with a structure as shown below:

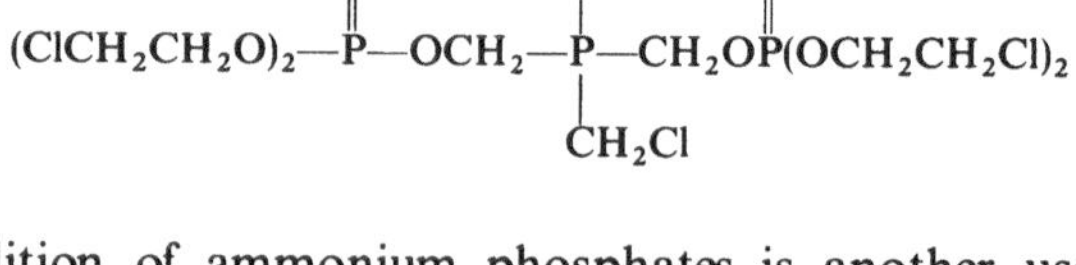

$$(ClCH_2CH_2O)_2-\overset{\overset{\displaystyle O}{\|}}{P}-OCH_2-\overset{\overset{\displaystyle CH_2Cl}{|}}{\underset{\underset{\displaystyle CH_2Cl}{|}}{P}}-CH_2O\overset{\overset{\displaystyle O}{\|}}{P}(OCH_2CH_2Cl)_2$$

The addition of ammonium phosphates is another useful, low-cost way to enhance flame retardance of polyurethanes. Diammonium and triammonium phosphates may be used, but they are highly water soluble and are susceptible to extraction during washing of coated fabrics. Ammonium polyphosphate is only slightly water soluble[130] and thus is less sensitive to contact with water and washing.

3.2.2. Reactive Flame-retardant Additives

The incorporation of reactive flame retardants in polyurethane coatings allows the flame-retarding compounds to be chemically bound to the polymer, and thus these materials are not susceptible to being removed by exposure to water, washing, or other cleaning solvents. These types of flame retardants are used only in thermoset coatings. Many flame-retardant polyols are available but only a few polyisocyanates are available commercially.

3.2.2.1. Polyisocyanates. Polyisocyanates are produced commercially by the reaction of phosgene with polyamines. This reaction involves complicated technology, and consequently there are few commercially available polyisocyanates. Their desirability as flame retardants, indicated by char-formation potential, lack of objectionable color formation upon weathering, and relative cost, is evaluated in the following paragraphs.

Toluene diisocyanate (TDI) is one of the lowest cost polyisocyanates. Although colorless coatings can be made, they tend to yellow upon aging. TDI is a pure difunctional material that can also be used to prepare thermoplastic polyurethanes. It has the least char formation potential of all aromatic polyisocyanates.

Crude methylene bis(phenyl isocyanate), (MDI), is relatively low in cost, but coatings produced from this highly colored material become even darker upon aging. Thermoplastic polyurethanes can not be made from this material. It has a medium char-formation potential.

Pure MDI is a high-cost material in comparison to the other aromatic isocyanates, but the colorless coatings produced tend to yellow upon aging. A pure difunctional material, MDI can be used to make thermoplastic polyurethanes. The product has a medium char-formation potential.

Polyaromatic polyisocyanate (PAPI) is derived from phenol–formaldehyde resins. It is a relatively low-cost, highly colored, crude polyfunctional material that produces colored coatings which deepen in color upon aging. Although it is not suitable for thermoplastic materials, it has the best char-formation potential among all common isocyanates. However, polyurethanes based on this material tend to produce large amounts of smoke when burning.[132]

Hydrogenated MDI (H_{12}MDI) or methylene bis(cyclohexyl) isocyanate is a high-cost aliphatic isocyanate, coatings of which do not yellow on exposure. It is a pure difunctional material and is suitable for producing thermoplastic polyurethanes. It has less char-formation potential than MDI.

Other polyisocyanates available in commercial quantities are isophorone diisocyanate, trimethyl hexamethylene diisocyanate, hexamethylene diisocyanate, and xylylene diisocyanate. Various halogen-containing polyisocyanates[133,134] and bromotoluene diisocyanate,[135,136] as well as phosphorus-containing polyisocyanates,[137-141] have been reported. Polyurethanes prepared from phosphorus polyisocyanates are usually sensitive to hydrolysis.[142] None of the above materials have had any significant usage in today's fabric coatings.

   3.2.2.2. Polyols and Amino Polyols. A wide variety of flame-retardant polyols is currently available because of the relatively simple processes involved in producing the two general types currently used, polyester and polyether polyols. Polyester polyols are prepared by reacting polyacids with low-molecular-weight polyols. Either or both components may be halogenated to increase flame retardance. For example, chlorendic anhydride may be reacted with excess glycerine to form a polyester polyol. Polyether polyols are made by reacting low-molecular-weight polyols with alkylene oxides or epichlorohydrin. Many flame-retardant polyether polyols generally are based on low-molecular-weight halogenated polyols. Phosphorus polyols are made in a similar manner by reacting phosphoric acid with alkylene oxides.

   It is difficult to evaluate the overall performance of flame-retardant polyols because of the large number and variety of proprietary polyurethane formulations. Information on them comes almost exclusively from patent disclosures and product bulletins. The general advantages and disadvantages of some flame-retardant polyols are summarized below.

   Polyester polyols, as a result of greater thermal stability, generally give polyurethanes with better flame retardance than similar polyurethanes prepared from polyether polyols.[143] Polyether urethanes are more sensitive to ultraviolet radiation, have better hydrolytic stability, and are usually more flexible. The best low-temperature flexibility is obtained with polytetramethylene ether glycols.[144]

   The amount and type of halogen present determines the extent of improvement of flame retardance of polyurethanes. Bromine generally is more effective than chlorine. Halogens attached to aliphatic carbon are more effective than those attached to aromatic groups. The position of bromine atoms in the polymer also influences flame retardance;[158,160] however, the heat stability, hydrolytic stability, and weather resistance of aliphatic bromides is usually not acceptable for fabric coatings.

   Polyester polyols containing chlorendic acid have high heat and hydrolytic stability and are generally suitable polyols for flame-retardant polyurethane coatings.[161-164] A relatively new method[144,165,166] of preparing chlorinated polyether polyols uses trichloromethylethylene oxide as the halogen source and may be useful in polyurethane coatings.

   As with other materials, the synergistic effect in using halogens and antimony oxide[160,162] and the less pronounced synergism between phosphorus compounds and halogens[164,167] provides increased flame retardance. In addition to the use of phosphorus and halogenated polyols, the use of amine-containing polyols will also increase flame retardancy because amines are reduced to NH_3 instead of being oxidized in the presence of flaming conditions. This avoids the heat of combustion change of hydrogen into water that most organic materials display.

Polyurethanes based on low-molecular-weight polyethers give better flame retardation[145] than high-molecular-weight polyethers, probably because low-molecular-weight polyethers require more polyisocyanate for curing. The incorporation of a higher amount of polyisocyanate increases nitrogen content and aromatic ring content when aromatic polyisocyanates are used, both of which increase flame retardance. The same should be true of high- and low-molecular-weight polyester polyols.

Incorporating or incrasing the cyclic components, such as the addition of hydroxyalkylated bisphenolic products[146] and trimellitic anhydride polyesters,[147] in the polyol increases flame retardancy. Cyclic polyethers are more effective flame-retardant additives than open-chain polyethers;[131,143,148] however, materials containing large amounts of aromatic rings may not be flexible for fabric coatings.

Increased flame retardancy may be obtained by incorporating phosphorus into the polyurethane polymer. Polyols derived from phosphoric acid[124,149-153] are inexpensive but usually have relatively poor hydrolytic stability.[140] Phosphorus compounds containing phosphorus–carbon bonds have better hydrolytic stability[154] but generally cost more. Phosphate-containing polyols have poor hydrolytic stability and are of limited value in producing durable flame-retardant polyurethane coatings.

When a phosphorus diol is used to make polyurethanes, there is some evidence that it is more flame retardant if the phosphorus atom is not in the polyurethane backbone, but rather on a side chain. Thus diethyl-*N,N*-bis(hydroxyethyl)aminomethyl phosphate is a better flame-retardant additive than its isomer, bis(hydroxyethyl)-*N,N*-diethylaminomethyl phosphonate.[155]

$$(CH_3CH_2O)_2\overset{\displaystyle O}{\overset{\displaystyle \|}{P}}-CH_2N(CH_2CH_2OH)_2 > (HOCH_2CH_2-O)_2\overset{\displaystyle O}{\overset{\displaystyle \|}{P}}-CH_2N(CH_2CH_3)_2$$

The chemical incorporation of phosphorus into polyurethanes gives better flame retardancy than simple blending of nonreactive additives.[130,143] The following polyols containing phosphorus

$$(HOCH_2)_3P$$

$$(HOCH_2)_3P{=}O$$

$$(HOCH_2)_4P-Cl$$

$$(HOCH_2)_2P-CH_2CH_2CH_2CH_3$$

$$(HOCH_2)_2\overset{\displaystyle O}{\overset{\displaystyle \|}{P}}\,CH_2CH_2CH_2CH_3$$

$$H_2N \langle\!\!-\!\!-\!\!-\!\rangle\!\!-\!\!P\!\!-\!\!\langle\!\!-\!\!-\!\!-\!\rangle\!\!-\!NH_2$$

have been incorporated into polyurethane polymers,[154,156-158] but their usefulness for polyurethane coatings has not been established. In addition to providing acceptable performance properties, flame-retardant materials must also be economically competitive. Since many of the above compounds are relatively expensive, there is probably only limited growth potential for these materials based on current economics.

Polyols with high OH functionality[143] per unit weight increase flame retardancy by creating high cross-link densities. This creates a hard rigid material that resists burning. However, the flexibility of coated fabrics is often seriously decreased.

3.3. Other Coatings Suitable for Fabrics

Many polymeric materials capable of forming flexible films can be used to make flame-retardant coatings for fabrics. Most elastomeric polymers, the more common of which have been reviewed by Fabris and Summer,[168] can also be used, including those coating systems that can be flexibilized by plasticizers. However, because of cost vs. performance properties and the limited market, very few polymeric materials have been developed for use as fabric coatings. The polymers that have found commercial utility can be divided into self-extinguishing, nonburning, combustible, and nonelastomeric materials groups.

3.3.1. Self-Extinguishing Elastomers

Chlorinated rubbers, such as chloroprene and chloroprene–acrylonitrile copolymers, have been suggested as excellent materials to be used in formulating flame-retardant products, since they are relatively low in cost[169] and, when extreme flame retardation is needed, may have added up to 30 phr of antimony oxide and 30 phr zinc borate. Hydrated alumina and kaolin may also be added to improve the flame retardancy of these materials.[171,172]

Chlorosulfonated polyethylenes are produced from the combined reaction of sulfur dioxide and chlorine on polyethylene and cured with metal oxides. The use of clay and carbon black as fillers increases the flame retardancy,[173] and antimony oxide or hydrated alumina may be added to the coating when extreme flame retardation is needed. Tricresyl phosphate and chlorinated paraffin have been suggested as plasticizers for chlorosulfonated polyethylene to increase the flexibility and flame retardance.[168]

Silicone elastomers such as methyl vinyl siloxane (Silastic 745 U, Dow-Corning) and methyl vinyl phenyl siloxane (Silastic 446 Base, Dow Corning) are polymers which are rated as self-extinguishing by Trexler.[174] The main advantage of these polymers is that they maintain their flexibility and strength over temperatures ranging from -110 to $+310°C$, but are degraded at very high temperatures. The average char obtained on exposing the polymer to an arc jet can be as high as 90% of the original weight.[175] In order to further increase flame retardancy, halogenated organic compounds, antimony oxide, aluminum silicate, and other metal oxides can be used.[176-179]

Chlorinated polyethylene (Dow MX2245-25), tested by Trexler[174] along with 34 other commercial elastomers, was the only fluorine-free material found to be nonburning. It is made by chlorination of polyethylene and is similar in characteristics[180] to PVC. Because of this similarity and because it is low in cost, it has been suggested as a substitute for PVC.

Many fluoroelastomers, such as 3M's Fluorel L-3203-6[181,182] and Du Pont's Viton,[183] are copolymers of hexafluoropropane and vinylidene fluoride; Kel-F (Spencer Kellog Co.) is a copolymer of vinylidene fluoride and trifluorochloroethylene. All are nonburning polymers. Fluorel L-3203-6 and Fluorel KF-2140 have been found to have good flame retardance under the extremely severe conditions encountered in space exploration,[181,184] and some of the protective coatings derived from this family of polymers are self-extinguishing in an atmosphere of pure oxygen.[185]

Nitroso rubbers, copolymers of trifluoronitrosomethane and tetrafluoroethylene or trifluoroethylene, are shown in the following formula:

$$\left[N\!-\!O\!\!+\!\!CF_2\!-\!CF)_{\!n}\right]_{\!m}$$
$$\underset{CF_3}{|} \qquad \underset{X}{|} \qquad X = H \text{ or } F$$

These polymers, which are nonburning by the most rigid standards, have been investigated primarily for use in aerospace applications.[186]

3.3.2. Combustible Elastomers

Almost all film-forming elastomers can be used as fabric coatings and almost all polymers and elastomers can be formulated to obtain some degree of flame retardance. Numerous articles and patents report techniques used for increasing flame retardance, such as the incorporation of flame-retardant comonomers in the vehicle polymer or the use of additives in the coating formulation. An interesting study recently reported[174] that 34 elastomers were compounded to achieve a "nonburning" product by using only four flame-retardant additives, namely, antimony oxide; Chloro-

TABLE 3

Summary of Nonburning Compounds

Polymer	Class	Unplasticized Compound No.	Additives[a]	Class	Plasticized Compound No.	Additives[a]
Natural poly(isoprene)	NB	NR-110	1, 2			
Synthetic poly(isoprene)	NB	IR-38	1, 2			
Styrene–butadiene	NB	SBR-26	1, 2			
Poly(butadiene)	NB	BR-36	1, 2			
Ethylene–propylene terpolymer	NB	EPDM-27	1, 2	NB	EPDM-29	1, 2, 4
Isobutylene–isoprene	NB	IIR-48	1, 2	NB	IIR-50	1, 2, 4
Chloroisobutylene–isoprene	NB	IIR-38B	1, 2	NB	IIR-26B	1, 2, 4
Epichlorohydrin	NB	CO-13	1, 2			
Epichlorohydrin–ethylene oxide	NB	CO-28	1, 3			
Chloroprene	NB	CR-13	1, 3			
Chlorosulfonated poly(ethylene)	NB	CSM-13	1, 3			
Polyester urethane	NB	AU-18	1, 3			
Butadiene acrylonitrile (33% AN)	NB	NBR-18A	1, 2			
Butadiene acrylonitrile (44% AN)	NB	NBR-71B	1, 2			
Carboxylated butadiene acrylonitrile	NB	NBR-27	1, 2			
Nitrile polyvinyl chloride (70-30)	NB	NBR-54	1, 2			
Nitrile polyvinyl chloride (50-50)	NB	NBR-63	1, 2			
Nitrile polyvinyl chloride (50-50 plasticized)	NB	NBR-44	1, 2			
Polyacrylate (Hycar 4021)	NB	ABR-14	1, 3			
Polyacrylate (Hycar 4032)	NB	ABR-28B	1, 3			
Polyacrylate (Thiacril 76)	NB	ABR-64	1, 3			
Polyacrylate (Cyanacryl R)	NB	ABR-45	1, 3			
Polyacrylate (Cyanacryl LT 3)	NB	ABR-55	1, 3			

[a] 1 = antimony oxide; 2 = Chlorowax 70, chlorinated paraffin, 70% chlorine; 3 = Dechlorane 515 (perchloropentacyclodecane); 4 = Fyrol FR-2[tris(dichloropropyl)phosphate].

was 70, a chlorinated paraffin containing 70% chlorine; Dechlorane 515, perchloropentacyclodecane; and Fyrol FR-2 tris(dichloropropyl phosphate).

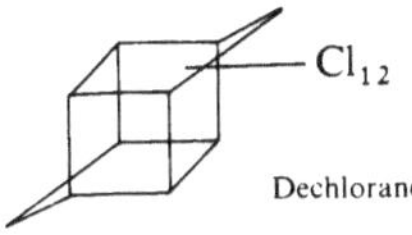

Dechlorane 515

$$O=P-(OCH_2\overset{\displaystyle Cl}{\overset{\displaystyle |}{C}H}-\overset{\displaystyle Cl}{\overset{\displaystyle |}{C}H_2})_3$$

Fyrol FR-2

The results of this study are summarized in Table 3.

The physical properties of elastomers compounded with flame-retardant additives usually change. As an example, Table 4 shows the composition and the effects of adding antimony oxide and Chlorowax 70 to polyisoprene and natural rubber.

In Trexler's study there are also four materials which could not be formulated to obtain a nonburning classification. The four materials are: bromoisobutylene–isoprene (Hycar 2202, Goodrich Rubber Co.), methyl vinyl phenyl siloxane (Silastic 446 Base, Dow Corning), propylene oxide (Dynagen XP-139, from an unknown source), and polyether urethane (Adiprene L-100, Du Pont Chemical Co.). This probably resulted from

TABLE 4
Natural and Synthetic Poly(isoprene)

Ingredients	Natural		Synthetic	
	Control	NR-110	Control	IR-38
Smoked sheet	100	100		
Shell isoprene 305			100	100
Zinc oxide	3	3	3	3
Stearic acid	3	3	3	3
N 301 carbon black	43	43		
N 330 carbon black			50	50
Captax	1	1		
Santocure			0.5	0.5
Sulfur	3	3	2.4	2.4
Antimony oxide		25		20
Chlorowax 70		50		40
Press cure: min/°F	30/300		25/300	
Flammability				
Match, in./min	C7.6	NB	C7.2	NB
Mil 417, in./min	C4.6	NB	C3.5	NB
Tensile strength, psi	3295	1760	3230	2350
Elongation, %	515	590	425	640
Durometer	63	53	50	56
Tear strength, lb/in.	595	160	420	410

the inherent instability of the polymer in the presence of heat, as well as incompatibility of additives with the polymer. However, these materials did have enough flame retardance to be used as fabric coatings in many areas.

3.3.3. Nonelastomeric Fabric Coatings

Any coating material that can be plasticized to produce a flexible film could be used for fabric coatings. However, very few coatings of this type offer any major advantages over currently available PVC coatings. Hence, there is little incentive to formulate coatings of this type and little information is available. Hajela *et al.*,[187] reported that a modified alkyd resin based on linseed oil photochlorinated in carbon tetrachloride showed good flame retardance on cotton fabric when formulated with antimony oxide. Another report[188] disclosed that when polyester and nylon-6 textiles were back-coated with self-cross-linking polyacrylate, urea resin, or silicone resin, they were resistant to thermal shrinking and did not form holes on direct contact with lighted cigarettes.

4. Heat-Resistant Organic Coatings

Heat-resistant organic coatings are expensive. As a result, they are usually not used for applications primarily requiring flame retardance. Some are so stable that they may be used as exterior coatings for cooking wares which will be in direct contact with flame. This fact is consistent with the current theories of flame propagation: Combustible materials decompose in the presence of heat to produce combustible gases. The exothermic oxidation of these gases generates more heat which, in turn, produces more combustible gases, and so forth. Since heat-resistant coatings are stable at high temperatures, little or no combustible gases are produced. Thus ignition and flame propagation are retarded or do not occur.

The flame retardance of heat-resistant coatings is often not tested, but the performance of these coatings is evaluated by measuring thermal decomposition. The extent of decomposition is dependent upon temperature and time, as well as the nature of the surrounding atmosphere (oxygen rich, inert, vacuum, etc.) and is expressed in terms of a temperature–time limit or a temperatue–time curve. For example, the thermostability of some silicone resins is 1000 hr at 250°C or 710 yr at 180°C.

Thermogravimetric analysis (TGA) is most often used to determine heat stability of heat-resistant coating materials because it is a simple and rapid method of determining the weight loss of a material as a function of temperature and time. Heat stability is expressed in terms of the time and/or temperature required to reach a specified percent weight loss. Other methods such as the isoteniscope test, differential thermal analysis (DTA), and torsional braid analysis (TBA) are less often used.

The heat-resistant organic coatings which have the largest use are aromatic polyimides, polyamidimide and polyesterimide, silicone resins, and fluorocarbon polymers.

Inorganic coatings are often heat and flame resistant, either because the material is in its highest oxidation state (silicates, borates, and cement coatings) or because the material does not produce any combustible gases when heated (graphite, various metal coatings, and hydrated metal oxides). Coatings such as zinc-rich silicate contain both types of materials. Since inorganic coatings are outside the scope of this chapter, further discussion of this area will be limited to organic and organometallic coatings.

4.1. Polyimide Coatings

Polymide coatings, some of the most heat-resistant organic coatings known, are widely used in applications requiring excellent performance properties at elevated temperatures. These coatings are flame retardant to the point of the films being noncombustible. A free polyimide film has zero strength at 800°F (20 psi load, 5 sec to failure), and no melting point. When the film is exposed to direct flame, it only chars.

The cured coating is almost indestructable by chemicals, moisture resistant, insensitive to solvents, grease, lubricants, and strong acids.[189] Further, these properties are retained over a wide temperature range from −250 to over 300°C. The coating also has excellent irradiative heat-transfer properties indicative of ability to transfer heat from the substrate to the environment. Such coatings find important application as coverings for magnet wires. Polyimide coatings exhibit good resistance to tear initiation,[190] high tensile and high impact strength, and they are only soluble in concentrated sulfuric and fuming nitric acids.

Polyimide polymers must be applied in a soluble form to produce coatings. One method of making polyimide coatings is to first form a high-molecular-weight polyamic acid solution by reacting a pure diacid anhydride (e.g., pyromellitic anhydride) with a diamine (e.g., 4,4′-diaminodiphenyl ether) in a suitable potent solvent (e.g., dimethyl formamide)[191-198] This solution is then used to coat the intended object, after which the solvent is removed and the coating dehydrated to form the polyimide structure by either a chemical[193-200] or thermal[201,202] method. In the thermal method, the coated article is heated at 300°C or heated in progressive stages up to 300°C for a period of up to 1 hr. In the chemical method, the

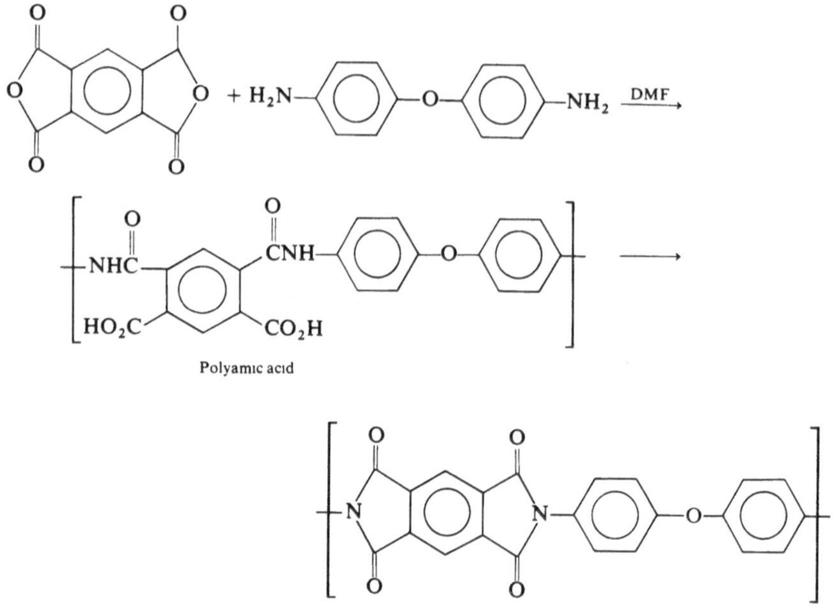

Polyamic acid

Polyimide

conversion is effected by utilizing a dehydrating agent such as acetic anhydride and a catalyst such as trimethylamine. Infrared analysis may be used to determine the completion of the dehydration by measuring the residual carbonyl absorptions of the acid and amide groups.

The use of different diacid anhydrides, diamines, solvents, and dehydrating agents will give coatings with somewhat different physical properties. All the coatings have excellent properties and heat resistance; however, they are highly colored and may range from light yellow to dark brown. The combination of good heat resistance and electrical properties —some have a service temperature in excess of 260°C—makes polyimides the best wire coatings available today. Decorative coatings for cookware are possible with polyimide coatings because of their ability to withstand direct contact with flame.

The high cost of these coatings, up to $45/lb for the finished coating, limits their use. Another limiting factor is the rather severe curing conditions which restricts the types of materials to which polyimide coatings can be applied. In an effort to overcome this handicap, many copolymers of polyimides have been developed that have poorer heat resistance but are less expensive and easier to apply. The copolymers, which may be polyesterimides, polyamidimides, etc., can be prepared by methods similar to that for the polyimides.

One such copolymer is prepared by reacting the diacid anhydride with phenol to form a soluble ester. This material is dissolved in a suitable solvent and mixed with a primary aromatic diamine and a catalyst such as a tertiary amine to give a clear, high-viscosity solution. This solution will produce a clear flexible coating when applied to a substrate and baked at an elevated temperature (ca. 350°C).[203-205]

4.2. Silicone Coatings

Heat-resistant and flame-retardant silicone coatings are based on polymers of disubstituted siloxanes of the following structure:

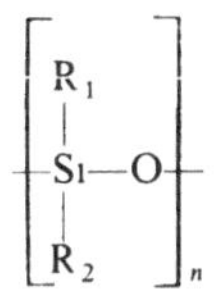

$$\left[\begin{matrix} R_1 \\ | \\ Si-O \\ | \\ R_2 \end{matrix}\right]_n$$

where R_1 and R_2 may be low-molecular-weight alkyl or aryl groups. For use as coatings, the polymers may be the product of several different monomers. Cross-linking can be introduced by including alkyl triacetoxy-silanes. After the coating is applied, hydrolysis by moisture creates Si—OH groups and the group cross-links to Si—O—Si linkages.

These coatings usually have low temperature flexibility, exterior durability, ozone- and corona-discharge resistance, minimum dependence of physical properties upon temperature, and good dielectric properties. In addition, the polymers have good film-forming properties and are hydrophobic and physiologically inert. Many silicone coatings can be heated in air at 180–200°C for a year without any appreciable change in their properties. They can also be subjected to temperatures as high as 350–450°C for shorter periods without significant loss of properties.

Under extreme conditions silicones are degraded either by oxidative rupture of the Si–C bond or by depolymerization. The oxidative degradation of polyorganosiloxanes begins with oxidation of the organic portions of the polymer. The degradation products of polymethyl siloxane in air have been found to be formaldehyde and formic acid.[206,207] The ease of oxidation increases with lengthening of the alkyl substituent. Thus the heat resistance of ethyl silicones is lower than the heat resistance of methyl silicones. However, phenyl-substituted siloxanes, when compared to methyl siloxanes, have superior oxidation resistance at elevated temperatures because they are less susceptible to oxidation than aliphatic groups. This is especially so below 400°C, owing to the resonance stabilization in the aromatic ring.

Thermal depolymerization, which leads to lower-molecular-weight cyclic products, is catalyzed by Lewis acids and/or water vapor. Here again phenyl-substituted siloxane polymers have more resistance because of increased $d\pi$-$p\pi$ interactions between silicon and oxygen.

The inert nature of polyorganosiloxanes and the absence of reactive sites, which make silicones resistant to thermal decomposition, also provide

TABLE 5
Recommended Uses of Silicone Resin Lacquers

Property determining the application	Type of silicone resin lacquer	Examples of application
Heat resistance		
Up to 230°C	Pure or modified silicone resin mixed with heat-resistant pigments. (Stowing enamels)	Smoothing irons Geysers Electrical heating appliances Domestic furnaces High-pressure steam conduits Furnaces and ovens, particularly oil furnaces Chimneys
	Unpigmented coatings or enamels colored with transparent pigments. (Stoving enamels)	Incandescent lamps Headlamp glasses Inspection glasses
Up to 400°C	Silicone lacquers with pigment (titanium dioxide)/binder ratio $\approx$ 3/1. (Stoving enamels with poor corrosion resistance)	Furnace tubes and doors Ovens
Up to 500°C, in special cases 600°C	Straight silicone resins pigmented with metallic pigments, especially Zn and Al, or with iron, mica, or or graphite. (Air drying)	Silencers and exhaust piepes Rotary kilns Furnace doors Aircraft parts Blast furnace units Engine parts Rockets Chimneys
Release effect	Clear or pigmented pure or modified silicone resins, including those containing relatively large amounts of silicone oil. (Stoving enamels)	Baking sheets and molds Frying pans Dust-removing planst Deicing units Molds for the plastics and rubber industry Drying pans and drums
Weather resistance	Mainly pigmented and modified silicone resins. (Stoving enamels)	Lighting units Reflectors Traffic signs Facade cladding sheets

resistance to weathering effects, ozone, glow discharge, and ultraviolet radiation. Resistance to weathering is enhanced by the high interfacial tension of all silicones against water, thereby minimizing any hydrolysis of the polymer.

The heat resistance of some silicones can be increased by adding certain pigments, such as aluminum bronze and zinc dust, to the coating. When a coating pigmented with zinc dust is covered with a coating pigmented with aluminum bronze, the composite coatings are stable at temperatures[208] in excess of the melting point of zinc ($419^\circ C$). Silicone enamel coatings pigmented with TiO_2 have been shown to have excellent gloss retention and do not yellow even after several hundred hours[209] at $400^\circ C$. However, these coatings do show a substantial loss of flexibility after extended exposure at high temperatures.

Silicone coatings may be used as substitutes for vitreous enamels in applications requiring operating temperatures below $400^\circ C$. Although vitreous enamels are harder and have a greater heat and chemical resistance, the silicone coatings have better flexibility, can be applied by conventional methods and baked on at relatively low temperatures. Table 5 illustrates the wide range of uses of silicone lacquer coatings.[210]

4.3. Fluoropolymers

Coatings based on fluoropolymers are effective in applications requiring good heat and chemical resistance. They have been used extensively for nonstick cookware, corrosion-resistant coatings, and in oxygen-rich environments where a high degree of flame retardance is necessary. Polytetrafluoroethylene (PTFE), polyvinylidene fluoride (PVF$_2$), and polyvinyl fluoride (PVF) are the fluoropolymers most widely used as heat-resistant coatings.

4.3.1. Polytetrafluoroethylene (PTFE)

PTFE, a fully fluorinated polymer has been found stable at high temperatures, in vacuum,[211] and in oxygen.[212] Coatings made from it are resistant to chemical attack, showing 3.7% increase in weight after 8 hr in CCl_4 at $200^\circ C$.[229,230] Although resistant to ultraviolet radiation, PTFE is somewhat degraded by high-energy radiation.[213]

Thermal decomposition occurs very slowly at $250^\circ C$ giving off small amounts of gaseous products. The initial weight loss for one PTFE coating is $0.0001\%/hr$ at $260^\circ C$, and $0.004\%/hr$ at $370^\circ C$. However, above $400^\circ C$, the rate of decomposition increases rapidly. The main decomposition product, tetrafluoroethylene monomer, is flammable when in a concentrated form. The polymer itself is nonflammable probably because of an insufficient concentration of the combustible monomer formation. This monomer, although more toxic than ethylene, still has a relatively low toxicity.[214] The coating can be used in contact with cooking food.

Coatings based on PTFE have many uses; however, its high cost ($3.20–4.60/lb) and high curing temperatures tend to limit its wider use. Currently the nonstick properties of PTFE coatings are in demand for kitchen cookware, steam irons, and high-temperature wire coatings.

PTFE coatings are "nonstick" because of their low surface energy (20 dyne/cm at 30°C). Since most liquids and liquid mixtures (foodstuffs, etc.) have a higher surface energy, they cannot wet or stick to the PTFE surface.

PTFE coating is cured by a very different technique than other coatings. Liquid dispersions of the powdered PTFE polymer are applied by spray, flow coating, dip, or electrodeposition. After the coating is applied, the volatile liquid is removed by heating at 70–80°C; then any wetting agents present are removed at 280–290°C. Finally, the powdered PTFE is sintered at 380°C to form a continuous film. The coating is usually 1.5 mils thick, and repeated applications are necessary for thicker coatings.

4.3.2. Polyvinylidene Fluoride

Polyvinylidene fluoride (PVF_2) is very similar to PTFE in that it is resistant to corrosive chemicals, oxygen, and, to a lesser extent, heat. The alternating methylene and difluoromethylene groups change the morphology of the polymer. Unlike PTFE, PVF_2 has a melting point (170°C), T_g value ($-40°C$), and is soluble or will form organosols with ω-butyrolactone, N,N-dimethylacetamide, dimethyl phthalate, 2-ethoxyethyl acetate, or Carbitol acetate. Although the material cannot be used above its melting point, PVF_2 is stable in air at 200°C, 30°C above its melting point. It decomposes in 15–60 min at 250°C. PVF_2, poorer in heat resistance than PTFE, is self-extinguishing and forms a char when exposed to a flame. The decomposition products are carbon and hydrogen fluoride. Its greater solubility makes it more useful in coatings than PTFE. Current applications include chemical- and corrosion-resistant coatings such as fuel tanks, pipelines, valves, and exterior coatings that are weather-, abrasion-, and chip-resistant. Various properties can be obtained by copolymerization of vinylidene fluoride with tetrafluoroethylene,[215] chlorotrifluoroethylene,[216-218] and hexafluoropropylene.[219-221] This coating, along with other fluoropolymer coatings, finds limited use because of its high cost (ca. $4.00/lb).

4.3.3. Polyvinyl Fluoride

This is another fluoropolymer coating with excellent toughness, flexibility, and chemical-resistant properties. The compound has a lower melting point (105°C) and poorer heat resistance than PTFE or PVF_2. Like those coatings, it has excellent weather resistance, so that a film of PVF required 7–8 yr exposure to reduce the elongation to break from 140% to 10%.[222]

PVF films can be made from solvent solutions[223] and by extrusion of the plasticized polymer.[224] The material is soluble above 100°C in a variety of solvents such as N-substituted amides, dinitriles, ketones, tetramethylene sulfones, and tetramethylurea.

Because they have excellent resistance to weathering, PVF coatings are used outdoors to decorate and protect a wide varity of substrates such as house sidings, etc. These coatings also do well in interior applications requiring good stain, dirt, and abrasion resistance. They are suitable as a mold release for molding epoxies and polyesters because of their low surface energy. Flame-retardant fabric coatings and free films of PVF are rated as slow burning or self-extinguishing, depending upon the exact formulation used.

4.3.4. Polychlorotrifluoroethylene

Polychlorotrifluoroethylene (PCTFE), like the other fluoropolymer coatings discussed, is relatively high in cost (\$4–15/lb depending on grade). Coatings of this material have a combination of high strength, low thermal expansion coefficient, good impact resistance, and low thermal conductivity. PCTFE does not carbonize nor does it support combustion when exposed to flame, and all forms of PCTFE are highly resistant to strong oxidizing agents. These properties make it useful for extended protection of substrates from the weather.[225-228]

5. References

1. H. W. Emmons, *Sci. Am.* **231**(1), 25 (1974).
1a. S. Steingiser, *J. Fire Flam.* **3**(3), 238 (1972).
2. Standard E-84, American Society for Testing and Materials, Philadelphia, Pa.
2a. ASTM Policy Defining Fire Hazard Standards, Limiting the Scope of Properties-Description Standards and Establishing a Committee on Fire Hazard Standards, American Society for Testing and Materials, Philadelphia, Pa. (Adopted Sept. 18, 1973).
3. W. J. Harland, *J. Paint Technol.* **44**(565), 43 (1972); **44**(575), 64 (1972).

4. C. J. Hilado and P. E. Burgess, *J. Fire Flam.* **3**(2), 154 (1972).
5. Standard E-286, American Society for Testing and Materials, Philadelphia, Pa.
6. C. P. Fenimore and F. J. Martin, *Mod. Plast.* **44**(3), 141 (1966).
7. C. P. Fenimore and G. W. Jones, *Combust. Flame* **10**(3), 295 (1966).
8. Standard D-2863, American Society for Testing and Materials, Philadelphia, Pa.
9. Standard D-1360, American Society for Testing and Materials, Philadelphia, Pa.
10. Standard D-1361, American Society for Testing and Materials, Philadelphia, Pa.
11. Standard E-162, American Society for Testing and Materials, Philadelphia, Pa.
12. C. A. Hafer, Final Report, Project No. 3-2947-26, Southwest Research Institute, San Antonio, Tex. (1970).
13. HET acid anhydride supplied by Hooker Chemical Corp., Niagara Falls, New York.
14. Cloran supplied by Universal Oil Products, Des Plaines, Ill.
15. R. F. Cleaver, *Polym., Paint Colour J.* **163**(3837), 107 (1973); **163**(3838), 152 (1973).
16. J. W. Lyons, *The Chemistry and Uses of Fire Retardants*, Chap. 6, Wiley-Interscience, New York (1970).
17. P. Thiery, *Fireproofing*, p. 113, from French by J. H. Coundry, Elsevier Pub. Co., Ltd., New York (1970).
18. V. M. Bhatnagar, *Fire Retardant Formulations Handbook*, Vol. 1, p. 1, Technomic Publishing Co., Westport, Conn. (1972).
19. T. F. Birkenhead, *Aust. Paint J.* **15**(6), 16 (1969).
20. I. Touval, *J. Fire Flam.* **3**(2), 130 (1972).
21. J. G. Bower, S. M. Draganov, and R. W. Sprague, *J. Fire Flam.* **3**(3), 181 (1972).
22. J. W. Lyons, *The Chemistry and Uses of Fire Retardants*, p. 244, Wiley-Interscience, New York (1970).
23. L. E. Newman and P. J. Silvester, *Am. Paint J.* **58**(24), 59 (1973).
24. W.-H. Chang, R. L. Scriven, J. R. Peffer, and S. Porter, *Ind. Eng. Chem., Prod. Res. Dev.* **12**, 278 (1973).
25. A. J. Papa and W. R. Proops, U.S. Pat. 3,779,953 (1973) (to Union Carbide Corp.).
26. J. P. Burns, J. Feltzin, and E. Koehn, U.S. Pat. 3,764,577 (1973) (to ICI America Inc.).
27. K. C. Frisch, Chap. 1 of this book.
28. J. H. Saunders and K. C. Frisch, *Polyurethanes: Chemistry and Technology*, Part 1, p. 94, Interscience, New York (1962).
29. B. D. Beitchman, *Ind. Chem., Prod. Res. Dev.* **5**, 35 (1966).
30. S. R. Sandler, *J. Appl. Polym. Sci.* **11**, 811 (1967).
31. J. Burkus, U.S. Pat. 2,979,485 (1961) (to U.S. Rubber Co.).
32. J. I. Jones and N. G. Savill, *J. Chem. Soc.* **1957**, 4392.
33. L. Nicholas and G. T. Gmitter, *J. Cell. Plast.* **1**, 85 (1965).
34. K. Ashida and T. Yagi, Brit. Pat. 1,155,768 (1969) (to Nisshin Spinning Co).
34a. N. Sasaki, T. Yokoyama, and T. Tanaka, *J. Polym. Sci., Polym. Chem. Ed.* **11**, 1765 (1973).
35. V. M. Bhatnagar, *Chem. Age India* **24**(1), 36 (1973).
36. J. W. Lyons, *J. Fire Flam.* **1**(4), 302 (1970).
37. J. W. Lyons, *The Chemistry and Uses of Fire Retardants*, p. 401, Wiley-Interscience, New York (1970).
37a. Anon., *Paint Varn. Prod.* **61**(11), 51 (1971).
38. J. H. Saunders and K. C. Frisch, *Polyurethanes: Chemistry and Technology*, Part I, Wiley-Interscience, New York (1963).
39. H. Tramm, C. Clar, P. Kuhnel, and W. Schuff, U.S. Pat. 2,106,938 (1938) (to Ruhrchemie Akt.).
40. H. L. Vandersall, *J. Fire Flam.* **2**, 97 (1971).
41. J. W. Lyons, *The Chemistry and Uses of Fire Retardants*, Wiley-Interscience, New York (1970).

42. D. J. Jones, *Intumescent Paints,* R. H. Chandler, Ltd., Braintree, Essex, England (1973).
43. H. F. Payne, *Organic Coating Technology,* Vol. I, Wiley, New York (1954).
44. H. L. Vandersall, paper presented at the Wayne State University Polymers Conference, Program 5, Detroit (1966).
45. R. E. Ellis, U.S. Pat. 3,102,821 (1963).
46. G. B. Verburg, D. A. Yeadon, E. T. Rayner, F. G. Dollear, H. P. Dupuy, L. L. Hopper, Jr., and E. York, *J. Paint Technol.* **38**, 407 (1966).
47. H. L. Vandersall, Phos-Chek P/30 Brand Fire Retardant—Its Use in Intumescent Paint, Monsanto Chemical Company, Inorganic Chemicals Div., Report No. 6521 (1965).
48. J. S. Schwartz and R. Pierrehumbert, *Am. Paint J.* **47**(55), 76 (1963).
49. F. J. Hahn and H. L. Vandersall, U.S. Pat 3,513,114 (1970) (to Monsanto Company).
50. F. J. Hahn and H. L. Vandersall, Can. Pat. 822,594 (1969) (to Monsanto Company).
51. F. Liberti and R. Pierrehumbert, *Am. Paint J.* **53**(7), 96 (1968).
52. C. A. Meyer, *Am. Paint J.* **53**(7), 78 (1968).
53. W. Juda, G. Jones, and N. Altman, U.S. Pat. 2,628,946 (1953) (to Albi Manufacturing Co., Inc.).
54. G. B. Verburg, E. T. Rayner, D. A. Yeadon, L. L. Hopper, Jr., L. A. Goldblatt, F. G. Dollear, and H. P. Dupuy, *J. Am. Oil Chem. Soc.* **41**, 670 (1964).
55. D. A. Yeadon, E. T. Rayner, G. B. Verburg, L. L. Hopper, Jr., F. G. Dollear, H. P. Dupuy, and H. Miller, *Off. Dig. Fed. Soc. Paint Technol.* **37**, 1095 (1965).
56. E. T. Rayner, D. A. Yeadon, G. B. Verburg, F. G. Dollear, H. P. Dupuy, L. L. Hopper, Jr., and H. Miller, *J. Paint Technol.* **38**, 105 (1966).
57. C. C. Clark and A. J. Krawczyk, U.S. Pat. 3,525,708 (1970) (to Textron, Inc.).
58. G. Quelle, C. A. Redfarn, and R. Thompson, Brit. Pat. 862,569 (1961) (to Alim Chemical Corp.); U.S. Pat. 3,077,458 (1963) (to Alim Chemical Corp.).
59. N. D. Blair, G. Witschard, and R. R. Hindersinn, *J. Paint Technol.* **44**, No. 573, 75 (1972).
60. Brit. Pat. 1,215,286 (1970) (to Rolls Royce, Ltd.).
61. G. J. Fleming, U.S. Pat. 3,634,036 (1972) (to U.S. Secretary of the Navy).
62. Nederlandse Organisatie voor Toegepast-Natuurwetenschappelijk Onderzoek ten behoeve van Nijverheid, Handel en Verkeer, Neth. Appl. 69-17, 423 (1970).
63. F. H. Thomas, H. M. Headrick, and E. L. Schulz, U.S. Pat. 3,422,046 (1969) (to Sherwin-Williams Company).
64. Brit. Pat. 1,280,543 (1972) (to Rolls Royce Ltd. and International Paint Co.).
65. Vsesjuzny Nauchno-Issledovatelsky Institute Protivopozharnoi Oborony Ministerstva Vnutrennikhdal SSSR, Ger. (East) Pat. 97,224 (1973).
66. A. J. Krawczyk, U.S. Pat. 3,627,726 (1971) (to Textron Inc.).
67. Brit. Pat. 925,570 (1963) (to Imperial Chemical Industries Ltd.).
68. K. Brack, U.S. Pat. 3,345,310 (1967) (to Hercules Inc.).
69. C. C. Clark and A. J. Krawczyk, U.S. Pat. 3,448,075 (1069) (to Textron Inc.).
70. C. C. Clark and A. J. Krawczyk, U.S. Pat. 3,365,420 (1968) (to Textron Inc.).
71. J. A. Parker, G. M. Fohlen, P. M. Sawko, and R. N. Griffin, *SAMPE J.,* p. 21 (Aug/Sept. 1968).
72. J. A. Parker and G. M. Fohlen, U.S. Pat. 3,535,130 (1970) (to U.S. National Aeronautics and Space Administration).
73. G. M. Fohlen, J. A. Parker, S. R. Riccitiello, and P. M. Sawko, *Proc. NASA Conf. Mater. Improv. Fire Saf.* **12**, 20 pp. (1970).
74. P. M. Sawko, U.S. Pat. 3,702,841 (1972) (to U.S. National Aeronautics and Space Administration).
75. Brit. Pat. 1,319,620 (1973) (to U.S. National Aeronautics and Space Administration).
76. P. M. Sawko, E. J. Fontes, and S. R. Riccitiello, *J. Paint Technol.* **44**(571), 51 (1972).
77. S. H. Roth, U.S. Pat. 3,703,409 (1972) (to Cities Service Company).

78. S. H. Roth, U.S. Pat. 3,703,410 (1972) (to Cities Service Company).
79. J. J. Seipel, S. H. Roth, and J. Green, U.S. Pat. 3,748,154 (1973) (to Cities Service Company).
80. J. Green, S. J. Roth, and J. J. Seipel, U.S. Pat. 3,703,487 (1972) (to Cities Service Company).
81. S. H. Roth, J. Green, and J. J. Seipel, U.S. Pat. 3,714,082 (1973) (to Cities Service Company).
82. R. F. Shannon, U.S. Pat. 3,630,764 (1971) (to Corning Fiberglas Corp.).
83. W. J. Clarke, S. Afr. Pat. 72-01,506 (1972) (to American Cyanamid Co.).
84. Brit. Pat. 1,095,857 (1967) (to Albi Manufacturing Co.).
85. W. L. Mackie, U.S. Pat. 3,697,422 (1972) (to U.S. Secretary of the Navy).
86. W. F. Moran and L. C. Kyrias, U.S. Pat. 3,663,267 (1972) (to Beatrice Foods Co.).
87. T. Okazaki, K. Asahara, and T. Kometani, Japan. Kokai 72-30,785 (1972) (to Nissan Chemical Industries Ltd.).
88. C. N. Berczi, S. Afr. Pat. 67-07,448 (1968) (to Deering Milliken Corp.).
89. D. G. Higgins, Coated Fabrics, in *Kirk-Othmer Encyclopedia of Chemical Technology*, A. Standen, ed., 2d ed., Vol. 5, p. 679, Wiley-Interscience, New York (1964).
90. G. C. Kantner, *Text. Chem. Color.* **5**(1), 21 (1973).
91. Fed. Spec. CCC-A-680a, Artificial Leather, (Cloth, Coated), Vinyl Resin, (Upholstery), U.S. General Services Administration, Washington, D.C. (May 26, 1966, amended Mar. 7, 1972).
92. *Fed. Regist.* **36**(5), 289 (1971).
93. M. M. O'Mara, W. Ward, D. P. Knechtges, and R. J. Meyer, in *Flame Retardancy of Polymeric Materials,* (W. C. Kuryla and A. J. Papa, eds.) Vol. 1, p. 250, Marcel Dekker, New York (1973).
94. DOT MVSS 302, Flame Retardance Standard for Motor Vehicle Interiors, U.S. Dept. of Transportation, Washington, D.C. (Sept. 1, 1972).
95. S. C. Stinson, *Plast. Technol.* **18**(3), 35 (1972).
96. G. Matthews, *Vinyl and Allied Polymers,* Vol. 2, p. 25, CRC Press, Cleveland, Ohio (1972).
97. C. J. Hilado, *Flammability Handbook for Plastics,* p. 164, Technomic Publishing Co., Stamford, Conn. (1969).
98. *Ibid.,* p. 39.
99. H. H. Cornish, *Arch. Environ. Health* **19**, 15 (1969).
100. *Fed. Regist.* **39**(67), 12342 (1974); *Chem. Eng. News* **52**(15), 6 (1974).
101. W. S. Penn, *PVC Technology,* p. 10, Maclaren & Sons Ltd., London (1966).
102. W. S. Penn, *High Polymeric Chemistry,* p. 71, Chapman and Hall Ltd., London (1949).
103. M. M. O'Mara, W. Ward, D. P. Knechtges, and R. J. Meyer, in *Flame Retardancy of Polymeric Materials,* (W. C. Kuryla and A. J. Papa, eds.) Vol. 1, p. 239, Marcel Dekker, New York (1973).
104. W. S. Penn, *PVC Technolgy,* p. 13, Maclaren & Sons Ltd., London (1966).
105. L. Holstejn, *Plast. Hmoty Kauc.* **9**(5), 134 (1972).
106. G. Pastuska, *Gummi, Asbest, Kunstst.* **19**(3), 275 (1966).
107. R. R. Hindersinn and G. M. Wagner, Fire Retardancy in *Encyclopedia of Polymer Science and Technology,* (N. M. Bikales, ed.) Vol. 7, p. 12, Wiley-Interscience, New York (1967).
108. W. S. Penn, *PVC. Technology,* p. 159, Maclaren & Sons Ltd., London (1966).
109. W. C. Kuryla in *Flame Retardancy of Polymeric Materials,* (W. C. Kuryla and A. J. Papa, eds.) Vol. 1, p. 1, Marcel Dekker, New York (1973).
110. NL Industries Inc., Product Bulletin F-2 (1968).
111. J. J. Pitts in *Flame Retardancy of Polymeric Materials,* (W. C. Kuryla and A. J. Papa, eds.) Vol. 1, p. 167, Marcel Dekker, New York (1973).
112. J. H. Saunders and J. K. Backus, *Rubber Chem. Technol.* **39**, 461 (1966).

113. R. W. Little, *Flameproofing Textile Fabrics*, Reinhold, New York (1947).
114. W. C. Kuryla in *Flame Retardancy of Polymeric Materials*, (W. C. Kuryla and A. J. Papa, eds) Vol. 1, p. 24, Marcel Dekker, New York (1973).
115. M. M. O'Mara, W. Ward, D. P. Knechtges, and R. J. Meyer in *Flame Retardancy of Polymeric Materials* (W. C. Kuryla and A. J. Papa, eds.) Vol. 1, p. 229, Marcel Dekker, New York (1973).
116. G. Matthews, *Vinyl and Allied Polymers*, Vol. 2, p. 111, CRC Press, Cleveland, Ohio (1972).
117. K. I. Freidgeim, L. F. Golutivina, I. V. Plotnikov, N. V. Syrokvashina, and A. N Kopyl, *Nauch.-Issled. Tr., Vses. Nauch.-Issled. Inst. Plencoh. Mater. Iskusstv. Kozhi No. 19*, 103 (1971).
118. W. G. Woods and J. G Bower, *Mod. Plast.* **47**(6), 140 (1970).
119. M. M. O'Mara, *J. Polym. Sci.* **9**, 1387 (1971).
120. G. Matthews, *Vinyl and Allied Polymers*, Vol. 2, p. 137, CRC Press, Cleveland, Ohio (1972).
121. Monsanto Co., Technical Bulletin O PL-342 (1966)
122. Diamond Shamrock Chemical Co., Technical Bulletin on Chlorowax (1971).
123. W. Carl and M. Wandek, Brit. Pat. 1,274,925 (May 17, 1972) (to Farbenfabriken Bayer A.G.).
124. J. W. Lyons, *The Chemistry and Uses of Fire Retardants*, p. 345, Wiley-Interscience, New York (1970).
125. Austral. Pat. 435,597 (1973) (to Imperial Chemical Industries Ltd.).
126. J. T. Harrington, U.S. Pat. 3,574,149 (1971) (to General Tire & Rubber Co.).
127. J. A. Parker and S. R. Riccitiello, U.S. Pat. 3,549,564 (1970) (to U.S. National Aeronautics and Space Administration).
128. A. J. Papa and W. R. Proops, *J. Appl. Polym. Sci.* **16**, 2361 (1972).
129. G. F. Baumann and J. F. Szabat, Fire Retardant Flexible Urethane Foam, in *Advances in Fire Retardants*, (V. M. Bhatnagar, ed.) Part I, p. 1, Technomic Publishing Co., Westport, Conn. (1972).
130. C. E. Miles and J. W. Lyons, *J. Cell. Plast.* **3**, 539 (1967).
131. E. A. Dickert and G. C. Toone, *Mod. Plast.* **42**(5), 197 (1965).
132. I. N. Einhorn and R. W. Nickelson, *Am. Chem. Soc., Div. Org. Coatings Plast. Chem., Pap.* 28, No. 1, 291 (1968).
133. P. E. Hoch, U.S. Pat. 3,151,143 (1964) (to Hooker Chemical Corp.).
134. P. A. Argabright and B. L. Phillips, U.S. Pat. 3,627,689 (1971) (to Marathon Oil Co.).
135. H. E. Holmquist, U.S. Pat. 3,479,304 (1969) (to E. I. duPont de Nemours & Co.).
136. R. C. Nametz, R. D. Deanin, and P. M. Lambert, *Mod. Plast.* **41**(1), 166 (1963).
137. A. C. Haven, *J. Am. Chem. Soc.* **78**, 842 (1956).
138. H. C. Fielding, Brit. Pat. 892,931 (1965) (to Imperial Chemical Industries Ltd.).
139. A. Holtschmidt and G. Oertel, *Angew. Chem., Intnatl. Ed. Eng.* **1**, 617 (1962).
140. L. M. Kindley, H. E. Podall, and N. Filipescu, *SPE Trans.* **2**, 122 (1962).
141. R. L. McConnell and H. W. Coover, U.S. Pat. 2,926,145 (1960) (to Eastman Kodak Co.).
142. H. C. Fielding, U.S. Pat. 3,144,302 (1964) (to Imperial Chemical Industries Ltd.).
143. J. J. Anderson, *Ind. Eng. Chem., Prod. Res. Dev.* **2**, 261 (1963).
144. N. E. Rustad and R. G. Krawiec, *Rubber Age (New York)* **105**(11), 45 (1973).
145. J. K. Backus, D. L. Bernard, W. C. Darr, and J. H. Saunders, *J. Appl. Polym. Sci.* **12**, 1053 (1968).
146. A. Heslinga and P. J. Napjus, Ger. Pat. 1,197,614 (1964) (to Chemische Fabrik Kalk G.m.b.H.).
147. Standard Oil Co., Neth. Appl. 6,604,906 (1966); Neth. Appl. 6,516,709 (1966).
148. K. C. Frisch, J. Cell. Plast, *1*, 3 (1965).

149. Brit. Pat. 951,792 (1964) (to Virginia-Carolina Chemical Corp.).
150. J. J. Anderson, Brit. Pat. 1,034,489 (1967) (to Mobil Oil Corp.).
151. P. G. Pope, J. E. Sawyer, and R. C. Nametz, *J. Cell. Plast.* **4**, 438 (1968).
152. F. D. Popp and W. E. McEwen, *Chem. Rev.* **58**, 321 (1958).
153. D. Shaw and B. W. Greenwald, U.S. Pat. 3,291,867 (1966) (to Merck and Co.).
154. E. Dyer and R. A. Dunbar, *J. Polym. Sci., Part A-1* **8**, 629 (1970).
155. L. Friedman, U.S. Pat. 3,309,342 (1967) (to Union Carbide Corp.).
156. P. A. T. Hoye and H. Coates, Brit. Pat. 974,033 (1964) (to Albright & Wilson Manufacturing Ltd.).
157. M. I. Bakhitov, E. V. Kuznetsov, and M. Ya. Obryadina, in *Soviet Urethane Technology,* (A. M. Schiller, ed.) Vol. 1, p. 79, Technomic Publishing Co., Westport, Conn. (1973).
158. R. K. Valetdinov, E. V. Kuznetsov, M. Kh. Khasanov, and T. Kh. Valeeva, in *Soviet Urethane Technology,* (A. M. Schiller, ed.) Vol. 1, p. 94, Technomic Publishing Co., Westport, Conn. (1973).
159. A. J. Papa and W. R. Proops, *J. Appl. Polym. Sci.* **17**, 2463 (1973).
160. C. K. Lyon and T. H. Applewhite, *J. Cell. Plast.* **3**, 91 (1967).
161. P. Robitschek, U.S. Pat. 3,058,925 (1962) (to Hooker Chemical Corp.).
162. P. Robitschek and J. L. Olmstead, U.S. Pat. 2,909,501 (1959) (to Hooker Chemical Corp.).
163. P. Robitschek, *J. Cell. Plast.* **3**, 395 (1967).
164. R. R. Hindersinn and S. M. Creighton, U.S. Pat. 2,865,869 (1958) (to Hooker Chemical Corp.).
165. H. A. Bruson and J. S. Rose, U.S. Pat. 3,244,754 and U.S. Pat. 3,269,961 (1966) (to Olin Mathieson Chemical Corp.).
166. H. C. Bogt and P. Davis, *Polym. Eng. Sci.* **11**, 312 (1971).
167. E. V. Gouinlock, F. W. Long, and S. M. Creighton, *Plast. Technol.* **8**(12) 40 (1962).
168. H. J. Fabris and J. G. Sommer, in *Flame Retardancy of Polymeric Materials,* (W. C. Kuryla and A. J. Papa, eds.) Vol. 1, p. 135, Marcel Dekker, Inc., New York (1973).
169. K. Oplustil, Factors Affecting Flame Spread Rating of Neoprene Vulcanizates, E. I. duPont de Nemours & Co. Technical Bulletin 5D-113 (1968).
170. W. S. Penn, *Rubber J.* **151**, 23 (1969).
171. D. C. Thompson, J. G. Hagman, and N. N. Mueller, *Rubber Age (New York)* **83**, 819 (1958).
172. L. N. Kireenkova and Yu. S. Zuev, *Sov. Rubber Technol.* **27**(12), 23 (1968).
173. E. I. duPont de Nemours & Co., Hypalon in Wire and Cable, Hypalon Report No. 10, (1963), p. 9.
174. H. E. Trexler, *Rubber Chem. Technol.* **46**, 1114 (1973).
175. N. S. Vojvodich, *J. Macromol. Sci.-Chem.* **A3**, 367 (1969).
176. J. P. Gouzene, Ger. Pat. 1,298,877 (1969) (to Compagnie de Saint-Gobain).
177. Hooker Chemical Corp., Dechlorane 604, Preliminary Data Sheet No. 349 (1970).
178. E. E. Pepe, Ger. Pat. 1,936,345 (1970) (to Stauffer Chemical Co.).
179. J. Milgrom, J. T. Howarth, and J. G. Sheth, A Flame-Resistant Silicone Rubber, National Aeronautics and Space Administration N70-34148, Arthur D. Little, Inc., Boston, Mass. (1970).
180. F. H. Winslow, L. D. Loan, and W. Matreyek, *Am. Chem. Soc., Div. Org. Coatings Plast. Chem. Pap. 31,* No. 1, 124 (1971).
181. D. L. Supkis, Description and Application of Fluorel L-3203-6. Report No. MSC-01275, National Aeronautics and Space Administration, Manned Spacecraft Center, Houston.
182. Minnesota Mining and Manufacturing Co., Fluorel Elastomer, Technical Information Bulletin Y-TTDC (fluorel), p. 9.
183. H. J. Fabris and J. G. Sommer in *Flame Retardancy of Polymeric Materials,* (W. C. Kuryla and A. J. Papa, eds.) Vol. 1, p. 154, Marcel Dekker, New York (1973).

184. D. A. Stivers, Aerospace Adhesive Elastomers, Natl. SAMPE Tech. Conf. Proc., 2d, (1970) p. 363.
185. Anon., *Spinner Weber Textilveredl.* **69**, 88 (1971).
186. D. G. Sauers, NASA Conference on Materials for Improved Fire Safety, National Aeronautics and Space Administration, Manned Spacecraft Center, Houston (1970), p. 17-1.
187. B. P. Hajela, I. D. Singh, K. L. Bhatia, and S. S. Shukla, *Paintindia* **22**(5), 20 (1972).
188. S. Hikarida, Japan Pat. 71 37,434 (1971) (to Toyo Spinning Co. Ltd.).
189. *Mach. Des.* **36**, 234 (1964).
190. *Ibid.*, **36**, 232 (1964).
191. C. E. Sroog, A. L. Endrey, S. V. Abramo, C. E. Berr, W. M. Edwards, and K. L. Olivier, *Am. Chem. Soc., Div. Polym. Chem., Prepr.* **5**(1), 132 (1964).
192. C. E. Sroog, A. L. Endrey, S. V. Abramo, C. E. Berr, W. M. Edwards, and K. L. Olivier, *J. Polym. Sci., Part A* **3**, 1373 (1965).
193. C. E. Sroog, *J. Polym. Sci. Part C* **16**, 1191 (1967).
194. M. L. Wallach, *Am. Chem. Soc., Div. Polym. Chem., Prepr.* **6**(1), 53 (1965).
195. *Ibid.* **8**(1), 656 (1967).
196. *Ibid.* **8**(2), 1170 (1967).
197. M. L. Wallach, *J. Polym. Sci., Part A-2* **5**, 653 (1967).
198. R. Ikeda, *J. Polym. Sci., Part B* **4**, 353 (1966).
199. A. L. Endrey, U.S. Pat. 3,179,630 (1965) (to E. I. duPont de Nemours & Co.).
200. W. M. Edwards, U.S. Pat. 3,179,634 (1965) (to E. I. duPont de Nemours & Co.).
201. W. M. Edwards, U.S. Pat. 3,179,614 (1965) (to E. I. duPont de Nemours & Co.).
202. A. L. Endrey, U.S. Pat. 3,179,631 (1965) (to E. I. duPont de Nemours & Co.).
203. J. A. Kreuz, A. L. Endrey, F. P. Gay, and C. E. Sroog, *J. Polym. Sci., Part A-1* **4**, 2607 (1966).
204. Brit. Pat. 1,057,727 (1967) (to Dr. Beck & Co. G.m.b.H.).
205. F. H. Holub, U.S. Pat. 3,277,043 (1966) (to General Electric Co.).
206. D. C. Atkins, C. M. Murphy, and C. E. Saunders, *Ind. Eng. Chem.* **39**, 1395 (1947).
207. *Ibid.* **42**, 2462 (1950).
208. A. E. Durkin and A. H. Horner, *Mater. Methods* **38**(3), 114 (1953).
209. W. Noll, *Chemistry and Technology of Silicones*, p. 476, Academic Press, New York (1968).
210. *Ibid.*, p. 563.
211. J. M. Cox, B. A. Wright, and W. W. Wright, *J. Appl. Polym. Sci.* **8**, 2935 (1964).
212. *Ibid.*, 2951.
213. Fluon, *Eng. News* (Imperial Chemical Industries Ltd.), p. 5 (May 1964).
214. J. A. Zapp, *Arch. Environ. Health* **4**(3), 342 (1962).
215. T. A. Ford and W. E. Hanford, U.S. Pat. 2,435,537 (1948) (to E. I. duPont de Nemours and Co.).
216. T. A. Ford, U.S. Pat. 2,468,054 (1949) (to E. I. duPont de Nemours & Co.).
217. A. L. Dittman, H. J. Passino, and W. O. Teeters, U.S. Pat. 2,738,343 (1956) (to M. W. Kellogg Co.).
218. A. L. Dittman, H. J. Passino, and W. O. Teeters, U.S. Pat. 2,752,331 (1956) (to M. W. Kellogg Co.).
219. E. S. Lo, U.S. Pat. 3,023,187 (1962) (to Minnesota Mining and Manufacturing Co.).
220. D. R. Rexford, U.S. Pat. 3,051,677 (1962) (to E. I. duPont de Nemours and Co.).
221. E. S. Lo, Brit. Pat. 823,974 (1959) (to Minnesota Mining and Manufacturing Co.).
222. V. L. Simril and A. Curry, *Mod. Plast.* **36**(11), 121 (1959).
223. G. H. Kalb, D. D. Coffman, T. A. Ford, and F. L. Johnston, *J. Appl. Polym. Sci.* **4**, 55–61 (1960).
224. L. R. Bartron, U.S. Pat. 2,953,818 (1960) (to E. I. duPont de Nemours and Co.).

Index